Lecture Notes in Mathematics

Volume 1499

This series reports on new developments in all areas of mathematics and their applications – quickly, informally and at a high level. Mathematical texts analysing new developments in modelling and numerical simulation are welcome. The type of material considered for publication includes:

1. Research monographs 2. Lectures on a new field or presentations of a new angle in a classical field 3. Summer schools and intensive courses on topics of current research.

Texts which are out of print but still in demand may also be considered if they fall within these categories. The timeliness of a manuscript is sometimes more important than its form, which may be preliminary or tentative.

More information about this series at http://www.springer.com/series/304

Kazuaki Taira

Boundary Value Problems and Markov Processes

Functional Analysis Methods for Markov Processes

Third Edition

 Springer

Kazuaki Taira (iD)
Institute of Mathematics
University of Tsukuba
Tsukuba, Ibaraki
Japan

ISSN 0075-8434   ISSN 1617-9692   (electronic)
Lecture Notes in Mathematics
ISBN 978-3-030-48787-4   ISBN 978-3-030-48788-1   (eBook)
https://doi.org/10.1007/978-3-030-48788-1

Mathematics Subject Classification: 47D07, 35J25, 47D05, 60J35, 60J60

This Springer imprint is published by the registered company Springer Nature Switzerland AG.
The registered company address is: Gewerbestrasse 11, 6330 Cham, Switzerland

To the memory of my parents,
Yasunori Taira (1915–1990)
and
Yasue Taira (1918–2014)

Preface to the Third Edition

This book is an expanded and revised version of a set of lecture notes for the graduate courses given by the author at the University of Tsukuba (1988–1995 and 1998–2009), at Hiroshima University (1995–1998) and at Waseda University (2009–2016), which were addressed to advanced undergraduate and beginning-graduate students with an interest in functional analysis, partial differential equations and probability.

It provieds an easy-to-read reference describing a link between functional analysis, partial differential equations and probability via Markov processes.

The functional analytic approach to Markov processes is distinguished by its extensive use of ideas and techniques characteristic of recent developments in the theory of pseudo-differential operators, which may be considered as a modern version of the classical potential theory. It should be emphasized that pseudo-differential operators provide a constructive tool to deal with existence and smoothness of solutions of partial differential equations. The full power of this very refined theory is yet to be exploited.

Several recent developments in the theory of partial differential equations have made possible further progress in the (mesoscopic) study of elliptic boundary value problems and hence in the (microscopic) study of Markov processes, with a special emphasis on the (macroscopic) study of Feller semi-groups in functional analysis.

In addition to providing a comprehensive study of Markov processes, this book may also be considered as an accessible and careful introduction to two more advanced books:

(A) Diffusion processes and partial differential equations, Academic Press, Boston, Massachusetts, 1988.
ISBN: 0-12-682220-4

(B) Semigroups, boundary value problems and Markov processes, Second Edition, Springer Monographs in Mathematics (SMM), Springer-Verlag, 2014.
ISBN: 978-3-662-43696-7

Most mathematicians working in partial differential equations (the mesoscopic approach) are only vaguely familiar with the powerful ideas of stochastic analysis (the microscopic approach). However, the stochastic intuition which this book conveys may provide a profound insight into the study of three interrelated subjects in analysis: semigroups, elliptic boundary value problems and Markov processes. In this new edition we explain in detail this neglected microscopic side of functional analysis, boundary value problems and probability, which is the first purpose of the book. Indeed, we re-work and expand in a different spirit the material of the book (B), even though there is a lot of overlap between the table of contents of this new edition and that of the SMM volume. The following three chapters have been revised:

(1) Chapter 3: Markov Processes and Feller Semigroups. Section 3.6 is devoted to the local time on the boundary for the reflecting diffusion due to Paul Lévy.
(2) Chapter 4: L^p Theory of Pseudo-Differential Operators. Section 4.6 is devoted to the functional calculus for the Laplacian via the heat kernel.
(3) Chapter 6: L^p Theory of Elliptic Boundary Value Problems. Section 6.1 is a modern version of the classical potential theory in terms of pseudo-differential operators. Section 6.6 is a probabilistic approach to pseudo-differential operators via the reflecting Brownian motion.

Very recently, we have been able to solve several long-standing open problems in the spectral analysis of elliptic boundary value problems, such as the hypoelliptic Robin problem and the subelliptic oblique derivative problem. In the proofs we made essential use of the Boutet de Monvel calculus, which is one of the most influential ideas in the modern history of analysis. The presentation of the new aspect of the Boutet de Monvel calculus is the second purpose of this third edition. Indeed, we carefully re-worked the classical functional analytic methods for Markov processes (due to Sato–Ueno and Bony–Courrège–Priouret) from the viewpoint of the Boutet de Monvel calculus, which will provide a powerful method for future research in semigroups, boundary value problems and Markov processes. The following six chapters are included in this new edition:

(4) Chapter 5: Boutet de Monvel Calculus.
(5) Chapter 10: Elliptic Waldenfels Operators and Maximum Principles.
(6) Chapter 11: Boundary Operators and Boundary Maximum Principles.
(7) Chapter 14: Proofs of Theorems 1.8, 1.9, 1.10 and 1.11.
(8) Chapter 15: Path Functions of Markov Processes via Semigroup Theory.
(9) Chapter 16: Concluding Remarks.

In this third edition we mainly confined ourselves to simple but fundamental boundary conditions such as those in the Robin problem and the oblique derivative problem, which makes it possible to develop our basic machinery with a minimum of bother and also to present our principal ideas concretely and explicitly for advanced undergraduates and beginning-graduate students. Moreover, this edition is amply illustrated; 137 figures, 15 tables and 9 flowcharts of proofs are provided with appropriate captions. Having read this book, a broad spectrum of readers will be able to easily and effectively appreciate the mathematical crossroads of functional analysis, boundary value problems and probability developed in the more advanced books (A) and (B). This is the third purpose of the new edition.

Furthermore, this book provides a compendium for a large variety of facts from functional analysis, pseudo-differential operators and Markov processes – making it easy to quickly look up a theorem. Indeed, this book gives detailed coverage of important examples and applications. Bibliographical references are discussed primarily in the Notes and Comments at the end of each chapter. These notes are intended to supplement the text and place it in a better perspective.

In preparing this monograph, I am indebted to Professor Yasushi Ishikawa, who has read and commented on portions of various preliminary drafts from the viewpoint of probability. In particular, Section 3.6 is essentially due to him. I would like to extend my warmest thanks to Professors Francesco Altomare and Elmar Schrohe, who have showed constant interest in my work since 1996.

I would like to extend my hearty thanks to the staff of Springer-Verlag (Heidelberg), who have generously complied with all my wishes.

Last but not least, I owe a great debt of gratitude to my family, who gave me moral support during the preparation of this book.

Tsuchiura,   Kazuaki Taira
April 2020

Preface to the Second Edition

This monograph is an expanded and revised version of a set of lecture notes for the graduate courses given by the author both at Hiroshima University (1995–1997) and at the University of Tsukuba (1998–2000) which were addressed to the advanced undergraduates and beginning-graduate students with interest in functional analysis, partial differential equations and probability.

The first edition of this monograph, which was based on the lecture notes given at the University of Tsukuba (1988–1990), was published in 1991. This edition was found useful by a number of people, but it went out of print after a few years.

This second edition has been revised to streamline some of the analysis and to give better coverage of important examples and applications. The errors in the first printing are corrected thanks to kind remarks of many friends. In order to make the monograph more up-to-date, additional references have been included in the bibliography.

This second edition may be considered as a short introduction to the more advanced book *Semigroups, boundary value problems and Markov processes* which was published in the Springer Monographs in Mathematics series in 2004. For graduate students working in functional analysis, partial differential equations and probability, it may serve as an effective introduction to these three interrelated fields of analysis. For graduate students about to major in the subject and mathematicians in the field looking for a coherent overview, it will provide a method for the analysis of elliptic boundary value problems in the framework of L^p spaces.

This research was partially supported by Grant-in-Aid for General Scientific Research (No. 19540162), Ministry of Education, Culture, Sports, Science and Technology, Japan.

Last but not least, I owe a great debt of gratitude to my family who gave me moral support during the preparation of this book.

Tsukuba,   Kazuaki Taira
March 2009

Contents

Part II Pseudo-Differential Operators and Elliptic Boundary Value Problems

Introduction and Main Results

This book is an easy-to-read reference providing a link among functional analysis, partial differential equations and probability. In this introductory chapter, our problems and results are stated in such a fashion that a broad spectrum of readers could understand.

Table 1.1 below gives a bird's-eye view of Markov processes, Feller semigroups and elliptic boundary value problems and how these relate to each other.

1.1 Historical Perspective of Feller's Approach to Brownian Motion

In 1828 the Scottish botanist Robert Brown (1773–1858) observed that pollen grains suspended in water move chaotically, incessantly changing their direction of motion (see Figure 1.1 below). The physical explanation of this phenomenon is that a single grain suffers innumerable collisions with the randomly moving molecules of the surrounding water. A mathematical theory for Brownian motion was put forward by the German physicist Albert Einstein (1879–1955) in 1905 ([36]). Let $p(t, x, y)$ be the probability density function that a one-dimensional Brownian particle starting at position x will be found at position y at time t. Einstein derived the following formula from statistical mechanical considerations:

$$p(t, x, y) = \frac{1}{\sqrt{2\pi Dt}} \exp\left[-\frac{(y-x)^2}{2Dt}\right].$$

Here D is a positive constant determined by the radius of the particle, the interaction of the particle with surrounding molecules, temperature and the Boltzmann constant. This gives an accurate method of measuring Avogadro's

© Springer Nature Switzerland AG 2020
K. Taira, *Boundary Value Problems and Markov Processes*, Lecture Notes in Mathematics 1499,
https://doi.org/10.1007/978-3-030-48788-1_1

Probability (Microscopic) (approach)	Functional Analysis (Macroscopic approach)	Elliptic Boundary Value Problems (Mesoscopic approach)
Markov process $\mathcal{X} = (x_t)$	Feller semigroup $\{T_t\}_{t \geq 0}$	Infinitesimal generator $\overline{\mathcal{W}} = \mathfrak{W}$
Markov transition function $p_t(\cdot, dy)$	$T_t f = \int_{\overline{D}} p_t(\cdot, dy) f(y)$	$T_t = e^{t\overline{\mathcal{W}}} = e^{t\mathfrak{W}}$
Chapman and Kolmogorov equation	Semigroup property $T_{t+s} = T_t \cdot T_s$	Waldenfels operator $W = A + S$
Absorption and reflection phenomena	Function space $C_0 \left(\overline{D} \setminus M \right)$	Ventcel' (Wentzell) condition L

Table 1.1. A bird's-eye view of Markov processes, Feller semigroups and boundary value problems

number $N_A = 6{,}023 \cdot 10^{23}$ by observing particles. Einstein's theory was experimentally tested by the French physicist Jean Perrin (1870–1942) between 1906 and 1909 ([90]).

Brownian motion was put on a firm mathematical foundation for the first time by the American mathematician Nobert Wiener (1894–1964) in 1923 ([144], Knight [69]).

Let Ω be the space of continuous functions

$$\omega \colon [0, \infty) \longmapsto \mathbf{R} \quad \text{with coordinates } x_t(\omega) = \omega(t),$$

and let $\mathcal{F}$ be the smallest σ-algebra in Ω with respect to which all x_t, $t \geq 0$, are measurable. Namely, $\mathcal{F}$ is the smallest σ-algebra which contains all sets of the form

$$x_t^{-1}\left([a, b)\right) = \{\omega \in \Omega : a \leq x_t(\omega) < b\} \quad \text{for } t \geq 0 \text{ and } a < b.$$

Wiener constructed probability measures P_x, $x \in \mathbf{R}$, on $\mathcal{F}$ for which the following formula (1.1) holds true:

$$P_x\{\omega \in \Omega : a_1 \leq x_{t_1}(\omega) < b_1, a_2 \leq x_{t_2}(\omega) < b_2, \ldots, \tag{1.1}$$
$$a_n \leq x_{t_n}(\omega) < b_n\}$$

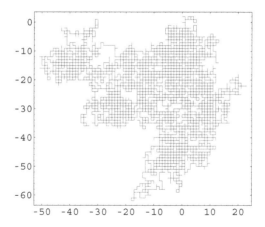

Fig. 1.1. Brownian motion

$$= \int_{a_1}^{b_1} \int_{a_2}^{b_2} \cdots \int_{a_n}^{b_n} p(t_1, x, y_1) p(t_2 - t_1, y_1, y_2) \cdots$$
$$p(t_n - t_{n-1}, y_{n-1}, y_n) \, dy_1 \, dy_2 \ldots dy_n \quad \text{for } 0 < t_1 < t_2 < \ldots < t_n < \infty.$$

This formula (1.1) expresses the "starting afresh" property of Brownian motion that if a Brownian particle reaches a position, then it behaves subsequently as though that position had been its initial position.

More precisely, let

$$p_t(x, E) = P_x \{\omega \in \Omega : x_0(\omega) = x, \ x_t(\omega) \in E\}$$

be the transition probability that a Brownian particle starting at position x will be found in the set E at time t (see Figure 1.2 below). Then the above formula (1.1) expresses the idea that a transition from the position x to the set E in time $t + s$ is composed of a transition from x to some position y in time t, followed by a transition from y to the set E in the remaining time s; the latter transition has probability $p_s(y, E)$ which depends only on y. Thus a Brownian particle "starts afresh"; this property is called the *Markov property* of Brownian motion. The measure P_x is called the *Wiener measure* starting at x.

Markov processes are an abstraction of the idea of Brownian motion. In the first works devoted to Markov processes, the most fundamental was the work of the Russian mathematician Andrey Nikolaevich Kolmogorov (1903–1987) in 1931 ([70]) where the general concept of a Markov transition function was introduced for the first time and an analytic method of describing Markov transition functions was proposed.

From the viewpoint of analysis, the transition function $p_t(x, \cdot)$ is something more convenient than the Markov process itself. In fact, it can be shown that the transition functions of Markov processes generate solutions of certain

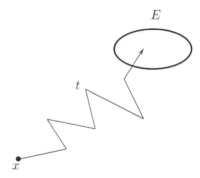

Fig. 1.2. An image of the transition probability $p_t(x, E)$

parabolic partial differential equations such as the classical diffusion equation; and, conversely, these differential equations can be used to construct and study the transition functions and the Markov processes themselves.

In the 1950s, the theory of Markov processes entered a new period of intensive development. We can associate with each transition function in a natural way a family of bounded linear operators acting on the space of continuous functions on the state space, and the Markov property implies that this family forms a *semigroup*. The Hille–Yosida theory of semigroups in functional analysis ([54], [147]) made possible further progress in the (microscopic) study of Markov processes. The (macroscopic) semigroup approach to Markov processes can be traced back to the pioneering work of the Croatian–American mathematician William Feller (1906–1970) in early 1950s. Feller [39], [40] characterized completely the analytic structure of one-dimensional diffusion processes; he gave an intrinsic (mesoscopic) representation of the infinitesimal generator of a one-dimensional diffusion process and determined all possible boundary conditions which describe the domain of the infinitesimal generator.

The functional analytic approach to one-dimensional Brownian motion can be visualized in Table 1.2 below (cf. Knight [69, Chapter 3, Section 3.1]):

The probabilistic meaning of Feller's work was clarified by Eugene Borisovich Dynkin (1924–2014) [33] and [34], Kiyosi Itô (1915–2008) and Henry P. McKean, Jr. [65], Daniel Ray [92] and others. One-dimensional diffusion processes are completely studied both from analytic and probabilistic viewpoints (see Ikeda–Watanabe [62], Revuz–Yor [94]).

The French mathematician Paul Lévy (1886–1971) found another construction of Brownian motion, and gave a profound description of (microscopic) qualitative properties of the individual Brownian path in his book ([76]): Processus stochastiques et mouvement brownien (1948).

Transition function $1/\sqrt{4\pi t}\,e^{-(x-y)^2/4t}\,dy$	Dynkin $\Longleftarrow\!\!\Longrightarrow$	Feller semigroup $T_t = e^{t\,\partial^2/\partial x^2}$
Laplace transform $\updownarrow$	$T_t f(x) =$ $\frac{1}{\sqrt{4\pi t}}\int_{\mathbf{R}} e^{-(x-y)^2/4t} f(y)\,dy$	Hille–Yosida $\updownarrow$
Green kernel $1/\sqrt{4\alpha}\,e^{-\sqrt{\alpha}\,|x-y|}$	$\Longleftarrow\!\!\Longrightarrow$ Riesz–Markov	Green operator $\left(\alpha - \partial^2/\partial x^2\right)^{-1}$

Table 1.2. Feller's approach to one-dimensional Brownian motion

1.2 Formulation of the Problem and Statement of Main Results

Now let D be a bounded domain of Euclidean space $\mathbf{R}^N$, with smooth boundary ∂D; its closure $\overline{D} = D \cup \partial D$ is an N-dimensional, compact smooth manifold with boundary (see Figure 1.3 below).

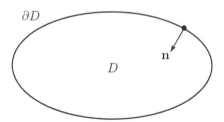

Fig. 1.3. The bounded domain D and the *inward normal* **n** to the boundary ∂D

In this section, we consider a second-order, *elliptic Waldenfels integro-differential operator* W with real coefficients such that

$$Wu(x) = Au(x) + Su(x) \tag{1.2}$$

$$:= \left(\sum_{i,j=1}^{N} a^{ij}(x) \frac{\partial^2 u}{\partial x_i \partial x_j}(x) + \sum_{i=1}^{N} b^i(x) \frac{\partial u}{\partial x_i}(x) + c(x)u(x) \right)$$

$$+ \int_{\mathbf{R}^N \setminus \{0\}} \left(u(x+z) - u(x) - \sum_{j=1}^{N} z_j \frac{\partial u}{\partial x_j}(x) \right) s(x,z)\, m(dz).$$

Here:

(1) $a^{ij} \in C^\infty(\overline{D})$, $a^{ij}(x) = a^{ji}(x)$ for all $1 \le i,j \le N$, and there exists a constant $a_0 > 0$ such that

$$\sum_{i,j=1}^{N} a^{ij}(x)\xi_i\xi_j \geq a_0|\xi|^2 \quad \text{for all } x \in \overline{D} \text{ and } \xi \in \mathbf{R}^N.$$

(2) $b^i \in C^\infty(\overline{D})$ for all $1 \leq i \leq N$.

(3) $c \in C^\infty(\overline{D})$, and $c(x) \leq 0$ in D, but $c(x) \not\equiv 0$ in D.

(4) $s(x, z) \in L^\infty(\mathbf{R}^N \times \mathbf{R}^N)$ and $0 \leq s(x, z) \leq 1$ almost everywhere in $\mathbf{R}^N \times \mathbf{R}^N$, and there exist constants $C_0 > 0$ and $0 < \theta_0 < 1$ such that

- $$|s(x, z) - s(y, z)| \leq C_0 |x - y|^{\theta_0} \qquad (1.3a)$$
  for all $x, y \in \overline{D}$ and almost all $z \in \mathbf{R}^N$,

- $$s(x, z) = 0 \quad \text{if } x \in D \text{ and } x + z \notin \overline{D}. \qquad (1.3b)$$

Probabilistically, the *support condition* (1.3b) implies that all jumps from D are within $\overline{D}$. Analytically, the support condition (1.3b) guarantees that the integral operator S may be considered as an operator acting on functions u defined on the closure $\overline{D}$ (see [48, Chapter II, Remark 1.19]).

(5) The measure $m(dz)$ is a Radon measure on $\mathbf{R}^N \setminus \{0\}$ which has a density with respect to the Lebesgue measure dz on $\mathbf{R}^N$, and satisfies the *moment condition*

$$\int_{\{0<|z|\leq 1\}} |z|^2 \, m(dz) + \int_{\{|z|>1\}} |z| \, m(dz) < \infty. \qquad (1.4)$$

(6) Finally, we assume that

$$(W1)(x) = (A1)(x) + (S1)(x) = c(x) \leq 0 \text{ and } c(x) \not\equiv 0 \text{ in } D. \qquad (1.5)$$

The operator $W = A + S$ is called a second-order, *Waldenfels integro-differential operator* or simply *Waldenfels operator* (cf. [15], [140]). The integro-differential operator S is called a second-order, *Lévy integro-differential operator* which is supposed to correspond to the jump phenomenon in the closure $\overline{D}$; a Markovian particle moves by jumps to a random point, chosen with kernel $s(x, z)$ and Radon measure $m(dz)$, in $\overline{D}$ (see [111], [11], [63]). The differential operator A describes analytically a strong Markov process with continuous paths in the interior D such as Brownian motion. The operator A is called a *diffusion operator*, and the functions $a^{ij}(x)$, $b^i(x)$ and $c(x)$ are called the *diffusion coefficients*, the *drift coefficients* and the *termination coefficient*, respectively. The Lévy integro-differential operator S is supposed to correspond to the jump phenomenon in the closure $\overline{D}$. Namely, a Markovian particle moves by jumps to a random point, chosen with kernel $s(x, z)$, in the interior D. The function $s(x, z)$ is called the *jump density*. Therefore, the Waldenfels integro-differential operator $W = A + S$ is supposed to correspond to such a diffusion phenomenon that a Markovian particle moves both by jumps and continuously in the state space D (see Figure 1.4 below).

In this context, the support condition (1.3b) implies that any Markovian particle does not move by jumps from the interior D into the outside of the

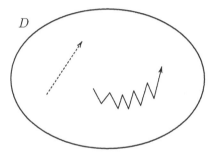

Fig. 1.4. A Markovian particle moves both by jumps and continuously in the state space

closure $\overline{D}$. On the other hand, the moment condition (1.4) imposes various conditions on the structure of jumps for the Lévy operator S. More precisely, the condition

$$\int_{\{0<|z|\leq 1\}} |z|^2 \, m(dz) < \infty$$

implies that the measure $m(\cdot)$ admits a singularity of order 2 at the origin, and this singularity at the origin is produced by the accumulation of *small jumps* of Markovian particles. The condition

$$\int_{\{|z|>1\}} |z| \, m(dz) < \infty$$

implies that the measure $m(\cdot)$ admits a singularity of order 1 at infinity, and this singularity at infinity is produced by the accumulation of *large jumps* of Markovian particles.

Example 1.1. A typical example of the Radon measure $m(dz)$ which satisfies the moment condition (1.4) is given by the formula

$$m(dz) = \begin{cases} \dfrac{1}{|z|^{N+2-\varepsilon}} \, dz & \text{for } 0 < |z| \leq 1, \\ \dfrac{1}{|z|^{N+\varepsilon}} \, dz & \text{for } |z| > 1, \end{cases}$$

where $\varepsilon > 0$.

Let L be a first-order, Ventcel' boundary condition such that

$$Lu(x') = \mu(x')\frac{\partial u}{\partial \mathbf{n}}(x') + \gamma(x')u(x') \quad \text{for } x' \in \partial D. \tag{1.6}$$

Here:

(1) $\mu \in C^\infty(\partial D)$ and $\mu(x') \geq 0$ on ∂D.
(2) $\gamma \in C^\infty(\partial D)$ and $\gamma(x') \leq 0$ on ∂D.

(3) $\mathbf{n} = (n_1, n_2, \ldots, n_N)$ is the unit inward normal to the boundary ∂D (see Figure 1.3).

Remark 1.1. Just as in Taira [123], we can study the boundary condition L_0 of the form

$$L_0 u(x') = \mu(x') \frac{\partial u}{\partial \boldsymbol{\nu}}(x') + \gamma(x') u(x'), \tag{1.6'}$$

where $\partial/\partial\boldsymbol{\nu}$ is the *conormal derivative* associated with the elliptic differential operator A

$$\frac{\partial}{\partial\boldsymbol{\nu}} = \sum_{j,k=1}^{N} a^{jk}(x') n_k \frac{\partial}{\partial x_j}.$$

We remark that if $\mu(x') \neq 0$ and $\gamma(x') \equiv 0$ on ∂D (resp. $\mu(x') \equiv 0$ and $\gamma(x') \neq 0$ on ∂D), then the boundary condition L is essentially the so-called Neumann (resp. Dirichlet) condition.

The terms $\mu(x')\partial u/\partial \mathbf{n}$ and $\gamma(x')u$ of the boundary condition L are supposed to correspond to reflection and absorption phenomena at the boundary ∂D, respectively. The situation may be represented schematically as in Figure 1.5 below.

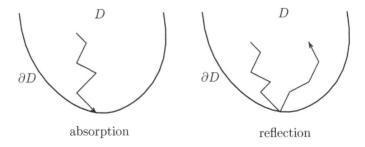

absorption   reflection

Fig. 1.5. The absorption phenomenon and the reflection phenomenon

1.2.1 The Differential Operator Case

First, we consider the case where $S \equiv 0$ in D, that is, the differential operator case:

$$\begin{cases} (A - \lambda)\, u = f & \text{in } D, \\ Lu = \varphi & \text{on } \partial D. \end{cases} \tag{$*$}_\lambda$$

Here

$$Au(x) = \sum_{i,j=1}^{N} a^{ij}(x) \frac{\partial^2 u}{\partial x_i \partial x_j}(x) + \sum_{i=1}^{N} b^i(x) \frac{\partial u}{\partial x_i}(x) + c(x) u(x)$$

and λ is a complex parameter.

We study the homogeneous problem $(*)_\lambda$ in the framework of L^p Sobolev spaces. If $1 \leq p < \infty$, we let

$L^p(D) =$ the space of (equivalence classes of) Lebesgue
measurable functions $u(x)$ on D such that
$|u(x)|^p$ is integrable on D.

The space $L^p(D)$ is a Banach space with the norm

$$\|u\|_p = \left(\int_D |u(x)|^p dx \right)^{1/p}.$$

If m is a non-negative integer, we define the usual Sobolev space

$W^{m,p}(D) =$ the space of (equivalence classes of) functions
$u \in L^p(D)$ whose derivatives $D^\alpha u(x), |\alpha| \leq m,$
in the sense of distributions are in $L^p(D)$.

The space $W^{m,p}(D)$ is a Banach space with the norm

$$\|u\|_{m,p} = \left(\sum_{|\alpha| \leq m} \int_D |D^\alpha u(x)|^p dx \right)^{1/p}.$$

We remark that

$$W^{0,p}(D) = L^p(D); \quad \| \cdot \|_{0,p} = \| \cdot \|_p.$$

Furthermore, we let

$B^{m-1/p,p}(\partial D) =$ the space of the boundary values $\varphi(x')$ of functions
$u \in W^{m,p}(D)$.

In the space $B^{m-1/p,p}(\partial D)$, we introduce a norm

$$|\varphi|_{m-1/p,p} = \inf \|u\|_{m,p},$$

where the infimum is taken over all functions $u \in W^{m,p}(D)$ that equal φ on the boundary ∂D. The space $B^{m-1/p,p}(\partial D)$ is a Banach space with respect to this norm $|\cdot|_{m-1/p,p}$; more precisely, it is a Besov space (cf. [2], [13], [112], [135]). Hence we have, by the trace theorem,

$$Lu = \mu(x')\frac{\partial u}{\partial \mathbf{n}} + \gamma(x')u \bigg|_{\partial D} \in B^{1-1/p,p}(\partial D) \quad \text{for } u \in W^{2,p}(D).$$

It should be emphasized that problem $(*)_\lambda$ is a *degenerate*, elliptic boundary value problem in the sense of Lopatinskii–Shapiro (see [26, Chapitre V, condition (4.5)]; [61, Chapter XX, Definition 20.1.1]; [93, Chapter 3, p. 194, Definition 1]; [146, Chapter II, Condition 11.1]). This is due to the fact that the so-called Lopatinskii–Shapiro complementary condition is violated at each point x' of the set

$$M = \{x' \in \partial D : \mu(x') = 0\}.$$

(see [123, Example 6.1]). More precisely, it is easy to see that the boundary value problem $(*)_\lambda$ is non-degenerate (or coercive) if and only if either $\mu(x') \neq 0$ on ∂D (the regular Robin case) or $\mu(x') \equiv 0$ and $\gamma(x') \neq 0$ on ∂D (the Dirichlet case). The generation theorem of analytic semigroups is well established in the non-degenerate case both in the L^p topology and in the topology of uniform convergence (cf. Friedman [43], Tanabe [132], Masuda [78], Stewart [110]).

In this book, under the condition that $\mu(x') \geq 0$ on ∂D we shall consider the problem of existence and uniqueness of solutions of the boundary value problem $(*)_\lambda$ in the framework of Sobolev spaces of L^p type, and generalize the generation theorem for analytic semigroups to the *degenerate* case.

Our fundamental conditions on L are formulated as follows:

(A) $\mu(x') \geq 0$ and $\gamma(x') \leq 0$ on ∂D.
(B) $\mu(x') - \gamma(x') = \mu(x') + |\gamma(x')| > 0$ on ∂D.

A probabilistic meaning of condition (B) is (see Figure 1.6 below) that absorption phenomenon occurs at each point of the boundary portion

$$M = \{x' \in \partial D : \mu(x') = 0\},$$

while reflection phenomenon occurs at each point of the boundary portion

$$\partial D \setminus M = \{x' \in \partial D : \mu(x') > 0\}.$$

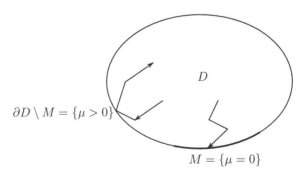

Fig. 1.6. A probabilistic meaning of condition (B)

Amann [8] studied the non-degenerate case; more precisely, he assumes that the boundary ∂D is the *disjoint union* of the two closed subsets M and $\partial D \setminus M$, each of which is an $(N-1)$ dimensional compact smooth manifold.

We give a simple example of the functions $\mu(x')$ and $\gamma(x')$ in a relatively compact domain D with smooth boundary ∂D in the plane $\mathbf{R}^2$ $(N = 2)$:

Example 1.2. Let $D = \{(x_1, x_2) \in \mathbf{R}^2 : x_1^2 + x_2^2 < 1\}$ be the *unit disk* with the boundary $\partial D = \{(x_1, x_2) \in \mathbf{R}^2 : x_1^2 + x_2^2 = 1\}$. For a local coordinate system $x_1 = \cos\theta$, $x_2 = \sin\theta$ with $\theta \in [0, 2\pi]$ on the unit circle ∂D, we define functions $\mu(x_1, x_2)$ and $\gamma(x_1, x_2)$ as follows:

$$\mu(x_1, x_2) = \mu(\cos\theta, \sin\theta)$$

$$= \begin{cases} e^{\frac{2}{\pi}-\frac{1}{\theta}}\left(1 - e^{\frac{2}{\pi}+\frac{1}{\theta-\frac{\pi}{2}}}\right) & \text{for } \theta \in \left[0, \frac{\pi}{2}\right], \\ 1 & \text{for } \theta \in \left[\frac{\pi}{2}, \pi\right], \\ e^{\frac{2}{\pi}+\frac{1}{\theta-\frac{3\pi}{2}}}\left(1 - e^{\frac{2}{\pi}-\frac{1}{\theta-\pi}}\right) & \text{for } \theta \in \left[\pi, \frac{3\pi}{2}\right], \\ 0 & \text{for } \theta \in \left[\frac{3\pi}{2}, 2\pi\right], \end{cases}$$

and

$$\gamma(x_1, x_2) = \mu(x_1, x_2) - 1 \quad \text{on } \partial D.$$

Here

$$M = \left\{(\cos\theta, \sin\theta) \in \mathbf{R}^2 : \theta \in \left[\frac{3\pi}{2}, 2\pi\right]\right\}.$$

Therefore, the crucial point in our approach is how to generalize the classical variational approach to the degenerate case.

We begin with the following *a priori* estimate (1.7) in the framework of L^p Sobolev spaces (see [121, Theorem 1.1]):

Theorem 1.2. *Let $1 < p < \infty$. Assume that conditions (A) and (B) are satisfied. Then, for any solution $u \in W^{2,p}(D)$ of problem $(*)_\lambda$ with $f \in L^p(D)$ and $\varphi \in B^{2-1/p,p}(\partial D)$ we have the* a priori *estimate*

$$\|u\|_{2,p} \leq C(\lambda)\left(\|f\|_p + |\varphi|_{2-1/p,p} + \|u\|_p\right), \tag{1.7}$$

with a positive constant $C(\lambda)$ depending on λ.

Remark 1.3. Some remarks are in order.

(1) It is worthwhile pointing out that the *a priori* estimate (1.7) is the same one for the Dirichlet condition: $\mu(x') \equiv 0$ and $\gamma(x') \neq 0$ on ∂D (cf. [4], [77]). This rather surprising result (elliptic estimates for a degenerate problem) works, since the degeneracy occurs only for the boundary data φ.

(2) More precisely, we can obtain an existence and uniqueness theorem for the *non-homogeneous* boundary value problem

$$\begin{cases} Au = f & \text{in } D, \\ Lu = \varphi & \text{on } \partial D \end{cases} \tag{*}$$

in the framework of L^p Sobolev spaces, if we take $S := 0$ in Theorem 1.8 below.

(I) Analytic Semigroups in the L^p Topology

First, we formulate a generation theorem for analytic semigroups in the L^p topology.

We associate with problem $(*)_\lambda$ an unbounded linear operator

$$\boxed{A_p \colon L^p(D) \longrightarrow L^p(D)}$$

in the Banach space $L^p(D)$ into itself as follows:

(a) The domain of definition $\mathcal{D}(A_p)$ is the set

$$\mathcal{D}(A_p) = \left\{ u \in W^{2,p}(D) : Lu = \mu(x')\frac{\partial u}{\partial \mathbf{n}} + \gamma(x')u = 0 \text{ on } \partial D \right\}. \quad (1.8)$$

(b) $A_p u = Au$ for every $u \in \mathcal{D}(A_p)$.

Then we can prove that the operator A_p generates an analytic semigroup in the Banach space $L^p(D)$ (see [121, Theorem 1.2]):

Theorem 1.4. *Let $1 < p < \infty$. Assume that conditions (A) and (B) are satisfied. Then we have the following two assertions:*

(i) For every positive number ε, there exists a positive constant $r_p(\varepsilon)$ such that the resolvent set of A_p contains the set

$$\Sigma_p(\varepsilon) = \left\{ \lambda = r^2 e^{i\theta} : r \geq r_p(\varepsilon),\ -\pi + \varepsilon \leq \theta \leq \pi - \varepsilon \right\},$$

and that the resolvent $(A_p - \lambda I)^{-1}$ satisfies the estimate

$$\left\| (A_p - \lambda I)^{-1} \right\| \leq \frac{c_p(\varepsilon)}{|\lambda|} \quad \text{for all } \lambda \in \Sigma_p(\varepsilon), \quad (1.9)$$

where $c_p(\varepsilon)$ is a positive constant depending on ε.
(ii) The operator A_p generates a semigroup $e^{z A_p}$ on $L^p(D)$ that is analytic in the sector

$$\Delta_\varepsilon = \left\{ z = t + is : z \neq 0, |\arg z| < \pi/2 - \varepsilon \right\}$$

for any $0 < \varepsilon < \pi/2$ (see Figure 1.7 below).

(II) Analytic Semigroups in the Topology of Uniform Convergence and Feller Semigroups

Secondly, we formulate a generation theorem for analytic semigroups in the topology of uniform convergence.

Let $C(\overline{D})$ be the space of real-valued, continuous functions $f(x)$ on $\overline{D}$. We equip the space $C(\overline{D})$ with the topology of uniform convergence on the whole $\overline{D}$; hence it is a Banach space with the maximum norm

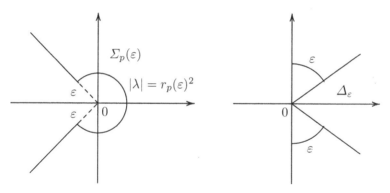

Fig. 1.7. The set $\Sigma_p(\varepsilon)$ and the sector Δ_ε

$$\|f\|_\infty = \max_{x \in \overline{D}} |f(x)|.$$

We introduce a subspace of $C(\overline{D})$ that is associated with the boundary condition L. We remark that the boundary condition

$$Lu = \mu(x')\frac{\partial u}{\partial \mathbf{n}} + \gamma(x')u\bigg|_{\partial D} = 0 \quad \text{on } \partial D$$

includes the condition

$$u = 0 \quad \text{on } M = \{x' \in \partial D : \mu(x') = 0\},$$

if $\gamma(x') \neq 0$ on M. With this fact in mind, we let

$$C_0\left(\overline{D} \setminus M\right) = \left\{ u \in C(\overline{D}) : u = 0 \text{ on } M \right\}.$$

The space $C_0\left(\overline{D} \setminus M\right)$ is a closed subspace of $C(\overline{D})$; hence it is a Banach space. Furthermore, we introduce an unbounded linear operator

$$\boxed{\mathfrak{A} \colon C_0\left(\overline{D} \setminus M\right) \longrightarrow C_0\left(\overline{D} \setminus M\right)}$$

as follows:

(a) The domain of definition $\mathcal{D}\left(\mathfrak{A}\right)$ is the set

$$\mathcal{D}\left(\mathfrak{A}\right) = \left\{ u \in C_0\left(\overline{D} \setminus M\right) : Au \in C_0\left(\overline{D} \setminus M\right), \ Lu = 0 \text{ on } \partial D \right\}. \quad (1.10)$$

(b) $\mathfrak{A}u = Au$ for every $u \in \mathcal{D}\left(\mathfrak{A}\right)$.

Here Au and Lu are taken in the sense of *distributions* (see Chapter 12).

Then we can prove that the operator $\mathfrak{A}$ generates an analytic semigroup in the Banach space $C_0\left(\overline{D} \setminus M\right)$ ([121, Theorem 1.3]):

Theorem 1.5. *Assume that conditions (A) and (B) are satisfied. Then we have the following two assertions:*

(i) For every positive number ε, there exists a positive constant $r(\varepsilon)$ such that the resolvent set of $\mathfrak{A}$ contains the set

$$\Sigma(\varepsilon) = \left\{\lambda = r^2 e^{i\theta} : r \geq r(\varepsilon),\ -\pi + \varepsilon \leq \theta \leq \pi - \varepsilon \right\},$$

and that the resolvent $(\mathfrak{A} - \lambda I)^{-1}$ satisfies the estimate

$$\left\| (\mathfrak{A} - \lambda I)^{-1} \right\| \leq \frac{c(\varepsilon)}{|\lambda|} \quad \text{for all } \lambda \in \Sigma(\varepsilon), \tag{1.11}$$

where $c(\varepsilon)$ is a positive constant depending on ε.

(ii) The operator $\mathfrak{A}$ generates a semigroup $e^{z\,\mathfrak{A}}$ on $C_0\left(\overline{D} \setminus M\right)$ that is analytic in the sector

$$\Delta_\varepsilon = \left\{ z = t + is : z \neq 0, |\arg z| < \pi/2 - \varepsilon \right\}$$

for any $0 < \varepsilon < \pi/2$ (see Figure 1.8 below).

Moreover, the operators $\left\{ e^{t\,\mathfrak{A}} \right\}_{t \geq 0}$ are non-negative and contractive on $C_0\left(\overline{D} \setminus M\right)$:

$$f \in C_0\left(\overline{D} \setminus M\right),\ 0 \leq f(x) \leq 1 \text{ on } \overline{D} \setminus M \tag{1.12}$$
$$\implies\ 0 \leq e^{t\,\mathfrak{A}} f(x) \leq 1 \text{ on } \overline{D} \setminus M.$$

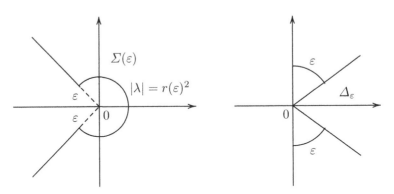

Fig. 1.8. The set $\Sigma(\varepsilon)$ and the sector Δ_ε

The main purpose of this book is devoted to the functional analytic approach to the problem of existence of Markov processes in probability theory.

A strongly continuous semigroup $\{T_t\}_{t \geq 0}$ on the space $C_0\left(\overline{D} \setminus M\right)$ is called a *Feller semigroup* on the state space $\overline{D} \setminus M$ if it is non-negative and contractive on the Banach space $C_0\left(\overline{D} \setminus M\right)$.

Therefore, we can reformulate assertion (1.12) of Theorem 1.5 as follows ([121, Theorem 1.4]):

Theorem 1.6. *If conditions (A) and (B) are satisfied, then the operator* $\mathfrak{A}$ *generates a Feller semigroup* $\left\{ e^{t\,\mathfrak{A}} \right\}_{t \geq 0}$ *on the state space* $\overline{D} \setminus M$.

Theorem 1.6 generalizes Bony–Courrège–Priouret [15, Théorème XIX] to the case where $\mu(x') \geq 0$ on the boundary ∂D (cf. [114, Theorem 10.1.3]).

It is known (cf. [34], [122, Chapter 9]) that if $\{T_t\}_{t>0}$ is a Feller semigroup on the state space $\overline{D} \setminus M$, then there exists a unique Markov transition function $p_t(x, \cdot)$ on the space $\overline{D} \setminus M$ such that

$$T_t f(x) = \int_{\overline{D} \setminus M} p_t(x, dy) f(y) \quad \text{for all } f \in C_0 \left(\overline{D} \setminus M \right).$$

Furthermore, it can be shown that the function $p_t(x, \cdot)$ is the transition function of some strong *Markov process* $\mathcal{X}$; hence the value $p_t(x, E)$ expresses the transition probability that a Markovian particle starting at position x will be found in the set E at time t.

The differential operator A describes analytically a strong Markov process with continuous paths in the interior D such as Brownian motion.

Rephrased, Theorem 1.6 asserts that there exists a Feller semigroup on the state space $\overline{D}$ corresponding to such a diffusion phenomenon that a Markovian particle moves continuously in the state space $\overline{D} \setminus M$ until it "dies" at the time when it reaches the set M where the particle is definitely absorbed (see Figure1.6).

Remark 1.7. It is worthwhile pointing out here that the condition

$$\mu(x') \geq 0 \text{ and } \gamma(x') \leq 0 \quad \text{on } \partial D$$

is necessary in order that the operator $\mathfrak{A}$ is the infinitesimal generator of a Feller semigroup $\{T_t\}_{t \geq 0}$ on the state space $\overline{D} \setminus M$ (cf. [114, Section 9.5]).

Our functional analytic approach to strong Markov processes may be visualized as in Table 1.3 below (see also Figure 1.9 below).

1.2.2 The Integro-Differential Operator Case

Now we consider the general case, that is, the integro-differential operator case:

$$\begin{cases} (W - \lambda)\, u = f & \text{in } D, \\ Lu = \varphi & \text{on } \partial D. \end{cases} \tag{$**$}_\lambda$$

Here

$$Wu(x) = Au(x) + Su(x) \tag{1.2}$$

$$:= \left(\sum_{i,j=1}^{N} a^{ij}(x) \frac{\partial^2 u}{\partial x_i \partial x_j}(x) + \sum_{i=1}^{N} b^i(x) \frac{\partial u}{\partial x_i}(x) + c(x)u(x) \right)$$

Transition function $p_t(x, dy)$	Dynkin $\Longleftarrow\Longrightarrow$	Feller semigroup $T_t = e^{t\mathfrak{A}}$
Laplace transform $\updownarrow$	$T_t f(x) = \int_{\overline{D}} p_t(x, dy) f(y)$	$\updownarrow$ Hille–Yosida
Green kernel $G_\alpha(x, y)$	$\Longleftarrow\Longrightarrow$ Riesz–Markov	Green operator $(\alpha I - \mathfrak{A})^{-1}$

Table 1.3. A semigroup approach to strong Markov processes

$$+ \int_{\mathbf{R}^N \setminus \{0\}} \left(u(x+z) - u(x) - \sum_{j=1}^{N} z_j \frac{\partial u}{\partial x_j}(x) \right) s(x, z) \, m(dz)$$

and λ is a complex parameter.

(I) Unique Solvability Theorems for Waldenfels Integro-Differential Operators

The first purpose of this book is to prove an existence and uniqueness theorem for the following non-homogeneous boundary value problem

$$\begin{cases} Wu = f & \text{in } D, \\ Lu = \varphi & \text{on } \partial D \end{cases} \tag{$**$}$$

in the framework of *Hölder spaces*.

Due to the *non-local character* of the Waldenfels integro-differential operator

$$W = A + S,$$

we find more difficulties in the bounded domain D than in the whole space $\mathbf{R}^N$. In fact, when considering the Dirichlet problem in D, it is natural to use the *zero extension* of functions in the interior D outside of the closure $\overline{D} = D \cup \partial D$. This extension has a probabilistic interpretation. Namely, this corresponds to stopping the diffusion process with jumps in the whole space $\mathbf{R}^N$ at the *first exit time* of the closure $\overline{D}$. However, the zero extension produces a singularity of solutions at the boundary ∂D. In order to remove this singularity, we introduce various conditions on the structure of jumps for the Waldenfels integro-differential operator $W = A + S$ such as conditions (1.3a), (1.3b) and (1.4).

On the other hand, we introduce a subspace of $C^{1+\theta}(\partial D)$ which is associated with the *degenerate boundary condition*

$$Lu = \mu(x')\frac{\partial u}{\partial \mathbf{n}} + \gamma(x')u \bigg|_{\partial D} \qquad \text{on } \partial D.$$

If conditions (A) and (B) are satisfied, we let

$$C_*^{1+\theta}(\partial D) := \mu(x')\, C^{1+\theta}(\partial D) + |\gamma(x')|\, C^{2+\theta}(\partial D)$$
$$= \left\{ \varphi = \mu(x')\varphi_1 - \gamma(x')\varphi_2 : \varphi_1 \in C^{1+\theta}(\partial D),\ \varphi_2 \in C^{2+\theta}(\partial D) \right\},$$

and define a norm

$$|\varphi|_{C_*^{1+\theta}(\partial D)}$$
$$= \inf \left\{ |\varphi_1|_{C^{1+\theta}(\partial D)} + |\varphi_2|_{C^{2+\theta}(\partial D)} : \varphi = \mu(x')\varphi_1 - \gamma(x')\varphi_2 \right\}.$$

Then it is easy to verify (see the proof of [123, Lemma 6.8]) that the space $C_*^{1+\theta}(\partial D)$ is a Banach space with respect to the norm $|\cdot|_{1+\theta}^*$. We remark that the space $C_*^{1+\theta}(\partial D)$ is an "interpolation space" between the spaces $C^{1+\theta}(\partial D)$ and $C^{2+\theta}(\partial D)$. More precisely, we have the assertions

$$C_*^{1+\theta}(\partial D)$$
$$= \begin{cases} C^{2+\theta}(\partial D) & \text{if } \mu(x') \equiv 0 \text{ on } \partial D \text{ (the Dirichlet case)}, \\ C_*^{1+\theta}(\partial D) = C^{1+\theta}(\partial D) & \text{if } \mu(x') > 0 \text{ on } \partial D \text{ (the Robin case)}, \end{cases}$$

and, for general $\mu(x')$, the continuous injections

$$C^{1+\theta}(\partial D) \subset C_*^{1+\theta}(\partial D) \subset C^{2+\theta}(\partial D).$$

The next theorem is a generalization of [117, Theorem 1] and [123, Theorem 1.1] to the integro-differential operator case :

Theorem 1.8. *Assume that the following three conditions (A), (B) and (H) are satisfied:*

(A) $\mu(x') \geq 0$ and $\gamma(x') \leq 0$ on ∂D.
(B) $\mu(x') - \gamma(x') = \mu(x') + |\gamma(x')| > 0$ on ∂D.
(H) The integral operator S satisfies conditions (1.3a), (1.3b), (1.4) and (1.5).

Then the mapping

$$\boxed{(W, L) : C^{2+\theta}(\overline{D}) \longrightarrow C^{\theta}(\overline{D}) \oplus C_*^{1+\theta}(\partial D)}$$

is an algebraic and topological isomorphism for all $0 < \theta \leq \theta_0$. In particular, for any $f \in C^{\theta}(\overline{D})$ and any $\varphi \in C_*^{1+\theta}(\partial D)$, there exists a unique solution $u \in C^{2+\theta}(\overline{D})$ of problem (**).

(II) Analytic Semigroups in the L^p Topology

The second purpose of this book is to study problem $(**)_\lambda$ from the point of view of analytic semigroup theory in functional analysis.

First, we formulate a generation theorem for analytic semigroups in the L^p topology. To do this, we associate with problem $(**)_\lambda$ an unbounded linear operator

$$\boxed{W_p \colon L^p(D) \longrightarrow L^p(D)}$$

in the Banach space $L^p(D)$ into itself as follows:

(a) The domain of definition $\mathcal{D}(W_p)$ is the set

$$\mathcal{D}(W_p) = \left\{ u \in W^{2,p}(D) : Lu = 0 \text{ on } \partial D \right\}. \tag{1.13}$$

(b) $W_p u = W u$ for every $u \in \mathcal{D}(W_p)$.

Here Wu and Lu are taken in the sense of *distributions* (see Chapters 12 and 13).

The next theorem is a generalization of Theorem 1.5 to the integro-differential operator case:

Theorem 1.9. *Let $1 < p < \infty$. Assume that conditions (A), (B) and (H) are satisfied. Then we have the following two assertions:*

(i) For every $\varepsilon > 0$, there exists a constant $r_p(\varepsilon) > 0$ such that the resolvent set of W_p contains the set

$$\Sigma_p(\varepsilon) = \left\{ \lambda = r^2 e^{i\vartheta} : r \geq r_p(\varepsilon), -\pi + \varepsilon \leq \vartheta \leq \pi - \varepsilon \right\},$$

and that the resolvent $(W_p - \lambda I)^{-1}$ satisfies the estimate

$$\left\| (W_p - \lambda I)^{-1} \right\| \leq \frac{c_p(\varepsilon)}{|\lambda|} \quad \text{for all } \lambda \in \Sigma_p(\varepsilon), \tag{1.14}$$

where $c_p(\varepsilon) > 0$ is a constant depending on ε.

(ii) The operator W_p generates a semigroup $e^{z W_p}$ on $L^p(D)$ which is analytic in the sector

$$\Delta_\varepsilon = \left\{ z = t + is : z \neq 0, |\arg z| < \pi/2 - \varepsilon \right\}$$

for any $0 < \varepsilon < \pi/2$.

(III) Analytic Semigroups in the Topology of Uniform Convergence and Feller Semigroups

Secondly, we state a generation theorem for analytic semigroups in the topology of uniform convergence. We introduce a linear operator

$$\boxed{\mathfrak{W} \colon C_0\left(\overline{D} \setminus M \right) \longrightarrow C_0\left(\overline{D} \setminus M \right)}$$

in the Banach space $C_0\left(\overline{D} \setminus M \right)$ as follows:

(a) The domain of definition $\mathcal{D}(\mathfrak{W})$ is the set

$$\mathcal{D}(\mathfrak{W}) \tag{1.15}$$
$$= \left\{ u \in C_0\left(\overline{D} \setminus M\right) \cap W^{2,p}(D) : Wu \in C_0\left(\overline{D} \setminus M\right), \; Lu = 0 \text{ on } \partial D \right\}.$$

(b) $\mathfrak{W}u = Wu$ for every $u \in \mathcal{D}(\mathfrak{W})$.

Here we remark that the domain $\mathcal{D}(\mathfrak{W})$ is *independent* of $N < p < \infty$ (see the proof of Lemma 14.12).

Then we can prove that the operator $\mathfrak{W}$ generates an analytic semigroup in the Banach space $C_0\left(\overline{D} \setminus M\right)$. In other words, Theorem 1.9 remains valid with $L^p(D)$ and W_p replaced by $C_0\left(\overline{D} \setminus M\right)$ and $\mathfrak{W}$, respectively:

Theorem 1.10. *Let $N < p < \infty$. Assume that conditions (A), (B) and (H) are satisfied. Then we have the following two assertions:*

(i) For every $\varepsilon > 0$, there exists a constant $r(\varepsilon) > 0$ such that the resolvent set of $\mathfrak{W}$ contains the set

$$\Sigma(\varepsilon) = \left\{ \lambda = r^2 e^{i\vartheta} : r \geq r(\varepsilon), -\pi + \varepsilon \leq \vartheta \leq \pi - \varepsilon \right\},$$

and that the resolvent $(\mathfrak{W} - \lambda I)^{-1}$ satisfies the estimate

$$\left\| (\mathfrak{W} - \lambda I)^{-1} \right\| \leq \frac{c(\varepsilon)}{|\lambda|} \quad \text{for all } \lambda \in \Sigma(\varepsilon), \tag{1.16}$$

where $c(\varepsilon) > 0$ is a constant depending on ε.
(ii) The operator $\mathfrak{W}$ generates a semigroup $e^{z\,\mathfrak{W}}$ on $C_0\left(\overline{D} \setminus M\right)$ that is analytic in the sector

$$\Delta_\varepsilon = \left\{ z = t + is : z \neq 0, |\arg z| < \pi/2 - \varepsilon \right\}$$

for any $0 < \varepsilon < \pi/2$.

Theorems 1.9 and 1.10 express a *regularizing effect* for the parabolic integro-differential operator $\partial/\partial t - W$ with the homogeneous boundary condition L (cf. [48, Chapter VIII, Theorem 3.1]).

As an application of Theorem 1.8, we consider the problem of existence of Markov processes in probability theory. To do this, we define a linear operator

$$\boxed{\mathcal{W} \colon C_0\left(\overline{D} \setminus M\right) \longrightarrow C_0\left(\overline{D} \setminus M\right)}$$

in the Banach space as follows:

(a) The domain of definition $\mathcal{D}(\mathcal{W})$ is the set

$$\mathcal{D}(\mathcal{W}) \tag{1.17}$$
$$= \left\{ u \in C^2(\overline{D}) \cap C_0\left(\overline{D} \setminus M\right) : Wu \in C_0\left(\overline{D} \setminus M\right), \; Lu = 0 \text{ on } \partial D \right\}.$$

(b) $\mathcal{W}u = Wu$ for every $u \in \mathcal{D}(\mathcal{W})$.

The next theorem asserts that there exists a Feller semigroup on the state space $\overline{D} \setminus M$ corresponding to such a diffusion phenomenon that a Markovian particle moves both by jumps and continuously in the state space $\overline{D} \setminus M$ until it "dies" at the time when it reaches the set M where the particle is definitely absorbed (cf. [72, Theorem 5.2], [111, Theorem 2.2], [48, Chapter VIII, Theorem 3.3]):

Theorem 1.11. *If conditions (A), (B) and (H) are satisfied, then the operator W is* closable *in the space $C_0(\overline{D} \setminus M)$, and its minimal closed extension $\overline{W}$ is the infinitesimal generator of some Feller semigroup $T_t = e^{t\overline{W}}$ on the state space $\overline{D} \setminus M$.*

Remark 1.12. By combining Theorems 1.11 and 1.10, we can prove that the operator $\mathfrak{W}$ coincides with the minimal closed extension $\overline{W}$ (see Section 14.6):

$$\mathfrak{W} = \overline{W}. \tag{14.45}$$

Our functional analytic approach to strong Markov processes may be visualized as in Figure 1.9 below.

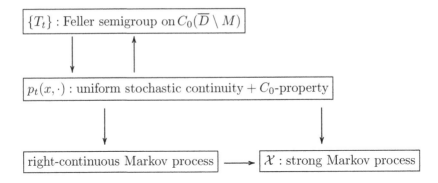

Fig. 1.9. A functional analytic approach to strong Markov processes

1.3 Summary of the Contents

This introductory chapter 1 is intended as a brief introduction to our problem and results in such a fashion that a broad spectrum of readers could understand. The contents of the book are divided into five principal parts.

Chapter 2 is devoted to a review of standard topics from the theory of analytic semigroups which forms a functional analytic background for the proof of Theorems 1.4 and 1.5.

Chapter 3 is devoted to a functional analytic approach to Markov processes in probability which forms a functional analytic background for the proof of Theorems 1.6 and 1.11.

Section 3.1 is devoted to the Riesz–Markov representation theorem which describes an intimate relationship between measures and linear functionals (Theorems 3.7 and 3.10).

Section 3.2 is devoted to a brief description of basic definitions and results about Markov processes. We give concrete important examples of Markov transition functions on the line $\mathbf{R} = (-\infty, \infty)$, and some examples of diffusion processes on the half line $\overline{\mathbf{R}^+} = [0, \infty)$ in which we must take account of the effect of the boundary point 0.

Section 3.3 provides a brief description of basic definitions and results about Markov processes and a class of semigroups (Feller semigroups) associated with Markov processes. The semigroup approach to Markov processes can be traced back to the pioneering work of Feller [39] and [40] in early 1950s (cf. [15], [98], [116]).

In Section 3.4 we prove various generation theorems of Feller semigroups by using the Hille–Yosida theory of semigroups (Theorems 3.34 and 3.36). We give two examples where it is difficult to begin with a Markov transition function and the infinitesimal generator is the basic tool of describing the process (Examples 3.16 and 3.17).

Section 3.5 is devoted to the reflecting diffusion that is associated with the homogeneous Neumann problem for the Laplacian.

In Section 3.6, following Sato–Tanaka [97] and Sato–Ueno [98] we introduce the notion of local time on the boundary for the reflecting diffusion constructed in Section 3.5.

In **Chapter 4** we present a brief description of the basic concepts and results of the L^p theory of pseudo-differential operators which may be considered as a modern version of the classical potential theory.

In Section 4.1 we define Hölder spaces $C^{k+\theta}(\overline{\Omega})$ and various Sobolev spaces $W^{s,p}(\Omega)$ and $H^{s,p}(\Omega)$ and also Besov spaces $B^{s,p}(\partial\Omega)$ on the boundary $\partial\Omega$ of a smooth domain Ω of Euclidean space $\mathbf{R}^n$. It is the imbedding characteristics of L^p Sobolev spaces that render these spaces so useful in the study of partial differential equations. In the proof of Theorem 1.5 we shall make use of some imbedding properties of L^p Sobolev spaces (Theorem 4.8). Moreover, we shall need the Rellich–Kondrachov compactness theorem for function spaces of L^p type (Theorem 4.10) in the proof of Theorem 1.4. The Rellich–Kondrachov theorem is a Sobolev space version of the Bolzano–Weierstrass theorem and the Ascoli–Arzelà theorem in calculus.

In Section 4.2 we formulate Seeley's extension theorem (Theorem 4.11), due to Seeley [103], which asserts that the functions in $C^\infty(\overline{\Omega})$ are the restrictions to Ω of functions in $C^\infty(\mathbf{R}^n)$.

It should be emphasized that Besov spaces $B^{s,p}(\partial\Omega)$ enter naturally in connection with boundary value problems in the framework of Sobolev spaces of L^p type. Indeed, we need to make sense of the restriction $u|_{\partial\Omega}$ to the boundary $\partial\Omega$ as an element of a Besov space on $\partial\Omega$ when u belongs to a Sobolev space on the domain Ω. In Section 4.3, we formulate an important trace theorem (Theorem 4.18) that will be used in the study of boundary value problems.

In Section 4.4, we present a brief description of basic concepts and results of the theory of Fourier integral operators and pseudo-differential operators. Pseudo-differential operators provide a constructive tool to deal with existence and smoothness of solutions of partial differential equations. The theory of pseudo-differential operators continues to be one of the most influential works in modern history of analysis, and is a very refined mathematical tool whose full power is yet to be exploited.

In Section 4.5 we present a brief description of the basic concepts and results of the L^p theory of pseudo-differential operators. We formulate the Besov space boundedness theorem due to Bourdaud [16] (Theorem 4.47) in Subsection 4.5.2 and we give a useful criterion for hypoellipticity due to Hörmander [59] (Theorem 4.49) in Subsection 4.5.4, which plays an essential role in the proof of our main results. In Subsection 4.5.5, following Coifman–Meyer [30, Chapitre IV, Proposition 1]) we state that the distribution kernel

$$s(x, y) = \frac{1}{(2\pi)^n} \int_{\mathbf{R}^n} e^{i(x-y)\cdot\xi}\, a(x, \xi)\, d\xi,$$

of a pseudo-differential operator $S \in L^m_{1,0}(\mathbf{R}^n)$ with symbol $a(x, \xi)$ satisfies the estimate (Theorem 4.51)

$$|s(x, y)| \leq \frac{C}{|x - y|^{m+n}} \quad \text{for all } x,\, y \in \mathbf{R}^n \text{ and } x \neq y.$$

In Section 4.6, by using the Riesz–Schauder theory we prove some of the most important results about elliptic pseudo-differential operators on a manifold and their indices in the framework of Sobolev spaces (Theorems 4.53 through 4.67). These results play an important role in the study of elliptic boundary value problems in Chapter 5.

The heat kernel

$$K_t(x) = \frac{1}{(4\pi t)^{n/2}}\, e^{-\frac{|x|^2}{4t}} \quad \text{for } t > 0$$

has many important and interesting applications in partial differential equations. In Section 4.7, by calculating various convolution kernels for the Laplacian via the Laplace transform we derive Newtonian, Riesz and Bessel potentials and also the Poisson kernel for the Dirichlet boundary value problem (Theorems 4.69, 4.70 and 4.73).

In **Chapter 5** we introduce the notion of *transmission property* due to Boutet de Monvel [19], which is a condition about symbols in the normal direction at the boundary. Elliptic boundary value problems cannot be treated directly by pseudo-differential operator methods. It was Boutet de Monvel [19] who brought in the operator-algebraic aspect with his calculus in 1971. He introduced a 2×2 matrix $\mathcal{A}$ of operators, and constructed a relatively small "algebra", called the *Boutet de Monvel algebra*, which contains the boundary value problems for elliptic differential operators as well as their parametrices.

We will take a close look at Boutet de Monvel's work.

Let Ω be a bounded, domain of Euclidean space $\mathbf{R}^n$ with smooth boundary $\partial\Omega$. Without loss of generality, we may assume that Ω is a relatively compact, open subset of an n-dimensional, compact smooth manifold $M = \widehat{\Omega}$ without boundary (see Figure 1.10 below). The manifold M is called the *double* of Ω.

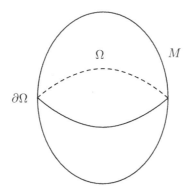

Fig. 1.10. The bounded domain Ω and the double M

Boutet de Monvel [19] introduced a 2×2 matrix $\mathcal{A}$ of operators

$$
\mathcal{A} = \begin{pmatrix} P_\Omega + G & K \\ T & S \end{pmatrix} : \begin{matrix} C^\infty(\overline{\Omega}) \\ \oplus \\ C^\infty(\partial\Omega) \end{matrix} \longrightarrow \begin{matrix} C^\infty(\overline{\Omega}) \\ \oplus \\ C^\infty(\partial\Omega) \end{matrix}
$$

Here:

(1) P is a pseudo-differential operator on the full manifold M and

$$
P_\Omega u = r^+ \left(P(e^+ u) \right) = P(e^+ u)\big|_\Omega \quad \text{for all } u \in C^\infty(\overline{\Omega}),
$$

where $e^+ u$ is the extension of u by zero to M

$$
e^+ u(x) = \begin{cases} u(x) & \text{for } x \in \overline{\Omega}, \\ 0 & \text{for } x \in M \setminus \overline{\Omega}, \end{cases}
$$

while $r^+ v$ is the restriction $v|_\Omega$ to Ω of a distribution v on M. In view of the *pseudo-local property* of P, we find that the operator P_Ω can be visualized as follows:

$$\boxed{P_\Omega \colon C^\infty(\overline{\Omega}) \xrightarrow{\ e^+\ } \mathcal{D}'(M) \xrightarrow{\ P\ } \mathcal{D}'(M) \xrightarrow{\ r^+\ } C^\infty(\Omega).}$$

The crucial requirement is that the symbol of P has the *transmission property* in order that P_Ω maps $C^\infty(\overline{\Omega})$ into itself. In Section 5.1 we introduce three basic function spaces H, H^+ and H^- (Proposition 5.1). In Section 5.2 we illustrate how the transmission property of the symbol ensures that the associated operator preserves smoothness up to the boundary (Theorems 5.6 and 5.7).

(2) S is a pseudo-differential operator on $\partial\Omega$.

(3) The potential operator K and trace operator T are generalizations of the potentials and trace operators known from the classical theory of elliptic boundary value problems, respectively.

(4) The entry G, a *singular Green operator*, is an operator which is smoothing in the interior Ω while it acts like a pseudo-differential operator in directions tangential to the boundary $\partial\Omega$. As an example, we may take the difference of two solution operators to (invertible) classical boundary value problems with the same differential part in the interior but different boundary conditions. In Section 5.3 we give typical examples of a potential operator K, a trace operator T and a singular Green operator G (see Examples 5.5 through 5.12).

Boutet de Monvel [19] proved that these operator matrices form an algebra in the following sense: Given another element of the calculus, say,

$$\mathcal{A}' = \begin{pmatrix} P'_\Omega + G' & K' \\ T' & S' \end{pmatrix} : \begin{array}{c} C^\infty(\overline{\Omega}) \\ \oplus \\ C^\infty(\partial\Omega) \end{array} \longrightarrow \begin{array}{c} C^\infty(\overline{\Omega}) \\ \oplus \\ C^\infty(\partial\Omega) \end{array}$$

the composition $\mathcal{A}'\mathcal{A}$ is again an operator matrix of the type described above. It is worth pointing out here that the product $P'_\Omega P_\Omega$ does not coincide with $(P'P)_\Omega$; in fact, the difference

$$P'_\Omega P_\Omega - (P'P)_\Omega$$

turns out to be a singular Green operator.

Section 5.4 is devoted to a brief historical perspective of the Wiener–Hopf technique. It should be emphasized that the Boutet de Monvel calculus is closely related to the classical Wiener–Hopf technique (see [145], [84]) that remains a source of inspiration to Mathematicians, Physicists and Engineers working in many diverse fields, and the areas of application continue to broaden.

In **Chapter 6** we consider the non-homogeneous general Robin problem

$$\begin{cases} Au = f & \text{in } \Omega, \\ B\gamma u := a(x')\,\dfrac{\partial u}{\partial \boldsymbol{\nu}} + b(x')u \Big|_{\partial\Omega} = \varphi & \text{on } \partial\Omega \end{cases} \qquad (6.1)$$

under the following two conditions (H.1) and (H.2) (corresponding to conditions (A) and (B) with $\mu(x') := a(x')$ and $\gamma(x') := -b(x')$):

(H.1) $a(x') \geq 0$ and $b(x') \geq 0$ on $\partial\Omega$.
(H.2) $a(x') + b(x') > 0$ on $\partial\Omega$.

Here $\nu = -\mathbf{n}$ is the unit *outward* normal to the boundary $\partial\Omega$.

We study the general Robin boundary value problem (6.1) in the framework of Sobolev spaces of L^p type, by using the L^p theory of pseudo-differential operators. The purpose of Section 6.1 is to describe, in terms of pseudo-differential operators, the classical surface and volume potentials arising in boundary value problems for elliptic differential operators. This calculus of pseudo-differential operators is applied to elliptic boundary value problems in Part III.

In Section 6.2 we consider the Dirichlet problem in the framework of Sobolev spaces of L^p type. This is a generalization of the classical potential approach to the Dirichlet problem. In Section 6.3 we formulate elliptic boundary value problems in the framework of Sobolev spaces of L^p type. The pseudo-differential operator approach to elliptic boundary value problems can be traced back to the pioneering work of Calderón [22] in early 1960s ([57], [105]). The idea of our approach is stated as follows.

In Section 6.4, we consider the following *homogeneous Neumann problem*:

$$\begin{cases} Av = f & \text{in } \Omega, \\ \dfrac{\partial v}{\partial \mathbf{n}}\bigg|_{\partial\Omega} = 0 & \text{on } \partial\Omega. \end{cases} \tag{1.18}$$

The existence and uniqueness theorem for the Neumann problem (1.18) is well established in the framework of Sobolev spaces of L^p type (cf. [4], [50], [77], [133], [146]). More precisely, in the Neumann problem for the differential operator A and its formal adjoint A^*, we have *parabolic* condensation of eigenvalues along the negative real axis, as discussed in Agmon [3, pp. 276–277].

We let

$$v := \mathcal{G}_N f.$$

The operator $\mathcal{G}_N$ is the *Green operator* for the homogeneous Neumann problem. Then it follows that a function $u(x)$ is a solution of the general Robin problem (6.1) if and only if the function

$$w(x) = u(x) - v(x) = u(x) - \mathcal{G}_N f(x)$$

is a solution of the problem

$$\begin{cases} Aw = 0 & \text{in } \Omega, \\ B\gamma w = -B\gamma v = - \left(a(x')\dfrac{\partial v}{\partial \nu} + b(x')v \right)\bigg|_{\partial\Omega} = -b(x')\,(v|_{\partial\Omega}) & \text{on } \partial\Omega. \end{cases}$$

However, we know that every solution w of the homogeneous equation

$$Aw = 0 \quad \text{in } \Omega$$

can be expressed by means of a single layer potential as follows:

$$w = \mathcal{P}\psi.$$

The operator $\mathcal{P}$ is the *Poisson operator* for the Dirichlet problem. Thus, by using the operators $\mathcal{G}_N$ and $\mathcal{P}$ we can reduce the study of the general Robin problem (6.1) to that of the equation

$$T\psi := B\gamma\left(\mathcal{P}\psi\right) = -b(x')\left((\mathcal{G}_N f)|_{\partial\Omega}\right) = -b(x')\gamma_0\left(\mathcal{G}_N f\right). \tag{1.19}$$

This is a modern version of the classical *Fredholm integral equation* in terms of pseudo-differential operators.

It is well known (cf. [26], [57], [61], [73], [93], [105], [133]) that the operator

$$T = B\gamma\mathcal{P} = a(x')\,\gamma_1\mathcal{P} + b(x')\,\gamma_0\mathcal{P} = -a(x')\,\Pi + b(x'),$$

is a pseudo-differential operator of first order on the boundary ∂D. More precisely, Π is called the *Dirichlet-to-Neumann operator* defined by the formula (cf. [138, p. 134, formula (4.13)])

$$\Pi\varphi = -\gamma_1\mathcal{P} = -\left.\frac{\partial}{\partial\boldsymbol{\nu}}\left(\mathcal{P}\varphi\right)\right|_{\partial\Omega} = \left.\frac{\partial}{\partial\mathbf{n}}\left(\mathcal{P}\varphi\right)\right|_{\partial\Omega} \quad \text{for all } \varphi \in C^\infty(\partial\Omega), \tag{6.3}$$

and further (cf. [26], [61], [73], [93], [105], [133]) that Π is a classical, *elliptic* pseudo-differential operator of first order on $\partial\Omega$.

The virtue of the boundary equation (1.19) is that there is no difficulty in taking adjoints or transposes after restricting the attention to the boundary, whereas boundary value problems in general do not have adjoints or transposes. This allows us to discuss the *existence* theory more easily. Here it should be emphasized that our reduction approach would break down if we use the Dirichlet problem (Theorem 6.15) as usual, instead of the Neumann problem (Theorem 6.16). The reader might be referred to Iwasaki [66, p. 563, Example].

The study of problem (6.1) can be expressed, in terms of the Boutet de Monvel calculus, in the following matrix formula:

$$\begin{pmatrix} A & 0 \\ B\gamma & 0 \end{pmatrix} \begin{pmatrix} \mathcal{G}_N & \mathcal{P} \\ 0 & 0 \end{pmatrix} = \begin{pmatrix} I & 0 \\ \gamma(x')\,(\gamma_0\mathcal{G}_N) & T \end{pmatrix}.$$

In Section 6.5 we study the non-homogeneous general Robin problem

$$\begin{cases} Au = f & \text{in } \Omega, \\ B\gamma u = a(x')\,\gamma_1 u + b(x')\,\gamma_0 u = \varphi & \text{on } \partial\Omega \end{cases} \tag{6.1}$$

and the non-homogeneous Neumann problem

$$\begin{cases} Av = g & \text{in } \Omega, \\ \left.\dfrac{\partial v}{\partial \mathbf{n}}\right|_{\partial D} = \psi & \text{on } \partial\Omega \end{cases} \tag{6.4}$$

in the framework of L^p Sobolev spaces from the viewpoint of the Boutet de Monvel calculus. Here

$$\gamma_1 u := \left.\frac{\partial u}{\partial \boldsymbol{\nu}}\right|_{\partial\Omega} = -\left.\frac{\partial u}{\partial \mathbf{n}}\right|_{\partial\Omega} \qquad \text{on } \partial\Omega.$$

Then we derive an index formula of Agranovič–Dynin type for the Neumann problem (6.4) and the general Robin problem (6.1) in the framework of L^p Sobolev spaces (Theorem 6.26).

In Section 6.6 we study an intimate relationship between the Dirichlet-to-Neumann operator Π and the reflecting diffusion in a bounded, domain Ω of Euclidean space $\mathbf{R}^N$ with smooth boundary ∂D (Theorems 6.27 and 6.28 and Remark 6.29).

In Section 6.7, following Mizohata [82, Chapter 3] and Wloka [146, Section 13] we prove that all the sufficiently large eigenvalues of the Dirichlet problem for the differential operator A lie in the *parabolic* type region (see assertion (6.52) and Figure 6.3).

Our subject proper starts with the third part (Chapters 7 and 9) of this book.

Chapter 7 is devoted to the proof of Theorem 1.2. The proof of Theorem 1.2 is flowcharted (Table 7.1).

The idea of our proof is stated as follows. First, we reduce the study of the boundary value problem $(*)_\lambda$ to that of a first-order pseudo-differential operator $T(\lambda) = LP(\lambda)$ on the boundary ∂D, just as in Section 6.4. Then we prove that conditions (A) and (B) are sufficient for the validity of the *a priori* estimate

$$\|u\|_{2,p} \le C(\lambda)\left(\|f\|_p + |\varphi|_{2-1/p,p} + \|u\|_p\right). \tag{1.7}$$

More precisely, we construct a *parametrix* $S(\lambda)$ for $T(\lambda)$ in the Hörmander class $L^0_{1,1/2}(\partial D)$ (Lemma 7.2), and apply the Besov-space boundedness theorem (Theorem 4.47) to $S(\lambda)$ to obtain the desired estimate (1.7) (Lemma 7.1).

Chapter 8 and the next Chapter 9 are devoted to the proof of Theorem 1.4. In this chapter we study the operator A_p, and prove *a priori* estimates for the operator $A_p - \lambda I$ (Theorem 8.3) which will play a fundamental role in the next chapter. In the proof we make good use of Agmon's method (Proposition 8.4). This is a technique of treating a spectral parameter λ as a second-order, elliptic differential operator of an extra variable and relating the old problem to a new problem with the additional variable.

In **Chapter 9** we prove Theorem 1.4 (Theorems 9.1 and 9.11). The proof of Theorem 1.4 is flowcharted (Table 9.1). Once again we make use of Agmon's method in the proof of Theorems 9.1 and 9.11. In particular, Agmon's method plays an important role in the proof of the *surjectivity* of the operator $A_p - \lambda I$ (Proposition 9.2).

Part IV (Chapters 10 and 11) is devoted to the general study of the maximum principles for second-order, elliptic Waldenfels operators.

In **Chapter 10**, following Bony–Courrège–Priouret [15] we prove various maximum principles for second-order, elliptic Waldenfels operators $W = P+S$ which play an essential role throughout the book. In Section 10.1 we give complete characterizations of linear operators W which satisfy the positive maximum principle (PM) closely related to condition (β') given in the Hille–Yosida–Ray theorem (Theorem 3.36):

$$x_0 \in \overset{\circ}{D}, \ v \in C_0^2(D) \ \text{ and } \ v(x_0) = \sup_{x \in D} v(x) \geq 0 \Longrightarrow Wv(x_0) \leq 0 \qquad \text{(PM)}$$

(Theorems 10.1, 10.2 and 10.4). In Section 10.2 we prove the weak and strong maximum principles and Hopf's boundary point lemma for second-order, elliptic Waldenfels operators $W = P + S$ (Theorems 10.5, 10.7 and Lemma 10.11) that play an important role in Chapters 12 and 13.

In **Chapter 11**, following Bony–Courrège–Priouret ([15, Chapter II]) we characterize Ventcel'–Lévy boundary operators T (Theorem 11.3) and general boundary operators $\Gamma = \Lambda + T$ (Theorem 11.4) defined on the compact smooth manifold $\overline{D}$ with boundary ∂D in terms of the positive boundary maximum principle (PMB):

$$x_0' \in \partial D, \ u \in C^2(\overline{D}) \ \text{ and } \ u(x_0') = \max_{x \in \overline{D}} u(x) \geq 0 \Longrightarrow \Gamma u(x_0') \leq 0. \qquad \text{(PMB)}$$

This chapter is very useful in the study of Markov processes with general Ventcel' boundary conditions in Chapter 14.

The fourth part (Chapters 12 through 14) of this book is devoted to the proofs of main theorems (Theorems 1.5 through 1.11).

Chapters 12 and 13 are devoted to the proofs of Theorem 1.5 and Theorem 1.6.

In **Chapter 12**, we prove part (i) of Theorem 1.5. The proof of part (i) of Theorem 1.5 is flowcharted (Table 12.1).

Part (i) of Theorem 1.5 follows from Theorem 1.4, by using Sobolev's imbedding theorems (Theorems 4.2 and 4.6) and a *λ-dependent localization* argument essentially due to Masuda [78] (Lemma 12.2).

Chapter 13 is devoted to the proofs of Theorem 1.6 and part (ii) of Theorem 1.5. This chapter is the heart of the subject.

In Section 13.1 we formulate general existence theorems for Feller semigroups in terms of elliptic boundary value problems with spectral parameter (Theorem 13.14).

In Section 13.2 we study Feller semigroups with reflecting barrier (Theorem 13.17) and then, by using these Feller semigroups we construct Feller semigroups corresponding to such a diffusion phenomenon that either absorption or reflection phenomenon occurs at each point of the boundary (Theorem

13.22). Our proof is based on the generation theorems of Feller semigroups discussed in Section 3.3.

In Section 13.3 we prove Theorem 1.6. To do so, we apply part (ii) of Theorem 3.34 to the operator $\mathfrak{A}$ defined by formula (1.11). The proof of Theorem 1.6 is flowcharted (Table 13.1).

Section 13.4 is devoted to the proof of part (ii) of Theorem 1.5. The proof is flowcharted (Table 13.2)

In **Chapter 14** we study a class of degenerate boundary value problems for second-order elliptic Waldenfels operators $W = A + S$, and generalize Theorems 1.4 and 1.5 (Theorems 1.8, 1.9, 1.10 and 1.11).

In Section 14.1, by using the Hölder space theory of pseudo-differential operators we study the non-homogeneous boundary value problem

$$\begin{cases} Au = f & \text{in } D, \\ Lu = \mu(x')\dfrac{\partial u}{\partial \mathbf{n}} + \gamma(x')u \Big|_{\partial D} = \varphi & \text{on } \partial D \end{cases} \tag{$*$}$$

in the framework of *Hölder spaces*, and prove an existence and uniqueness theorem (Theorem 14.1). More precisely, we prove that if conditions (A) and (B) are satisfied, then the mapping

$$\boxed{(A, L) : C^{2+\theta}(\overline{D}) \longrightarrow C^{\theta}(\overline{D}) \oplus C_*^{1+\theta}(\partial D)}$$

is an algebraic and topological *isomorphism* for all $0 < \theta < 1$.

In Section 14.2 we prove an existence and uniqueness theorem for the non-homogeneous boundary value problem

$$\begin{cases} Wu = (A+S)u = f & \text{in } D, \\ Lu = \varphi & \text{on } \partial D. \end{cases} \tag{$**$}$$

in the framework of Hölder spaces (Theorem 1.8). The proof of Theorem 1.8 is flowcharted (Table 14.1).

The essential point in the proof of Theorem 1.8 is to estimate the integral operator S in terms of Hölder norms (Lemmas 14.4 and 14.5). We show that if condition (H) is satisfied, then the operator

$$\boxed{(W, L) = (A, L) + (S, 0) : C^{2+\theta}(\overline{D}) \longrightarrow C^{\theta}(\overline{D}) \oplus C_*^{1+\theta}(\partial D)}$$

may be considered as a perturbation of a *compact* operator $(S, 0)$ to the operator (A, L) in the framework of Hölder spaces (Lemma 14.6).

In this way, the proof of Theorem 1.8 is reduced to the differential operator case (Theorem 14.1).

In Section 14.3 we prove Theorem 1.9 (Theorem 14.8). We estimate the integral operator S in terms of L^p norms, and show that S is an A_p-completely continuous operator in the sense of Gohberg–Kreĭn [51] (Lemmas 14.9 and 14.10). The proof of Theorem 1.9 is flowcharted (Table 14.2).

Section 14.4 is devoted to the proof of Theorem 1.10. Theorem 1.10 follows from Theorem 1.9 by using Sobolev's imbedding theorems and a λ-dependent localization argument. The proof is carried out in a chain of auxiliary lemmas (Lemmas 14.11, 14.12 and 14.15). The proof of Theorem 1.10 is flowcharted (Table 14.3).

In Section 14.5, as an application, we construct a Feller semigroup corresponding to such a diffusion phenomenon that a Markovian particle moves both by jumps and continuously in the state space until it "dies" at the time when it reaches the set where the particle is definitely absorbed, generalizing Theorem 1.6 (Theorem 1.11). The proof of Theorem 1.11 is flowcharted (Table 14.4).

In this book we have studied mainly Markov transition functions with only informal references to the random variables which actually form the Markov processes themselves. In **Chapter 15** we study this neglected side of our subject.

Section 15.1 is devoted to a review of the basic definitions and properties of Markov processes. In Section 15.2 we consider when the paths of a Markov process are actually continuous, and prove Theorem 3.19 (Corollary 15.7). In Section 15.3 we give a useful criterion for path-continuity of a Markov process $\{x_t\}$ in terms of the infinitesimal generator $\mathfrak{A}$ of the associated Feller semigroup $\{T_t\}$ (Theorem 15.9).

Section 15.4 is devoted to the study of three typical examples of multi-dimensional diffusion processes. More precisely, we prove that (1) the reflecting barrier Brownian motion (Theorem 15.11), (2) the reflecting and absorbing barrier Brownian motion (Theorem 15.14) and (3) the reflecting, absorbing and drifting barrier Brownian motion (Theorem 15.15) are multi-dimensional *diffusion processes*, namely, they are continuous strong Markov processes.

It should be emphasized that these three Brownian motions correspond to (1) the Neumann boundary value problem, (2) the Robin boundary value problem and (3) the oblique derivative boundary value problem for the Laplacian Δ in terms of elliptic boundary value problems, respectively.

In the final **Chapter 16**, as concluding remarks, we give an overview of general results on generation theorems for Feller semigroups proved mainly by the author using the theory of pseudo-differential operators ([57], [105], [106]) and the Calderón–Zygmund theory of singular integral operators ([23]).

In particular, we generalize Theorem 13.17 to the transversal case (Theorem 16.1) and Theorem 13.22 to the non-transversal case (Theorem 16.2), respectively.

1.4 An Overview of Main Theorems

Table 1.4 below gives an overview of analytic and Feller semigroups for the Waldenfels integro-differential operator

Closed operators	Function spaces	Generated semigroups
A_p (Theorem 1.4)	$L^p(D)$	$e^{z\,A_p}$ (analytic)
$\mathfrak{A}$ (Theorem 1.5)	$C_0\left(\overline{D}\setminus M\right)$	$e^{z\,\mathfrak{A}}$ (analytic)
$\mathfrak{A}$ (Theorem 1.6)	$C_0\left(\overline{D}\setminus M\right)$	$e^{t\,\mathfrak{A}}$ (Feller)
W_p (Theorem 1.9)	$L^p(D)$	$e^{z\,W_p}$ (analytic)
$\mathfrak{W}$ (Theorem 1.10)	$C_0\left(\overline{D}\setminus M\right)$	$e^{z\,\mathfrak{W}}$ (analytic)
$\overline{W}=\mathfrak{W}$ (Theorem 1.11)	$C_0\left(\overline{D}\setminus M\right)$	$e^{t\,\overline{W}}=e^{t\,\mathfrak{W}}$ (Feller)

Table 1.4. An overview of generation theorems for analytic and Feller semigroups (Theorems 1.4 through 1.11)

$$Wu(x) = Au(x) + Su(x) \tag{1.2}$$

$$:= \left(\sum_{i,j=1}^{N} a^{ij}(x)\frac{\partial^2 u}{\partial x_i \partial x_j}(x) + \sum_{i=1}^{N} b^i(x)\frac{\partial u}{\partial x_i}(x) + c(x)u(x) \right)$$

$$+ \int_{\mathbf{R}^N\setminus\{0\}} \left(u(x+z) - u(x) - \sum_{j=1}^{N} z_j \frac{\partial u}{\partial x_j}(x) \right) s(x,z)\, m(dz)$$

and the Ventcel' boundary condition

$$Lu(x') = \mu(x')\frac{\partial u}{\partial \mathbf{n}}(x') + \gamma(x')u(x'). \tag{1.6}$$

Both Theorems 1.4 and 1.9 are a generalization of Agranovich–Vishik [6, Theorem 5.1] to the *degenerate* case. Our approach here is distinguished by the extensive use of the ideas and techniques characteristic of the recent developments in the L^p theory of pseudo-differential operators which may be considered as a modern version of the classical potential theory. It should be emphasized that pseudo-differential operators provide a constructive tool to deal with existence and smoothness of solutions of partial differential equations.

Finally, Table 1.5 below gives a bird's-eye view of Markov processes and elliptic boundary value problems via the Boutet de Monvel calculus.

Field	Probability (Microscopic approach)	Partial Differential Equations (Mesoscopic approach)
Mathematical subject	Markov processes on the domain	Elliptic boundary value problems
Reduction to the boundary	Markov processes on the boundary	Fredholm integral equations on the boundary
Mathematical theory	Stochastic calculus	Boutet de Monvel calculus

Table 1.5. A bird's-eye view of Markov processes and elliptic boundary value problems via the Boutet de Monvel calculus

1.5 Notes and Comments

This chapter is mainly based on the lecture entitled *A mathematical study of diffusion* delivered at Mathematisch-Physikalisches Kolloquim, Leibniz Universität Hannover, Germany, on November 3, 2015.

For further study of Ventcel' boundary value problems for elliptic Waldenfels operators, the reader might be referred to [128, Theorems 1.1 and 1.2].

Analytic and Feller Semigroups and Markov
Processes

Analytic Semigroups

This chapter is devoted to a review of standard topics from the theory of analytic semigroups which forms a functional analytic background for the proof of Theorems 1.4 and 1.5.

2.1 Analytic Semigroups via the Cauchy Integral

Let E be a Banach space over the real or complex number field, and let $A \colon E \to E$ be a *densely defined*, closed linear operator with domain $\mathcal{D}(A)$.

Assume that the operator A satisfies the following two conditions (see Figure 2.1 below):

(1) The resolvent set of A contains the region

$$\Sigma_\omega = \{\lambda \in \mathbf{C} : \lambda \neq 0, |\arg \lambda| < \pi/2 + \omega\} \quad \text{for } 0 < \omega < \pi/2. \qquad (2.1)$$

(2) For each $\varepsilon > 0$, there exists a positive constant M_ε such that the resolvent $R(\lambda) = (A - \lambda I)^{-1}$ satisfies the estimate

$$\|R(\lambda)\| \leq \frac{M_\varepsilon}{|\lambda|} \qquad (2.2)$$

for all $\lambda \in \Sigma_\omega^\varepsilon = \{\lambda \in \mathbf{C} : \lambda \neq 0, \ |\arg \lambda| \leq \pi/2 + \omega - \varepsilon\}$.

Then we let

$$\boxed{U(t) = -\frac{1}{2\pi i} \int_\Gamma e^{\lambda t} R(\lambda) \, d\lambda = -\frac{1}{2\pi i} \int_\Gamma e^{\lambda t} (A - \lambda I)^{-1} \, d\lambda.} \qquad (2.3)$$

Here Γ is a path in the set $\Sigma_\omega^\varepsilon$ consisting of the following three curves (see Figure 2.2 below):

$$\Gamma^{(1)} = \left\{ r e^{-i(\pi/2 + \omega - \varepsilon)} : 1 \leq r < \infty \right\},$$

© Springer Nature Switzerland AG 2020
K. Taira, *Boundary Value Problems and Markov Processes*, Lecture Notes in Mathematics 1499,
https://doi.org/10.1007/978-3-030-48788-1_2

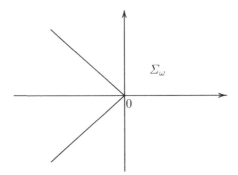

Fig. 2.1. The region Σ_ω in condition (1)

$$\Gamma^{(2)} = \left\{ e^{i\theta} : -(\pi/2 + \omega - \varepsilon) \le \theta \le \pi/2 + \omega - \varepsilon \right\},$$
$$\Gamma^{(3)} = \left\{ re^{i(\pi/2 + \omega - \varepsilon)} : 1 \le r < \infty \right\}.$$

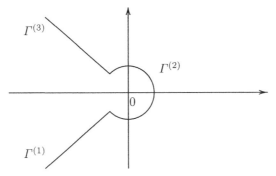

Fig. 2.2. The integral path Γ consisting of $\Gamma^{(1)}$, $\Gamma^{(2)}$ and $\Gamma^{(3)}$

It is easy to see that the integral

$$U(t) = -\frac{1}{2\pi i} \sum_{k=1}^{3} \int_{\Gamma^{(k)}} e^{\lambda t} R(\lambda) \, d\lambda$$

converges in the uniform operator topology of the Banach space $L(E, E)$ for all $t > 0$, and thus defines a bounded linear operator on E. Here $L(E, E)$ denotes the space of bounded linear operators on E.

Furthermore, we have the following proposition:

Proposition 2.1. *The operators $U(t)$, defined by formula (2.2), form a semi-group on E, that is, they enjoy the* semigroup property

$$U(t + s) = U(t) \cdot U(s) \quad \text{for all } t, s > 0.$$

Proof. By Cauchy's theorem, we may assume that

$$U(s) = -\frac{1}{2\pi i} \int_{\Gamma'} e^{\mu s} R(\mu) \, d\mu \quad \text{for } s > 0.$$

Here Γ' is a path obtained from the path Γ by translating each point of Γ to the right by a fixed small positive distance (see Figure 2.3 below).

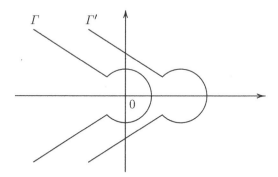

Fig. 2.3. The integral paths Γ and Γ'

Then we have, by Fubini's theorem,

$$
\begin{aligned}
U(t) \cdot U(s) &= \frac{1}{(2\pi i)^2} \int_{\Gamma} \int_{\Gamma'} e^{\lambda t} e^{\mu s} R(\lambda) R(\mu) \, d\lambda \, d\mu \\
&= \frac{1}{(2\pi i)^2} \int_{\Gamma} \int_{\Gamma'} e^{\lambda t} e^{\mu s} \frac{R(\lambda) - R(\mu)}{\lambda - \mu} \, d\lambda \, d\mu \\
&= \frac{1}{2\pi i} \int_{\Gamma} e^{\lambda t} R(\lambda) \left[\frac{1}{2\pi i} \int_{\Gamma'} \frac{e^{\mu s}}{\lambda - \mu} \, d\mu \right] d\lambda \\
&\quad - \frac{1}{2\pi i} \int_{\Gamma'} e^{\mu s} R(\mu) \left[\frac{1}{2\pi i} \int_{\Gamma} \frac{e^{\lambda t}}{\lambda - \mu} \, d\lambda \right] d\mu.
\end{aligned}
$$

We calculate the two terms in the last part.

(a) We let

$$f(\mu) = \frac{e^{\mu s}}{\lambda - \mu}, \quad \mu \in \mathbf{C}.$$

Then, by applying the residue theorem we obtain that (see Figure 2.4 below)

$$
\int_{\Gamma'^{(1)} \cap \{|\mu| \le r\}} f(\mu) \, d\mu + \int_{\Gamma'^{(2)}} f(\mu) \, d\mu + \int_{\Gamma'^{(3)} \cap \{|\mu| \le r\}} f(\mu) \, d\mu
$$
$$
+ \int_{-(\pi/2+\omega-\varepsilon)}^{\pi/2+\omega-\varepsilon} f(r \, e^{i\theta}) r \, i \, e^{i\theta} \, d\theta
$$

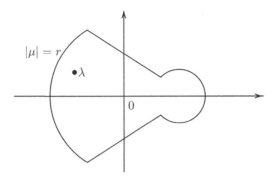

$|\mu| = r$

$\bullet\,\lambda$

0

Fig. 2.4. The truncated integral path Γ' for the residue theorem

$$= 2\pi i \operatorname{Res}\left[f(\mu)\right]_{\mu=\lambda}$$
$$= -2\pi\, i\, e^{\lambda s}.$$

However, we have, as $r \to \infty$,

$$\int_{\Gamma'^{(1)} \cap \{|\mu| \le r\}} f(\mu)\, d\mu \longrightarrow \int_{\Gamma'^{(1)}} f(\mu)\, d\mu,$$

$$\int_{\Gamma'^{(3)} \cap \{|\mu| \le r\}} f(\mu)\, d\mu \longrightarrow \int_{\Gamma'^{(3)}} f(\mu)\, d\mu,$$

and also

$$\left| \int_{-(\pi/2+\omega-\varepsilon)}^{\pi/2+\omega-\varepsilon} f\left(re^{i\theta}\right) r\, i\, e^{i\theta}\, d\theta \right| \le e^{-rs\cdot\sin(\omega-\varepsilon)} \int_{-(\pi/2+\omega-\varepsilon)}^{\pi/2+\omega-\varepsilon} \frac{d\theta}{\left|\frac{\lambda}{r} - e^{i\theta}\right|} \longrightarrow 0.$$

Therefore, we find that

$$\frac{1}{2\pi i} \int_{\Gamma'} \frac{e^{\mu s}}{\lambda - \mu}\, d\mu = -e^{\lambda s}.$$

(b) Similarly, since the path Γ lies to the left of the path Γ', we find that

$$\frac{1}{2\pi i} \int_{\Gamma} \frac{e^{\lambda t}}{\lambda - \mu}\, d\lambda = 0.$$

Summing up, we obtain that

$$U(t) \cdot U(s) = -\frac{1}{2\pi i} \int_{\Gamma} e^{\lambda(t+s)} R(\lambda)\, d\lambda = U(t+s) \quad \text{for all } t,\, s > 0.$$

The proof of Proposition 2.1 is complete. $\square$

2.2 Generation Theorem for Analytic Semigroups

The next theorem states that the semigroup $U(t)$ can be extended to an *analytic semigroup* in some sector containing the positive real axis ([123, p. 63, Theorem 3.2]):

Theorem 2.2. *Assume that the operator A satisfies conditions* (2.1) *and* (2.2). *The semigroup $U(t)$, defined by formula* (2.3), *can be extended to a semigroup $U(z)$ that is* analytic *in the sector*

$$\Delta_\omega = \{z = t + is : z \neq 0, |\arg z| < \omega\},$$

and enjoys the following three properties:

(a) The operators $AU(z)$ and $\frac{dU}{dz}(z)$ are bounded operators on E for each $z \in \Delta_\omega$, and satisfy the relation

$$\frac{dU}{dz}(z) = AU(z) \quad \text{for all } z \in \Delta_\omega. \tag{2.4}$$

(b) For each $0 < \varepsilon < \omega/2$, there exist positive constants $\widetilde{M}_0(\varepsilon)$ and $\widetilde{M}_1(\varepsilon)$ such that

$$\|U(z)\| \leq \widetilde{M}_0(\varepsilon) \qquad \text{for all } z \in \Delta_\omega^{2\varepsilon}, \tag{2.5}$$

$$\|AU(z)\| \leq \frac{\widetilde{M}_1(\varepsilon)}{|z|} \qquad \text{for all } z \in \Delta_\omega^{2\varepsilon}, \tag{2.6}$$

where (see Figure 2.5 below)

$$\Delta_\omega^{2\varepsilon} = \{z \in \mathbf{C} : z \neq 0, |\arg z| \leq \omega - 2\varepsilon\}.$$

(c) For each $x \in E$, we have, as $z \to 0$, $z \in \Delta_\omega^{2\varepsilon}$,

$$U(z)x \longrightarrow x \quad \text{in } E.$$

Proof. (i) The *analyticity* of $U(z)$: If $\lambda \in \Gamma^{(3)}$ and $z \in \Delta_\omega^{2\varepsilon}$, that is, if we have the formulas

$$\lambda = |\lambda|e^{i\theta}, \quad \theta = \pi/2 + \omega - \varepsilon,$$
$$z = |z|e^{i\varphi}, \quad |\varphi| \leq \omega - 2\varepsilon,$$

then it follows that

$$\lambda z = |\lambda| |z|e^{i(\theta+\varphi)},$$

with

$$\frac{\pi}{2} + \varepsilon \leq \theta + \varphi \leq \frac{\pi}{2} + 2\omega - 3\varepsilon < \frac{3\pi}{2} - 3\varepsilon.$$

Note that

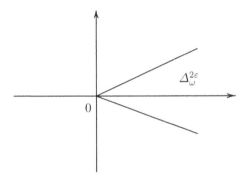

Fig. 2.5. The sector $\Delta_\omega^{2\varepsilon}$ in Theorem 2.2

$$\cos(\theta + \varphi) \le \cos(\pi/2 + \varepsilon) = -\sin\varepsilon.$$

Hence we have the inequality

$$|e^{\lambda z}| \le e^{-|\lambda|\,|z|\sin\varepsilon} \quad \text{for all } \lambda \in \Gamma^{(3)} \text{ and } z \in \Delta_\omega^{2\varepsilon}. \tag{2.7}$$

Similarly, we have the inequality

$$|e^{\lambda z}| \le e^{-|\lambda|\,|z|\sin\varepsilon} \quad \text{for all } \lambda \in \Gamma^{(1)} \text{ and } z \in \Delta_\omega^{2\varepsilon}. \tag{2.8}$$

For each small $\varepsilon > 0$, we let

$$K_\omega^\varepsilon = \Delta_\omega^{2\varepsilon} \cap \{z \in \mathbf{C} : |z| \ge \varepsilon\} = \{z \in \mathbf{C} : |z| \ge \varepsilon, |\arg z| \le \omega - 2\varepsilon\}.$$

Then, by combining estimates (2.2), (2.7) and (2.8) we obtain that

$$\left\| e^{\lambda z} R(\lambda) \right\| \le \frac{M_\varepsilon}{|\lambda|} e^{-\varepsilon \sin\varepsilon\cdot|\lambda|} \quad \text{for all } \lambda \in \Gamma^{(1)} \cup \Gamma^{(3)} \text{ and } z \in K_\omega^\varepsilon. \tag{2.9}$$

On the other hand, we have the estimate

$$\left\| e^{\lambda z} R(\lambda) \right\| \le M_\varepsilon e^{|z|} \quad \text{for all } \lambda \in \Gamma^{(2)} \text{ and } z \in K_\omega^\varepsilon. \tag{2.10}$$

Therefore, we find that the integral

$$U(z) = -\frac{1}{2\pi i} \int_\Gamma e^{\lambda z} R(\lambda)\, d\lambda = -\frac{1}{2\pi i} \sum_{k=1}^3 \int_{\Gamma^{(k)}} e^{\lambda z} R(\lambda)\, d\lambda \tag{2.11}$$

converges in the Banach space $L(E, E)$, uniformly in $z \in K_\omega^\varepsilon$, for every $\varepsilon > 0$. This proves that the operator $U(z)$ is analytic in the domain

$$\Delta_\omega = \bigcup_{\varepsilon > 0} K_\omega^\varepsilon.$$

By the analyticity of $U(z)$, it follows that the operators $U(z)$ also enjoy the semigroup property

$$U(z + w) = U(z) \cdot U(w) \quad \text{for all } z, \, w \in \Delta_\omega.$$

(ii) We prove that the operators $U(z)$ enjoy properties (a), (b) and (c).

(b) First, by using Cauchy's theorem we obtain that

$$U(z) = -\frac{1}{2\pi i} \int_\Gamma e^{\lambda z} \, R(\lambda) \, d\lambda = -\frac{1}{2\pi i} \int_{\Gamma_{|z|}} e^{\lambda z} \, R(\lambda) \, d\lambda,$$

where $\Gamma_{|z|}$ is a path consisting of the following three curves (see Figure 2.6 below):

$$\Gamma_{|z|}^{(1)} = \left\{ re^{-i(\pi/2 + \omega - \varepsilon)} : \frac{1}{|z|} \le r < \infty \right\},$$

$$\Gamma_{|z|}^{(2)} = \left\{ \frac{1}{|z|} e^{i\theta} : -(\pi/2 + \omega - \varepsilon) \le \theta \le \pi/2 + \omega - \varepsilon \right\},$$

$$\Gamma_{|z|}^{(3)} = \left\{ re^{i(\pi/2 + \omega - \varepsilon)} : \frac{1}{|z|} \le r < \infty \right\}.$$

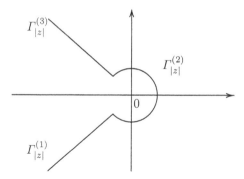

Fig. 2.6. The integral path $\Gamma_{|z|}$ consisting of $\Gamma_{|z|}^{(1)}$, $\Gamma_{|z|}^{(2)}$ and $\Gamma_{|z|}^{(3)}$

However, by estimates (2.2), (2.7) and (2.8), it follows that

$$\left\| e^{\lambda z} R(\lambda) \right\| \le \frac{M_\varepsilon}{|\lambda|} e^{-|\lambda| \, |z| \, \sin \varepsilon} \quad \text{for all } \lambda \in \Gamma_{|z|}^{(1)} \cup \Gamma_{|z|}^{(3)} \text{ and } z \in \Delta_\omega^{2\varepsilon}.$$

Hence we have, for $k = 1, \, 3$,

$$\int_{\Gamma_{|z|}^{(k)}} \left\| e^{\lambda z} R(\lambda) \right\| \, |d\lambda| \le M_\varepsilon \int_{\frac{1}{|z|}}^{\infty} e^{-\rho |z| \, \sin \varepsilon} \, \rho^{-1} \, d\rho$$

$$= M_\varepsilon \int_1^\infty e^{-\sin \varepsilon \cdot s} s^{-1} \, ds.$$

We have also, for $k = 2$,

$$\int_{\Gamma_{|z|}^{(2)}} \left\| e^{\lambda z} R(\lambda) \right\| |d\lambda| \le M_\varepsilon \int_{-(\pi/2+\omega-\varepsilon)}^{\pi/2+\omega-\varepsilon} e \, d\theta$$

$$= 2e \, M_\varepsilon (\pi/2 + \omega - \varepsilon)$$

$$\le 2\pi \, e \, M_\varepsilon.$$

Summing up, we obtain the following estimate:

$$\|U(z)\| \le \frac{1}{2\pi} \sum_{k=1}^{3} \int_{\Gamma_{|z|}^{(k)}} \left\| e^{\lambda z} R(\lambda) \right\| |d\lambda|$$

$$\le \frac{1}{2\pi} \left(2M_\varepsilon \int_1^\infty s^{-1} e^{-\sin \varepsilon \cdot s} \, ds + 2\pi \, e \, M_\varepsilon \right)$$

$$= \frac{M_\varepsilon}{\pi} \left(\int_1^\infty s^{-1} e^{-\sin \varepsilon \cdot s} \, ds + \pi e \right).$$

This proves the desired estimate (2.5), with

$$\widetilde{M_0}(\varepsilon) = \frac{M_\varepsilon}{\pi} \left(\int_1^\infty s^{-1} e^{-\sin \varepsilon \cdot s} \, ds + \pi \, e \right).$$

To prove estimate (2.6), we note that

$$AR(\lambda) = (A - \lambda I + \lambda I) R(\lambda) = I + \lambda R(\lambda),$$

so that

$$\|AR(\lambda)\| \le 1 + M_\varepsilon \quad \text{for all } \lambda \in \Sigma_\omega^\varepsilon.$$

Hence, by arguing just as in the proof of estimate (2.5) we obtain that

$$\left\| \int_\Gamma e^{\lambda z} AR(\lambda) \, d\lambda \right\| \le 2 \int_{\frac{1}{|z|}}^\infty e^{-\rho |z| \sin \varepsilon} (1 + M_\varepsilon) \, d\rho \qquad (2.12)$$

$$+ \int_{-(\pi/2+\omega-\varepsilon)}^{\pi/2+\omega-\varepsilon} (1 + M_\varepsilon) \, e \, \frac{1}{|z|} \, d\theta$$

$$\le 2 (1 + M_\varepsilon) \left(\int_1^\infty e^{-\sin \varepsilon \cdot s} \, ds + \pi \, e \right) \frac{1}{|z|}$$

$$\text{for all } z \in \Delta_\omega^{2\varepsilon}.$$

This proves that the integral

$$\int_\Gamma e^{\lambda z} AR(\lambda) \, d\lambda$$

is convergent in the Banach space $L(E, E)$, for every $z \in \Delta_\omega^{2\varepsilon}$. By the closedness of A, it follows that

$$U(z) \in \mathcal{D}(A) \quad \text{for all } z \in \Delta_\omega^{2\varepsilon},$$

and

$$AU(z) = -\frac{1}{2\pi i} \int_\Gamma e^{\lambda z} AR(\lambda) \, d\lambda \quad \text{for all } z \in \Delta_\omega^{2\varepsilon}. \tag{2.13}$$

Therefore, the desired estimate (2.6) follows from estimate (2.12), with

$$\widetilde{M_1}(\varepsilon) = \frac{1 + M_\varepsilon}{\pi} \left(\int_1^\infty e^{-\sin \varepsilon \cdot s} \, ds + \pi e \right).$$

We remark that formula (2.12) remains valid for all $z \in \Delta_\omega$, since $\Delta_\omega = \bigcup_{\varepsilon > 0} \Delta_\omega^{2\varepsilon}$.

(a) By estimates (2.9) and (2.10), we can differentiate formula (2.11) under the integral sign to obtain that

$$\frac{dU}{dz}(z) = -\frac{1}{2\pi i} \int_\Gamma e^{\lambda z} \lambda R(\lambda) \, d\lambda \quad \text{for all } z \in \Delta_\omega. \tag{2.14}$$

On the other hand, it follows from formula (2.13) that

$$AU(z) = -\frac{1}{2\pi i} \int_\Gamma e^{\lambda z} AR(\lambda) \, d\lambda \tag{2.15}$$

$$= -\frac{1}{2\pi i} \int_\Gamma e^{\lambda z} \left(I + \lambda R(\lambda) \right) \, d\lambda$$

$$= -\frac{1}{2\pi i} \int_\Gamma e^{\lambda z} \lambda R(\lambda) \, d\lambda \quad \text{for all } z \in \Delta_\omega,$$

since we have, by Cauchy's theorem,

$$\int_\Gamma e^{\lambda z} \, d\lambda = 0.$$

Therefore, the desired formula (2.4) follows immediately from formulas (2.14) and (2.15).

(c) Now let x_0 be an arbitrary element of $\mathcal{D}(A)$. By the residue theorem, it follows that

$$x_0 = \frac{1}{2\pi i} \int_\Gamma \frac{e^{\lambda z}}{\lambda} x_0 \, d\lambda.$$

Hence we have the formula

$$U(z)x_0 - x_0 = -\frac{1}{2\pi i} \int_\Gamma e^{\lambda z} \left(R(\lambda) + \frac{1}{\lambda} \right) x_0 \, d\lambda$$

$$= -\frac{1}{2\pi i} \int_\Gamma \frac{e^{\lambda z}}{\lambda} R(\lambda) A x_0 \, d\lambda.$$

Here we remark that

- $\left\| \dfrac{1}{\lambda} R(\lambda) \right\| \leq \dfrac{M_\varepsilon}{|\lambda|^2}$ for all $\lambda \in \Gamma$,

bullet $\left| e^{\lambda z} \right| \leq 2 e^{-|\lambda|\,|z|\,\sin\varepsilon} + e^{|z|}$ for all $z \in \Delta_\omega^{2\varepsilon}$ and $\lambda \in \Gamma$.

Thus it follows from an application of the Lebesgue dominated convergence theorem that, as $z \to 0$, $z \in \Delta_\omega^{2\varepsilon}$,

$$U(z)x_0 - x_0 \longrightarrow -\frac{1}{2\pi i} \int_\Gamma \frac{1}{\lambda} R(\lambda)\, A x_0 \, d\lambda.$$

However, we have the assertion

$$\int_\Gamma \frac{1}{\lambda} R(\lambda)\, A x_0 \, d\lambda = 0.$$

Indeed, by Cauchy's theorem it follows that

$$\begin{aligned}
\int_\Gamma \frac{1}{\lambda} R(\lambda) A x_0 \, d\lambda &= \lim_{r \to \infty} \int_{\Gamma \cap \{|\lambda| \leq r\}} \frac{1}{\lambda} R(\lambda)\, A x_0 \, d\lambda \\
&= - \lim_{r \to \infty} \int_{C_r} \frac{1}{\lambda} R(\lambda)\, A x_0 \, d\lambda \\
&= 0,
\end{aligned}$$

where C_r is a closed path shown in Figure 2.7 below.

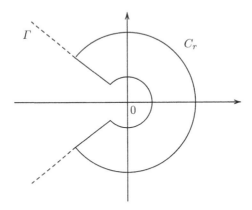

Fig. 2.7. The closed path C_r for Cauchy's theorem

Summing up, we have proved that

$$U(z)x_0 \longrightarrow x_0 \quad \text{as } z \to 0, \ z \in \Delta_\omega^{2\varepsilon},$$

for each $x_0 \in \mathcal{D}(A)$.

Since the domain $\mathcal{D}(A)$ is dense in E and

$$\|U(z)\| \leq \widetilde{M_0}(\varepsilon) \quad \text{for all } z \in \Delta_\omega^{2\varepsilon},$$

it follows that, for each $x \in E$,

$$U(z)x \longrightarrow x \quad \text{as } z \to 0, \, z \in \Delta_\omega^{2\varepsilon}.$$

The proof of Theorem 2.2 is now complete. $\quad\square$

2.3 Remark on the Resolvent Estimate (2.2)

Assume that the operator A satisfies a stronger condition than condition (2.2):

(1) The resolvent set of A contains the region Σ shown in Figure 2.8 below.
(2) There exists a constant $M > 0$ such that the resolvent $R(\lambda) = (A - \lambda I)^{-1}$ satisfies the estimate

$$\|R(\lambda)\| \leq \frac{M}{1 + |\lambda|} \quad \text{for all } \lambda \in \Sigma. \tag{2.16}$$

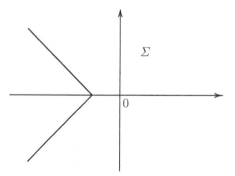

Fig. 2.8. The region Σ in condition (1)

We define the semigroup $U(t)$ by the formula (by replacing formula (2.3))

$$U(t) = -\frac{1}{2\pi i} \int_\Gamma e^{\lambda t} R(\lambda) \, d\lambda = -\frac{1}{2\pi i} \int_\Gamma e^{\lambda t} (A - \lambda I)^{-1} \, d\lambda. \tag{2.3'}$$

Here the path Γ runs in the set Σ from $\infty e^{-i\omega}$ to $\infty e^{i\omega}$, avoiding the positive real axis and the origin (see Figure 2.9 below).
   Then we can prove the following remark:

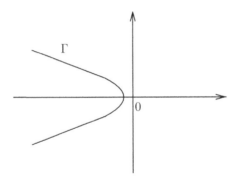

Fig. 2.9. The integral path Γ in formula $(2.3')$

Remark 2.3. The estimates (2.5) and (2.6) can be replaced as follows:

$$\|U(z)\| \leq \widetilde{M_0}(\varepsilon)\, e^{-\delta \cdot \mathrm{Re}\, z} \qquad \text{for all } z \in \Delta_\omega^{2\varepsilon}, \tag{2.17}$$

$$\|AU(z)\| \leq \frac{\widetilde{M_1}(\varepsilon)}{|z|}\, e^{-\delta \cdot \mathrm{Re}\, z} \qquad \text{for all } z \in \Delta_\omega^{2\varepsilon}, \tag{2.18}$$

with some constant $\delta > 0$.

Proof. Take a real number δ such that

$$0 < \delta < \frac{1}{M_\varepsilon}.$$

Then we have, by estimate (2.15),

$$\delta \left\|(A - \lambda I)^{-1}\right\| \leq \frac{\delta M_\varepsilon}{|\lambda| + 1} \leq \delta M_\varepsilon < 1 \quad \text{for all } \lambda \in \Sigma_\omega^\varepsilon.$$

Hence it follows that the operator $(A + \delta I) - \lambda I$ has the inverse

$$((A + \delta I) - \lambda I)^{-1} = \left(I + \delta\,(A - \lambda I)^{-1}\right)^{-1} (A - \lambda I)^{-1},$$

and

$$
\begin{aligned}
\left\|((A + \delta I) - \lambda I)^{-1}\right\| &\leq \left\|\left(I + \delta\,(A - \lambda I)^{-1}\right)^{-1}\right\| \cdot \left\|(A - \lambda I)^{-1}\right\| \\
&\leq \frac{M_\varepsilon}{|\lambda| + 1} \frac{1}{1 - \|\delta(A - \lambda I)^{-1}\|} \\
&\leq \frac{M_\varepsilon}{|\lambda| + 1} \frac{1}{1 - \delta M_\varepsilon} \\
&\leq \left(\frac{M_\varepsilon}{1 - \delta M_\varepsilon}\right) \frac{1}{|\lambda|}.
\end{aligned}
$$

This proves that the operator $A + \delta I$ satisfies condition (2.2), so that estimates (2.4) and (2.5) remain valid for the operator $A + \delta I$.

$$\|V(z)\| \leq \widetilde{M}_0(\varepsilon) \qquad \text{for all } z \in \Delta_\omega^{2\varepsilon}, \qquad (2.19)$$

$$\|(A + \delta I)\, V(z)\| \leq \frac{\widetilde{M}_1(\varepsilon)}{|z|} \quad \text{for all } z \in \Delta_\omega^{2\varepsilon}, \qquad (2.20)$$

where

$$V(z) = -\frac{1}{2\pi i} \int_\Gamma e^{\lambda z} \, (A + \delta I - \lambda I)^{-1} \, d\lambda.$$

However, we have, by Cauchy's theorem,

$$V(z) = -\frac{1}{2\pi i} \int_\Gamma e^{\lambda z} \, (A + \delta I - \lambda I)^{-1} \, d\lambda \qquad (2.21)$$

$$= -\frac{1}{2\pi i} \int_{\Gamma + \delta} e^{\lambda z} \, (A + \delta I - \lambda I)^{-1} \, d\lambda$$

$$= -\frac{1}{2\pi i} \int_\Gamma e^{\mu z} \, e^{\delta z} \, (A - \mu I)^{-1} \, d\mu$$

$$= e^{\delta z} U(z) \quad \text{for all } z \in \Delta_\omega^{2\varepsilon}.$$

In view of formula (2.21), the desired estimates (2.17) and (2.18) follow from estimates (2.19) and (2.20).

The proof of Remark 2.3 is complete.   □

2.4 Notes and Comments

This chapter is adapted from Taira [123, Chapter 3]. For more leisurely treatments of analytic semigroups, the reader is referred to Friedman [43], Pazy [87], Tanabe [131] and Yosida [147].

3

Markov Processes and Feller Semigroups

This chapter is devoted to a functional analytic approach to Markov processes in probability which forms a functional analytic background for the proof of Theorems 1.6 and 1.11.

Section 3.1 is devoted to the Riesz–Markov representation theorem which describes an intimate relationship between measures and linear functionals (Theorems 3.7 and 3.10).

Section 3.2 is devoted to a brief description of basic definitions and results about Markov processes. We give concrete important examples of Markov transition functions on the line $\mathbf{R} = (-\infty, \infty)$, and some examples of diffusion processes on the half line $\overline{\mathbf{R}^+} = [0, \infty)$ in which we must take account of the effect of the boundary point 0.

Section 3.3 provides a brief description of basic definitions and results about Markov processes and a class of semigroups (Feller semigroups) associated with Markov processes. The semigroup approach to Markov processes can be traced back to the pioneering work of Feller [39] and [40] in early 1950s (cf. [15], [98], [116]).

In Section 3.4 we prove various generation theorems of Feller semigroups by using the Hille–Yosida theory of semigroups (Theorems 3.34 and 3.36). We give two examples where it is difficult to begin with a Markov transition function and the infinitesimal generator is the basic tool of describing the process (Examples 3.16 and 3.17).

Section 3.5 is devoted to the reflecting diffusion that is associated with the homogeneous Neumann problem for the Laplacian

$$\begin{cases} \Delta u = f & \text{in } D, \\ \dfrac{\partial u}{\partial \mathbf{n}} = 0 & \text{on } \partial D. \end{cases}$$

In Section 3.6, following Sato–Tanaka [97] and Sato–Ueno [98] we introduce the notion of local time on the boundary for the reflecting diffusion constructed in Section 3.5.

© Springer Nature Switzerland AG 2020
K. Taira, *Boundary Value Problems and Markov Processes*, Lecture Notes in Mathematics 1499,
https://doi.org/10.1007/978-3-030-48788-1_3

3.1 Continuous Functions and Measures

One of the fundamental theorems in analysis is the Riesz–Markov representation theorem which describes an intimate relationship between measures and linear functionals.

3.1.1 Space of Continuous Functions

A topological space is said to be *locally compact* if every point has a compact neighborhood. Let (X, ρ) be a locally compact metric space. Let $C(X)$ be the collection of real-valued, continuous functions on X. We define in the set $C(X)$ addition and scalar multiplication of functions in the usual way:

$$(f + g)(x) = f(x) + g(x) \quad \text{for all } x \in X.$$
$$(\alpha f)x = \alpha f(x) \quad \text{for all } \alpha \in \mathbf{R} \text{ and } x \in X.$$

Then $C(X)$ is a real linear space.

The next two results show that locally compact metric spaces have a rich supply of continuous functions that vanish outside compact sets (see [42, Chapter 4, Lemma 4.15 and Theorem 4.16]):

Lemma 3.1 (Urysohn). *Let (X, ρ) be a locally compact metric space. If $K \subset U \subset X$ where K is compact and U is open, then there exists a real-valued, continuous function $f \in C(X)$ such that (see Figure 3.1 below)*

$$\begin{cases} 0 \le f(x) \le 1 & \text{on } X, \\ f(x) = 1 & \text{on } K, \\ f(x) = 0 & \text{outside a compact subset of } U. \end{cases}$$

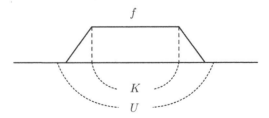

Fig. 3.1. The function f in Lemma 3.1

Theorem 3.2 (Tietze's extension theorem). *Let (X, ρ) be a locally compact metric space and K a compact subset of X. If $f \in C(K)$, then there exists a real-valued, continuous function $F \in C(X)$ such that $F = f$ on K. Moreover, the function $F(x)$ may be taken to vanish outside a compact set.*

If $f \in C(X)$, the *support* of f, denoted by supp f, is the smallest closed set outside of which $f(x)$ vanishes, that is, the closure of the set $\{x \in X : f(x) \neq 0\}$. If supp f is compact, we say that f is *compactly supported*. We define a subspace of $C(X)$ as follows:

$$C_c(X) = \{f \in C(X) : \text{supp } f \text{ is compact}\}.$$

Namely, $C_c(X)$ is the space of compactly supported, continuous functions on X. The notation 'C_c' is used only for the moment in this Section 3.5 (later it is mostly 'C_0').

If (X, ρ) is a non-compact, locally compact metric space, then we can make X into a compact space by adding a single "point at infinity" in such a way that the functions in $C_0(X)$ are precisely those continuous functions f such that $f(x) \to 0$ as x approaches the point at infinity.

More precisely, let ∂ denote a point that is not an element of X and let $X_\partial = X \cup \{\partial\}$. Then we have the following proposition:

Proposition 3.3. *Let (X, ρ) be a locally compact metric space and let $\mathcal{T}$ be the collection of all subsets U of $X_\partial = X \cup \{\partial\}$ such that either (i) U is an open subset of X or (ii) $\partial \in U$ and $U^c = X_\partial \setminus U$ is a compact subset of X. Then the space $(X_\partial, \mathcal{T})$ is a compact space and the inclusion map $i \colon X \to X_\partial$ is an embedding. Furthermore, if $f \in C(X)$, then $f(x)$ extends continuously to X_∂ if and only if $f(x) = g(x) + c$ where $g \in C_0(X)$ and $c \in \mathbf{R}$, in which case the continuous extension is given by $f(\partial) = c$ (see Figure 3.2 below).*

The space $X_\partial = X \cup \{\partial\}$ is called the *one-point compactification* of X and the point ∂ is called the *point at infinity*.

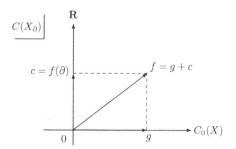

Fig. 3.2. The function $f = g + c$ in Proposition 3.3

Furthermore, if $f \in C(X)$, we say that f is *vanishes at infinity* if the set $\{x \in X : |f(x)| \geq \varepsilon\}$ is compact for every $\varepsilon > 0$, and we write

$$\lim_{x \to \partial} f(x) = 0.$$

We define a subspace $C_0(X)$ of $C(X)$ as follows:

$$C_0(X) = \left\{ f \in C(X) : \lim_{x \to \partial} f(x) = 0 \right\}.$$

It is easy to see that $C_0(X)$ is a Banach space with the supremum (maximum) norm

$$\|f\|_\infty = \sup_{x \in X} |f(x)|.$$

The next proposition asserts that $C_0(X)$ is the uniform closure of $C_c(X)$ (see [42, Chapter 4, Proposition 4.35]):

Proposition 3.4. *Let (X, ρ) be a locally compact metric space. The space $C_0(X)$ is the closure of $C_c(X)$ in the topology of uniform convergence.*

Proof. Assume that $\{f_n\}$ is a sequence in $C_c(X)$ which converges uniformly to some function $f \in C(X)$. Then, for any given $\varepsilon > 0$ there exists a number $n \in \mathbf{N}$ such that

$$\|f - f_n\|_\infty < \varepsilon.$$

Hence we have the assertion

$$|f(x)| < \varepsilon \quad \text{if } x \in X \setminus \operatorname{supp} f_n.$$

This proves that the set $\{x \in X : |f(x)| \geq \varepsilon\}$ is compact for every $\varepsilon > 0$, so that $f \in C_0(X)$.

Conversely, if $f \in C_0(X)$, we let

$$K_n = \left\{ x \in X : |f(x)| \geq \frac{1}{n} \right\} \quad \text{for } n \in \mathbf{N}.$$

Since K_n is compact, by applying Urysohn's lemma (Lemma 3.1) we can find a function $g_n \in C_c(X)$ such that $0 \leq g_n \leq 1$ and $g_n = 1$ on K_n. Then it follows that $f_n = g_n f \in C_c(X)$ and that

$$\|f - f_n\|_\infty = \|(1 - g_n)f\|_\infty \leq \frac{1}{n} \quad \text{for all } n \in \mathbf{N}.$$

This proves that $\{f_n\}$ converges uniformly to $f \in C(X)$.

The proof of Proposition 3.4 is complete. $\quad \square$

3.1.2 Space of Signed Measures

Let $(X, \mathcal{M})$ be a measurable space. A real-valued function μ defined on the σ-algebra $\mathcal{M}$ is called a *signed measure* or *real measure* if it is countably additive, that is,

$$\mu \left(\sum_{i=1}^\infty A_i \right) = \sum_{i=1}^\infty \mu(A_i)$$

for any disjoint countable collection $\{A_i\}_{i=1}^\infty$ of members of $\mathcal{M}$. It should be noticed that every rearrangement of the series $\sum_i \mu(A_i)$ also converges, since the disjoint union $\sum_{i=1}^{\infty} A_i$ is not changed if the subscripts are permuted. A signed measure takes its values in $(-\infty, \infty)$, but a non-negative measure may take ∞; hence the non-negative measures do not form a subclass of the signed measures.

If μ and λ are signed measures on $\mathcal{M}$, we define the sum $\mu + \lambda$ and the scalar multiple $\alpha \mu$ as follows:

$$(\mu + \lambda)(A) = \mu(A) + \lambda(A) \quad \text{for all } A \in \mathcal{M},$$
$$(\alpha \mu)(A) = \alpha \mu(A) \quad \text{for all } \alpha \in \mathbf{R} \text{ and } A \in \mathcal{M}.$$

Then it is clear that $\mu + \lambda$ and $\alpha \mu$ are signed measures.

If μ is a signed measure, we define a function $|\mu|$ on $\mathcal{M}$ by the formula

$$|\mu|(A) = \sup \left\{ \sum_{i=1}^{n} |\mu(A_i)| \right\} \quad \text{for all } A \in \mathcal{M},$$

where the supremum is taken over all finite partitions $\{A_i\}$ of A into members of $\mathcal{M}$. Then the function $|\mu|$ is a finite non-negative measure on $\mathcal{M}$. The measure $|\mu|$ is called the *total variation measure* of μ, and the quantity $|\mu|(X)$ is called the *total variation* of μ. We remark that

$$|\mu(A)| \leq |\mu|(A) \leq |\mu|(X) \quad \text{for all } A \in \mathcal{M}. \tag{3.1}$$

Furthermore, we can verify that the quantity $|\mu|(X)$ satisfies the axioms of a norm. Thus the totality of signed measures on $\mathcal{M}$ is a normed linear space by the norm $\|\mu\| := |\mu|(X)$.

If we define two functions μ^+ and μ^- on $\mathcal{M}$ by the formulas

$$\mu^+ = \frac{1}{2}(|\mu| + \mu),$$
$$\mu^- = \frac{1}{2}(|\mu| - \mu),$$

then it follows from inequalities (3.31) that both μ^+ and μ^- are finite non-negative measures on $\mathcal{M}$. It should be emphasized that the measures μ^+ and μ^- are the positive and negative variation measures of μ, respectively. We also have the *Jordan decomposition* of μ:

$$\mu = \mu^+ - \mu^-.$$

3.1.3 The Riesz–Markov Representation Theorem

First, we characterize the non-negative linear functionals on $C(K)$. We show that non-negative linear functionals on the spaces of continuous functions are given by integration against Radon measures. This fact constitutes an essential link between measure theory and functional analysis, providing a powerful tool for constructing measures.

Let (X, ρ) be a locally compact metric space and K a compact subset of X, and

$$C_c(X) = \{f \in C(X) : \operatorname{supp} f \text{ is compact}\}.$$
$$C_0(X) = \{f \in C(X) : f \text{ vanishes at infinity}\}.$$

First, we characterize the non-negative linear functionals on $C_c(X)$. A linear functional I on $C_c(X)$ is said to be *non-negative* if $I(f) \geq 0$ whenever $f \geq 0$, that is, if it satisfies the condition

$$f \in C_c(X), \ f(x) \geq 0 \text{ on } X \implies I(f) \geq 0.$$

Then we can prove that non-negativity implies a rather strong continuity property (see [42, Chapter 7, Proposition 7.1]):

Proposition 3.5. *If I is a non-negative linear functional on $C_c(X)$, then, for every compact set $K \subset X$ there exists a positive constant C_K such that*

$$|I(f)| \leq C_K \|f\|_\infty \quad \text{for all } f \in C_c(X) \text{ with } \operatorname{supp} f \subset K.$$

Here

$$\|f\|_\infty = \sup_{x \in X} |f(x)|.$$

If μ is Borel measure on X such that $\mu(K) < \infty$ for every compact set $K \subset X$, then it follows that

$$C_c(X) \subset L^1(X, \mu).$$

Hence the map

$$I_\mu : f \longmapsto \int_X f(x)\, d\mu(x)$$

is a non-negative linear functional on the space $C_c(X)$. The purpose of this subsection is to prove that every non-negative linear functional on $C_c(X)$ arises in this fashion. In doing this, we impose some additional conditions on μ, subject to which μ is uniquely determined.

A *Radon measure* on X is a Borel measure that is finite on all compact sets in X, and is outer regular on all Borel sets in X and inner regular on all open sets in X.

If U is an open set in X and $f \in C_c(X)$, then we write

$$f \prec U$$

to mean that (see Figure 3.3 below)

$$\begin{cases} 0 \leq f(x) \leq 1 & \text{on } X, \\ \operatorname{supp} f \subset U. \end{cases}$$

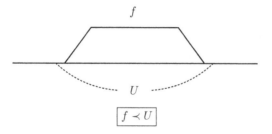

Fig. 3.3. The notation $f \prec U$

On the other hand, if K is a subset of X, we let

$$\chi_K(x) = \begin{cases} 1 & \text{if } x \in K, \\ 0 & \text{if } x \in X \setminus K, \end{cases}$$

and we write

$$f \geq \chi_K$$

to mean that (see Figure 3.4 below)

$$f(x) \geq \chi_K(x) \quad \text{on } X.$$

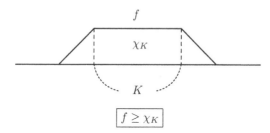

Fig. 3.4. The notation $f \geq \chi_K$

The next theorem asserts that non-negative linear functionals on the space $C_c(X)$ are given by integration against Radon measures (see [42, Chapter 7, Theorem 7.2]):

Theorem 3.6 (the Riesz representation theorem). *Let (X, ρ) be a locally compact metric space. If I is a non-negative linear functional on the space $C_c(X)$, then there is a unique Radon measure μ on X such that*

$$I(f) = \int_X f(x) \, d\mu(x) \quad \text{for all } f \in C_c(X). \tag{3.2}$$

Furthermore, the Radon measure μ enjoys the following two properties (3.3) and (3.4):

$$\bullet \ \mu(U) = \sup\{I(f) : f \in C_c(X), \ f \prec U\} \tag{3.3}$$

for every open set $U \subset X$.

$$\bullet \ \mu(K) = \inf\{I(f) : f \in C_c(X), \ f \geq \chi_K\} \tag{3.4}$$

for every compact set $K \subset X$.

Proof. The proof is divided into two steps.

Step 1: First, we prove the uniqueness of a Radon measure. More precisely, we show that a Radon measure μ is determined by I on all Borel subsets of X.

Assume that μ is a Radon measure such that

$$I(f) = \int_X f(x)\,d\mu \quad \text{for all } f \in C_c(X).$$

Let U be an arbitrary open subset of X. Then we have, for every function $f \prec U$,

$$I(f) = \int_X f(x)\,d\mu \leq \int_U d\mu = \mu(U).$$

On the other hand, if K is a compact subset of U, then it follows from an application of Urysohn's lemma (Lemma 3.1) that there exists a function $f \in C_c(X)$ such that $f \prec U$ and $f = 1$ on K. Hence we have the inequality

$$\mu(K) = \int_X \chi_K(x)\,d\mu \leq \int_X f(x)\,d\mu = I(f).$$

However, since μ is inner regular, we obtain that

$$\begin{aligned}
\mu(U) &= \sup\{\mu(K) : K \subset U, \ K \text{ is compact}\} \\
&\leq \sup\{I(f) : f \in C_c(X), \ f \prec U\} \leq \mu(U).
\end{aligned}$$

Therefore, we have, for every open set $U \subset X$,

$$\mu(U) = \sup\{I(f) : f \in C_c(X), \ f \prec U\}.$$

This proves that the Radon measure μ is determined by I on open subsets U of X, and hence by I on all Borel subsets of X, since it is outer regular on all Borel sets.

Step 2: The proof of the uniqueness suggests how to construct a Radon measure μ. More precisely, we begin by defining $\mu(U)$ for an arbitrary open set $U \subset X$ by the formula

$$\mu(U) = \sup\{I(f) : f \in C_c(X), \ f \prec U\},$$

and then define $\mu^*(E)$ for an arbitrary set $E \subset X$ by the formula

$$\mu^*(E) = \inf \{\mu(U) : U \supset X, \ U \text{ is open}\}.$$

It should be noticed that $\mu^*(U) = \mu(U)$ if U is open, since we have $\mu(U) \leq \mu(V)$ for $U \subset V$.

The idea of the proof may be explained as follows:

(i) First, we prove that μ^* is an outer measure:

$$\mu^*(\emptyset) = 0.$$
$$\mu^*(E) \leq \mu^*(F) \quad \text{if } E \subset F.$$
$$\mu^*\left(\bigcup_{j=1}^{\infty} E_j\right) \leq \sum_{j=1}^{\infty} \mu^*(E_j).$$

(ii) Secondly, we show that every open subset U of X is μ^*-measurable:

$$\mu^*(E) = \mu^*(E \cap U) + \mu^*(E \setminus U)$$
$$\text{for all } E \subset X \text{ such that } \mu^*(E) < \infty.$$

It follows from an application of Carathéodory's theorem ([42, Theorem 1.11]) that every Borel set is μ^*-measurable and further that the restriction μ of μ^* to the σ-algebra $\mathcal{B}_X$ is a Borel measure. It should be emphasized that $\mu^*(U) = \mu(U)$ if U is open and that the measure μ is outer regular and satisfies condition (3.3).

(iii) Thirdly, we prove that the measure μ satisfies the condition (3.3). This implies that μ is finite on compact subsets of X and is *inner regular* on open subsets U of X:

$$\mu(U) = \sup \{\mu(K) : K \subset U, \ K \text{ is compact}\}.$$

Indeed, if U is open and if α is an arbitrary number satisfying the condition $\alpha < \mu(U)$, then we can choose a function $f \in C_c(X)$ such that $f \prec U$ and that $I(f) > \alpha$. We let

$$K = \operatorname{supp} f.$$

If g is a function in $C_c(X)$ satisfying the condition $g \geq \chi_K$, then it follows that

$$g - f \geq 0$$

so that, by the positivity of I,

$$I(g) \geq I(f) > \alpha.$$

However, we have, by formula (3.34),

$$\mu(K) > \alpha.$$

This proves that μ is inner regular on open sets, since α is an arbitrary number satisfying the condition $\alpha < \mu(U)$.

(iv) Finally, we prove formula (3.2).

Proof of Assertion (i): It suffices to show that if $\{U_j\}$ is a sequence of open sets in X and if $U = \cup_{j=1}^{\infty} U_j$, then we have the inequality

$$\mu(U) \leq \sum_{j=1}^{\infty} \mu(U_j).$$

Indeed, it follows from this inequality that we have, for any subset $E \subset X$,

$$\mu^*(E) = \inf\left\{\sum_{j=1}^{\infty} \mu(U_j) : E \subset \bigcup_{j=1}^{\infty} U_j, \ U_j \text{ is open}\right\},$$

and further ([42, Proposition 1.10]) that the expression of the right-hand side defines an outer measure.

If $U = \cup_{j=1}^{\infty} U_j$ and if $f \in C_c(X)$ such that $f \prec U$, then we let

$$K = \operatorname{supp} f.$$

Since K is compact, it follows that, for some finite n,

$$K \subset \bigcup_{j=1}^{n} U_j.$$

Moreover, we can find functions $g_1, g_2, \ldots, g_n \in C_c(X)$ such that $g_j \prec U_j$ and $\sum_{j=1}^{n} g_j = 1$ on K (a partition of unity subordinate to the covering $\{U_j\}$). However, since we have the formula

$$f = \sum_{j=1}^{n} f g_j, \quad f g_j \prec U_j,$$

we obtain that, for any function $f \prec U$,

$$I(f) = \sum_{j=1}^{n} I(f g_j) \leq \sum_{j=1}^{n} \mu(U_j) \leq \sum_{j=1}^{\infty} \mu(U_j),$$

so that

$$\mu(U) = \sup\{I(f) : f \in C_c(X), \ f \prec U\} \leq \sum_{j=1}^{\infty} \mu(U_j).$$

Proof of Assertion (ii): It suffices to show that

$$\mu^*(E) \geq \mu^*(E \cap U) + \mu^*(E \setminus U) \tag{3.5}$$
$$\text{for all } E \subset X \text{ such that } \mu^*(E) < \infty.$$

First, we consider the case where E is open. Then, for any given $\varepsilon > 0$ we can find a function $f \in C_c(X)$ such that

$$\begin{cases} f \prec E \cap U, \\ I(f) > \mu(E \cap U) - \varepsilon. \end{cases}$$

Moreover, since the set $E \setminus \operatorname{supp} f$ is also open, we can find a function $g \in C_c(X)$ such that

$$\begin{cases} g \prec E \setminus \operatorname{supp} f, \\ I(g) > \mu(E \setminus \operatorname{supp} f) - \varepsilon. \end{cases}$$

However, we have the assertions

$$f + g \prec E, \quad \operatorname{supp} f \subset U,$$

and so

$$\begin{aligned} \mu(E) \geq I(f) + I(g) &> \mu(E \cap U) + \mu(E \setminus \operatorname{supp} f) - 2\varepsilon \\ &\geq \mu^*(E \cap U) + \mu^*(E \setminus U) - 2\varepsilon. \end{aligned}$$

Therefore, by letting $\varepsilon \downarrow 0$ in this inequality we obtain the desired inequality (3.5).

Secondly, we consider the general case where $\mu^*(E) < \infty$. Then, for any given $\varepsilon > 0$ we can find an open subset $V \supset E$ such that

$$\mu(V) < \mu^*(E) + \varepsilon.$$

Hence it follows that

$$\begin{aligned} \mu^*(E) + \varepsilon > \mu(V) &\geq \mu^*(V \cap U) + \mu^*(V \setminus U) \\ &\geq \mu^*(E \cap U) + \mu^*(E \setminus U). \end{aligned}$$

Therefore, by letting $\varepsilon \downarrow 0$ in this inequality we obtain the desired inequality (3.5).

Proof of Assertion (iii): Let K be an arbitrary compact subset of X, and let $f \in C_c(X)$ such that $f \geq \chi_K$. If ε is an arbitrary positive number, we define an open set U_ε as follows:

$$U_\varepsilon = \{x \in X : f(x) > 1 - \varepsilon\}.$$

Then it follows that we have, for any function $g \prec U_\varepsilon$,

$$\frac{1}{1 - \varepsilon} f - g \geq 0,$$

and so, by the positivity of I,

$$I(g) \leq \frac{1}{1 - \varepsilon} I(f).$$

Hence we have the inequality

$$\mu(K) \leq \mu(U_\varepsilon) = \sup\{I(g) : g \in C_c(X),\ g \prec U_\varepsilon\} \leq \frac{1}{1-\varepsilon} I(f).$$

Therefore, by letting $\varepsilon \downarrow 0$ in this inequality we obtain that

$$\mu(K) \leq I(f).$$

This proves that we have, for every compact set $K \subset X$,

$$\mu(K) \leq \inf\{I(f) : f \in C_c(X),\ f \geq \chi_K\}. \tag{3.6}$$

On the other hand, for any open set $U \supset K$, by using Urysohn's lemma (Lemma 3.1) we can find a function $h \in C_c(X)$ such that $h \geq \chi_K$ and $h \prec U$. Hence we have the inequality

$$I(h) \leq \mu(U) = \sup\{I(f) : f \in C_c(X),\ f \prec U\}.$$

However, since μ is outer regular on K, it follows that

$$\mu(K) = \inf\{\mu(U) : U \supset K,\ U \text{ is open}\}.$$

Hence we have proved that

$$I(h) \leq \mu(K).$$

This proves that

$$\inf\{I(f) : f \in C_c(X),\ f \geq \chi_K\} \leq I(h) \leq \mu(K). \tag{3.7}$$

Therefore, the desired formula (3.4) follows by combining inequalities (3.6) and (3.7).

Proof of Assertion (iv): To do this, we have only to show that

$$I(f) = \int_X f(x)\,d\mu \quad \text{for all } f \in C_c(X, [0, 1]).$$

Indeed, it suffices to note that the space $C_c(X)$ is the linear span of functions in the space $C_c(X, [0, 1])$.

For any positive integer $N \in \mathbf{N}$, we let

$$K_j = \left\{x \in X : f(x) \geq \frac{j}{N}\right\} \quad \text{for } 1 \leq j \leq N,$$

and

$$K_0 = \operatorname{supp} f.$$

Moreover, we define functions $f_1,\ f_2,\ \ldots,\ f_N \in C_c(X, [0, 1])$ by the formulas

$$f_j(x) = \min\left\{\max\left\{f(x) - \frac{j-1}{N}, 0\right\}, \frac{1}{N}\right\} \quad \text{for } 1 \leq j \leq N.$$

Here it should be noticed that

$$f_j(x) = \begin{cases} 0 & \text{if } x \notin K_{j-1}, \\ f(x) - \frac{j-1}{N} & \text{if } x \in K_{j-1} \setminus K_j, \\ \frac{1}{N} & \text{if } x \in K_j. \end{cases}$$

Then it follows that

$$\frac{1}{N}\chi_{K_j} \le f_j \le \frac{1}{N}\chi_{K_{j-1}},$$

so that

$$\frac{1}{N}\mu(K_j) = \frac{1}{N}\int_X \chi_{K_j}(x)\,d\mu \le \int_X f_j(x)\,d\mu \tag{3.8}$$

$$\le \frac{1}{N}\int_X \chi_{K_{j-1}}(x)\,d\mu = \frac{1}{N}\mu(K_{j-1}).$$

Also, if U is an open set containing K_{j-1}, then we have the condition

$$N f_j \prec U,$$

and the inequality

$$I(f_j) \le \frac{\mu(U)}{N}.$$

Hence, by formula (3.4) and outer regularity of μ it follows that

$$\frac{1}{N}\mu(K_j) = \frac{1}{N}\inf\{I(f) : f \in C_c(X),\ f \ge \chi_{K_j}\} \le I(f_j) \tag{3.9}$$

$$\le \frac{1}{N}\inf\{\mu(U) : U \supset K_{j-1},\ U \text{ is open}\} = \frac{1}{N}\mu(K_{j-1}).$$

However, since we have the formula

$$f = \sum_{j=1}^{N} f_j,$$

it follows from inequalities (3.8) and (3.9) that

$$\frac{1}{N}\sum_{j=1}^{N}\mu(K_j) \le \sum_{j=1}^{N}\int_X f_j(x)\,d\mu = \int_X f(x)\,d\mu \le \frac{1}{N}\sum_{j=0}^{N-1}\mu(K_j),$$

$$\frac{1}{N}\sum_{j=1}^{N}\mu(K_j) \le \sum_{j=1}^{N}I(f_j) = I(f) \le \frac{1}{N}\sum_{j=0}^{N-1}\mu(K_j).$$

Hence we have the inequalities

$$\left| I(f) - \int_X f(x)\,d\mu \right| \le \frac{\mu(K_0) - \mu(K_N)}{N} \le \frac{\mu(\text{supp } f)}{N} \quad \text{for all } N \in \mathbf{N}.$$

Therefore, by letting $N \to \infty$ in this inequality we obtain the desired formula (3.2), since $\mu(\text{supp } f) < \infty$.

Now the proof of Theorem 3.6 is complete. $\quad\square$

We recall (Proposition 3.4) that $C_0(X)$ is the uniform closure of $C_c(X)$. Hence we find that if μ is a Radon measure on X, then the linear functional

$$I_\mu : f \longmapsto \int_X f(x)\, d\mu(x)$$

extends continuously to $C_0(X)$ if and only if it is bounded with respect to the uniform norm. This happens only when

$$\mu(X) = \sup\left\{ I(f) : f \in C_c(X),\ 0 \le f \le 1 \text{ on } X \right\} < \infty,$$

in which case $\mu(X)$ is the operator norm $\|I\|$ of F.

Therefore, we have the following *locally compact version* of the Riesz–Markov representation theorem (see [42, Chapter 7, Theorem 7.17 and Corollary 7.18]):

Theorem 3.7 (the Riesz–Markov representation theorem). *Let (X, ρ) be a locally compact metric space. If F is a non-negative linear functional on the space $C_0(X)$, then there exists a unique Radon measure μ on X such that*

$$F(f) = \int_X f(x)\, d\mu(x) \quad \text{for all } f \in C_0(X),$$

and we have the formula

$$\mu(X) = \sup\left\{ \int_X f(x)\, d\mu(x) : f \in C_0(X),\ 0 \le f \le 1 \text{ on } X \right\} = \|F\|.$$

Corollary 3.8. *Let (K, ρ) be a* compact *metric space. Then we have the following two assertions (i) and (ii):*

(i) *To each non-negative linear functional T on $C(K)$, there corresponds a unique Radon measure μ on K such that*

$$T(f) = \int_K f(x)\, d\mu(x) \quad \text{for all } f \in C(K), \tag{3.10}$$

and we have the formula
$$\|T\| = \mu(K). \tag{3.11}$$

(ii) *Conversely, every finite Radon measure μ on K defines a non-negative linear functional T on $C(K)$ through formula (3.10), and relation (3.11) holds true.*

Remark 3.9. It is easy to see that every open set in a compact metric space is a σ-compact. Thus we find that every finite Radon measure μ is regular.

Now we can characterize the space of all bounded linear functionals on $C(K)$, that is, the dual space $C(K)'$ of $C(K)$. Remark that the dual space $C(K)'$ is a Banach space with the operator norm

$$\|T\| = \sup_{\substack{f \in C(K) \\ \|f\| \leq 1}} |Tf|.$$

The next theorem is a *compact* version of the Riesz–Markov representation theorem:

Theorem 3.10 (Riesz–Markov). *Let (K, ρ) be a* compact *metric space. Then we have the following two assertions (i) and (ii):*

(i) To each $T \in C(K)'$, there corresponds a unique real Borel measure μ on K such that formula (3.40) holds true for all $f \in C(K)$, and we have the formula

$$\|T\| = \text{the total variation } |\mu|(K) \text{ of } \mu. \tag{3.12}$$

(ii) Conversely, every real Borel measure μ on K defines a bounded linear functional $T \in C(K)'$ through formula (3.40), and relation (3.42) holds true.

Remark 3.11. The positive and negative variation measures μ^+, μ^- of a real Borel measure μ are both regular.

We recall that the space of all real Borel measures μ on K is a normed linear space by the norm

$$\|\mu\| = \text{the total variation } |\mu|(K) \text{ of } \mu. \tag{3.13}$$

Therefore, we can restate Theorem 3.10 as follows:

Theorem 3.12. *The dual space $C(K)'$ of $C(K)$ can be identified with the space of all real Borel measures on K normed by formula (3.43).*

3.1.4 Weak Convergence of Measures

Let K be a compact metric space and let $C(K)$ be the Banach space of real-valued continuous functions on K with the supremum (maximum) norm

$$\|f\|_\infty = \sup_{x \in K} |f(x)|.$$

A sequence $\{\mu_n\}_{n=1}^\infty$ of real Borel measures on K is said to *converge weakly* to a real Borel measure μ on K if it satisfies the condition

$$\lim_{n \to \infty} \int_K f(x)\, d\mu_n(x) = \int_K f(x)\, d\mu(x) \quad \text{for every } f \in C(K). \tag{3.14}$$

Theorem 3.12 asserts that the space of all real Borel measures on K normed by formula (3.13) can be identified with the strong dual space $C(K)'$ of $C(K)$. Thus the weak convergence (3.14) of real Borel measures is just the weak* convergence of $C(K)'$.

Two more results are important when studying the weak convergence of measures (see [20, Chapter 3, Corollary 3.30], [44, Chapter 4, Theorem 4.12.3], [100, Chapter 8, Theorem 8.13], [147, Chapter V, Section 1, Theorem 10]):

Theorem 3.13. *The Banach space $C(K)$ is separable, that is, it contains a countable, dense subset.*

Corollary 3.14. *Let X be a separable Banach space. Every bounded sequence $\{f_n\}_{n=1}^{\infty}$ in the strong dual space X^* has a subsequence which converges weakly* to an element f of X^*.*

The next theorem is one of the fundamental theorems in measure theory:

Theorem 3.15. *Every sequence $\{\mu_n\}_{n=1}^{\infty}$ of real Borel measures on K satisfying the condition*

$$\sup_{n \geq 1} |\mu_n| (K) < +\infty \tag{3.15}$$

has a subsequence which converges weakly to a real Borel measure μ on K. Furthermore, if the measures μ_n are all non-negative, then the measure μ is also non-negative.

Proof. By virtue of Theorem 3.13, we can apply Corollary 3.14 with $X := C(K)$ to obtain the first assertion, since condition (3.15) implies the boundedness of the Borel measures μ_n. The second assertion is an immediate consequence of the first assertion of Corollary 3.8.

The proof of Theorem 3.15 is complete.   □

3.2 Elements of Markov Processes

This section provides a brief description of basic definitions and results about Markov processes.

3.2.1 Definition of Markov Processes

Let $(\Omega, \mathcal{F})$ be a measurable space. A non-negative measure P on $\mathcal{F}$ is called a *probability measure* if $P(\Omega) = 1$. The triple $(\Omega, \mathcal{F}, P)$ is called a *probability space*. The elements of Ω are known as sample points, those of $\mathcal{F}$ as events and the values $P(A)$, $A \in \mathcal{F}$, are their probabilities. An extended real-valued, $\mathcal{F}$-measurable function X on Ω is called a *random variable*. The integral

$$\int_{\Omega} X(\omega) \, P(d\omega)$$

(if it exists) is called the *expectation* of X, and is denoted by $E(X)$.

We begin with a review of conditional probabilities and conditional expectations (see [122, Chapter 2, Section 2.6]). Let $\mathcal{G}$ be a σ-algebra contained in $\mathcal{F}$. If X is an integrable random variable, then the *conditional expectation* of X for given $\mathcal{G}$ is any random variable Y which satisfies the following two conditions (CE1) and (CE2):

(CE1) The function Y is $\mathcal{G}$-measurable.
(CE2) $\int_A Y(\omega) \, P(d\omega) = \int_A X(\omega) \, P(d\omega)$ for all $A \in \mathcal{G}$.

We recall that conditions (CE1) and (CE2) determine Y up to a set in $\mathcal{G}$ of measure zero. We shall write

$$Y = E\left(X \mid \mathcal{G}\right).$$

Hence we have, for all $A \in \mathcal{G}$,

$$\int_A X(\omega) \, P(d\omega) = \int_A E(X \mid \mathcal{G}) \, P(d\omega).$$

When X is the characteristic function χ_B of a set $B \in \mathcal{F}$, we shall write

$$P(B \mid \mathcal{G}) = E\left(\chi_B \mid \mathcal{G}\right).$$

The function $P\left(B \mid \mathcal{G}\right)$ is called the *conditional probability* of B for given $\mathcal{G}$. This function can also be characterized as follows:

(CP1) The function $P\left(B \mid \mathcal{G}\right)$ is $\mathcal{G}$-measurable.
(CP2) $P(A \cap B) = E\left(P\left(B \mid \mathcal{G}\right); A\right)$ for every $A \in \mathcal{G}$. Namely, we have, for every $A \in \mathcal{G}$,

$$P(A \cap B) = \int_A P\left(B \mid \mathcal{G}\right)(\omega) \, P(d\omega).$$

It should be emphasized that the function $P\left(B \mid \mathcal{G}\right)$ is determined up to a set in $\mathcal{G}$ of P-measure zero, that is, it is an equivalence class of $\mathcal{G}$-measurable functions on Ω with respect to the measure P.

Markov processes are an abstraction of the idea of Brownian motion. Let K be a locally compact, separable metric space and let $\mathcal{B}$ be the σ-algebra of all Borel sets in K, that is, the smallest σ-algebra containing all open sets in K. Let $(\Omega, \mathcal{F}, P)$ be a probability space. A function $X(\omega)$ defined on Ω taking values in K is called a *random variable* if it satisfies the condition

$$X^{-1}(E) = \{\omega \in \Omega : X(\omega) \in E\} \in \mathcal{F} \quad \text{for all } E \in \mathcal{B}.$$

We express this by saying that X is $\mathcal{F}/\mathcal{B}$-measurable. A family $\{x_t\}_{t \geq 0}$ of random variables is called a *stochastic process*, and it may be thought of as the motion in time of a physical particle. The space K is called the *state space* and Ω the *sample space*. For a fixed $\omega \in \Omega$, the function $x_t(\omega)$, $t \geq 0$, defines in the state space K a *trajectory* or *path* of the process corresponding to the sample point ω.

In this generality the notion of a stochastic process is of course not so interesting. The most important class of stochastic processes is the class of Markov processes which is characterized by the Markov property. Intuitively, this is the principle of the lack of any "memory" in the system. More precisely, (temporally homogeneous) *Markov property* is that the prediction of subsequent motion of a particle, knowing its position at time t, depends neither on the value of t nor on what has been observed during the time interval $[0, t)$; that is, a particle "starts afresh".

Now we introduce a class of Markov processes which we will deal with in this book (see [34, Chapter III, Section 1], [14], [94]).

Definition 3.16. *Assume that we are given the following:*

(1) A locally compact, separable metric space K and the σ-algebra $\mathcal{B}$ of all Borel sets in K. A point ∂ is adjoined to K as the point at infinity if K is not compact, and as an isolated point if K is compact (see Figure 3.5 below). We let

$$K_\partial = K \cup \{\partial\},$$
$$\mathcal{B}_\partial = \text{ the } \sigma\text{-algebra in } K_\partial \text{ generated by } \mathcal{B}.$$

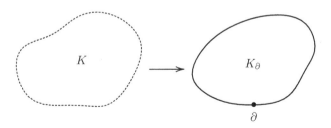

Fig. 3.5. The compactification K_∂ of K

(2) The space Ω of all mappings $\omega \colon [0, \infty] \to K_\partial$ such that $\omega(\infty) = \partial$ and that if $\omega(t) = \partial$ then $\omega(s) = \partial$ for all $s \geq t$. Let ω_∂ be the constant map $\omega_\partial(t) = \partial$ for all $t \in [0, \infty]$.
(3) For each $t \in [0, \infty]$, the coordinate map x_t defined by $x_t(\omega) = \omega(t)$ for all $\omega \in \Omega$.
(4) For each $t \in [0, \infty]$, a mapping $\varphi_t \colon \Omega \to \Omega$ defined by $(\varphi_t \omega)(s) = \omega(t + s)$ for all $\omega \in \Omega$. Note that $\varphi_\infty \omega = \omega_\partial$ and $x_t \circ \varphi_s = x_{t+s}$ for all $t, s \in [0, \infty]$.

(5) A σ-algebra $\mathcal{F}$ in Ω and an increasing family $\{\mathcal{F}_t\}_{0 \leq t \leq \infty}$ of sub-σ-algebras of $\mathcal{F}$.
(6) For each $x \in K_\partial$, a probability measure P_x on $(\Omega, \mathcal{F})$.

We say that these elements define a (temporally homogeneous) Markov process $\mathcal{X} = (x_t, \mathcal{F}, \mathcal{F}_t, P_x)$ if the following four conditions (i) through (iv) are satisfied:

(i) For each $0 \le t < \infty$, the function x_t is $\mathcal{F}_t/\mathcal{B}_\partial$-measurable, that is,

$$\{x_t \in E\} = \{\omega \in \Omega : x_t(\omega) \in E\} \in \mathcal{F}_t \quad \text{for all } E \in \mathcal{B}_\partial.$$

(ii) For all $0 \le t < \infty$ and $E \in \mathcal{B}$, the function

$$p_t(x, E) = P_x\{x_t \in E\} \tag{3.16}$$

is a Borel measurable function of $x \in K$.

(iii) $P_x\{\omega \in \Omega : x_0(\omega) = x\} = 1$ for each $x \in K_\partial$.

(iv) For all $t, h \in [0, \infty]$, $x \in K_\partial$ and $E \in \mathcal{B}_\partial$, we have the formula

$$P_x\{x_{t+h} \in E \mid \mathcal{F}_t\} = p_h(x_t, E) \quad \text{a. e.,}$$

or equivalently,

$$\boxed{P_x\left(A \cap \{x_{t+h} \in E\}\right) = \int_A p_h\left(x_t(\omega), E\right) dP_x(\omega) \quad \text{for all } A \in \mathcal{F}_t.}$$

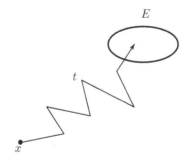

Fig. 3.6. The transition probability $p_t(x, E)$

Here is an intuitive way of thinking about the above definition of a Markov process. The sub-σ-algebra $\mathcal{F}_t$ may be interpreted as the collection of events which are observed during the time interval $[0, t]$. The value $P_x(A)$, $A \in \mathcal{F}$, may be interpreted as the probability of the event A under the condition that a particle starts at position x; hence the value $p_t(x, E)$ expresses the transition probability that a particle starting at position x will be found in the set E at time t (see Figure 3.6 above). The function $p_t(x, \cdot)$ is called the *transition function* of the process $\mathcal{X}$. The transition function $p_t(x, \cdot)$ specifies the probability structure of the process. The intuitive meaning of the crucial condition (iv) is that the future behavior of a particle, knowing its history up to time t, is the same as the behavior of a particle starting at $x_t(\omega)$, that is, a particle starts afresh.

A Markovian particle moves in the space K until it "dies" or "disappear" at the time when it reaches the point ∂; hence the point ∂ is called the *terminal point* or *cemetery*. With this interpretation in mind, we let

$$\zeta(\omega) = \inf\{t \in [0,\infty] : x_t(\omega) = \partial\}.$$

The random variable ζ is called the *lifetime* of the process $\mathcal{X}$. The process $\mathcal{X}$ is said to be *conservative* if it satisfies the condition

$$P_x\{\zeta = \infty\} = 1 \quad \text{for all } x \in K.$$

3.2.2 Markov Processes and Markov Transition Functions

In the first works devoted to Markov processes, the most fundamental was A. N. Kolmogorov's work ([70]) where the general concept of a Markov transition function was introduced for the first time and an analytic method of describing Markov transition functions was proposed. From the point of view of analysis, the transition function is something more convenient than the Markov process itself. In fact, it can be shown that the transition functions of Markov processes generate solutions of certain parabolic partial differential equations such as the classical diffusion equation; and, conversely, these differential equations can be used to construct and study the transition functions and the Markov processes themselves.

In the 1950s, the theory of Markov processes entered a new period of intensive development. We can associate with each transition function in a natural way a family of bounded linear operators acting on the space of continuous functions on the state space, and the Markov property implies that this family forms a semigroup. The Hille–Yosida theory of semigroups in functional analysis made possible further progress in the study of Markov processes.

Our first job is thus to give the precise definition of a transition function adapted to the theory of semigroups (see [34, Chapter III, Section 2]):

Definition 3.17. *Let* (K, ρ) *be a locally compact, separable metric space and let* $\mathcal{B}$ *be the* σ-*algebra of all Borel sets in* K. *A function* $p_t(x, E)$, *defined for all* $t \geq 0$, $x \in K$ *and* $E \in \mathcal{B}$, *is called a (temporally homogeneous) Markov transition function on* K *if it satisfies the following four conditions (a) through (d):*

(a) *$p_t(x, \cdot)$ is a non-negative measure on $\mathcal{B}$ and $p_t(x, K) \leq 1$ for all $t \geq 0$ and $x \in K$.*
(b) *$p_t(\cdot, E)$ is a Borel measurable function for all $t \geq 0$ and $E \in \mathcal{B}$.*
(c) *$p_0(x, \{x\}) = 1$ for all $x \in K$.*
(d) *(The* Chapman–Kolmogorov *equation) For all $t, s \geq 0$, $x \in K$ and $E \in \mathcal{B}$, we have the equation*

$$p_{t+s}(x, E) = \int_K p_t(x, dy)\, p_s(y, E). \tag{3.17}$$

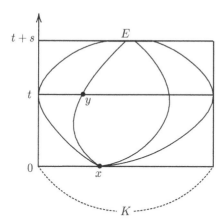

Fig. 3.7. An intuitive meaning of the Chapman–Kolmogorov equation (3.17)

Here is an intuitive way of thinking about the above definition of a Markov transition function. The value $p_t(x, E)$ expresses the transition probability that a physical particle starting at position x will be found in the set E at time t. The Chapman–Kolmogorov equation (3.17) expresses the idea that a transition from the position x to the set E in time $t + s$ is composed of a transition from x to some position y in time t, followed by a transition from y to the set E in the remaining time s; the latter transition has probability $p_s(y, E)$ which depends only on y (see Figure 3.7 above). Thus a particle "starts afresh"; this property is called the *Markov property*.

The Chapman–Kolmogorov equation (3.17) asserts that $p_t(x, K)$ is monotonically increasing as $t \downarrow 0$, so that the limit

$$p_{+0}(x, K) = \lim_{t \downarrow 0} p_t(x, K)$$

exists.

A transition function $p_t(x, \cdot)$ is said to be *normal* if it satisfies the condition

$$p_{+0}(x, K) = 1 \quad \text{for all } x \in K.$$

The next theorem, due to Dynkin [33, Chapter 4, Section 2] (or [34, p. 85, Theorem 3.2]), justifies the definition of a transition function, and hence it will be fundamental for our further study of Markov processes:

Theorem 3.18 (Dynkin). *For every Markov process, the function $p_t(x, \cdot)$, defined by formula (3.16), is a Markov transition function. Conversely, every normal Markov transition function corresponds to some Markov process.*

Here are some important examples of normal transition functions on the line $\mathbf{R} = (-\infty, \infty)$:

Example 3.1 (Uniform motion). If $t \geq 0$, $x \in \mathbf{R}$ and $E \in \mathcal{B}$, we let

$$p_t(x, E) = \chi_E(x + vt),$$

where v is a constant, and $\chi_E(y) = 1$ if $y \in E$ and $= 0$ if $y \notin E$.

This process, starting at x, moves deterministically with constant velocity v.

Example 3.2 (Poisson process). If $t \geq 0$, $x \in \mathbf{R}$ and $E \in \mathcal{B}$, we let

$$p_t(x, E) = e^{-\lambda t} \sum_{n=0}^{\infty} \frac{(\lambda t)^n}{n!} \chi_E(x + n),$$

where λ is a positive constant.

This process, starting at x, advances one unit by jumps, and the probability of n jumps during the time 0 and t is equal to $e^{-\lambda t}(\lambda t)^n/n!$.

Example 3.3 (Brownian motion). If $t > 0$, $x \in \mathbf{R}$ and $E \in \mathcal{B}$, we let

$$p_t(x, E) = \frac{1}{\sqrt{2\pi t}} \int_E \exp\left[-\frac{(y - x)^2}{2t}\right] dy,$$

and

$$p_0(x, E) = \chi_E(x).$$

This is a mathematical model of one-dimensional Brownian motion. Its character is quite different from that of the Poisson process; the transition function $p_t(x, E)$ satisfies the condition

$$p_t(x, [x - \varepsilon, x + \varepsilon]) = 1 - o(t) \quad \text{as } t \downarrow 0,$$

for all $\varepsilon > 0$ and $x \in \mathbf{R}$. This means that the process never stands still, as does the Poisson process. Indeed, this process changes state not by jumps but by *continuous* motion. A Markov process with this property is called a *diffusion process*.

Example 3.4 (Brownian motion with constant drift). If $t > 0$, $x \in \mathbf{R}$ and $E \in \mathcal{B}$, we let

$$p_t(x, E) = \frac{1}{\sqrt{2\pi t}} \int_E \exp\left[-\frac{(y - mt - x)^2}{2t}\right] dy,$$

and

$$p_0(x, E) = \chi_E(x),$$

where m is a constant.

This represents Brownian motion with a constant drift of magnitude m superimposed; the process can be represented as $\{x_t + mt\}$, where $\{x_t\}$ is Brownian motion on $\mathbf{R}$.

Example 3.5 (Cauchy process). If $t > 0$, $x \in \mathbf{R}$ and $E \in \mathcal{B}$, we let

$$p_t(x, E) = \frac{1}{\pi} \int_E \frac{t}{t^2 + (y - x)^2} \, dy,$$

and

$$p_0(x, E) = \chi_E(x).$$

This process can be thought of as the "trace" on the real line of trajectories of two-dimensional Brownian motion, and it moves by jumps (see [69, Lemma 2.12]). More precisely, if $B_1(t)$ and $B_2(t)$ are two independent Brownian motions and if T is the first passage time of $B_1(t)$ to x, then $B_2(T)$ has the Cauchy density

$$\frac{1}{\pi} \frac{|x|}{x^2 + y^2}, \quad -\infty < y < \infty.$$

Here are three more examples of diffusion processes on the half line $\overline{\mathbf{R}^+} = [0, \infty)$ in which we must take account of the effect of the boundary point 0:

Example 3.6 (Reflecting barrier Brownian motion). If $t > 0$, $x \in \overline{\mathbf{R}^+}$ and $E \in \mathcal{B}$, we let

$$p_t(x, E) = \frac{1}{\sqrt{2\pi t}} \left(\int_E \exp\left[-\frac{(y - x)^2}{2t} \right] dy + \int_E \exp\left[-\frac{(y + x)^2}{2t} \right] dy \right),$$

and

$$p_0(x, E) = \chi_E(x).$$

This represents Brownian motion with a reflecting barrier at $x = 0$; the process may be represented as $\{|x_t|\}$, where $\{x_t\}$ is Brownian motion on $\mathbf{R}$. Indeed, since $\{|x_t|\}$ goes from x to y if $\{x_t\}$ goes from x to $\pm y$ due to the symmetry of the transition function in Example 3.3 about $x = 0$, it follows that (see Figure 3.8 below)

$$p_t(x, E) = P_x\{|x_t| \in E\}$$

$$= \frac{1}{\sqrt{2\pi t}} \left(\int_E \exp\left[-\frac{(y - x)^2}{2t} \right] dy + \int_E \exp\left[-\frac{(y + x)^2}{2t} \right] dy \right).$$

Example 3.7 (Absorbing barrier Brownian motion). If $t > 0$, $x \in K = [0, \infty)$ and $E \in \mathcal{B}$, we let

$$p_t(x, E) = \frac{1}{\sqrt{2\pi t}} \left(\int_E \exp\left[-\frac{(y - x)^2}{2t} \right] dy - \int_E \exp\left[-\frac{(y + x)^2}{2t} \right] dy \right),$$

and

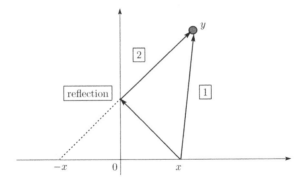

Fig. 3.8. The reflecting barrier Brownian motion

$$p_0(x, E) = \chi_E(x).$$

This represents Brownian motion with an absorbing barrier at $x = 0$; a Brownian particle dies at the first moment when it hits the boundary point $x = 0$ (see Figure 3.9 below). Namely, the boundary point 0 of K is the terminal point.

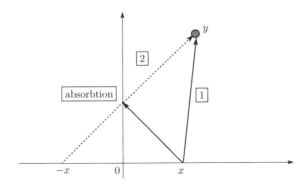

Fig. 3.9. The absorbing barrier Brownian motion

Example 3.8 (Sticking barrier Brownian motion). If $t > 0$, $x \in \overline{\mathbf{R}^+}$ and $E \in \mathcal{B}$, we let

$$p_t(x, E) = \frac{1}{\sqrt{2\pi t}} \left(\int_E \exp\left[-\frac{(y-x)^2}{2t}\right] dy - \int_E \exp\left[-\frac{(y+x)^2}{2t}\right] dy \right)$$
$$+ \left(1 - \frac{1}{\sqrt{2\pi t}} \int_{-x}^{x} \exp\left[-\frac{z^2}{2t}\right] dz\right) \chi_E(0),$$

and

$$p_0(x, E) = \chi_E(x).$$

This represents Brownian motion with a sticking barrier at $x = 0$. When a Brownian particle reaches the boundary point 0 for the first time, instead of reflecting it sticks there forever; in this case the state 0 is called a *trap*.

3.2.3 Path Functions of Markov Processes

It is naturally interesting and important to ask the following problem:

Problem. Given a Markov transition function $p_t(x, \cdot)$, under which conditions on $p_t(x, \cdot)$ does there exist a Markov process with transition function $p_t(x, \cdot)$ whose paths are almost surely continuous ?

A Markov process $\mathcal{X} = (x_t, \mathcal{F}, \mathcal{F}_t, P_x)$ is said to be *right continuous* provided that we have, for each $x \in K$,

$P_x\{\omega \in \Omega : \text{the mapping } t \mapsto x_t(\omega) \text{ is a right continuous function}$
$\text{from } [0, \infty) \text{ into } K_\partial\} = 1.$

Furthermore, we say that $\mathcal{X}$ is *continuous* provided that we have, for each $x \in K$,

$P_x\{\omega \in \Omega : \text{the mapping } t \mapsto x_t(\omega) \text{ is a continuous function}$
$\text{from } [0, \zeta(\omega)) \text{ into } K_\partial\} = 1,$

where ζ is the lifetime of the process $\mathcal{X}$.

Now we give some useful criteria for path continuity in terms of Markov transition functions (see Dynkin [33, Chapter 6], [34, p. 91, Theorem 3.5]):

Theorem 3.19. *Let (K, ρ) be a locally compact, separable metric space and let $p_t(x, \cdot)$ be a normal Markov transition function on K.*
(i) Assume that the following two conditions (L) and (M) are satisfied:

(L) For each $s > 0$ and each compact $E \subset K$, we have the condition

$$\lim_{x \to \partial} \sup_{0 \le t \le s} p_t(x, E) = 0.$$

(M) For each $\varepsilon > 0$ and each compact $E \subset K$, we have the condition

$$\lim_{t \downarrow 0} \sup_{x \in E} p_t(x, K \setminus U_\varepsilon(x)) = 0,$$

where $U_\varepsilon(x) = \{y \in K : \rho(y, x) < \varepsilon\}$ is an ε-neighborhood of x.

Then there exists a Markov process $\mathcal{X}$ with transition function $p_t(x, \cdot)$ whose paths are right continuous on $[0, \infty)$ and have left-hand limits on $[0, \zeta)$ almost surely.
(ii) Assume that condition (L) and the following condition (N) (replacing condition (M)) are satisfied:

(N) For each $\varepsilon > 0$ and each compact $E \subset K$, we have the condition

$$\lim_{t \downarrow 0} \frac{1}{t} \sup_{x \in E} p_t(x, K \setminus U_\varepsilon(x)) = 0,$$

or equivalently

$$\sup_{x \in E} p_t(x, K \setminus U_\varepsilon(x)) = o(t) \quad \text{as } t \downarrow 0.$$

Then there exists a Markov process X with transition function $p_t(x, \cdot)$ whose paths are almost surely continuous on $[0, \zeta)$.

Remark 3.20. It is known (see Dynkin [33, Lemma 6.2]) that if the paths of a Markov process are right continuous, then the transition function $p_t(x, \cdot)$ satisfies the condition

$$\lim_{t \downarrow 0} p_t(x, U_\varepsilon(x)) = 1 \quad \text{for all } x \in K.$$

3.2.4 Stopping Times

In this subsection we formulate the starting afresh property for suitable random times τ, that is, the events $\{\omega \in \Omega : \tau(\omega) < a\}$ should depend on the process $\{x_t\}$ only "up to time a", but not on the "future" after time a. This idea leads us to the following definition (see [34, Chapter III, Section 3]):

Definition 3.21. *Let $\{\mathcal{F}_t : t \geq 0\}$ be an increasing family of σ-algebras in a probability space $(\Omega, \mathcal{F}, P)$. A mapping $\tau : \Omega \to [0, \infty]$ is called a* stopping time *or* Markov time *with respect to $\{\mathcal{F}_t\}$ if it satisfies the condition*

$$\{\tau < a\} = \{\omega \in \Omega : \tau(\omega) < a\} \in \mathcal{F}_a \quad \text{for all } a > 0. \tag{3.18}$$

If we introduce another condition

$$\{\tau \leq a\} = \{\omega \in \Omega : \tau(\omega) \leq a\} \in \mathcal{F}_a \quad \text{for all } a > 0, \tag{3.19}$$

then condition (3.19) implies condition (3.18); hence we obtain a smaller family of stopping times.

Conversely, we can prove the following lemma (see [122, Lemma 9.23]):

Lemma 3.22. *Assume that the family $\{\mathcal{F}_t\}$ is* right-continuous, *that is,*

$$\mathcal{F}_t = \bigcap_{s > t} \mathcal{F}_s \quad \text{for each } t \geq 0.$$

Then condition (3.18) implies condition (3.19).

Therefore, we find that conditions (3.18) and (3.19) are equivalent provided that the family $\{\mathcal{F}_t\}$ is right-continuous.

If τ is a stopping time with respect to the right-continuous family $\{\mathcal{F}_t\}$ of σ-algebras, we let

$$\mathcal{F}_\tau = \{A \in \mathcal{F} : A \cap \{\tau \le a\} \in \mathcal{F}_a \quad \text{for all } a > 0\}.$$

Intuitively, we may think of $\mathcal{F}_\tau$ as the "past" up to the random time τ. Then we have the following lemma (see [122, Lemma 9.24]):

Lemma 3.23. $\mathcal{F}_\tau$ *is a σ-algebra.*

Now we list some elementary properties of stopping times and their associated σ-algebras:

(i) Any non-negative constant mapping is a stopping time. More precisely, if $\tau \equiv t_0$ for some constant $t_0 \ge 0$, then it follows that τ is a stopping time and that $\mathcal{F}_\tau$ reduces to $\mathcal{F}_{t_0}$.

(ii) If $\{\tau_n\}$ is a finite or denumerable collection of stopping times for the family $\{\mathcal{F}_t\}$, then it follows that

$$\tau = \inf_n \tau_n$$

is also a stopping time.

(iii) If $\{\tau_n\}$ is a finite or denumerable collection of stopping times for the family $\{\mathcal{F}_t\}$, then it follows that

$$\tau = \sup_n \tau_n$$

is also a stopping time.

(iv) If τ is a stopping time and t_0 is a positive constant, then it follows that $\tau + t_0$ is also a stopping time.

(v) Let τ_1 and τ_2 be stopping times for the family $\{\mathcal{F}_t\}$ such that $\tau_1 \le \tau_2$ on Ω. Then it follows that

$$\mathcal{F}_{\tau_1} \subset \mathcal{F}_{\tau_2}.$$

This is a generalization of the monotonicity of the family $\{\mathcal{F}_t\}$.

(vi) Let $\{\tau_n\}_{n=1}^\infty$ be a sequence of stopping times for the family $\{\mathcal{F}_t\}$ such that $\tau_{n+1} \le \tau_n$ on Ω. Then it follows that the limit

$$\tau = \lim_{n \to \infty} \tau_n = \inf_{n \ge 1} \tau_n$$

is a stopping time and further that

$$\mathcal{F}_\tau = \bigcap_{n \ge 1} \mathcal{F}_{\tau_n}.$$

This property generalizes the right-continuity of the family $\{\mathcal{F}_t\}$.

3.2.5 Definition of Strong Markov Processes

A Markov process is called a *strong Markov process* if the "starting afresh" property holds not only for every fixed moment but also for suitable random times. In this subsection, following Dynkin [34, Chapter III, Section 3] we formulate precisely this "strong" Markov property, and give a useful criterion for the strong Markov property.

Let (K, ρ) be a locally compact, separable metric space, and let $K_\partial = K \cup \{\partial\}$ be its *one-point compactification*. Namely, we add a new point ∂ to the locally compact space K as the *point at infinity* if K is not compact, and as an isolated point if K is compact.

Let $\mathcal{X} = (x_t, \mathcal{F}, \mathcal{F}_t, P_x)$ be a Markov process. For each $t \in [0, \infty]$, we define a mapping

$$\Phi_t \colon [0, t] \times \Omega \longrightarrow K_\partial$$

by the formula

$$\Phi_t(s, \omega) = x_s(\omega).$$

A Markov process $\mathcal{X} = (x_t, \mathcal{F}, \mathcal{F}_t, P_x)$ is said to be *progressively measurable* with respect to $\{\mathcal{F}_t\}$ if the mapping Φ_t is $\mathcal{B}_{[0,t]} \times \mathcal{F}_t/\mathcal{B}_\partial$-measurable for each $t \in [0, \infty]$, that is, if we have the condition

$$\Phi_t^{-1}(E) = \{\Phi_t \in E\} \in \mathcal{B}_{[0,t]} \times \mathcal{F}_t \quad \text{for all } E \in \mathcal{B}_\partial.$$

Here $\mathcal{B}_{[0,t]}$ is the σ-algebra of all Borel sets in the interval $[0, t]$ and $\mathcal{B}_\partial$ is the σ-algebra in K_∂ generated by $\mathcal{B}$. It should be noticed that if $\mathcal{X}$ is progressively measurable and if τ is a stopping time, then the mapping $x_\tau \colon \omega \mapsto x_{\tau(\omega)}(\omega)$ is $\mathcal{F}_\tau/\mathcal{B}_\partial$- measurable.

Definition 3.24. *We say that a progressively measurable Markov process* $\mathcal{X} = (x_t, \mathcal{F}, \mathcal{F}_t, P_x)$ *has the* strong Markov property *with respect to* $\{\mathcal{F}_t\}$ *if the following condition is satisfied:*

For all $h \geq 0$, $x \in K_\partial$, $E \in \mathcal{B}_\partial$ and all stopping times τ, we have the formula

$$P_x\{x_{\tau+h} \in E \mid \mathcal{F}_\tau\} = p_h(x_\tau, E),$$

or equivalently,

$$\boxed{P_x(A \cap \{x_{\tau+h} \in E\}) = \int_A p_h(x_{\tau(\omega)}(\omega), E)\, dP_x(\omega) \quad \text{for all } A \in \mathcal{F}_\tau.}$$

This expresses the idea of "starting afresh" at random times (cf. Definition 3.16).

We shall state a simple criterion for the strong Markov property in terms of transition functions.

Let (K, ρ) be a locally compact, separable metric space. We add a point ∂ to the metric space K as the point at infinity if K is not compact, and as an isolated point if K is compact; so the space $K_\partial = K \cup \{\partial\}$ is compact (see Figure 3.5). Let $C(K)$ be the space of real-valued, bounded continuous functions $f(x)$ on K; the space $C(K)$ is a Banach space with the supremum norm

$$\|f\|_\infty = \sup_{x \in K} |f(x)|.$$

We say that a function $f \in C(K)$ converges to zero as $x \to \partial$ if, for each $\varepsilon > 0$, there exists a compact subset E of K such that

$$|f(x)| < \varepsilon \quad \text{for all } x \in K \setminus E,$$

and we then write $\lim_{x \to \partial} f(x) = 0$. We let

$$C_0(K) = \left\{ f \in C(K) : \lim_{x \to \partial} f(x) = 0 \right\}.$$

The space $C_0(K)$ is a closed subspace of $C(K)$; hence it is a Banach space. Note that $C_0(K)$ may be identified with $C(K)$ if K is compact.

We introduce a useful convention as follows:

Any real-valued function $f(x)$ on K is extended to the space $K_\partial = K \cup \{\partial\}$ by setting $f(\partial) = 0$.

From this point of view, the space $C_0(K)$ is identified with the subspace of $C(K_\partial)$ which consists of all functions $f(x)$ satisfying the condition $f(\partial) = 0$:

$$C_0(K) = \{ f \in C(K_\partial) : f(\partial) = 0 \}.$$

Furthermore, we can extend a Markov transition function $p_t(x, \cdot)$ on K to a Markov transition function $p'_t(x, \cdot)$ on K_∂ by the formulas:

$$\begin{cases} p'_t(x, E) = p_t(x, E) & \text{for all } x \in K \text{ and } E \in \mathcal{B}, \\ p'_t(x, \{\partial\}) = 1 - p_t(x, K) & \text{for all } x \in K, \\ p'_t(\partial, K) = 0, \quad p'_t(\partial, \{\partial\}) = 1. \end{cases}$$

Intuitively, this means that a Markovian particle moves in the space K until it "dies" at the time it reaches the point ∂; hence the point ∂ is called the *terminal point*.

Now we introduce some conditions on the measures $p_t(x, \cdot)$ related to continuity in $x \in K$, for fixed $t \geq 0$ (see [34, Chapter II, Section 5]):

Definition 3.25. *(i) A Markov transition function $p_t(x, \cdot)$ is called a* Feller *function if the function*

$$T_t f(x) = \int_K p_t(x, dy) f(y)$$

is a continuous function of $x \in K$ whenever f is in $C(K)$, that is, if we have the condition

$$f \in C(K) \implies T_t f \in C(K).$$

(ii) We say that $p_t(x, \cdot)$ is a C_0-function if the space $C_0(K)$ is an invariant subspace of $C(K)$ for the operators T_t:

$$f \in C_0(K) \implies T_t f \in C_0(K).$$

Remark 3.26. The Feller property is equivalent to saying that the measures $p_t(x, \cdot)$ depend continuously on $x \in K$ in the usual weak topology, for every fixed $t \geq 0$.

The next result gives a useful criterion for the strong Markov property ([33, Theorem 5.10], [34, p. 99, Theorem 3.10])):

Theorem 3.27. *Assume that the transition function of some right-continuous Markov process has the C_0-property. Then it is a strong Markov process.*

3.2.6 Strong Markov Property and Uniform Stochastic Continuity

In this subsection, following Dynkin [33] and [34] we introduce the basic notion of uniform stochastic continuity of transition functions, and give simple criteria for the strong Markov property in terms of transition functions (see Figure 3.10 below).

Let (K, ρ) be a locally compact, separable metric space. We begin with the following definition (see [34, p. 92, Condition $M(\Gamma)$]):

Definition 3.28. *A transition function p_t on K is said to be uniformly stochastically continuous on K if it satisfies the following condition:*

For each $\varepsilon > 0$ and each compact $E \subset K$, we have the assertion

$$\lim_{t \downarrow 0} \sup_{x \in E} \left[1 - p_t(x, U_\varepsilon(x)) \right] = 0, \qquad (3.20)$$

where $U_\varepsilon(x) = \{ y \in K : \rho(y, x) < \varepsilon \}$ is an ε-neighborhood of x.

It should be noticed that every uniformly stochastically continuous transition function p_t is normal and satisfies condition (M) in Theorem 3.19. Therefore, by combining part (i) of Theorem 3.19 and Theorem 3.27 we obtain the following theorem (see [34, p. 92, Theorem 3.7], [122, Theorem 9.28]):

Theorem 3.29. *If a uniformly stochastically continuous, C_0 transition function satisfies condition (L), then it is the transition function of some strong Markov process whose paths are right-continuous and have no discontinuities other than jumps.*

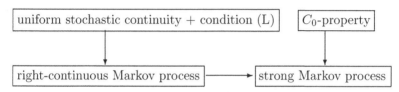

Fig. 3.10. A functional analytic approach to strong Markov processes in Theorems 3.19, 3.27 and 3.29

Theorems 3.19, 3.27 and 3.29 can be visualized as follows.

A continuous strong Markov process is called a *diffusion process*. The next result states a sufficient condition for the existence of a diffusion process with a prescribed transition function (see [34, p. 91, Theorem 3.5], [122, Theorem 9.29]):

Theorem 3.30. *If a uniformly stochastically continuous, C_0 transition function satisfies conditions (L) and (N), then it is the transition function of some diffusion process.*

This theorem is an immediate consequence of part (ii) of Theorem 3.19 and Theorem 3.29.

3.3 Markov Transition Functions and Feller Semigroups

This section provides a brief description of basic definitions and results about Markov processes and a class of semigroups (Feller semigroups) associated with Markov processes. The semigroup approach to Markov processes can be traced back to the pioneering work of Feller [39] and [40] in early 1950s (cf. [15], [98], [116]).

The Feller or C_0-property deals with continuity of a Markov transition function $p_t(x, E)$ in x, and does not, by itself, have no concern with continuity in t. We give a necessary and sufficient condition on $p_t(x, E)$ in order that its associated operators $\{T_t\}_{t \geq 0}$, defined by the formula

$$T_t f(x) = \int_K p_t(x, dy) \, f(y) \quad \text{for } f \in C_0(K), \tag{3.21}$$

is *strongly continuous* in t on the space $C_0(K)$:

$$\lim_{s \downarrow 0} \|T_{t+s} f - T_t f\|_\infty = 0 \quad \text{for every } f \in C_0(K). \tag{3.22}$$

Then we have the following theorem (cf. [114, Theorem 9.2.3]; [122, Theorem 9.33]):

Theorem 3.31. *Let $p_t(x, \cdot)$ be a C_0-transition function on a locally compact, separable metric space K. Then the associated operators $\{T_t\}_{t \geq 0}$, defined by formula (3.21) is strongly continuous in t on $C_0(K)$ if and only if $p_t(x, \cdot)$ is uniformly stochastically continuous on K and satisfies condition (L).*

Proof. (i) The "if" part: Since continuous functions with compact support are dense in $C_0(K)$, it suffices to prove the strong continuity of $\{T_t\}$ at $t = 0$:

$$\lim_{t \downarrow 0} \|T_t f - f\|_\infty = 0 \tag{3.23}$$

for all such functions f.

For any compact subset E of K containing the support $\operatorname{supp} f$ of f, we have the inequality

$$\|T_t f - f\|_\infty \leq \sup_{x \in E} |T_t f(x) - f(x)| + \sup_{x \in K \setminus E} |T_t f(x)| \tag{3.24}$$

$$\leq \sup_{x \in E} |T_t f(x) - f(x)| + \|f\|_\infty \cdot \sup_{x \in K \setminus E} p_t(x, \operatorname{supp} f).$$

However, condition (L) implies that, for each $\varepsilon > 0$ we can find a compact subset E of K such that, for all sufficiently small $t > 0$,

$$\sup_{x \in K \setminus E} p_t(x, \operatorname{supp} f) < \varepsilon. \tag{3.25}$$

On the other hand, we have, for each $\delta > 0$,

$$T_t f(x) - f(x)$$
$$= \int_{U_\delta(x)} p_t(x, dy)(f(y) - f(x))$$
$$+ \int_{K \setminus U_\delta(x)} p_t(x, dy)\,(f(y) - f(x)) - f(x)\,(1 - p_t(x, K)),$$

and hence

$$\sup_{x \in E} |T_t f(x) - f(x)|$$
$$\leq \sup_{\rho(x,y) < \delta} |f(y) - f(x)| + 3\|f\|_\infty \sup_{x \in E} [1 - p_t(x, U_\delta(x))].$$

Since the function $f(x)$ is uniformly continuous, we can choose a positive constant δ such that

$$\sup_{\rho(x,y) < \delta} |f(y) - f(x)| < \varepsilon.$$

Furthermore, it follows from condition (3.20) with $\varepsilon := \delta$ (the uniform stochastic continuity of $p_t(x, \cdot)$) that, for all sufficiently small $t > 0$,

$$\sup_{x \in E} [1 - p_t(x, U_\delta(x))] < \varepsilon.$$

Hence we have, for all sufficiently small $t > 0$,

$$\sup_{x \in E} |T_t f(x) - f(x)| < \varepsilon(1 + 3\|f\|_\infty). \tag{3.26}$$

Therefore, by carrying inequalities (3.25) and (3.26) into inequality (3.24) we obtain that, for all sufficiently small $t > 0$,

$$\|T_t f - f\|_\infty < \varepsilon(1 + 4\|f\|_\infty).$$

This proves the desired formula (3.23), that is, the strong continuity of $\{T_t\}$.

(ii) The "only if" part: For any $x \in K$ and $\varepsilon > 0$, we define a continuous function $f_x(y)$ by the formula (see Figure 3.11 below)

$$f_x(y) = \begin{cases} 1 - \dfrac{1}{\varepsilon}\rho(x, y) & \text{if } \rho(x, y) \leq \varepsilon, \\ 0 & \text{if } \rho(x, y) > \varepsilon. \end{cases} \tag{3.27}$$

Let E be an arbitrary compact subset of K. Then, for all sufficiently small

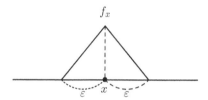

Fig. 3.11. The function f_x

$\varepsilon > 0$, the functions f_x, $x \in E$, are in $C_0(K)$ and satisfy the condition

$$\|f_x - f_z\|_\infty \leq \frac{1}{\varepsilon}\rho(x, z) \quad \text{for all } x, z \in E. \tag{3.28}$$

However, for any $\delta > 0$, by the compactness of E we can find a finite number of points x_1, x_2, ..., x_n of E such that

$$E = \bigcup_{k=1}^{n} U_{\delta\varepsilon/4}(x_k),$$

and hence

$$\min_{1 \leq k \leq n} \rho(x, x_k) \leq \frac{\delta\varepsilon}{4} \quad \text{for all } x \in E.$$

Thus, by combining this inequality with inequality (3.28) with $z := x_k$ we obtain that

$$\min_{1 \leq k \leq n} \|f_x - f_{x_k}\|_\infty \leq \frac{\delta}{4} \quad \text{for all } x \in E. \tag{3.29}$$

Now we have, by formula (3.27),

$$0 \leq 1 - p_t(x, U_\varepsilon(x)) \leq 1 - \int_{K_\partial} p_t(x, dy) f_x(y)$$
$$= f_x(x) - T_t f_x(x)$$
$$\leq \|f_x - T_t f_x\|_\infty$$
$$\leq \|f_x - f_{x_k}\|_\infty + \|f_{x_k} - T_t f_{x_k}\|_\infty$$
$$+ \|T_t f_{x_k} - T_t f_x\|_\infty$$
$$\leq 2\|f_x - f_{x_k}\|_\infty + \|f_{x_k} - T_t f_{x_k}\|_\infty \quad \text{for all } x \in E.$$

In view of inequality (3.29), the first term on the last inequality is bounded by $\delta/2$ for the right choice of k. Furthermore, it follows from the strong continuity (3.23) of $\{T_t\}$ that the second term tends to zero as $t \downarrow 0$ for each $k = 1, 2, \ldots, n$.

Consequently, we have, for all sufficiently small $t > 0$,

$$\sup_{x \in E} [1 - p_t(x, U_\varepsilon(x))] \leq \delta.$$

This proves the desired condition (3.20), that is, the uniform stochastic continuity of $p_t(x, \cdot)$.

Finally, it remains to verify condition (L). Our proof is based on a reduction to absurdity. Assume, to the contrary, that:

For some $s > 0$ and some compact $E \subset K$, there exist a positive constant ε_0, a sequence $\{t_k\}$, $t_k \downarrow t$ $(0 \leq t \leq s)$ and a sequence $\{x_k\}$, $x_k \to \partial$, such that

$$p_{t_k}(x_k, E) \geq \varepsilon_0. \tag{3.30}$$

Now we take a relatively compact subset U of K containing E, and let (see Figure 3.12 below)

$$f(x) = \frac{\rho(x, K \setminus U)}{\rho(x, E) + \rho(x, K \setminus U)}.$$

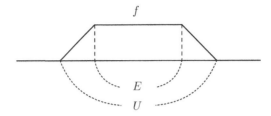

Fig. 3.12. The function $f(x)$

Then it follows that the function $f(x)$ is in $C_0(K)$ and satisfies the condition

$$(T_t f)(x) = \int_K p_t(x, dy) f(y) \geq p_t(x, E) \geq 0.$$

Therefore, by combining this inequality with inequality (3.30) we obtain that

$$(T_{t_k} f)(x_k) \geq p_{t_k}(x_k, E) \geq \varepsilon_0 \quad \text{for all } k \in \mathbf{N}. \tag{3.31}$$

However, we have the inequality

$$(T_{t_k} f)(x_k) \leq \|T_{t_k} f - T_t f\|_\infty + (T_t f)(x_k) \quad \text{for all } k \in \mathbf{N}. \tag{3.32}$$

Since the semigroup $\{T_t\}$ is strongly continuous and since we have the assertion

$$\lim_{k \to \infty} (T_t f)(x_k) = \int_{K_\partial} p'_t(\partial, dy) f(y) = f(\partial) = 0,$$

we can let $k \to \infty$ in inequality (3.32) to obtain that

$$\limsup_{k \to \infty} (T_{t_k} f)(x_k) = 0.$$

This contradicts inequality (3.31).

The proof of Theorem 3.31 is now complete. $\square$

Now we are in a position to define the following definition:

Definition 3.32. *A family* $\{T_t\}_{t \geq 0}$ *of bounded linear operators acting on the Banach space* $C_0(K)$ *is called a* Feller semigroup *on the state space* K *if it satisfies the following three conditions (i), (ii) and (iii):*

(i) $T_{t+s} = T_t \cdot T_s$, *t*, *s* ≥ 0 *(the semigroup property);* $T_0 = I$.
(ii) The family $\{T_t\}$ *is strongly continuous in t for all* $t \geq 0$:

$$\lim_{s \downarrow 0} \|T_{t+s} f - T_t f\|_\infty = 0 \quad \text{for each } f \in C_0(K).$$

(iii) The family $\{T_t\}$ *is non-negative and contractive on* $C_0(K)$:

$$f \in C_0(K), \, 0 \leq f(x) \leq 1 \quad \text{on } K \implies 0 \leq T_t f(x) \leq 1 \quad \text{on } K.$$

Rephrased, Theorem 3.31 gives a characterization of Feller semigroups in terms of Markov transition functions (see Figure 3.13 below):

Theorem 3.33 (Dynkin). *If* $p_t(x, \cdot)$ *is a uniformly stochastically continuous,* C_0-transition function on a locally compact, separable metric space K and satisfies condition (L), then its associated operators $\{T_t\}_{t \geq 0}$, defined by formula (3.21), form a Feller semigroup on the state space K.

Conversely, if $\{T_t\}_{t \geq 0}$ *is a Feller semigroup on the state space* K, *there exists a uniformly stochastically continuous,* C_0-transition function $p_t(x, \cdot)$ on K, satisfying condition (L), such that formula (3.21) holds true.

The most important applications of Theorem 3.33 are of course in the second statement.

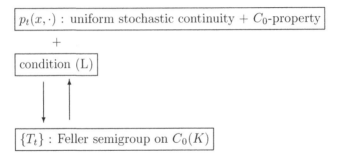

$p_t(x, \cdot)$: uniform stochastic continuity + C_0-property

+

condition (L)

$\{T_t\}$: Feller semigroup on $C_0(K)$

Fig. 3.13. A functional analytic approach to strong Markov processes in Theorems 3.31 and 3.33

3.4 Generation Theorems for Feller Semigroups

In this section we prove various generation theorems of Feller semigroups by using the Hille–Yosida theory of semigroups (Theorems 3.34 and 3.36) which form a functional analytic background for the proof of Theorem 1.6.

If $\{T_t\}_{t\geq 0}$ is a Feller semigroup on the state space K, we define its *infinitesimal generator* A by the formula

$$Au = \lim_{t\downarrow 0} \frac{T_t u - u}{t} \quad \text{for } u \in C_0(K), \tag{3.33}$$

provided that the limit (3.33) exists in the space $C_0(K)$. More precisely, the generator A is a linear operator from $C_0(K)$ into itself defined as follows:

(1) The domain $\mathcal{D}(A)$ of A is the set

$$\mathcal{D}(A) = \{u \in C_0(K) : \text{the limit (3.33) exists}\}.$$

(2) $Au = \lim_{t\downarrow 0} \dfrac{T_t u - u}{t}$ for every $u \in \mathcal{D}(A)$.

The next theorem is a version of the Hille–Yosida theorem adapted to the present context (cf. [114, Theorem 9.3.1 and Corollary 9.3.2]; [122, Theorem 9.35]):

Theorem 3.34 (Hille–Yosida). *(i) Let $\{T_t\}_{t\geq 0}$ be a Feller semigroup on the state space K and let A be its infinitesimal generator. Then we have the following four assertions (a) through (d):*

(a) The domain $\mathcal{D}(A)$ is dense in the space $C_0(K)$.
(b) For each $\alpha > 0$, the equation $(\alpha I - A)u = f$ has a unique solution u in $\mathcal{D}(A)$ for any $f \in C_0(K)$. Hence, for each $\alpha > 0$ the Green operator $(\alpha I - A)^{-1} : C_0(K) \to C_0(K)$ can be defined by the formula

$$u = (\alpha I - A)^{-1}f \quad \text{for } f \in C_0(K).$$

(c) For each $\alpha > 0$, the operator $(\alpha I - A)^{-1}$ is non-negative on $C_0(K)$:

$$f \in C_0(K), \ f(x) \geq 0 \quad \text{on } K \implies (\alpha I - A)^{-1} f(x) \geq 0 \quad \text{on } K.$$

(d) For each $\alpha > 0$, the operator $(\alpha I - A)^{-1}$ is bounded on $C_0(K)$ with norm

$$\|(\alpha I - A)^{-1}\| \leq \frac{1}{\alpha}.$$

(ii) Conversely, if A is a linear operator from $C_0(K)$ into itself satisfying condition (a) and if there is a non-negative constant α_0 such that, for all $\alpha > \alpha_0$, conditions (b), (c) and (d) are satisfied, then A is the infinitesimal generator *of some Feller semigroup $\{T_t\}_{t \geq 0}$ on the state space K.*

Proof. In view of the Hille–Yosida theory (see [147, Chapter IX, Section 7]), it suffices to show that the semigroup $\{T_t\}_{t \geq 0}$ is non-negative if and only if its resolvents (Green operators) $\{(\alpha I - A)^{-1}\}_{\alpha > \alpha_0}$ are non-negative.

The "only if" part is an immediate consequence of the following expression of the resolvent $(\alpha I - A)^{-1}$ in terms of the semigroup $\{T_t\}$:

$$(\alpha I - A)^{-1} = \int_0^\infty \exp[-\alpha t] \, T_t \, dt \quad \text{for } \alpha > 0.$$

On the other hand, the "if" part follows from the expression of the semigroup $T_t(\alpha)$ in terms of the Yosida approximation $J_\alpha = \alpha(\alpha I - A)^{-1}$:

$$T_t(\alpha) = \exp[-\alpha t] \, \exp[\alpha t \, J_\alpha] = \exp[-\alpha t] \sum_{n=0}^\infty \frac{(\alpha t)^n}{n!} \, J_\alpha^n,$$

and the definition of the semigroup T_t:

$$T_t = \lim_{\alpha \to \infty} T_t(\alpha).$$

The proof of Theorem 3.34 is complete. $\square$

Corollary 3.35. *Let K be a* compact *metric space and let A be the infinitesimal generator of a Feller semigroup on the state space K. Assume that the constant function 1 belongs to the domain $\mathcal{D}(A)$ of A and that we have, for some constant c,*

$$(A1)(x) \leq -c \quad \text{on } K. \tag{3.34}$$

Then the operator $A' = A + cI$ is the infinitesimal generator of some Feller semigroup on the state space K.

Proof. It follows from an application of part (i) of Theorem 3.34 that the operators

$$(\alpha I - A')^{-1} = ((\alpha - c) I - A)^{-1}$$

are defined and non-negative on the whole space $C(K)$, for all $\alpha > c$. However, in view of inequality (3.34) we obtain that

$$\alpha \leq \alpha - (A1 + c) = (\alpha I - A') 1 \quad \text{on } K,$$

so that

$$\alpha(\alpha I - A')^{-1} 1 \leq (\alpha I - A')^{-1} ((\alpha I - A')1) = 1 \quad \text{on } K.$$

Hence we have, for all $\alpha > c$,

$$\|(\alpha I - A')^{-1}\| = \|(\alpha I - A')^{-1} 1\|_\infty \leq \frac{1}{\alpha}.$$

Therefore, by applying part (ii) of Theorem 3.34 to the operator A' we find that A' is the infinitesimal generator of some Feller semigroup on the state space K.

The proof of Corollary 3.35 is complete. □

The Hille–Yosida theory of semigroups via the *Laplace transform* can be visualized as in Figure 3.14 below.

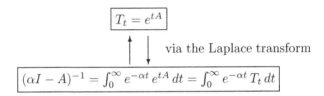

Fig. 3.14. The Hille–Yosida theory of semigroups via the Laplace transform

Now we write down explicitly the infinitesimal generators of Feller semigroups associated with the transition functions in Examples 3.1 through 3.8 (cf. [35]).

Example 3.9 (Uniform motion). $K = \mathbf{R}$ and

$$\begin{cases} \mathcal{D}(A) = \{f \in C_0(K) : f' \in C_0(K)\}, \\ Af = vf' \quad \text{for every } f \in \mathcal{D}(A). \end{cases}$$

Example 3.10 (Poisson process). $K = \mathbf{R}$ and

$$\begin{cases} \mathcal{D}(A) = C_0(K), \\ Af(x) = \lambda(f(x+1) - f(x)) \quad \text{for every } f \in \mathcal{D}(A). \end{cases}$$

The operator A is not "local"; the value $Af(x)$ depends on the values $f(x)$ and $f(x+1)$. This reflects the fact that the Poisson process changes state by jumps.

Example 3.11 (Brownian motion). $K = \mathbf{R}$ and

$$\begin{cases} \mathcal{D}(A) = \{f \in C_0(K) : f' \in C_0(K), \ f'' \in C_0(K)\}, \\ Af = \dfrac{1}{2}f'' \quad \text{for every } f \in \mathcal{D}(A). \end{cases}$$

The operator A is "local", that is, the value $Af(x)$ is determined by the values of f in an arbitrary small neighborhood of x. This reflects the fact that Brownian motion changes state by continuous motion.

Example 3.12 (Brownian motion with constant drift). $K = \mathbf{R}$ and

$$\begin{cases} \mathcal{D}(A) = \{f \in C_0(K) : f' \in C_0(K), \ f'' \in C_0(K)\}, \\ Af = \dfrac{1}{2}f'' + mf' \quad \text{for every } f \in \mathcal{D}(A). \end{cases}$$

Example 3.13 (Cauchy process). $K = \mathbf{R}$ and, the domain $\mathcal{D}(A)$ contains C^2 functions on K with compact support, and the infinitesimal generator A is of the form

$$Af(x) = \frac{1}{\pi} \int_{-\infty}^{+\infty} (f(x+y) - f(x)) \frac{dy}{y^2}.$$

The operator A is not "local", which reflects the fact that the Cauchy process changes state by jumps. More precisely, see Example 4.8 and Table 4.4 with $n := 1$ and $\alpha := 1$ in Chapter 4.

Example 3.14 (Reflecting barrier Brownian motion). $K = [0, \infty)$ and

$$\begin{cases} \mathcal{D}(A) = \{f \in C_0(K) : f' \in C_0(K), \ f'' \in C_0(K), \ f'(0) = 0\}, \\ Af = \dfrac{1}{2}f'' \quad \text{for every } f \in \mathcal{D}(A). \end{cases}$$

Example 3.15 (Sticking barrier Brownian motion). $K = [0, \infty)$ and

$$\begin{cases} \mathcal{D}(A) = \{f \in C_0(K) : f' \in C_0(K), \ f'' \in C_0(K), \ f''(0) = 0\}, \\ Af = \dfrac{1}{2}f'' \quad \text{for every } f \in \mathcal{D}(A). \end{cases}$$

Finally, here are two more examples where it is difficult to begin with a transition function and the infinitesimal generator is the basic tool of describing the process.

Example 3.16 (Sticky barrier Brownian motion). $K = [0, \infty)$ and

$$\begin{cases} \mathcal{D}(A) = \{f \in C_0(K) : f' \in C_0(K), \ f'' \in C_0(K), \ f'(0) - \alpha f''(0) = 0\}, \\ Af = \dfrac{1}{2}f'' \quad \text{for every } f \in \mathcal{D}(A). \end{cases}$$

Here α is a positive constant.

This process may be thought of as a "combination" of the reflecting and sticking Brownian motions. The reflecting and sticking cases are obtained by letting $\alpha \to 0$ and $\alpha \to \infty$, respectively.

Example 3.17 (Absorbing barrier Brownian motion). $K = [0, \infty)$ where the boundary point 0 is identified with the point at infinity ∂.

$$\begin{cases} \mathcal{D}(A) = \{f \in C_0(K) : f' \in C_0(K), \ f'' \in C_0(K), f(0) = 0\}, \\ Af = \dfrac{1}{2}f'' \quad \text{for every } f \in \mathcal{D}(A). \end{cases}$$

This represents Brownian motion with an absorbing barrier at $x = 0$; a Brownian particle "dies" at the first moment when it hits the boundary $x = 0$. Namely, the point 0 is the terminal point.

It is worth pointing out here that a strong Markov process cannot stay at a single position for a positive length of time and then leave that position by continuous motion; it must either jump away or leave instantaneously.

We give a simple example of a strong Markov process which changes state not by continuous motion but by jumps when the motion reaches the boundary.

Example 3.18. $K = [0, \infty)$.

$$\begin{cases} \mathcal{D}(A) = \{f \in C_0(K) \cap C^2(K) : f' \in C_0(K), f'' \in C_0(K), \\ \quad f''(0) = 2c \int_0^\infty (f(y) - f(0))dF(y)\}, \\ Af = \dfrac{1}{2}f'' \quad \text{for every } f \in \mathcal{D}(A). \end{cases}$$

Here c is a positive constant and F is a distribution function on the interval $(0, \infty)$.

This process may be interpreted as follows. When a Brownian particle reaches the boundary $x = 0$, it stays there for a positive length of time and then jumps back to a random point, chosen with the function F, in the interior $(0, \infty)$. The constant c is the parameter in the "waiting time" distribution at the boundary $x = 0$. We remark that the boundary condition

$$f''(0) = 2c \int_0^\infty (f(y) - f(0)) \, dF(y)$$

depends on the values of f far away from the boundary $x = 0$, unlike the boundary conditions in Examples 3.14 through 3.17.

Although Theorem 3.34 asserts precisely when a linear operator A is the infinitesimal generator of some Feller semigroup, it is usually difficult to verify conditions (b) through (d). So we give useful criteria in terms of the *maximum principle* (see [15], [98], [92], [114, Theorem 9.3.3 and Corollary 9.3.4]; [122, Theorem 9.50]):

Theorem 3.36 (Hille–Yosida–Ray). *If K is a* compact *metric space, then we have the following two assertions (i) and (ii):*

(i) Let B be a linear operator from $C(K) = C_0(K)$ into itself, and assume that:

    (α) The domain $\mathcal{D}(B)$ of B is dense *in the space $C(K)$.*

    (β) There exists an open and dense subset K_0 of K such that if a function $u \in \mathcal{D}(B)$ takes a positive *maximum at a point x_0 of K_0, then we have the inequality*

$$Bu(x_0) \leq 0.$$

Then the operator B is closable *in the space $C(K)$.*

(ii) Let B be as in part (i), and further assume that:

    (β′) If a function $u \in \mathcal{D}(B)$ takes a positive *maximum at a point x' of K, then we have the inequality*

$$Bu(x') \leq 0.$$

    (γ) For some $\alpha_0 \geq 0$, the range $\mathcal{R}(\alpha_0 I - B)$ of $\alpha_0 I - B$ is dense *in the space $C(K)$.*

Then the minimal closed extension *$\overline{B}$ of B is the* infinitesimal generator *of some Feller semigroup on the state space K.*

Proof. (i) It suffices to show that:

$$\{u_n\} \subset \mathcal{D}(B), \ u_n \to 0 \text{ and } Bu_n \to v \quad \text{in } C(K) \implies v = 0.$$

Our proof is based on a reduction to absurdity. By replacing v by $-v$ if necessary, we assume, to the contrary, that:

The function $v(x)$ takes a positive value at some point of K.

Then, since K_0 is open and dense in K, we can find a point x_0 of K_0, a neighborhood U of x_0 contained in K_0 and a positive constant ε such that we have, for all sufficiently large n,

$$Bu_n(x) > \varepsilon \quad \text{for all } x \in U. \tag{3.35}$$

On the other hand, by condition (α) there exists a function $h \in \mathcal{D}(B)$ such that

$$\begin{cases} h(x_0) > 1, \\ h(x) < 0 \quad \text{for all } x \in K \setminus U. \end{cases}$$

Therefore, since $u_n \to 0$ in $C(K)$, it follows that the function

$$u_n'(x) = u_n(x) + \frac{\varepsilon h(x)}{1 + \|Bh\|_\infty}$$

satisfies the conditions

$$u'_n(x_0) = u_n(x_0) + \frac{\varepsilon h(x_0)}{1 + \|Bh\|_\infty} > 0,$$

$$u'_n(x) = u_n(x) + \frac{\varepsilon h(x)}{1 + \|Bh\|_\infty} < 0 \quad \text{for all } x \in K \setminus U,$$

if n is sufficiently large. This implies that the function $u'_n \in \mathcal{D}(B)$ takes its positive maximum at a point x'_n of $U \subset K_0$. Hence we have, by condition (β),

$$(Bu'_n)(x'_n) \leq 0.$$

However, it follows from inequality (3.35) that

$$(Bu'_n)(x'_n) = (Bu_n)(x'_n) + \varepsilon \frac{(Bh)(x'_n)}{1 + \|Bh\|_\infty} > (Bu_n)(x'_n) - \varepsilon > 0.$$

This is a contradiction.

(ii) We apply part (ii) of Theorem 3.34 to the minimal closed extension $\overline{B}$ of B. The proof is divided into six steps.

Step 1: First, we show that

$$u \in \mathcal{D}(B), \ (\alpha_0 I - B) u \geq 0 \quad \text{on } K \implies u \geq 0 \quad \text{on } K. \tag{3.36}$$

By condition (γ), we can find a function $v \in \mathcal{D}(B)$ such that

$$(\alpha_0 I - B) v \geq 1 \quad \text{on } K. \tag{3.37}$$

Then we have, for any $\varepsilon > 0$,

$$\begin{cases} u + \varepsilon v \in \mathcal{D}(B), \\ (\alpha_0 I - B)(u + \varepsilon v) \geq \varepsilon \quad \text{on } K. \end{cases}$$

In view of condition (β'), this implies that the function $-(u(x) + \varepsilon v(x))$ does not take any positive maximum on K, so that

$$u(x) + \varepsilon v(x) \geq 0 \quad \text{on } K.$$

Thus, by letting $\varepsilon \downarrow 0$ in this inequality we obtain that

$$u(x) \geq 0 \quad \text{on } K.$$

This proves the desired assertion (3.36).

Step 2: It follows from assertion (3.36) that the inverse $(\alpha_0 I - B)^{-1}$ of $\alpha_0 I - B$ is defined and non-negative on the range $\mathcal{R}(\alpha_0 I - B)$. Moreover, it is bounded with norm

$$\left\|(\alpha_0 I - B)^{-1}\right\| \leq \|v\|_\infty. \tag{3.38}$$

Here $v(x)$ is the function that satisfies condition (3.37).

Indeed, since $g = (\alpha_0 I - B)v \geq 1$ on K, it follows that, for all $f \in C(K)$,

$$-\|f\|_\infty g \leq f \leq \|f\|_\infty g \quad \text{on } K.$$

Hence, by the non-negativity of $(\alpha_0 I - B)^{-1}$ we have, for all $f \in \mathcal{R}(\alpha_0 I - B)$,

$$-\|f\|_\infty v \leq (\alpha_0 I - B)^{-1} f \leq \|f\|_\infty v \quad \text{on } K.$$

This proves the desired inequality (3.38).

Step 3: Next we show that

$$\mathcal{R}\left(\alpha_0 I - \overline{B}\right) = C(K). \tag{3.39}$$

Let $f(x)$ be an arbitrary element of $C(K)$. By condition (γ), we can find a sequence $\{u_n\}$ in $\mathcal{D}(B)$ such that $f_n = (\alpha_0 I - B)u_n \to f$ in $C(K)$. Since the inverse $(\alpha_0 I - B)^{-1}$ is bounded, it follows that $u_n = (\alpha_0 I - B)^{-1} f_n$ converges to some function $u \in C(K)$, and hence $Bu_n = \alpha_0 u_n - f_n$ converges to $\alpha_0 u - f$ in $C(K)$. Thus we have, by the closedness of $\overline{B}$,

$$\begin{cases} u \in \mathcal{D}\left(\overline{B}\right), \\ \overline{B}u = \alpha_0 u - f, \end{cases}$$

so that

$$(\alpha_0 I - \overline{B})u = f.$$

This proves the desired assertion (3.39).

Step 4: Furthermore, we show that

$$u \in \mathcal{D}\left(\overline{B}\right), \ (\alpha_0 I - \overline{B})u \geq 0 \quad \text{on } K \implies u \geq 0 \quad \text{on } K. \tag{3.40}$$

Since $\mathcal{R}\left(\alpha_0 I - \overline{B}\right) = C(K)$, in view of the proof of assertion (3.40) it suffices to show the following assertion:

If a function $u \in \mathcal{D}\left(\overline{B}\right)$ takes a positive maximum at a point x' of K, then we have the inequality

$$\left(\overline{B}u\right)(x') \leq 0. \tag{3.41}$$

Our proof of assertion (3.41) is based on a reduction to absurdity. Assume, to the contrary, that

$$\left(\overline{B}u\right)(x') > 0.$$

Since there exists a sequence $\{u_n\}$ in $\mathcal{D}(B)$ such that $u_n \to u$ and $Bu_n \to \overline{B}u$ in $C(K)$, we can find a neighborhood U of x' and a positive constant ε such that, for all sufficiently large n,

$$(Bu_n)(x) > \varepsilon \quad \text{for all } x \in U. \tag{3.42}$$

Furthermore, by condition (α) we can find a function $h \in \mathcal{D}(B)$ such that

$$\begin{cases} h(x') > 1, \\ h(x) < 0 \quad \text{for all } x \in K \setminus U. \end{cases}$$

Then it follows that the function

$$u'_n(x) = u_n(x) + \frac{\varepsilon h(x)}{1 + \|Bh\|_\infty}$$

satisfies the condition

$$\begin{cases} u'_n(x') > u(x') > 0, \\ u'_n(x) < u(x') \quad \text{for all } x \in K \setminus U, \end{cases}$$

if n is sufficiently large. This implies that the function $u'_n \in \mathcal{D}(B)$ takes its positive maximum at a point x'_n of U. Hence we have, by condition (β'),

$$(Bu'_n)(x'_n) \leq 0 \quad \text{for } x'_n \in U.$$

However, it follows from inequality (3.42) that

$$(Bu'_n)(x'_n) = (Bu_n)(x'_n) + \varepsilon \frac{(Bh)(x'_n)}{1 + \|Bh\|_\infty} > (Bu_n)(x'_n) - \varepsilon > 0.$$

This is a contradiction.

Step 5: In view of Steps 3 and 4, we obtain that the inverse $(\alpha_0 I - \overline{B})^{-1}$ of $\alpha_0 I - \overline{B}$ is defined on the whole space $C(K)$, and is bounded with norm

$$\left\| (\alpha_0 I - \overline{B})^{-1} \right\| = \left\| (\alpha_0 I - \overline{B})^{-1} 1 \right\|_\infty .$$

Step 6: Finally, we show that:

For *all* $\alpha > \alpha_0$, the inverse $(\alpha I - \overline{B})^{-1}$ of $\alpha I - \overline{B}$ is defined on the whole space $C(K)$, and is non-negative and bounded with norm

$$\left\| (\alpha I - \overline{B})^{-1} \right\| \leq \frac{1}{\alpha}. \tag{3.43}$$

We let

$$G_{\alpha_0} = (\alpha_0 I - \overline{B})^{-1}.$$

First, we choose a constant $\alpha_1 > \alpha_0$ such that

$$(\alpha_1 - \alpha_0) \|G_{\alpha_0}\| < 1,$$

and let

$$\alpha_0 < \alpha \leq \alpha_1.$$

Then the C. Neumann series (see [147, Chapter II, Section 1, Theorem 2])

$$u = \left(I + \sum_{n=1}^{\infty} (\alpha_0 - \alpha)^n G_{\alpha_0}^n \right) G_{\alpha_0} f$$

converges in $C(K)$, and is a solution of the equation

$$u - (\alpha_0 - \alpha) G_{\alpha_0} u = G_{\alpha_0} f \quad \text{for any } f \in C(K).$$

Hence we have the assertions

$$\begin{cases} u \in \mathcal{D}\left(\overline{B}\right), \\ \left(\alpha I - \overline{B}\right) u = f. \end{cases}$$

This proves that

$$\mathcal{R}\left(\alpha I - \overline{B}\right) = C(K), \quad \alpha_0 < \alpha \le \alpha_1. \tag{3.44}$$

Thus, by arguing just as in the proof of Step 1 we obtain that, for any $\alpha_0 < \alpha \le \alpha_1$,

$$u \in \mathcal{D}\left(\overline{B}\right), \; \left(\alpha I - \overline{B}\right) u \ge 0 \text{ on } K \implies u \ge 0 \text{ on } K. \tag{3.45}$$

By combining assertions (3.44) and (3.45), we find that, for any $\alpha_0 < \alpha \le \alpha_1$, the inverse $(\alpha I - \overline{B})^{-1}$ is defined and non-negative on the whole space $C(K)$.
    We let

$$G_\alpha = \left(\alpha I - \overline{B}\right)^{-1} \quad \text{for } \alpha_0 < \alpha \le \alpha_1.$$

Then it follows that the operator G_α is bounded with norm

$$\|G_\alpha\| \le \frac{1}{\alpha}. \tag{3.46}$$

Indeed, in view of assertion (3.41) it follows that if a function $u \in \mathcal{D}\left(\overline{B}\right)$ takes a positive maximum at a point x' of K, then we have the inequality

$$\left(\overline{B}u\right)(x') \le 0,$$

so that

$$\max_{x \in K} u(x) = u(x') \le \frac{1}{\alpha}(\alpha I - \overline{B})u(x') \le \frac{1}{\alpha} \left\|\left(\alpha I - \overline{B}\right) u\right\|_\infty. \tag{3.47}$$

Similarly, if the function $u \in \mathcal{D}\left(\overline{B}\right)$ takes a negative minimum at a point of K, then (replacing $u(x)$ by $-u(x)$), we have the inequality

$$-\min_{x \in K} u(x) = \max_{x \in K}\left(-u(x)\right) \le \frac{1}{\alpha} \left\|\left(\alpha I - \overline{B}\right) u\right\|_\infty. \tag{3.48}$$

The desired inequality (3.46) follows from inequalities (3.47) and (3.48).
    Summing up, we have proved assertion (3.43) for all $\alpha_0 < \alpha \le \alpha_1$.
    Now we assume that assertion (3.43) is proved for all $\alpha_0 < \alpha \le \alpha_{n-1}$, $n = 2, 3, \ldots$. Then, by taking

$$\alpha_n = 2\alpha_{n-1} - \frac{\alpha_1}{2}, \quad n \ge 2,$$

or equivalently

$$\alpha_n = \left(2^{n-2} + \frac{1}{2}\right)\alpha_1 \quad \text{for } n \geq 2,$$

we have, for all $\alpha_{n-1} < \alpha \leq \alpha_n$,

$$(\alpha - \alpha_{n-1})\|G_{\alpha_{n-1}}\| \leq \frac{\alpha - \alpha_{n-1}}{\alpha_{n-1}} \leq \frac{\alpha_n - \alpha_{n-1}}{\alpha_{n-1}}$$

$$= \frac{1}{1 + 2^{2-n}}$$

$$< 1.$$

Hence assertion (3.43) for $\alpha_{n-1} < \alpha \leq \alpha_n$ is proved just as in the proof of assertion (3.43) for $\alpha_0 < \alpha \leq \alpha_1$. This proves the desired assertion (3.43) for all $\alpha > \alpha_0$.

Consequently, by applying part (ii) of Theorem 3.34 to the operator $\overline{B}$ we obtain that $\overline{B}$ is the infinitesimal generator of some Feller semigroup on the state space K.

The proof of Theorem 3.36 is now complete. □

The next corollary gives a sufficient condition in order that the infinitesimal generators of Feller semigroups are stable under *bounded perturbations*:

Corollary 3.37. *Let A be the infinitesimal generator of a Feller semigroup on a* compact *metric space K and let M be a bounded linear operator on the Banach space $C(K)$ into itself. Assume that either M or $A' = A + M$ satisfies condition (β'). Then the operator A' is the infinitesimal generator of some Feller semigroup on the state space K.*

Proof. We apply part (ii) of Theorem 3.36 to the operator A'.

First, we remark that $A' = A + M$ is a densely defined, closed linear operator from $C(K)$ into itself. Since the semigroup $\{T_t\}_{t \geq 0}$ is non-negative and contractive on $C(K)$, it follows that if a function $u \in \mathcal{D}(A)$ takes a positive maximum at a point x' of K, then we have the inequality

$$Au(x') = \lim_{t \downarrow 0} \frac{T_t u(x') - u(x')}{t} \leq 0.$$

This implies that if M satisfies condition (β'), so does $A' = A + M$.

We let

$$G_{\alpha_0} = (\alpha_0 I - A)^{-1}, \quad \alpha_0 > 0.$$

If α_0 is so large that

$$\|G_{\alpha_0} M\| \leq \|G_{\alpha_0}\| \cdot \|M\| \leq \frac{\|M\|}{\alpha_0} < 1,$$

then the Neumann series (see [147, Chapter II, Section 1, Theorem 2])

$$u = \left(I + \sum_{n=1}^{\infty} (G_{\alpha_0} M)^n \right) G_{\alpha_0} f$$

converges in $C(K)$, and is a solution of the equation

$$u - G_{\alpha_0} M u = G_{\alpha_0} f \quad \text{for any } f \in C(K).$$

Hence we have the assertions

$$\begin{cases} u \in \mathcal{D}(A) = \mathcal{D}(A'), \\ (\alpha_0 I - A')u = f. \end{cases}$$

This proves that

$$\mathcal{R}(\alpha_0 I - A') = C(K).$$

Therefore, by applying part (ii) of Theorem 3.36 to the operator A' we obtain that A' is the infinitesimal generator of some Feller semigroup on the state space K.

The proof of Corollary 3.37 is complete. □

3.5 Reflecting Diffusion

Let D be a bounded domain in Euclidean space $\mathbf{R}^N$ with smooth boundary ∂D, and a fixed abstract point ∂ is adjoined to the closure $\overline{D}$ as an isolated point. We consider the following homogeneous Neumann problem for the Laplacian Δ:

$$\begin{cases} \Delta u = f & \text{in } D, \\ \dfrac{\partial u}{\partial \mathbf{n}} = 0 & \text{on } \partial D. \end{cases}$$

Sato and Ueno [98] constructed the following Markov process

$$\mathcal{X} = (x_t, W, \mathcal{B}_t, \mathcal{B}, P_x, x \in \overline{D} \cup \{\partial\})$$

on the state space $\overline{D} \cup \{\partial\}$ (see [98, Theorem 7.1]):

(1) Let W be the space of right-continuous with left limits functions

$$w \colon [0, +\infty] \longrightarrow \overline{D} \cup \{\partial\}$$

with coordinates $x_t(w) = w(t)$.

(2) Let

$$\mathcal{B}_t = \sigma\left(x_s : 0 \le s \le t\right)$$

be the smallest σ-field of subsets of W with respect to which all mappings $\{x_s : 0 \le s \le t\}$ are measurable and let

$$\mathcal{B} = \sigma\left(x_s : 0 \le s < \infty\right)$$

be the smallest σ-field of subsets of W with respect to which all mappings $\{x_s : 0 \le s < \infty\}$ are measurable, respectively.

(3) A random variable $\zeta \colon W \to [0, +\infty]$, called the *lifetime*, such that

$$x_t(w) \begin{cases} \in \overline{D} & \text{for } 0 \leq t < \zeta(w), \\ = \partial & \text{for } \zeta(w) \leq t \leq +\infty. \end{cases}$$

The sample path $x_t(w)$ is continuous for $0 \leq t < \zeta(w)$.

(4) For each $t \in [0, +\infty]$, a pathwise *shift mapping* $\theta_t \colon W \to W$ defined by the formula $\theta_t w(s) = w(t+s)$ for all $w \in W$.

(5) The system of measures $\{P_x, x \in \overline{D} \cup \{\partial\}\}$ on $(W, \mathcal{B})$ such that:

(a) $P_x(B)$ is $\mathcal{B}(\overline{D} \cup \{\partial\})$-measurable in x, for each $B \in \mathcal{B}$.

(b) $P_x(\{w \in W : x_0(w) = x\}) = 1$ for each $x \in \overline{D}$.

(c) $P_x\left(\theta_t^{-1}(B) \mid \mathcal{B}_t\right) = P_{x_t}(B)$ for each $B \in \mathcal{B}$. More precisely, the formula

$$E_x\left(f \circ \theta_t \mid \mathcal{B}_t\right) = E_{x_t}(f)$$

holds true for every $\mathcal{B}$-measurable, bounded function f on $\overline{D}$, with P_x probability one.

(d) The system

$$P_x\left(\{w \in W : x_t(w) = y\}\right) = p(t, x, dy)$$

is called the system of *transition probabilities*.

(6) $P_x(\{w \in W : \zeta(w) = +\infty\}) = 1$ for each $x \in \overline{D}$. Namely, this process is *conservative*.

(7) A transition semigroup of linear operators

$$T_t f(x) := E_x\left(f(x_t)\right) = \int_{\overline{D}} p(t, x, dy)\, f(y)$$

forms a *Feller semigroup* on the state space $\overline{D}$, and its infinitesimal generator $\mathfrak{A}$ is equal to the *minimal closed extension* $\overline{\Delta}_N$ in the Banach space $C(\overline{D})$ of the operator Δ_N defined as follows:

(a) The domain $\mathcal{D}(\Delta_N)$ of Δ_N is the space

$$\mathcal{D}(\Delta_N) = \left\{ u \in C^2(\overline{D}) : \frac{\partial u}{\partial \mathbf{n}} = 0 \text{ on } \partial D \right\}. \tag{3.49}$$

(b) $\Delta_N u = \Delta u$ for every $u \in \mathcal{D}(\Delta_N)$.

The Markov process

$$\boxed{\mathcal{X} = \left(x_t, W, \mathcal{B}_t, \mathcal{B}, P_x, x \in \overline{D} \cup \{\partial\}\right)}$$

is called the *reflecting diffusion* on $\overline{D}$.

3.6 Local Time on the Boundary for the Reflecting Diffusion

Following P. Lévy [76], Sato and Ueno [98] constructed the *local time* $\tau(t)$ on the boundary for the reflecting diffusion $\mathcal{X} = (x_t, W, \mathcal{B}_t, \mathcal{B}, P_x, x \in \overline{D} \cup \{\partial\})$ (see [98, Theorem 7.2]):

Roughly speaking, the local time $\tau(t)$ can be defined by the formula

$$
\tau(t, w) = \lim_{\rho \downarrow 0} \frac{1}{\rho} \int_0^t \chi_{D_\rho}(x_s(w))\, ds \qquad \text{for every } w \in W,
$$

where

$$
D_\rho = \{x \in \overline{D} : \operatorname{dist}(x, \partial D) < \rho\} \quad \text{for } \rho > 0,
$$

$$
\chi_{D_\rho}(x) = \begin{cases} 1 & \text{if } x \in D_\rho, \\ 0 & \text{otherwise.} \end{cases}
$$

The local time $\tau(t)$ enjoys the following four properties (1) through (4):

(1) $P_x(\{w \in W : \lim_{t \to +\infty} \tau(t, w) = +\infty\}) = 1$ if $x \in \overline{D}$.
(2) $P_x(\{w \in W : \tau(t, w) > 0 \text{ for all } t > 0\}) = 1$ if and only if $x \in \partial D$.
(3) $\tau(t, w)$ is a continuous, non-negative additive functional of $\mathcal{X}$. More precisely, we have the assertions:
  (3a) $\tau(t, w)$ is measurable on the sample space $\{x_s : 0 \le s \le t\}$, for each $t \ge 0$.
  (3b) $0 = \tau(0, \omega) \le \tau(t, w)$ and $\tau(t, w)$ is continuous in t.
  (3c) For all $x \in \overline{D}$, we have the formula
$$
P_x(\{\omega \in W : \tau(t + s, w) = \tau(t, w) + \tau(s, \theta_t(w)) \\ \text{for each } t, s \ge 0\}) = 1.
$$

(4) $\tau(t, w)$ increases at t only when $x_t(w)$ is on the boundary ∂D.

Intuitively, $\tau(t, w)$ is the *sojourn time* (or occupation time) of a sample path $x_s(w)$ on the boundary ∂D up to time t.

Let $\tau^{-1}(t)$ is the right-continuous inverse of the local time $\tau(t)$ on the boundary for the reflecting diffusion

$$
\mathcal{X} = (x_t, W, \mathcal{B}_t, \mathcal{B}, P_x, x \in \overline{D} \cup \{\partial\}),
$$

where

$$
\tau^{-1}(t, w) = \inf\{s \ge 0 : \tau(s, w) \ge t\} \quad \text{for every } w \in W.
$$

The inverse local time $\tau^{-1}(t, w)$ is the amount of real time spent by a Markovian particle in the space $\overline{D}$ necessary to realize that its sojourn time on the boundary ∂D exceeds t.

Table 3.1 below gives a bird's-eye view of the standard time and the local time via Figure 3.15 below.

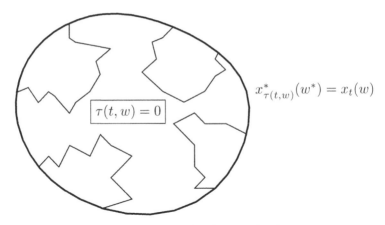

Fig. 3.15. The local time $\tau(t, w)$

State space	Trajectory	Watch
Interior D	$x_t(w)$	t (standard time)
Boundary ∂D	$x^*_{\tau(t,w)}(w^*) = x_t(w)$	$\tau(t, w)$ (local time)

Table 3.1. A world watch due to P. Lévy

3.7 Notes and Comments

The material of this chapter is adapted from Blumenthal–Getoor [14], Dynkin [33], [34], Dynkin–Yushkevich [35], Feller [39], [40], Ikeda–Watanabe [62], Itô–McKean, Jr. [65], Lamperti [74], Revuz–Yor [94] and also Taira [114, Chapter 9] and [122, Chapters 2 and 9]. In particular, our treatment of temporally homogeneous Markov processes follows the expositions of Dynkin [33], [34] and Blumenthal–Getoor [14].

However, unlike many other books on Markov processes, this chapter focuses on the relationship among three subjects: Feller semigroups, transition functions and Markov processes. Our semigroup approach to the problem of construction of Markov processes with Ventcel' boundary conditions is distinguished by the extensive use of the ideas and techniques characteristic of the recent developments in functional analysis methods.

Hille–Phillips [54] and Yosida [147] are the classics for semigroup theory of linear operators. The reader might be referred to Engel–Nagel [37], Goldstein [52] and Pazy [87] for the modern theory of semigroups of linear operators.

Section 3.1: This section is adapted from Folland [42, Chapter 7] and Rudin [95, Chapter 2]. Proposition 3.4 is taken from Folland [42, Proposition 4.35] and the proof of Theorem 3.6 is adapted from Folland [42, Theorem 7.2]. A locally compact version of the Riesz–Markov representation theorem (Theorem 3.7) is taken from Folland [42, Theorem 7.17] and a compact version of the Riesz–Markov representation theorem (Theorem 3.10) is taken from Folland [42, Corollary 7.18], respectively. See also Kolmogorov–Fomin [71], Friedman [44] and Schechter [100].

Theorem 3.15 is called *Prokhorov's theorem* in probability ([38, p. 104, Theorem 2.2], [94, p. 10, Theorem (5.4)]).

Section 3.2: The results here are adapted from Blumenthal–Getoor [14], Dynkin [34], Lamperti [74] and Revuz–Yor [94]. Theorem 3.18 is taken from Dynkin [33, Chapter 4, Section 2], while Theorem 3.19 is taken from Dynkin [33, Chapter 6] and [34, Chapter 3, Section 2]. Theorem 3.27 is due to Dynkin [33, Theorem 5.10] and Theorem 3.29 is due to Dynkin [33, Theorem 6.3], respectively. Theorem 3.29 is a non-compact version of Lamperti [74, Chapter 8, Section 3, Theorem 1]. Subsection 3.2.6 is adapted from Lamperti [74, Chapter 9, Section 2].

Section 3.3: The semigroup approach to Markov processes can be traced back to the work of Kolmogorov [70]. It was substantially developed in the early 1950s, with Feller [39] and [40] doing the pioneering work. Our presentation here follows the book of Dynkin [34] and also part of Lamperti's [74]. Theorem 3.31 is a non-compact version of Lamperti [74, Chapter 7, Section 7, Theorem 1].

Altomare et al. prove that a class of Feller semigroups can be approximated by iterates of modified Bernstein–Schnabl operators ([7]), which allows to infer the preservation of such properties as the Lipschitz continuity and convexity of the semigroups.

Section 3.4: Theorem 3.36 is due to Sato–Ueno [98, Theorem 1.2] and Bony–Courrège–Priouret [15, Théorème de Hille–Yosida–Ray] (cf. [64], [92], [122]).

Sections 3.5 and 3.6: These two sections are based on the talk entitled *Probabilistic approach to pseudo-differential operators* delivered at Special Seminar, Leibniz Universität Hannover, Germany, on October 16, 2019. The material is adapted from Sato [96], Sato–Tanaka [97] and Sato–Ueno [98].

This chapter is an expanded and revised version of Chapter 2 of the second edition [121].

Part II

Pseudo-Differential Operators and Elliptic Boundary Value Problems

L^p Theory of Pseudo-Differential Operators

In this chapter we present a brief description of the basic concepts and results of the L^p theory of pseudo-differential operators, which may be considered as a modern version of the classical potential approach.

In Section 4.1 we define Hölder spaces $C^{k+\theta}(\overline{\Omega})$ and various Sobolev spaces $W^{s,p}(\Omega)$ and $H^{s,p}(\Omega)$ and also Besov spaces $B^{s,p}(\partial\Omega)$ on the boundary $\partial\Omega$ of a smooth domain Ω of Euclidean space $\mathbf{R}^n$. It is the imbedding characteristics of L^p Sobolev spaces that render these spaces so useful in the study of partial differential equations. In the proof of Theorem 1.5 we shall make use of some imbedding properties of L^p Sobolev spaces (Theorem 4.8). Moreover, we shall need the Rellich–Kondrachov compactness theorem for function spaces of L^p type (Theorem 4.10) in the proof of Theorem 1.4. The Rellich–Kondrachov theorem is a Sobolev space version of the Bolzano–Weierstrass theorem and the Ascoli–Arzelà theorem in calculus.

In Section 4.2 we formulate Seeley's extension theorem (Theorem 4.11), due to Seeley [103], which asserts that the functions in $C^{\infty}(\overline{\Omega})$ are the restrictions to Ω of functions in $C^{\infty}(\mathbf{R}^n)$.

It should be emphasized that Besov spaces $B^{s,p}(\partial\Omega)$ enter naturally in connection with boundary value problems in the framework of Sobolev spaces of L^p type. Indeed, we need to make sense of the restriction $u|_{\partial\Omega}$ to the boundary $\partial\Omega$ as an element of a Besov space on $\partial\Omega$ when u belongs to a Sobolev space on the domain Ω. In Section 4.3, we formulate the trace theorem (Theorem 4.18) and the sectional trace theorem (Theorem 4.19) that play an important role in the study of elliptic boundary value problems.

In Section 4.4, we present a brief description of basic concepts and results of the theory of Fourier integral operators and pseudo-differential operators. The L^p theory of pseudo-differential operators may be considered as a modern version of the classical potential theory. Especially, pseudo-differential operators provide a constructive tool to deal with existence and smoothness of solutions of partial differential equations. The theory of pseudo-differential operators continues to be one of the most influential works in modern history

© Springer Nature Switzerland AG 2020
K. Taira, *Boundary Value Problems and Markov Processes*, Lecture Notes in Mathematics 1499,
https://doi.org/10.1007/978-3-030-48788-1_4

of analysis, and is a very refined mathematical tool whose full power is yet to be exploited.

In Section 4.5 we present a brief description of the basic concepts and results of the L^p theory of pseudo-differential operators. We formulate the Besov space boundedness theorem due to Bourdaud [16] (Theorem 4.47) in Subsection 4.5.2 and we give a useful criterion for hypoellipticity due to Hörmander [59] (Theorem 4.49) in Subsection 4.5.4, which will play an essential role in the proof of our main results. In Subsection 4.5.5, following Coifman–Meyer [30, Chapitre IV, Proposition 1]) we state that the distribution kernel $s(x, y)$ of a pseudo-differential operator $S \in L_{1,0}^m(\mathbf{R}^n)$ satisfies the estimate (Theorem 4.51)

$$|s(x, y)| \leq \frac{C}{|x - y|^{m+n}} \quad \text{for all } x, y \in \mathbf{R}^n \text{ and } x \neq y.$$

In Section 4.6, by using the Riesz–Schauder theory we prove some of the most important results about elliptic pseudo-differential operators on a manifold and their indices in the framework of Sobolev spaces (Theorems 4.53 through 4.67). These results play an important role in the study of elliptic boundary value problems in Chapter 5.

The heat kernel has many important and interesting applications in partial differential equations. In Section 4.7, by calculating various convolution kernels for the Laplacian via the Laplace transform we derive Newtonian, Riesz and Bessel potentials and also the Poisson kernel for the Dirichlet boundary value problem (Theorems 4.69, 4.70 and 4.73).

4.1 Function Spaces

Let Ω be a bounded domain of Euclidean space $\mathbf{R}^n$ with smooth boundary $\partial\Omega$. Its closure $\overline{\Omega} = \Omega \cup \partial\Omega$ is an n-dimensional, compact smooth manifold with boundary. We may assume the following three conditions (a), (b) and (c) (see Theorems 4.15 and 4.16 in Section 4.3):

(a) The domain Ω is a relatively compact, open subset of an n-dimensional, compact smooth manifold M without boundary in which Ω has a smooth boundary $\partial\Omega$ (see Figures 4.1 below).
(b) In a neighborhood W of $\partial\Omega$ in M a normal coordinate t is chosen so that the points of W are represented as (x', t), $x' \in \partial\Omega$, $-1 < t < 1$; $t > 0$ in Ω, $t < 0$ in $M \setminus \overline{\Omega}$ and $t = 0$ only on $\partial\Omega$ (see Figures 4.2 below).
(c) The manifold M is equipped with a strictly positive density μ which, on W, is the product of a strictly positive density ω on $\partial\Omega$ and the Lebesgue measure dt on $(-1, 1)$. This manifold M is called the *double* of Ω (see [83]).

The function spaces we shall treat are the following (cf. [2], [13], [21], [43], [112], [135]):

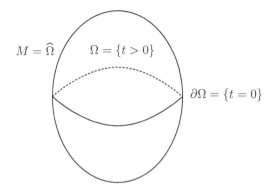

Fig. 4.1. The double M of Ω

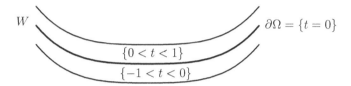

Fig. 4.2. The tubular neighborhood W of the boundary $\partial\Omega$

(i) The generalized Sobolev spaces $H^{s,p}(\Omega)$ and $H^{s,p}(M)$, consisting of all potentials of order s of L^p functions. When s is integral, these spaces coincide with the usual Sobolev spaces $W^{s,p}(\Omega)$ and $W^{s,p}(M)$, respectively.
(ii) The Besov spaces $B^{s,p}(\partial\Omega)$. These are functions spaces defined in terms of the L^p modulus of continuity, and enter naturally in connection with boundary value problems.

4.1.1 Hölder Spaces

Let Ω be a subset of $\mathbf{R}^n$ and let $0 < \theta < 1$. A function φ defined on Ω is said to be *Hölder continuous* with exponent θ if the quantity

$$[\varphi]_{\theta;\Omega} = \sup_{\substack{x,y \in \Omega \\ x \neq y}} \frac{|\varphi(x) - \varphi(y)|}{|x - y|^{\theta}}$$

is finite. We say that φ is *locally Hölder continuous* with exponent θ if it is Hölder continuous with exponent θ on compact subsets of D. Hölder continuity may be viewed as a fractional differentiability.

Let Ω be an open subset of $\mathbf{R}^n$ and $0 < \theta < 1$. We let

$$C^{\theta}(\Omega) = \text{the space of functions in } C(\Omega) \text{ which are locally Hölder}$$
$$\text{continuous with exponent } \theta \text{ on } \Omega.$$

If k is a positive integer, we let

$$C^{k+\theta}(\Omega) = \text{the space of functions in } C^k(\Omega) \text{ all of whose}$$
$$k\text{-th order derivatives are locally Hölder}$$
$$\text{continuous with exponent } \theta \text{ on } \Omega.$$

If K is a compact subset of Ω, we define a seminorm $q_{K,k}$ on $C^{k+\theta}(\Omega)$ by the formula

$$C^{k+\theta}(\Omega) \ni \varphi \longmapsto q_{K,k}(\varphi) = \sup_{\substack{x \in K \\ |\alpha| \le k}} |\partial^\alpha \varphi(x)| + \sup_{|\alpha|=k} [\partial^\alpha \varphi]_{\theta;K} .$$

It is easy to see that the Hölder space $C^{k+\theta}(\Omega)$ is a Fréchet space. Furthermore, we let

$$C^\theta(\overline{\Omega}) = \text{the space of functions in } C(\overline{\Omega}) \text{ which are Hölder}$$
$$\text{continuous with exponent } \theta \text{ on } \overline{\Omega}.$$

If k is a positive integer, we let

$$C^{k+\theta}(\overline{\Omega}) = \text{the space of functions in } C^k(\overline{\Omega}) \text{ all of whose } k\text{-th order}$$
$$\text{derivatives are Hölder continuous with exponent } \theta \text{ on } \overline{\Omega}.$$

Let m be a non-negative integer. We equip the space $C^{m+\theta}(\overline{\Omega})$ with the topology defined by the family $\{q_{K,k}\}$ of seminorms where K ranges over all compact subsets of $\overline{\Omega}$. It is easy to see that the Hölder space $C^{m+\theta}(\overline{\Omega})$ is a Fréchet space.

If Ω is a *bounded* domain of $\mathbf{R}^n$, then $C^{m+\theta}(\overline{\Omega})$ is a Banach space with the norm

$$\|\varphi\|_{C^{m+\theta}(\overline{\Omega})} = \|\varphi\|_{C^m(\overline{\Omega})} + \sup_{|\alpha|=m} [\partial^\alpha \varphi]_{\theta;\overline{\Omega}}$$
$$= \sup_{\substack{x \in \overline{\Omega} \\ |\alpha| \le m}} |\partial^\alpha \varphi(x)| + \sup_{|\alpha|=m} [\partial^\alpha \varphi]_{\theta;\overline{\Omega}} .$$

4.1.2 L^p Spaces

Throughout this subsection, Ω will denote a bounded domain in Euclidean space $\mathbf{R}^n$. By a measurable function on Ω, we shall mean an equivalence class of measurable functions on Ω which differ only on a subset of measure zero. Any pointwise property attributed to a measurable function will thus be understood to hold true in the usual sense for some function in the same equivalence class. The supremum and infimum of a measurable function will then be understood as the essential supremum and infimum.

First, if $1 \le p < \infty$, we let

$$L^p(\Omega) = \text{the space of (equivalence classes of) Lebesgue}$$

measurable functions $u(x)$ on Ω such that $|u(x)|^p$
is integrable on Ω.

The space $L^p(\Omega)$ is a Banach space with the norm

$$\|u\|_p = \left(\int_\Omega |u(x)|^p \, dx \right)^{1/p}.$$

For $p = \infty$, we let

$L^\infty(\Omega) =$ the space of (equivalence classes of) essentially bounded,
Lebesgue measurable functions $u(x)$ on Ω.

The space $L^\infty(\Omega)$ is a Banach space with the norm

$$\|u\|_\infty = \operatorname{ess\,sup}_{x \in \Omega} |u(x)|.$$

4.1.3 Fourier Transforms

If $f \in L^1(\mathbf{R}^n)$, we define its (direct) *Fourier transform* $\widehat{f}$ by the formula

$$\widehat{f}(\xi) = \int_{\mathbf{R}^n} e^{-i\,x\cdot\xi} f(x)\, dx \quad \text{for } \xi = (\xi_1, \xi_2, \ldots, \xi_n) \in \mathbf{R}^n, \qquad (4.1)$$

where $x \cdot \xi = x_1\xi_1 + x_2\xi_2 + \ldots + x_n\xi_n$. It follows from an application of
the Lebesgue dominated convergence theorem ([42, Theorem 2.24]) that the
function $\widehat{f}(\xi)$ is continuous on $\mathbf{R}^n$, and further we have the inequality

$$\|\widehat{f}\|_\infty = \sup_{\xi \in \mathbf{R}^n} |\widehat{f}(\xi)| \le \|f\|_1.$$

We also denote $\widehat{f}$ by $\mathcal{F}f$.

Similarly, if $g \in L^1(\mathbf{R}^n)$, we define the function $\check{g}(x)$ by the formula

$$\check{g}(x) = \frac{1}{(2\pi)^n} \int_{\mathbf{R}^n} e^{i\,x\cdot\xi} g(\xi)\, d\xi \quad \text{for } x = (x_1, x_2, \ldots, x_n) \in \mathbf{R}^n.$$

The function $\check{g}(x)$ is called the *inverse Fourier transform* of g. We also denote
$\check{g}$ by $\mathcal{F}^*g$.

Now we introduce a subspace of $L^1(\mathbf{R}^n)$ which is invariant under the
Fourier transform. We let

$\mathcal{S}(\mathbf{R}^n) =$ the space of smooth functions $\varphi(x)$ on $\mathbf{R}^n$ such that,
for any non-negative integer j, the quantity

$$p_j(\varphi) = \sup_{\substack{x \in \mathbf{R}^n \\ |\alpha| \le j}} \left\{ (1 + |x|^2)^{j/2} |\partial^\alpha \varphi(x)| \right\}$$

is finite.

The space $\mathcal{S}(\mathbf{R}^n)$ is called the *Schwartz space* or *space of smooth functions on* $\mathbf{R}^n$ *rapidly decreasing at infinity.* We equip the space $\mathcal{S}(\mathbf{R}^n)$ with the topology defined by the countable family $\{p_j\}$ of seminorms. It is easy to verify that $\mathcal{S}(\mathbf{R}^n)$ is complete; so it is a Fréchet space.

Now we give typical examples of functions in $\mathcal{S}(\mathbf{R})$:

Example 4.1. (1) For every $a > 0$, we let

$$\varphi_a(x) = e^{-a^2 x^2} \in \mathcal{S}(\mathbf{R}),$$
$$\psi_a(x) = x^2 e^{-a^2 x^2} \in \mathcal{S}(\mathbf{R}).$$

The Fourier transform $\widehat{\varphi}_a(\xi)$ of $\varphi_a(x)$ is given by the formula

$$\widehat{\varphi}_a(\xi) = \int_{\mathbf{R}} e^{-i x \cdot \xi} e^{-a^2 x^2} \, dx = \frac{\sqrt{\pi}}{a} e^{-\frac{\xi^2}{4a^2}} \quad \text{for } \xi \in \mathbf{R}.$$

Hence the Fourier transform $\widehat{\psi}_a(\xi)$ of $\psi_a(x)$ is given by the formula

$$\begin{aligned}
\widehat{\psi}_a(\xi) &= \int_{\mathbf{R}} e^{-i x \cdot \xi} x^2 e^{-a^2 x^2} \, dx \\
&= -\frac{\partial^2}{\partial \xi^2} \left(\int_{\mathbf{R}} e^{-i x \cdot \xi} e^{-a^2 x^2} \, dx \right) = -\frac{\partial^2 \widehat{\varphi}_a}{\partial \xi^2} \\
&= \frac{\sqrt{\pi}}{2a^3} \left(1 - \frac{\xi^2}{2a^2} \right) e^{-\frac{\xi^2}{4a^2}} \quad \text{for } \xi \in \mathbf{R}.
\end{aligned}$$

(2) The Fourier transform $\widehat{K}_t(\xi)$ of the heat kernel

$$K_t(x) = \frac{1}{\sqrt{4\pi t}} e^{-\frac{x^2}{4t}} \in \mathcal{S}(\mathbf{R}) \quad \text{for } x \in \mathbf{R}^n \text{ and } t > 0,$$

is given by the formula

$$\begin{aligned}
\widehat{K}_t(\xi) &= \frac{1}{\sqrt{4\pi t}} \int_{\mathbf{R}} e^{-i x \cdot \xi} e^{-\frac{x^2}{4t}} \, dx = \frac{1}{\sqrt{4\pi t}} \widehat{\varphi}_{1/(2\sqrt{t})}(\xi) \\
&= e^{-t \xi^2} \quad \text{for } \xi \in \mathbf{R} \text{ and } t > 0.
\end{aligned}$$

The next theorem summarizes the basic properties of the Fourier transform ([42, Section 8.3], [146, Section 1.9]):

Theorem 4.1. *(i) The Fourier transforms $\mathcal{F}$ and $\mathcal{F}^*$ map $\mathcal{S}(\mathbf{R}^n)$ continuously into itself. Furthermore, we have, for all multi-indices α and β,*

$$\widehat{D^\alpha \varphi}(\xi) = \xi^\alpha \widehat{\varphi}(\xi) \quad \text{for every } \varphi \in \mathcal{S}(\mathbf{R}^n),$$

$$D^\beta \widehat{\varphi}(\xi) = \widehat{(-x)^\beta \varphi}(\xi) \quad \text{for every } \varphi \in \mathcal{S}(\mathbf{R}^n).$$

(ii) The Fourier transforms $\mathcal{F}$ and $\mathcal{F}^*$ are isomorphisms of $\mathcal{S}(\mathbf{R}^n)$ onto itself; more precisely, $\mathcal{F}\mathcal{F}^* = \mathcal{F}^*\mathcal{F} = I$ on $\mathcal{S}(\mathbf{R}^n)$. In particular, we have the formula

$$\varphi(x) = \frac{1}{(2\pi)^n} \int_{\mathbf{R}^n} e^{i\,x\cdot\xi} \widehat{\varphi}(\xi)\, d\xi \quad \text{for every } \varphi \in \mathcal{S}(\mathbf{R}^n). \tag{4.2}$$

Formula (4.2) is called the Fourier inversion formula
  (iii) If φ, $\psi \in \mathcal{S}(\mathbf{R}^n)$, we have the formulas

$$\int_{\mathbf{R}^n} \varphi(x)\widehat{\psi}(x)\, dx = \int_{\mathbf{R}^n} \widehat{\varphi}(\xi)\psi(\xi)\, d\xi, \tag{4.3a}$$

$$\int_{\mathbf{R}^n} \varphi(x)\psi(x)\, dx = \frac{1}{(2\pi)^n} \int_{\mathbf{R}^n} \varphi(\xi)\psi(\xi)\, d\xi. \tag{4.3b}$$

Formulas (4.3a) and (4.3b) are called the Parseval formulas.

4.1.4 Tempered Distributions

For the spaces $C_0^\infty(\mathbf{R}^n)$, $\mathcal{S}(\mathbf{R}^n)$ and $C^\infty(\mathbf{R}^n)$, we have the following two inclusions (i) and (ii):

(i) The injection of $C_0^\infty(\mathbf{R}^n)$ into $\mathcal{S}(\mathbf{R}^n)$ is continuous and the space $C_0^\infty(\mathbf{R}^n)$ is dense in $\mathcal{S}(\mathbf{R}^n)$.
(ii) The injection of $\mathcal{S}(\mathbf{R}^n)$ into $C^\infty(\mathbf{R}^n)$ is continuous and the space $\mathcal{S}(\mathbf{R}^n)$ is dense in $C^\infty(\mathbf{R}^n)$.
    For any given function $\varphi \in \mathcal{S}(\mathbf{R}^n)$ (resp. $\varphi \in C^\infty(\mathbf{R}^n)$), it is easy to verify that

$$\psi_j \varphi \longrightarrow \varphi \quad \text{in } \mathcal{S}(\mathbf{R}^n) \text{ (resp. in } C^\infty(\mathbf{R}^n)) \text{ as } j \to \infty.$$

Hence the dual space $\mathcal{S}'(\mathbf{R}^n) = \mathcal{L}(\mathcal{S}(\mathbf{R}^n), \mathbf{C})$ can be identified with a linear subspace of $\mathcal{D}'(\mathbf{R}^n)$ containing $\mathcal{E}'(\mathbf{R}^n)$, by the identification of a continuous linear functional on $\mathcal{S}(\mathbf{R}^n)$ with its restriction to $C_0^\infty(\mathbf{R}^n)$. Namely, we have the inclusions

$$\mathcal{E}'(\mathbf{R}^n) \subset \mathcal{S}'(\mathbf{R}^n) \subset \mathcal{D}'(\mathbf{R}^n).$$

The elements of $\mathcal{S}'(\mathbf{R}^n)$ are called *tempered distributions* on $\mathbf{R}^n$. In other words, the tempered distributions are precisely those distributions on $\mathbf{R}^n$ that have continuous extensions to $\mathcal{S}(\mathbf{R}^n)$.

Roughly speaking, the tempered distributions are those which grow at most polynomially at infinity, since the functions in $\mathcal{S}(\mathbf{R}^n)$ die out faster than any power of x at infinity. In fact, we have the following four examples (1) through (4) of tempered distributions:

(1) The functions in $L^p(\mathbf{R}^n)$ $(1 \leq p \leq \infty)$ are tempered distributions.
(2) A locally integrable function on $\mathbf{R}^n$ is a tempered distribution if it grows at most polynomially at infinity.
(3) If $u \in \mathcal{S}'(\mathbf{R}^n)$ and $f(x)$ is a smooth function on $\mathbf{R}^n$ all of whose derivatives grow at most polynomially at infinity, then the product fu is a tempered distribution.
(4) Any derivative of a tempered distribution is also a tempered distribution.

Now we give some concrete and important examples of distributions in the space $\mathcal{S}'(\mathbf{R}^n)$:

Example 4.2. (a) The Dirac measure: $\delta(x)$.
(b) Riesz potentials:

$$R_\alpha(x) = \frac{\partial \Omega((n-\alpha)/2)}{2^\alpha \pi^{n/2} \Gamma(\alpha/2)} \frac{1}{|x|^{n-\alpha}} \quad \text{for } 0 < \alpha < n.$$

(c) Newtonian potentials:

$$N(x) = \frac{\Gamma((n-2)/2)}{4\,\pi^{n/2}} \frac{1}{|x|^{n-2}} \quad \text{for } n \geq 3.$$

(d) Riesz kernels:

$$R_j(x) = \sqrt{-1}\,\frac{\Gamma((n+1)/2)}{\pi^{(n+1)/2}} \,\text{v. p.}\,\frac{x_j}{|x|^{n+1}} \quad \text{for } 1 \leq j \leq n.$$

The distribution v. p. $(x_j/|x|^{n+1})$ is an extension of v. p. $(1/x)$ to the n-dimensional case.

The importance of tempered distributions lies in the fact that they have Fourier transforms.

If $u \in \mathcal{S}'(\mathbf{R}^n)$, we define its (direct) *Fourier transform* $\mathcal{F}u = \hat{u}$ by the formula

$$\langle \mathcal{F}u, \varphi \rangle = \langle u, \mathcal{F}\varphi \rangle \quad \text{for all } \varphi \in \mathcal{S}(\mathbf{R}^n). \tag{4.4}$$

Then we have $\mathcal{F}u \in \mathcal{S}'(\mathbf{R}^n)$, since the Fourier transform

$$\mathcal{F}: \mathcal{S}(\mathbf{R}^n) \longrightarrow \mathcal{S}(\mathbf{R}^n)$$

is an isomorphism. Furthermore, in view of formulas (4.3a) and (4.3b) it follows that the above definition (4.4) agrees with definition (4.1) if $u \in \mathcal{S}(\mathbf{R}^n)$. We also denote $\mathcal{F}u$ by $\hat{u}$.

Similarly, if $v \in \mathcal{S}'(\mathbf{R}^n)$, we define its *inverse Fourier transform* $\mathcal{F}^*v = \check{v}$ by the formula

$$\langle \mathcal{F}^*v, \psi \rangle = \langle v, \mathcal{F}^*\psi \rangle \quad \text{for all } \psi \in \mathcal{S}(\mathbf{R}^n).$$

The next theorem, which is a consequence of Theorem 4.1, summarizes the basic properties of Fourier transforms in the space $\mathcal{S}'(\mathbf{R}^n)$ (see [26, Chapitre I, Théorème 2.10]):

Theorem 4.2. *(i) The Fourier transforms $\mathcal{F}$ and $\mathcal{F}^*$ map $\mathcal{S}'(\mathbf{R}^n)$ continuously into itself. Furthermore, we have, for all multi-indices α and β,*

$$\mathcal{F}(D^\alpha u)(\xi) = \xi^\alpha \mathcal{F}u(\xi) \quad \text{for every } u \in \mathcal{S}'(\mathbf{R}^n),$$
$$D_\xi^\beta(\mathcal{F}u(\xi)) = \mathcal{F}((-x)^\beta u)(\xi) \quad \text{for every } u \in \mathcal{S}'(\mathbf{R}^n).$$

(ii) The Fourier transforms $\mathcal{F}$ and $\mathcal{F}^*$ are isomorphisms of $\mathcal{S}'(\mathbf{R}^n)$ onto itself; more precisely, $\mathcal{F}\mathcal{F}^* = \mathcal{F}^*\mathcal{F} = I$ on $\mathcal{S}'(\mathbf{R}^n)$.

(iii) The transforms $\mathcal{F}$ and $\mathcal{F}^*$ are norm-preserving operators on $L^2(\mathbf{R}^n)$ and $\mathcal{F}\mathcal{F}^* = \mathcal{F}^*\mathcal{F} = I$ on $L^2(\mathbf{R}^n)$. This assertion is referred to as the Plancherel theorem.

The situation of Theorems 4.1 and 4.2 can be visualized as in Figures 4.3 and 4.4 below.

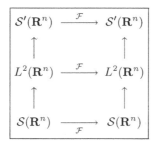

Fig. 4.3. The mapping properties of the Fourier transform $\mathcal{F}$

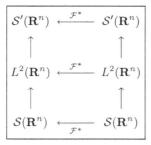

Fig. 4.4. The mapping properties of the inverse Fourier transform $\mathcal{F}^*$

Finally, we give a typical example of tempered distributions from the viewpoint of distribution kernels in the theory of pseudo-differential operators:

Example 4.3. Let $0 < \alpha < 2$. Then we can define a tempered distribution v.p. $\dfrac{1}{|x|^{n+\alpha}}$ as follows:

(1) The case where $0 < \alpha < 1$:

$$\left\langle \text{v.p.}\, \frac{1}{|x|^{n+\alpha}}, \varphi \right\rangle = \lim_{\varepsilon \downarrow 0} \int_{|y| \geq \varepsilon} \frac{\varphi(y) - \varphi(0)}{|y|^{n+\alpha}} \, dy$$

for all $\varphi \in \mathcal{S}(\mathbf{R}^n)$.

(2) The case where $1 \leq \alpha < 2$:

$$\left\langle \text{v.p.}\, \frac{1}{|x|^{n+\alpha}}, \varphi \right\rangle = \lim_{\varepsilon \downarrow 0} \int_{|y| \geq \varepsilon} \frac{\varphi(y) + \varphi(-y) - 2\varphi(0)}{|y|^{n+\alpha}} \, dy$$

for all $\varphi \in \mathcal{S}(\mathbf{R}^n)$.

Here "v.p." stands for Cauchy's "valeur principale" in French. Moreover, its Fourier transform is given by the formula (see [122, Example 5.28])

$$\mathcal{F}\left(\text{v.p.}\, \frac{1}{|x|^{n+\alpha}} \right) = \frac{\pi^{n/2}\, \Gamma(-\alpha/2)}{2^{\alpha}\, \Gamma((\alpha + n)/2)} \, |\xi|^{\alpha}.$$

4.1.5 Sobolev Spaces

If $s \in \mathbf{R}$, we define a linear map

$$J^s \colon \mathcal{S}'(\mathbf{R}^n) \longrightarrow \mathcal{S}'(\mathbf{R}^n)$$

by the formula

$$J^s u = \mathcal{F}^* \left(\left(1 + |\xi|^2\right)^{-s/2} \mathcal{F} u \right) \quad \text{for all } u \in \mathcal{S}'(\mathbf{R}^n).$$

This formula can be visualized as in Figure 4.5 below.

$$
\begin{array}{ccc}
u \in \mathcal{S}'(\mathbf{R}^n) & \xrightarrow{\;J^s = (1-\Delta)^{-s/2}\;} & \mathcal{S}'(\mathbf{R}^n) \ni J^s u \\[2pt]
{\scriptstyle \mathcal{F}}\downarrow & & \uparrow{\scriptstyle \mathcal{F}^*} \\[2pt]
\mathcal{F} u \in \mathcal{S}'(\mathbf{R}^n) & \xrightarrow[\;(1+|\xi|^2)^{-s/2}\;]{} & \mathcal{S}'(\mathbf{R}^n) \ni (1+|\xi|^2)^{-s/2}\mathcal{F} u
\end{array}
$$

Fig. 4.5. The definition of the Bessel potential J^s

Then it is easy to see that the map J^s is an isomorphism of $\mathcal{S}'(\mathbf{R}^n)$ onto itself and that its inverse is the map J^{-s}. The function $J^s u$ is called the *Bessel potential* of order s of u.

It should be noticed that if we define a function $G_s(x)$ by the formula

$$G_s(x) := \frac{1}{\Gamma(s/2)} \frac{1}{(4\pi)^{n/2}} \int_0^\infty e^{-t - \frac{|x|^2}{4t}} t^{\frac{s-n}{2}} \frac{dt}{t} \quad \text{for } s > 0,$$

then we have the important formula (see Stein [108, Chapter V, Section 3])

$$\widehat{G_s}(\xi) = \left(1 + |\xi|^2\right)^{-s/2} \quad \text{for } s > 0.$$

It is known (see Aronszajn–Smith [12]) that the function $G_s(x)$ is represented in the form

$$G_s(x) = \frac{1}{2^{(n+s-2)/2} \pi^{n/2} \Gamma(s/2)} K_{(n-s)/2}(|x|) |x|^{\frac{s-n}{2}},$$

where $K_{(n-s)/2}(z)$ is the *modified Bessel function* of the third kind (see Watson [142]).

First, we define Sobolev spaces of fractional order or Bessel potential spaces (see Figure 4.6):

Definition 4.3. *If $s \in \mathbf{R}$ and $1 < p < \infty$, we let*

$$H^{s,p}(\mathbf{R}^n) = \text{the image of } L^p(\mathbf{R}^n) \text{ under the mapping } J^s.$$

We equip $H^{s,p}(\mathbf{R}^n)$ with the norm

$$\|u\|_{s,p} = \left\|J^{-s}u\right\|_p \quad \text{for } u \in H^{s,p}(\mathbf{R}^n).$$

The space $H^{s,p}(\mathbf{R}^n)$ is called the Sobolev space of fractional order s *or* Bessel potential space of order s.

The situation can be visualized as in Figure 4.6 below.

$$\boxed{u = J^s v \in H^{s,p}(\mathbf{R}^n) \xleftarrow{\ J^s\ } L^p(\mathbf{R}^n) \ni v = J^{-s}u}$$

Fig. 4.6. The Sobolev spaces $H^{s,p}(\mathbf{R}^n)$ and $L^p(\mathbf{R}^n)$

We list some basic topological properties of $H^{s,p}(\mathbf{R}^n)$ ([2], [135]):

(1) The Schwartz space $\mathcal{S}(\mathbf{R}^n)$ is dense in each $H^{s,p}(\mathbf{R}^n)$.
(2) The space $H^{-s,p'}(\mathbf{R}^n)$ is the dual space of $H^{s,p}(\mathbf{R}^n)$, where $p' = p/(p-1)$ is the exponent conjugate to p ([2, p. 62, Theorem 3.9], [135, p. 178, Theorem 2.11.2]).

(3) If $s > t$, then we have the inclusions

$$S(\mathbf{R}^n) \subset H^{s,p}(\mathbf{R}^n) \subset H^{t,p}(\mathbf{R}^n) \subset S'(\mathbf{R}^n),$$

with continuous injections.

(4) If s is a non-negative integer, then the space $H^{s,p}(\mathbf{R}^n)$ is isomorphic to the usual Sobolev space $W^{s,p}(\mathbf{R}^n)$, that is, the space $H^{s,p}(\mathbf{R}^n)$ coincides with the space of functions $u \in L^p(\mathbf{R}^n)$ such that $D^\alpha u \in L^p(\mathbf{R}^n)$ for $|\alpha| \le s$, and the norm $\| \cdot \|_{s,p}$ is equivalent to the norm

$$\left(\sum_{|\alpha| \le s} \int_{\mathbf{R}^n} |D^\alpha u(x)|^p dx \right)^{1/p}.$$

4.1.6 Besov Spaces

Secondly, we define Besov spaces:

Definition 4.4. *(1) If $1 < p < \infty$, we let*

$B^{1,p}(\mathbf{R}^{n-1}) =$ *the space of (equivalence classes of) functions*
$\varphi(x') \in L^p(\mathbf{R}^{n-1})$ *for which the integral*

$$\iint_{\mathbf{R}^{n-1} \times \mathbf{R}^{n-1}} \frac{|\varphi(x'+y') - 2\varphi(x') + \varphi(x'-y')|^p}{|y'|^{n-1+p}} \, dy' \, dx'$$

is finite.

The space $B^{1,p}(\mathbf{R}^{n-1})$ is a Banach space with respect to the norm

$$|\varphi|_{1,p} = \left(\int_{\mathbf{R}^{n-1}} |\varphi(x')|^p \, dx' \right.$$
$$\left. + \iint_{\mathbf{R}^{n-1} \times \mathbf{R}^{n-1}} \frac{|\varphi(x'+y') - 2\varphi(x') + \varphi(x'-y')|^p}{|y'|^{n-1+p}} \, dy' \, dx' \right)^{1/p}.$$

(2) If $p = \infty$, we let

$B^{1,\infty}(\mathbf{R}^{n-1}) =$ *the space of (equivalence classes of) functions*
$\varphi(x') \in L^\infty(\mathbf{R}^{n-1})$ *for which the quantity*

$$\sup_{|y'|>0} \frac{\|\varphi(\cdot + y') - 2\varphi(\cdot) + \varphi(\cdot - y')\|_\infty}{|y'|}$$

is finite.

The space $B^{1,\infty}(\mathbf{R}^{n-1})$ is a Banach space with respect to the norm

$$|\varphi|_{1,\infty} = \|\varphi\|_\infty + \sup_{|y'|>0} \frac{\|\varphi(\cdot + y') - 2\varphi(\cdot) + \varphi(\cdot - y')\|_\infty}{|y'|}.$$

(3) If $s \in \mathbf{R}$ and $1 < p \leq \infty$, we let

$$B^{s,p}(\mathbf{R}^{n-1}) = \textit{the image of } B^{1,p}(\mathbf{R}^{n-1}) \textit{ under the mapping}$$
$$J'^{s-1} \textit{ where } J'^{s-1} \textit{ is the Bessel potential}$$
$$\textit{of order } s - 1 \textit{ on } \mathbf{R}^{n-1}.$$

We equip the space $B^{s,p}(\mathbf{R}^{n-1})$ with the norm

$$|\varphi|_{s,p} = \left| J'^{-s+1}\varphi \right|_{1,p} \quad \textit{for } \varphi(x') \in B^{s,p}(\mathbf{R}^{n-1}).$$

The space $B^{s,p}(\mathbf{R}^{n-1})$ is called the Besov space *of order s.*

The situation can be visualized as in Figure 4.7 below.

$$\boxed{\varphi = J'^{s-1}\psi \in B^{s,p}(\mathbf{R}^{n-1}) \xleftarrow{J'^{s-1}} B^{1,p}(\mathbf{R}^{n-1}) \ni \psi = J'^{-s+1}\varphi}$$

Fig. 4.7. The Besov spaces $B^{s,p}(\mathbf{R}^{n-1})$ and $B^{1,p}(\mathbf{R}^{n-1})$

We list some basic topological properties of $B^{s,p}(\mathbf{R}^{n-1})$ (see [135]):

(1) The Schwartz space $\mathcal{S}(\mathbf{R}^{n-1})$ is dense in each $B^{s,p}(\mathbf{R}^{n-1})$.
(2) The space $B^{-s,p'}(\mathbf{R}^{n-1})$ is the dual space of $B^{s,p}(\mathbf{R}^{n-1})$, where $p' = p/(p-1)$ is the exponent conjugate to p ([135, p. 178, Theorem 2.11.2]).
(3) If $s > t$, then we have the inclusions

$$\mathcal{S}(\mathbf{R}^{n-1}) \subset B^{s,p}(\mathbf{R}^{n-1}) \subset B^{t,p}(\mathbf{R}^{n-1}) \subset \mathcal{S}'(\mathbf{R}^{n-1}),$$

with continuous injections.
(4) If $s = m + \sigma$ where m is a non-negative integer and $0 < \sigma < 1$, then the Besov space $B^{s,p}(\mathbf{R}^{n-1})$ coincides with the space of functions $\varphi(x') \in H^{m,p}(\mathbf{R}^{n-1})$ such that, for $|\alpha| = m$ the integral (Slobodeckiĭ seminorm)

$$\iint_{\mathbf{R}^{n-1} \times \mathbf{R}^{n-1}} \frac{|D^{\alpha}\varphi(x') - D^{\alpha}\varphi(y')|^p}{|x' - y'|^{n-1+p\sigma}} \, dx' \, dy' < \infty.$$

Furthermore, the norm $|\varphi|_{s,p}$ is equivalent to the norm ([135, p. 36, part (iv)])

$$\left(\sum_{|\alpha| \leq m} \int_{\mathbf{R}^{n-1}} |D^{\alpha}\varphi(x')|^p \, dx' \right.$$
$$\left. + \sum_{|\alpha|=m} \iint_{\mathbf{R}^{n-1} \times \mathbf{R}^{n-1}} \frac{|D^{\alpha}\varphi(x') - D^{\alpha}\varphi(y')|^p}{|x' - y'|^{n-1+p\sigma}} \, dx' \, dy' \right)^{1/p}.$$

4.1.7 General Sobolev and Besov Spaces

Now we define the Sobolev spaces $H^{s,p}(\Omega)$ and $H^{s,p}(M)$ of fractional order and the Besov spaces $B^{s,p}(\partial\Omega)$ for a bounded domain Ω of $\mathbf{R}^n$ with smooth boundary $\partial\Omega$ and the double M of Ω (see Figure 4.1):

Definition 4.5. *If $s \in \mathbf{R}$ and $1 < p < \infty$, we define*

$$H^{s,p}(\Omega) = the\ space\ of\ distributions\ u \in \mathcal{D}'(\Omega)\ such\ that$$
$$there\ exists\ a\ function\ U \in H^{s,p}(\mathbf{R}^n)\ with\ U|_\Omega = u,$$

and equip the space $H^{s,p}(\Omega)$ with the norm

$$\|u\|_{s,p} = \inf \|U\|_{s,p},$$

where the infimum is taken over all such U. The space $H^{s,p}(\Omega)$ is a Banach space with respect to the norm $\|\cdot\|_{s,p}$. We remark that

$$H^{0,p}(\Omega) = L^p(\Omega); \quad \|\cdot\|_{0,p} = \|\cdot\|_p.$$

Then we have the following important relationships between the L^p Sobolev spaces $H^{s,p}(\Omega)$ and $W^{s,p}(\Omega)$ for all $s \geq 0$ and $1 < p < \infty$ (see [81, Theorem]):

Theorem 4.6 (Meyers–Serrin). *If Ω is a bounded, Lipschitz domain, then we have, for all $s \geq 0$ and $1 < p < \infty$,*

$$H^{s,p}(\Omega) = W^{s,p}(\Omega).$$

We introduce a space of distributions on Ω which behave locally just like the distributions in $H^{s,p}(\mathbf{R}^n)$:

Definition 4.7. *(1) If $s \in \mathbf{R}$ and $1 < p < \infty$, we define*

$$H^{s,p}_{\mathrm{loc}}(\Omega) = the\ space\ of\ distributions\ u \in \mathcal{D}'(\Omega)\ such\ that$$
$$\varphi u \in H^{s,p}(\mathbf{R}^n)\ for\ all\ \varphi \in C^\infty_0(\Omega).$$

We equip the localized Sobolev space $H^{s,p}_{\mathrm{loc}}(\Omega)$ with the topology defined by the seminorms

$$u \longmapsto \|\varphi u\|_{s,p}$$

as φ ranges over $C^\infty_0(\Omega)$. It is easy to verify that $H^{s,p}_{\mathrm{loc}}(\Omega)$ is a Fréchet space.
(2) Similarly, the localized Besov space $B^{s,p}_{\mathrm{loc}}(\partial\Omega)$ is defined for $s \in \mathbf{R}$ and $1 < p \leq \infty$, with $H^{s,p}(\mathbf{R}^n)$ replaced by $B^{s,p}(\mathbf{R}^{n-1})$.

If M is the double of Ω, then the Sobolev spaces $H^{s,p}(M)$ of fractional order are defined to be locally the spaces $H^{s,p}(\mathbf{R}^n)$, upon using local coordinate systems flattening out M, together with a partition of unity. Similarly, the Besov spaces $B^{s,p}(\partial\Omega)$ are defined with $H^{s,p}(\mathbf{R}^n)$ replaced by $B^{s,p}(\mathbf{R}^{n-1})$.

The norms of $H^{s,p}(M)$ and $B^{s,p}(\partial\Omega)$ will be denoted by $\|\cdot\|_{s,p}$ and $|\cdot|_{s,p}$, respectively.

4.1.8 Sobolev's Imbedding Theorems

It is the imbedding characteristics of L^p Sobolev spaces that render these spaces so useful in the study of partial differential equations. In the proof of Theorem 1.5 (see [121, Theorem 1.3]) we need the following imbedding properties of L^p Sobolev spaces (see the proof of Lemma 12.5):

Theorem 4.8 (Sobolev). *Let Ω be a bounded domain in the Euclidean space $\mathbf{R}^n$ with boundary $\partial\Omega$ of class C^2. Let j and m be non-negative integers and $1 \le p < \infty$. Then we have the following three assertions (i), (ii) and (iii):*

(i) If $1 \le p < n$, we have the continuous injection

$$H^{2,p}(\Omega) \subset H^{1,q}(\Omega) \quad for \quad \frac{1}{p} - \frac{1}{n} \le \frac{1}{q} \le \frac{1}{p}.$$

(ii) If $(m-1)p < n < mp$, we have the continuous injection

$$H^{j+m,p}(\Omega) \subset C^{j+\nu}(\overline{\Omega}) \quad for \quad 0 < \nu \le m - \frac{n}{p}.$$

(iii) If $(m-1)p = n$, we have the continuous injection

$$H^{j+m,p}(\Omega) \subset C^{j+\nu}(\overline{\Omega}) \quad for \quad 0 < \nu < 1.$$

For a proof of Theorem 4.8, see Adams–Fournier [2, Theorem 4.12], Friedman [43, Part I, Theorems 10.2 and 11.1] and also [123, Theorems 4.17 and 4.19].

Since the elements of $H^{k,p}(\Omega)$ are, strictly speaking, not functions defined everywhere in Ω but rather equivalence classes of such functions defined and equal up to sets of measure zero, we must clarify what it meant by an imbedding of $H^{j+m,p}(\Omega)$ into $C^{j+\nu}(\overline{\Omega})$. The imbedding $H^{j+m,p}(\Omega) \to C^{j+\nu}(\overline{\Omega})$ means that each $u \in H^{j+m,\nu}(\Omega)$ can, when considered as a function, be redefined on a set of zero measure in Ω in such a way that the modified function $\tilde{u}$ (which equals u in $H^{j+m,\nu}(\Omega)$) belongs to $C^{j+\nu}(\overline{\Omega})$ and satisfies the inequality

$$\|\tilde{u}\|_{C^{j+\nu}(\overline{\Omega})} \le C\|u\|_{H^{j+m,\nu}(\Omega)},$$

with some constant $C > 0$ depending on m, p, n and j.

In the proof of Theorem 1.5 (see [121, Theorem 1.3]) we make use of the following inequality (see the proof of Lemma 12.1):

Theorem 4.9 (Gagliardo–Nirenberg). *Let Ω be a bounded domain in $\mathbf{R}^n$ with boundary of class C^2, and $1 \le p, r \le \infty$. Then we have the following two assertions (i) and (ii):*

(i) If $p \ne n$ and if the inequality

$$\frac{1}{q} = \frac{1}{n} + \theta\left(\frac{1}{p} - \frac{2}{n}\right) + (1-\theta)\frac{1}{r} \quad for \quad \frac{1}{2} \le \theta \le 1$$

holds true, then we have, for all functions $u \in H^{2,p}(\Omega) \cap L^r(\Omega)$,

$$\|u\|_{1,q} \leq C_1 \|u\|_{2,p}^{\theta} \|u\|_r^{1-\theta}, \tag{4.5}$$

with a positive constant $C_1 = C_1(\Omega, p, r, \theta)$.
(ii) If $n/2 < p < \infty$, $p \neq n$ and if the inequality

$$0 \leq \nu < \theta \left(2 - \frac{n}{p} \right) - (1 - \theta)\frac{n}{r}$$

holds true, then we have, for all functions $u \in H^{2,p}(\Omega) \cap L^r(\Omega)$,

$$\|u\|_{C^{\nu}(\overline{\Omega})} \leq C_2 \|u\|_{2,p}^{\theta} \|u\|_r^{1-\theta}, \tag{4.6}$$

with a positive constant $C_2 = C_2(\Omega, p, r, \theta)$.

For a proof of Theorem 4.9, see Gagliardo [47], Friedman [43, Part I, Theorem 9.3] and also [123, Theorem 9.3].

4.1.9 The Rellich–Kondrachov theorem

The next *compactness theorem* for function spaces of L^p type plays an essential role in the proof of Theorem 1.4 ([121, Theorem 1.2]):

Theorem 4.10 (Rellich–Kondrachov). *Let Ω be a bounded domain in $\mathbf{R}^n$ with smooth boundary $\partial\Omega$. If $1 < p < \infty$ and $s > t$, then the injections*

$$H^{s,p}(\Omega) \longrightarrow H^{t,p}(\Omega),$$
$$B^{s,p}(\partial\Omega) \longrightarrow B^{t,p}(\partial\Omega)$$

are both compact (or completely continuous).

The Rellich–Kondrachov theorem is an L^p Sobolev space version of the Bolzano–Weierstrass theorem and the Ascoli–Arzelà theorem in calculus (see Table 4.1 below):

For a proof of Theorem 4.10, see Adams–Fournier [2, Theorem 6.3 and Paragraph 7.32], Friedman [43, Part I, Theorem 11.2], Gilbarg–Trudinger [50, Theorem 7.22] and also Triebel [135, p. 233, Remark 1].

4.2 Seeley's Extension Theorem

The next theorem, due to Seeley [103], asserts that the functions in $C^{\infty}(\overline{\Omega})$ are the restrictions to Ω of functions in $C^{\infty}(\mathbf{R}^n)$ (see [103], [2, Theorems 5.21 and 5.22]):

Subjects	Sequences	Compactness theorems
Theory of real numbers	Sequences of real numbers	The Bolzano–Weierstrass theorem
Calculus	Sequences of continuous functions	The Ascoli–Arzerà theorem
Theory of distributions	Sequences of distributions	The Rellich–Kondrachov theorem

Table 4.1. A bird's-eye view of three compactness theorems in calculus

Theorem 4.11 (Seeley's extension theorem). *Let Ω be either the half space $\mathbf{R}_+^n$ or a smooth domain in $\mathbf{R}^n$ with bounded boundary $\partial\Omega$. Then there exists a continuous linear extension operator*

$$E \colon C^\infty(\overline{\Omega}) \longrightarrow C^\infty(\mathbf{R}^n).$$

Furthermore, the restriction to the space $C_0^\infty(\overline{\Omega})$ of E is a continuous linear extension operator on $C_0^\infty(\overline{\Omega})$ into $C_0^\infty(\mathbf{R}^n)$. Here

$$C_0^\infty(\overline{\Omega}) = \text{the space of restrictions to } \Omega \text{ of functions in } C_0^\infty(\mathbf{R}^n).$$

Proof. The proof of Theorem 4.11 is divided into Steps I and II.

Step I: First, we consider the case where

$$\Omega = \mathbf{R}_+^n.$$

The proof is based on the following elementary lemma (see [123, Lemma 4.22]):

Lemma 4.12. *There exists a function $w(t)$ in the space $\mathcal{S}(\mathbf{R})$ such that*

$$\begin{cases} \operatorname{supp} w \subset [1, \infty), \\ \int_1^\infty t^n \, w(t) \, dt = (-1)^n, \quad n = 0, 1, 2, \ldots. \end{cases}$$

In this book, assuming Lemma 4.12 we prove Theorem 4.11.

By using the function w in Lemma 4.12, we can define a linear operator

$$E \colon C^\infty(\overline{\mathbf{R}_+^n}) \longrightarrow C^\infty(\mathbf{R}^n)$$

by the formula

$$E\varphi(x', x_n) = \begin{cases} \varphi(x', x_n) & \text{if } x_n \geq 0, \\ \int_1^\infty w(s)\, \theta(-x_n s)\, \varphi(x', -sx_n)\, ds & \text{if } x_n < 0, \end{cases}$$

where

$$x = (x', x_n) \in \mathbf{R}^n, \quad x' = (x_1, x_2, \ldots, x_{n-1}) \in \mathbf{R}^{n-1},$$

and

$$\begin{cases} \theta \in C_0^\infty(\mathbf{R}), \\ \operatorname{supp} \theta \subset [-2, 2], \\ \theta(t) = 1 \text{ for } |t| \leq 1. \end{cases}$$

Then it is easy to verify the following three assertions (1), (2) and (3):

(1) $E\varphi \in C^\infty(\mathbf{R}^n)$.
(2) The operator E maps $C^\infty(\overline{\mathbf{R}_+^n})$ continuously into $C^\infty(\mathbf{R}^n)$.
(3) If $\operatorname{supp} \varphi \subset \{x \in \mathbf{R}^n : |x'| \leq r, 0 \leq x_n \leq a\}$ for some $r > 0$ and $a > 0$, then it follows that $\operatorname{supp} E\varphi \subset \{x \in \mathbf{R}^n : |x'| \leq r, |x_n| \leq a\}$.

This proves Theorem 4.11 for the half space $\mathbf{R}_+^n$.

Step II: Now we assume that Ω is a smooth domain in $\mathbf{R}^n$ with bounded boundary $\partial\Omega$. Then we can choose a finite covering $\{V_j\}_{j=1}^N$ of $\partial\Omega$ by open subsets of $\mathbf{R}^n$ and C^∞ diffeomorphisms χ_j of V_j onto the unit ball

$$B(0, 1) = \{x \in \mathbf{R}^n : |x| < 1\}$$

(see Figures 4.8 and 4.9 below) such that the open sets

$$V_j = \chi_j^{-1}\left(\left\{x \in \mathbf{R}^n : |x'| < \frac{1}{2}, |x_n| < \frac{\sqrt{3}}{2}\right\}\right), \quad 1 \leq j \leq N,$$

form an open covering of the tubular neighborhood

$$\Omega_\delta = \{x \in \Omega : \operatorname{dist}(x, \partial\Omega) < \delta\}$$

for some $\delta > 0$. Furthermore, we can choose an open set V_0 in Ω, bounded away from $\partial\Omega$ (see Figure 4.10 below) such that

$$\Omega \subset V_0 \cup \left(\cup_{j=1}^N V_j\right).$$

Let $\{\omega_j\}_{j=0}^N$ be a *partition of unity* subordinate to the open covering $\{V_j\}_{j=0}^N$. Namely, the family $\{\omega_j\}_{j=0}^N$ satisfies the following three conditions (PU1), (PU2) and (PU3):

(PU1) $0 \leq \omega_j(x) \leq 1$ for all $x \in \overline{\Omega}$ and $0 \leq j \leq N$.
(PU2) $\operatorname{supp} \omega_j \subset V_j$ for each $0 \leq j \leq N$.
(PU3) The functions $\{\omega_j\}_{j=0}^N$ satisfy the condition

$$\sum_{j=0}^N \omega_j(x) = 1 \quad \text{for every } x \in \overline{\Omega}.$$

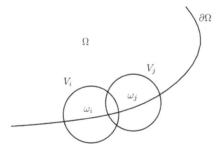

Fig. 4.8. The open covering $\{V_j\}$ and the partition of unity $\{\omega_j\}$

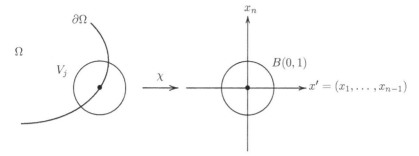

Fig. 4.9. The coordinate transformation χ_j maps V_j onto $B(0,1)$

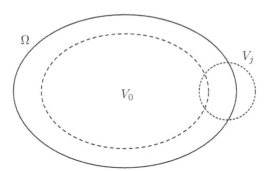

Fig. 4.10. The open sets V_0 and V_j

If $\varphi \in C^{\infty}(\overline{\Omega})$, we define a linear operator

$$E\varphi = \omega_0\varphi + \sum_{j=1}^{N} \chi_j^* \left(E \left((\chi_j^{-1})^*(\omega_j\varphi) \right) \right).$$

Then it is easy to verify that this operator E enjoys the desired properties. Theorem 4.11 is proved, apart from the proof of Lemma 4.12. $\square$

4.3 Trace Theorems

First, we study the restrictions to the hyperplane $\{x_n = 0\}$ of functions in the L^p Sobolev space $H^{s,p}(\mathbf{R}^n_+)$ on the half-space $\mathbf{R}^n_+$.

If $x = (x_1, \ldots, x_n)$ is a point of $\mathbf{R}^n$, we write

$$x = (x', x_n), \quad x' = (x_1, \ldots, x_{n-1}) \in \mathbf{R}^{n-1}.$$

If j is a non-negative integer, we define the trace map

$$\gamma_j \colon C_0^\infty(\overline{\mathbf{R}^n_+}) \longrightarrow C_0^\infty(\mathbf{R}^{n-1})$$

by the formula

$$\gamma_j u(x') = \lim_{x_n \downarrow 0} D_n^j u(x', x_n) \quad \text{for } u \in C_{(0)}^\infty(\overline{\mathbf{R}^n_+}).$$

Then we have the following theorem:

Theorem 4.13. *Let* $1 < p < \infty$. *If* $0 \le j < s - 1/p$, *then the trace map*

$$\gamma_j \colon C_{(0)}^\infty(\overline{\mathbf{R}^n_+}) \longrightarrow C_0^\infty(\mathbf{R}^{n-1})$$

extends uniquely to a continuous linear map

$$\gamma_j \colon H^{s,p}(\mathbf{R}^n_+) \longrightarrow B^{s-j-1/p,p}(\mathbf{R}^{n-1}).$$

Furthermore, if $u \in H^{s,p}(\mathbf{R}^n_+)$, *then the mapping*

$$x_n \longmapsto D_n^j u(\cdot, x_n)$$

is a continuous function on $[0, \infty)$ *with values in* $B^{s-j-1/p,p}(\mathbf{R}^{n-1})$.

The next theorem shows that the result of Theorem 4.13 is sharp:

Theorem 4.14 (the trace theorem). *Let* $1 < p < \infty$. *If* $0 \le j < s - 1/p$, *then the trace map*

$$\gamma \colon H^{s,p}(\mathbf{R}^n_+) \longrightarrow \prod_{0 \le j < s - 1/p} B^{s-j-1/p,p}(\mathbf{R}^{n-1})$$

$$u \longmapsto (\gamma_j u)_{0 \le j < s - 1/p}$$

is continuous and surjective.

Example 4.4. We consider the case where $p = 2$, $s = 1$ and $j = 0$. For any given function $\varphi \in C_0^\infty(\mathbf{R}^{n-1})$, the function

$$u(x', x_n)$$
$$= \frac{2}{(2\pi)^n} \int_{\mathbf{R}^n} e^{i\,x' \cdot \xi'} \, e^{i\,x_n\,\xi_n} \, \frac{(1 + |\xi'|^2)^{1/2}}{1 + |\xi'|^2 + \xi_n^2} \, \widehat{\varphi}(\xi') \, dxi' \, d\xi_n$$

satisfies the conditions:

$$\|u\|^2_{H^{1,2}(\mathbf{R}^n)} = 2 \, \|\varphi\|^2_{B^{1/2,2}(\mathbf{R}^{n-1})},$$
$$(\gamma_0 u)(x') = \varphi(x') \quad \text{for all } x' \in \mathbf{R}^{n-1}.$$

Now let Ω be a bounded domain in $\mathbf{R}^n$ with smooth boundary $\partial\Omega$. Its closure $\overline{\Omega} = \Omega \cup \partial\Omega$ is an n-dimensional, compact smooth manifold with boundary. Then we may assume the following three conditions (a), (b) and (c) (see Figures 4.1 and 4.2):

(a) The domain Ω is a relatively compact open subset of an n-dimensional, compact smooth manifold M without boundary. The manifold $M = \widehat{\Omega}$ is the *double* of Ω.

(b) In a neighborhood W of $\partial\Omega$ in M a normal coordinate t is chosen so that the points of W are represented as (x', t), $x' \in \partial\Omega$, $-1 < t < 1$; $t > 0$ in Ω, $t < 0$ in $M \setminus \overline{\Omega}$ and $t = 0$ only on $\partial\Omega$.

(c) The manifold M is equipped with a strictly positive density μ which, on W, is the product of a strictly positive density ω on $\partial\Omega$ and the Lebesgue measure dt on $(-1, 1)$.

More precisely, we state two fundamental theorems on smooth manifolds with boundary. Let M be an n-dimensional, paracompact smooth manifold with boundary ∂M. The first theorem states that the boundary ∂M has an open neighborhood in M which is diffeomorphic to $\partial M \times [0, 1)$:

Theorem 4.15 (the product neighborhood theorem). *There exists a C^∞ diffeomorphism φ of $\partial M \times [0, 1)$ onto an open neighborhood W of ∂M in M which is the identity map on ∂M.*

The diffeomorphism φ is called a *collar* for M and the neighborhood W is called a *product neighborhood* of ∂M, respectively.

The second theorem states that M is a submanifold of some n-dimensional smooth manifold without boundary. Let $M_0 = M \times \{0\}$ and $M_1 = M \times \{1\}$ be two copies of M. The *double* $\widehat{M}$ of M is the topological space obtained from the union $M_0 \cup M_1$ by identifying $(x, 0)$ with $(x, 1)$ for each x in ∂M (see Figure 4.11 below). By using the product neighborhood theorem, we have the following theorem:

Theorem 4.16. *The double $\widehat{M}$ of M is an n-dimensional smooth manifold without boundary, and is uniquely determined up to C^∞ diffeomorphisms.*

If j is a non-negative integer, we can define the trace map

$$\gamma_j \colon C^\infty(\overline{\Omega}) \longrightarrow C^\infty(\partial\Omega)$$

by the formula

$$\gamma_j u(x') = \lim_{t\downarrow 0} D_t^j u(x', t) \quad \text{for all } u \in C^\infty(\overline{\Omega}).$$

Then Theorems 4.13 and 4.14 extend to this case as follows (see [2], [13], [107], [135]):

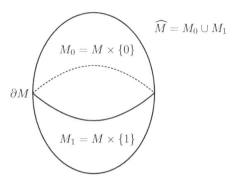

Fig. 4.11. The double $\widehat{M}$ of M

Theorem 4.17. *Let $1 < p < \infty$. If $0 \le j < s - 1/p$, then the trace map*

$$\gamma_j \colon C^\infty(\overline{\Omega}) \longrightarrow C^\infty(\partial\Omega)$$

extends uniquely to a continuous linear map

$$\gamma_j \colon H^{s,p}(\Omega) \longrightarrow B^{s-j-1/p,p}(\partial\Omega).$$

Furthermore, if $u \in H^{s,p}(\Omega)$, then the mapping

$$t \longmapsto D_t^j u(\cdot, t)$$

is a continuous function on $[0,1)$ with values in $B^{s-j-1/p,p}(\partial\Omega)$.

Theorem 4.18 (the trace theorem). *Let Ω be a bounded domain in $\mathbf{R}^n$ with smooth boundary $\partial\Omega$, and $1 < p < \infty$. If $0 \le j < s - 1/p$, then the trace map*

$$\gamma \colon H^{s,p}(\Omega) \longrightarrow \prod_{0 \le j < s - 1/p} B^{s-j-1/p,p}(\partial\Omega)$$

$$u \longmapsto (\gamma_j u)_{0 \le j < s - 1/2}$$

is continuous and surjective.

4.3.1 Sectional Traces

Let k be a non-negative integer. A distribution $u \in \mathcal{D}'(\Omega)$ is said to have *sectional traces* on $\partial\Omega$ up to order k if the mapping

$$t \longmapsto u(\cdot, t)$$

is a C^k function on $[0,1)$ with values in $\mathcal{D}'(\partial\Omega)$. This is equivalent to saying that, for every $\phi \in C^\infty(\partial\Omega)$, the function $\langle u(\cdot, t), \phi \cdot \omega \rangle$ is of class C^k on $[0,1)$.

Then we define the sectional trace $\gamma_j u$ on $\partial\Omega$ of order j, $0 \leq j \leq k$, by the formula

$$\gamma_j u = \lim_{t \downarrow 0} D_t^j u(\cdot, t) \quad \text{in } \mathcal{D}'(\partial\Omega).$$

Theorem 4.18 asserts that if $u \in H^{s,p}(\Omega)$ with $s > 1/p$, then it has sectional Traces

$$\gamma_j u \in B^{s-j-1/p,p}(\partial\Omega) \quad \text{up to order } j < s - 1/p.$$

The next result is useful for the interpretation and study of boundary conditions in terms of distributions:

Theorem 4.19 (the sectional trace theorem). *Let Ω be a bounded domain in $\mathbf{R}^n$ with smooth boundary $\partial\Omega$, and let*

$$A = \sum_{i,j=1}^n a^{ij}(x)\frac{\partial^2}{\partial x_i \partial x_j} + \sum_{i=1}^n b^i(x)\frac{\partial}{\partial x_i} + c(x)$$

be a second-order, differential operator with real C^∞ coefficients. Assume that the boundary $\partial\Omega$ is non-characteristic with respect to the operator A, that is,

$$\sum_{i,j=1}^n a^{ij}(x')\nu_i \nu_j > 0 \quad \text{on } \partial\Omega,$$

where $\nu = (\nu_1, \ldots, \nu_n)$ is the unit normal to $\partial\Omega$ at x'.

If $u \in H^{\sigma,p}(\Omega)$ and $Au \in H^{s,p}(\Omega)$ with $\sigma \leq s + 2$, then u has sectional traces $\gamma_j u$ on $\partial\Omega$ up to order $k < s + 2 - 1/p$. Moreover, it follows that $\gamma_j u \in B^{\sigma-j-1/p,p}(\partial\Omega)$ and we have the inequality

$$|\gamma_j u|_{B^{\sigma-j-1/p,p}(\partial\Omega)} \leq C_{\sigma,s}\left(\|Au\|_{H^{s,p}(\Omega)} + \|u\|_{H^{\sigma,p}(\Omega)}\right) \tag{4.7}$$

with some constant $C\sigma, s > 0$.

Remark 4.20. Theorem 4.19 is an expression of the fact that if we know about the derivatives of the solution u of $Au = f$ in tangential directions, then we can derive information about the normal derivatives $\gamma_j u$ by means of the equation $Au = f$.

Proof. By using local coordinate systems flattening out the boundary $\partial\Omega$, together with a partition of unity, we may assume that

$$\Omega = \mathbf{R}_+^n,$$
$$u \in \mathcal{E}'(\mathbf{R}^n),$$

and that

$$u = 0 \quad \text{in } \mathbf{R}_+^n.$$

The proof of Theorem 4.19 is divided into seven steps.

Step (1): First, we can find a constant $C > 0$ and a integer $\ell \geq 0$ such that

$$|\widehat{u}(\xi)| \leq C \left(1 + |\xi|\right)^{\ell} \quad \text{for all } \xi \in \mathbf{R}^n.$$

Hence we have, for every $\varepsilon > 0$,

$$u \in H^{-\ell - n/2 - \varepsilon}(\mathbf{R}^n).$$

If $B \in L_{\mathrm{cl}}^m(\mathbf{R}^n)$ for $m < 0$, it follows that

$$Bu \in H^{-\ell - m - n/2 - \varepsilon}(\mathbf{R}^n).$$

By Sobolev's imbedding theorem, we obtain that

$$Bu \in C^k(\mathbf{R}^n) \quad \text{for } k < -m - n - \ell. \tag{4.8}$$

Take a function $\chi(\xi) \in C^\infty(\mathbf{R}^n)$ such that

$$\chi(\xi) = \begin{cases} 0 & \text{in a neighborhood of } \xi = 0, \\ 1 & \text{for } |\xi| \geq 1. \end{cases}$$

If $a(x, \xi) \in S_{\mathrm{cl}}^\mu(\mathbf{R}^n \times \mathbf{R}^n)$ is a symbol of $A \in L_{\mathrm{cl}}^\mu(\mathbf{R}^n)$, we can express it as follows:

$$a(x, \xi) = \chi(\xi)\, a(x, \xi) + (1 - \chi(\xi))\, a(x, \xi),$$

where

$$(1 - \chi(\xi))\, a(x, \xi) \in S^{-\infty}(\mathbf{R}^n \times \mathbf{R}^n).$$

By applying assertion (4.8) with $m := -\infty$, we find that

$$(I - \chi(D))\, a(x, D) u \in C^\infty(\mathbf{R}^n).$$

Hence we are reduced to the following case:

$$\begin{cases} \widetilde{A} = \widetilde{a}(x, D) \in L_{\mathrm{cl}}^\mu(\mathbf{R}^n), \\ \widetilde{a}(x, \xi) = \chi(\xi)\, a(x, \xi), \\ \widetilde{a}(x, t\,\xi) = t^\mu\, a(x, \xi) \quad \text{for all } t > 0. \end{cases}$$

Step (2): Let $\Phi(x)$ be an arbitrary function in $C_0^\infty(\mathbf{R}^n)$. Then we have the formula

$$
\begin{aligned}
\left\langle \widetilde{A}u, \Phi \right\rangle &= \frac{1}{(2\pi)^n} \int_{\mathbf{R}^n} \widehat{u}(\xi) \left(\int_{\mathbf{R}^n} e^{i\, x \cdot \xi} \widetilde{a}(x, \xi)\, \Phi(x)\, dx \right) d\xi \\
&= \frac{1}{(2\pi)^n} \int_{|\xi| \geq 1} \widehat{u}(\xi) \left(\int_{\mathbf{R}^n} e^{i\, x \cdot \xi} a(x, \xi)\, \Phi(x)\, dx \right) d\xi \\
&\quad + \frac{1}{(2\pi)^n} \int_{|\xi| \leq 1} \widehat{u}(\xi) \left(\int_{\mathbf{R}^n} e^{i\, x \cdot \xi} \chi(\xi)\, a(x, \xi)\, \Phi(x)\, dx \right) d\xi
\end{aligned}
$$

$$:= \langle Bu, \Phi \rangle + \langle Ru, \Phi \rangle .$$

However, by using Fubini's theorem we can write down the last term as follows:

$$\langle Ru, \Phi \rangle = \frac{1}{(2\pi)^n} \int_{|\xi| \leq 1} \widehat{u}(\xi) \left(\int_{\mathbf{R}^n} e^{i\,x \cdot \xi} \chi(\xi)\, a(x, \xi)\, \Phi(x)\, dx \right) d\xi$$

$$= \frac{1}{(2\pi)^n} \int_{\mathbf{R}^n} \left(\int_{|\xi| \leq 1} \left\langle e^{i(x-y) \cdot \xi}, u(y) \right\rangle \chi(\xi)\, a(x, \xi)\, d\xi \right) \Phi(x)\, dx$$

$$= \int_{\mathbf{R}^n} \langle K_R(x, y), u(y) \rangle\, \Phi(x)\, dx,$$

where $K_R(x, y)$ is the distribution kernel of R, given by the formula

$$K_R(x, y) = \frac{1}{(2\pi)^n} \int_{|\xi| \leq 1} e^{i(x-y) \cdot \xi}\, \chi(\xi)\, a(x, \xi)\, d\xi.$$

Since we have the assertion

$$K_R(x, y) \in C^\infty(\mathbf{R}^n \times \mathbf{R}^n),$$

it follows that

$$Ru \in C^\infty(\mathbf{R}^n).$$

Therefore, we are reduced to the study of the term

$$Bu := \widetilde{A}u - Ru,$$

where

$$\langle Bu, \Phi \rangle = \frac{1}{(2\pi)^n} \int_{|\xi| \geq 1} \widehat{u}(\xi) F(\xi)\, d\xi,$$

$$F(\xi) = \int_{\mathbf{R}^n} e^{i\,x \cdot \xi}\, a(x, \xi)\, \Phi(x)\, dx.$$

Step (3): Since $a(x, \xi)$ is a rational function of ξ, we find that the poles of the function $a(x, \xi', \xi_n)$ of ξ_n remains in some compact subset of the complex place $\mathbf{C}$ when x belongs to a compact subset of $\mathbf{R}^n$ and ξ' belongs to a compact subset of $\mathbf{R}^{n-1}$, respectively. Hence, for every fixed $\xi' \in \mathbf{R}^{n-1}$ the function

$$\mathbf{C} \ni \xi_n \longmapsto F(\xi', \xi_n) = \int_{\mathbf{R}^n} e^{i\,x' \cdot \xi'}\, e^{i\,x_n\, \xi_n}\, a(x, \xi', \xi_n)\, \Phi(x)\, dx$$

is holomorphic, for $|\xi_n|$ sufficiently large.

Moreover, we have the following claim:

Claim 4.21. *Assume that* $\Phi(x) \in C_0^\infty(\mathbf{R}^n)$ *satisfies the condition*

$$\operatorname{supp} \Phi \subset \mathbf{R}_+^n.$$

Then the function

$$F(\xi', \xi_n) = \int_{\mathbf{R}^n} e^{i\,x \cdot \xi}\, a(x, \xi)\, \Phi(x)\, dx$$

is rapidly decreasing with respect to the variable $\xi = (\xi', \xi_n)$, *for* $\xi' \in \mathbf{R}^{n-1}$ *and* $\operatorname{Im} \xi_n \geq 0$.

Proof. By integration by parts, we have, for any multi-index α,

$$
\begin{aligned}
\xi^\alpha F(\xi) &= \int_{\mathbf{R}^n} D_x^\alpha e^{i\,x \cdot \xi} \cdot a(x, \xi', \xi_n)\, \Phi(x)\, dx \\
&= (-1)^{|\alpha|} \int_{\mathbf{R}^n} e^{i\,x \cdot \xi} \cdot D_x^\alpha \left(a(x, \xi', \xi_n)\, \Phi(x) \right) dx.
\end{aligned}
$$

However, we remark that

- $e^{i\,x \cdot \xi} = e^{i\,x' \cdot \xi'}\, e^{i\,x_n\,\xi_n} = e^{i\,x' \cdot \xi'}\, e^{i\,x_n\,\operatorname{Re}\,\xi_n} \cdot e^{-x_n\,\operatorname{Im}\,\xi_n}$,
- $x \in \operatorname{supp} \Phi \Longrightarrow x_n > 0$.

Hence we have, for some constant $C_\alpha > 0$,

$$|\xi^\alpha F(\xi)| \leq C\,(1 + |\xi|)^\mu \quad \text{for all } \xi' \in \mathbf{R}^{n-1} \text{ and } \operatorname{Im} \xi_n \geq 0.$$

Therefore, for any multi-index α we can find a constant $C_\alpha' > 0$ such that

$$|F(\xi)| \leq C_\alpha'\,(1 + |\xi|)^{\mu - |\alpha|} \quad \text{for all } \xi' \in \mathbf{R}^{n-1} \text{ and } \operatorname{Im} \xi_n \geq 0.$$

The proof of Claim 4.21 is complete. $\square$

On the other hand, we have the following claim:

Claim 4.22. *Assume that* $u = 0$ *in* $\mathbf{R}_+^n$, *that is,*

$$\operatorname{supp} u \subset \{x = (x', x_n) \in \mathbf{R}^n : x_n \leq 0\}.$$

Then the function $\widehat{u}(\xi', \xi_n)$ *is slowly increasing with respect to the variable* $\xi = (\xi', \xi_n)$, *for* $\xi' \in \mathbf{R}^{n-1}$ *and* $\operatorname{Im} \xi_n \geq 0$.

Proof. Since $u \in \mathcal{E}'(\mathbf{R}^n)$, we can find a constant $C > 0$ and a non-negative integer ℓ such that

$$|\langle u, \varphi \rangle| \leq C \sup_{x \in \operatorname{supp} u} \left(\sum_{|\alpha| \leq \ell} |\partial^\alpha \varphi(x)| \right) \quad \text{for all } \varphi \in C^\infty(\mathbf{R}^n). \tag{4.9}$$

Thus, by taking

$$\varphi(x) := e^{-i\,x \cdot \xi},$$

we have the formula

$$\widehat{u}(\xi',\xi_n) = \left\langle u, e^{-i\,x\cdot\xi} \right\rangle = \left\langle u, e^{-i\,x'\cdot\xi'}\, e^{-i\,x_n\,\xi_n} \right\rangle$$

$$= \left\langle u, e^{-i\,x'\cdot\xi'}\, e^{-i\,x_n\,\mathrm{Re}\,\xi_n}\, e^{x_n\,\mathrm{Im}\,\xi_n} \right\rangle.$$

However, since $u = 0$ in $\mathbf{R}_+^n$, it follows that

$$e^{x_n\,\mathrm{Im}\,\xi_n} \le 1 \quad \text{for all } x \in \operatorname{supp} u \text{ and } \mathrm{Im}\,\xi_n \ge 0.$$

Therefore, we have, by inequality (4.9),

$$|\widehat{u}(\xi)| \le C\,(1+|\xi|)^{\ell} \quad \text{for all } x \in \operatorname{supp} u \text{ and } \mathrm{Im}\,\xi_n \ge 0,$$

just as in the proof of the Paley–Wiener–Schwartz theorem.
The proof of Claim 4.22 is complete. $\quad\square$

Step (4): Now we make a contour deformation in the integral

$$\langle Bu, \Phi \rangle = \frac{1}{(2\pi)^n} \int_{|\xi|\ge 1} \widehat{u}(\xi) F(\xi)\, d\xi$$

$$= \frac{1}{(2\pi)^n} \int_{|\xi|\ge 1} \widehat{u}(\xi',\xi_n) F(\xi',\xi_n)\, d\xi'd\xi_n.$$

We consider the two cases for $|\xi| \ge 1$:
(a) $|\xi'| \ge 1$ and $-\infty < \xi_n < \infty$.
(b) $|\xi'| < 1$ and $\xi_n \ge \sqrt{1-|\xi'|^2}$ and $\xi_n \le -\sqrt{1-|\xi'|^2}$.
Case (a): In this case, we take a positive contour $\Gamma_{\xi'}$ in the upper half-plane $\{\mathrm{Im}\,\xi_n > 0\}$ enclosing the poles of $a(x,\xi',\xi_n)$ of ξ_n. Then we have, by Cauchy's theorem,

$$\int_{-R}^{R} \widehat{u}(\xi',\xi_n) F(\xi',\xi_n)\, d\xi_n + \int_{C_R} \widehat{u}(\xi',\xi_n) F(\xi',\xi_n)\, d\xi_n$$

$$= \int_{\Gamma_{\xi'}} \widehat{u}(\xi',\xi_n)\, F(\xi',\xi_n)\, d\xi_n.$$

However, by Claims 4.21 and 4.22 it follows that

$$\lim_{R\to\infty} \int_{C_R} \widehat{u}(\xi',\xi_n)\, F(\xi',\xi_n)\, d\xi_n = 0.$$

Hence, by passing to the limit we obtain that

$$\int_{-\infty}^{\infty} \widehat{u}(\xi',\xi_n)\, F(\xi',\xi_n)\, d\xi_n = \int_{\Gamma_{\xi'}} \widehat{u}(\xi',\xi_n) F(\xi',\xi_n)\, d\xi_n \quad \text{for all } |\xi' \ge 1.$$

Case (b): In this case, we take a positive contour $\Gamma_{\xi'}$ in the upper half-plane $\{\mathrm{Im}\,\xi_n > 0\}$ enclosing the poles of $a(x,\xi',\xi_n)$ of ξ_n, completed by the segment

$$\left[-\sqrt{1-|\xi'|^2},\sqrt{1-|\xi'|^2}\right].$$

Similarly, we have, by Cauchy's theorem,

$$\int_{\sqrt{1-|\xi'|^2}}^{R}\widehat{u}(\xi',\xi_n)F(\xi',\xi_n)\,d\xi_n + \int_{C_R}\widehat{u}(\xi',\xi_n)F(\xi',\xi_n)\,d\xi_n$$

$$+ \int_{-R}^{-\sqrt{1-|\xi'|^2}}\widehat{u}(\xi',\xi_n)F(\xi',\xi_n)\,d\xi_n$$

$$= \int_{\Gamma_{\xi'}}\widehat{u}(\xi',\xi_n)F(\xi',\xi_n)\,d\xi_n.$$

Hence, by passing to the limit we obtain that

$$\int_{-\infty}^{-\sqrt{1-|\xi'|^2}}\widehat{u}(\xi',\xi_n)F(\xi',\xi_n)\,d\xi_n + \int_{\sqrt{1-|\xi'|^2}}^{\infty}\widehat{u}(\xi',\xi_n)F(\xi',\xi_n)\,d\xi_n$$

$$= \int_{\Gamma_{\xi'}}\widehat{u}(\xi',\xi_n)F(\xi',\xi_n)\,d\xi_n \quad \text{for all } |\xi'| < 1.$$

Summing up, we have proved the formula

$$\langle Bu,\Phi\rangle = \frac{1}{(2\pi)^n}\int_{|\xi|\geq 1}\widehat{u}(\xi',\xi_n)F(\xi',\xi_n)\,d\xi'd\xi_n \tag{4.10}$$

$$= \frac{1}{(2\pi)^n}\int_{|\xi'|\geq 1}\left(\int_{-\infty}^{\infty}\widehat{u}(\xi',\xi_n)\,F(\xi',\xi_n)\,d\xi_n\right)d\xi'$$

$$+ \frac{1}{(2\pi)^n}\int_{|\xi'|<1}\left(\int_{-\infty}^{-\sqrt{1-|\xi'|^2}}\widehat{u}(\xi',\xi_n)\,F(\xi',\xi_n)\,d\xi_n\right)d\xi'$$

$$+ \frac{1}{(2\pi)^n}\int_{|\xi'|<1}\left(\int_{\sqrt{1-|\xi'|^2}}^{\infty}\widehat{u}(\xi',\xi_n)F(\xi',\xi_n)\,d\xi_n\right)d\xi'$$

$$= \frac{1}{(2\pi)^n}\int_{\mathbf{R}^{n-1}}\int_{\Gamma_{\xi'}}\widehat{u}(\xi',\xi_n)\,F(\xi',\xi_n)\,d\xi_nd\xi'.$$

Step (5): We remark that the symbol $a(x,\xi)$ is positively homogeneous of order μ with respect to the variable ξ. Hence, without loss of generality we may assume that

$$\Gamma_{\xi'} = |\xi'|\,\Gamma_1 \quad \text{for all } |\xi'| \geq 1,$$

or equivalently, we have, for all $|\xi'| \geq 1$,

$$\xi_n \in \Gamma_{\xi'} \Longleftrightarrow \xi_n = |\xi'|\,\eta_n, \ \eta_n = \frac{\xi_n}{|\xi'|} \in \Gamma_1.$$

This implies that there exists a constant $c_0 > 0$ such that

$$|\xi_n| \le c_0 \left(1 + |\xi'|\right) \quad \text{for all } \xi_n \in \Gamma_{\xi'} \text{ and } |\xi'| \ge 1.$$

Step (6): If we let

$$\Phi(x', x_n) = \varphi(x') \otimes \psi(x_n) \in C_0^\infty(\mathbf{R}_+^n)$$

with

$$\varphi(x') \in C_0^\infty(\mathbf{R}^{n-1}), \quad \psi(x_n) \in C_0^\infty(\mathbf{R}^+),$$

we obtain from formula (4.10) that

$$\langle Bu, \varphi(x') \otimes \psi(x_n) \rangle \tag{4.11}$$

$$= \frac{1}{(2\pi)^n} \int_{\mathbf{R}^{n-1}} \left(\int_{\Gamma_{\xi'}} \widehat{u}(\xi', \xi_n) \right.$$

$$\times \left. \left(\int_{\mathbf{R}^n} e^{i x \cdot \xi} a\,(x', x_n, \xi', \xi_n)\, \varphi(x')\, \psi(x_n)\, dx'\, dx_n \right) d\xi_n \right) d\xi'.$$

Moreover, we have the following claim:

Claim 4.23. *The formula* (4.11) *can be explicitly expressed in the form*

$$\langle Bu, \varphi(x') \otimes \psi(x_n) \rangle = \frac{1}{(2\pi)^{n-1}} \int_0^\infty \psi(x_n) \left(\int_{\mathbf{R}^{n-1}} G(x_n, \xi')\, d\xi' \right) dx_n$$

where the function

$$G(x_n, \xi')$$

$$= \frac{1}{2\pi} \int_{\Gamma_{\xi'}} e^{i x_n \xi_n}\, \widehat{u}(\xi', \xi_n) \left(\int_{\mathbf{R}^{n-1}} e^{i x' \cdot \xi'} a\,(x', x_n, \xi', \xi_n)\, \varphi(x')\, dx' \right) d\xi_n$$

is rapidly decreasing with respect to the variable ξ'.

Proof. We consider Friedrichs' mollifiers (see [122, Subsection 5.2.6], [136, Subsection 1.3.2]). Let $\rho(x)$ be a non-negative, bell-shaped C^∞ function on $\mathbf{R}^n$ satisfying the following two conditions (4.12a) and (4.12b):

$$\operatorname{supp} \rho = \{x \in \mathbf{R}^n : |x| \le 1\}. \tag{4.12a}$$

$$\int_{\mathbf{R}^n} \rho(x)\, dx = 1. \tag{4.12b}$$

For example, we may take

$$\rho(x) = \begin{cases} C \exp[-1/(1 - |x|^2)] & \text{if } |x| < 1, \\ 0 & \text{if } |x| \ge 1, \end{cases}$$

where the constant factor C is so chosen that condition (4.12b) is satisfied.

For each $\varepsilon > 0$, we define

$$\rho_\varepsilon(x) = \frac{1}{\varepsilon^n} \rho\left(\frac{x}{\varepsilon}\right),$$

then $\rho_\varepsilon(x)$ is a non-negative, C^∞ function on $\mathbf{R}^n$, and satisfies the conditions

$$\operatorname{supp} \rho_\varepsilon = \{x \in \mathbf{R}^n : |x| \leq \varepsilon\}; \tag{4.13a}$$

$$\int_{\mathbf{R}^n} \rho_\varepsilon(x)\, dx = 1. \tag{4.13b}$$

The functions $\{\rho_\varepsilon\}$ are called *Friedrichs' mollifiers* (see Figure 4.12 below).

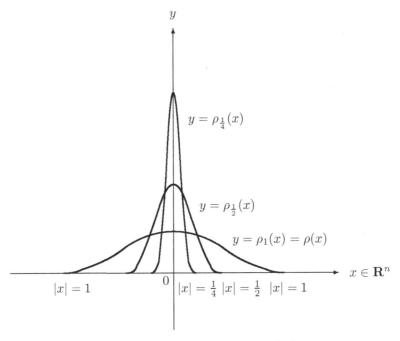

Fig. 4.12. Friedrichs' mollifiers $\{\rho_\varepsilon\}$

Now we let

$$\rho_k(x) := k^n\, \rho(kx) \quad \text{for all integer } k > 0.$$

Since $u \in \mathcal{E}'(\mathbf{R}^n)$ satisfies the condition

$$\{x = (x', x_n) \in \mathbf{R}^n : x_n \leq 0\},$$

it follows that

$$\begin{cases} \rho_k * u \in C_0^\infty(\mathbf{R}^n), \\ \operatorname{supp}(\rho_k * u) \subset \{x = (x', x_n) \in \mathbf{R}^n : x_n \leq 0\}, \\ \rho_k * u \longrightarrow u \quad \text{in } \mathcal{E}'(\mathbf{R}^n) \text{ as } k \to \infty. \end{cases}$$

Hence we have the assertion

$$B\left(\rho_k * u\right) \longrightarrow Bu \quad \text{in } \mathcal{D}'(\mathbf{R}^n) \text{ as } k \to \infty,$$

and so

$$
\begin{aligned}
&\langle Bu, \varphi(x') \otimes \psi(x_n)\rangle \\
&= \lim_{k \to \infty} \langle B\left(\rho_k * u\right), \varphi(x') \otimes \psi(x_n)\rangle \\
&= \lim_{k \to \infty} \frac{1}{(2\pi)^n} \int_{\mathbf{R}^{n-1}} \left(\int_{\Gamma_{\xi'}} \widehat{\rho_k}(\xi)\, \widehat{u}(\xi', \xi_n) \right. \\
&\quad \times \left. \left(\int_{\mathbf{R}^n} e^{i\,x\cdot\xi} a\left(x', x_n, \xi', \xi_n\right) \varphi(x')\psi(x_n)\, dx'\, dx_n \right) d\xi_n \right) d\xi'.
\end{aligned}
$$

Here we remark the following three facts (a), (b) and (c):

(a) $\widehat{\rho_k}(\xi) = \widehat{\rho}(\xi/k) \in \mathcal{S}(\mathbf{R}^n)$ for all integer $k > 0$.
(b) $|\xi_n| \le c_0\left(1 + |\xi'|\right)$ for all $\xi_n \in \Gamma_{\xi'}$ and $|\xi'| \ge 1$.
(c) $\Gamma_{\xi'} = |\xi'|\, \Gamma_1$ for all $|\xi'| \ge 1$.

Hence, by using Fubini's theorem we obtain that

$$
\begin{aligned}
&\frac{1}{(2\pi)^n} \int_{\mathbf{R}^{n-1}} \left(\int_{\Gamma_{\xi'}} \widehat{\rho_k}(\xi', \xi_n)\, \widehat{u}(\xi', \xi_n) \right. \\
&\quad \times \left. \left(\int_{\mathbf{R}^n} e^{i\,x\cdot\xi} a\left(x', x_n, \xi', \xi_n\right) \varphi(x')\, \psi(x_n)\, dx'\, dx_n \right) d\xi_n \right) d\xi' \\
&= \frac{1}{(2\pi)^n} \int_0^\infty \psi(x_n) \left(\int_{\mathbf{R}^{n-1}} \left(\int_{\Gamma_{\xi'}} e^{i\,x_n\,\xi_n}\, \widehat{\rho}\left(\frac{(\xi', \xi_n)}{k} \right) \widehat{u}(\xi', \xi_n) \right. \right. \\
&\quad \times \left. \left. \left(\int_{\mathbf{R}^{n-1}} e^{i\,x'\cdot\xi'}\, a(x', x_n, \xi', \xi_n)\varphi(x')\, dx' \right) d\xi_n \right) d\xi' \right) dx_n.
\end{aligned}
$$

However, we have the following assertion (d):

(d) The function

$$\int_{\mathbf{R}^{n-1}} e^{i\,x'\cdot\xi'}\, a\left(x', x_n, \xi', \xi_n\right) \varphi(x')\, dx'$$

is rapidly decreasing with respect to ξ', since $\varphi(x') \in C_0^\infty(\mathbf{R}^{n-1})$.

Indeed, by integration by parts it suffices to note that

$$
\begin{aligned}
&\xi'^{\alpha'} \int_{\mathbf{R}^{n-1}} e^{i\,x'\cdot\xi'}\, a(x', x_n, \xi', \xi_n)\, \varphi(x')\, dx' \\
&= \int_{\mathbf{R}^{n-1}} D_{x'}^{\alpha'}\left(e^{i\,x'\cdot\xi'} \right) a\left(x', x_n, \xi', \xi_n\right) \varphi(x')\, dx' \\
&= (-1)^{|\alpha'|} \int_{\mathbf{R}^{n-1}} \left(e^{i\,x'\cdot\xi'} \right) D_{x'}^{\alpha'}\left(a\left(x', x_n, \xi', \xi_n\right) \varphi(x')\right) dx'.
\end{aligned}
$$

Hence, we find that the function

$$
\begin{aligned}
G_k(x_n, \xi') = \frac{1}{2\pi} \int_{\Gamma_{\xi'}} e^{i\,x_n\,\xi_n}\, \widehat{\rho}\left(\frac{(\xi', \xi_n)}{k}\right)\, \widehat{u}(\xi', \xi_n) \\
\times \left(\int_{\mathbf{R}^{n-1}} e^{i\,x'\cdot\xi'}\, a\,(x', x_n, \xi', \xi_n)\, \varphi(x')\, dx'\right) d\xi_n
\end{aligned}
$$

is rapidly decreasing with respect to the variable ξ'. Moreover, we have the inequality

$$
\left|\widehat{\rho}\left(\frac{(\xi', \xi_n)}{k}\right)\right| \le \widehat{\rho}(0) = \int_{\mathbf{R}^n} \rho(x)\, dx = 1 \quad \text{for all integer } k > 0.
$$

Therefore, by applying Lebesgue's dominated convergence theorem we obtain that

$$
\begin{aligned}
&\langle Bu, \varphi(x') \otimes \psi(x_n)\rangle \\
&= \lim_{k\to\infty} \frac{1}{(2\pi)^n} \int_{\mathbf{R}^{n-1}} \left(\int_{\Gamma_{\xi'}} \widehat{\rho_k}(\xi)\, \widehat{u}(\xi', \xi_n)\right. \\
&\qquad\left. \times \left(\int_{\mathbf{R}^n} e^{i\,x\cdot\xi}\, a\,(x', x_n, \xi', \xi_n)\, \varphi(x')\, \psi(x_n)\, dx'\, dx_n\right) d\xi_n\right) d\xi' \\
&= \lim_{k\to\infty} \frac{1}{(2\pi)^n} \int_0^\infty \psi(x_n) \left(\int_{\mathbf{R}^{n-1}} \left(\int_{\Gamma_{\xi'}} e^{i\,x_n\,\xi_n}\, \widehat{\rho}\left(\frac{(\xi', \xi_n)}{k}\right)\, \widehat{u}(\xi', \xi_n)\right.\right. \\
&\qquad\left.\left. \times \left(\int_{\mathbf{R}^{n-1}} e^{i\,x'\cdot\xi'}\, a\,(x', x_n, \xi', \xi_n)\, \varphi(x')\, dx'\right) d\xi_n\right) d\xi'\right) dx_n \\
&= \frac{1}{(2\pi)^n} \int_0^\infty \psi(x_n) \left(\int_{\mathbf{R}^{n-1}} \left(\int_{\Gamma_{\xi'}} e^{i\,x_n\,\xi_n}\, \widehat{u}(\xi', \xi_n)\right.\right. \\
&\qquad\left.\left. \times \left(\int_{\mathbf{R}^{n-1}} e^{i\,x'\cdot\xi'}\, a\,(x', x_n, \xi', \xi_n)\, \varphi(x')\, dx'\right) d\xi_n\right) d\xi'\right) dx_n \\
&= \frac{1}{(2\pi)^{n-1}} \int_0^\infty \psi(x_n) \left(\int_{\mathbf{R}^{n-1}} G(x_n, \xi')\, d\xi'\right) dx_n,
\end{aligned}
$$

where

$$
\begin{aligned}
&G(x_n, \xi') \\
&= \frac{1}{2\pi} \int_{\Gamma_{\xi'}} e^{i\,x_n\,\xi_n}\, \widehat{u}(\xi', \xi_n) \left(\int_{\mathbf{R}^{n-1}} e^{i\,x'\cdot\xi'}\, a\,(x', x_n, \xi', \xi_n)\, \varphi(x')\, dx'\right) d\xi_n.
\end{aligned}
$$

The proof of Claim 4.23 is complete. $\square$

Step (7): By virtue of Claim 4.23, we find that the distribution

$$
[0, \infty) \ni x_n \longmapsto Bu(\cdot, x_n)|_{\mathbf{R}_+^n}
$$

is equal to a function in the space $C^\infty\left([0,\infty); \mathcal{D}'(\mathbf{R}^{n-1})\right)$, given by the formula

$$C_0^\infty(\mathbf{R}^{n-1}) \ni \varphi \longmapsto \frac{1}{(2\pi)^{n-1}} \int_{\mathbf{R}^{n-1}} \left(\int_{\Gamma_{\xi'}} e^{i\,x_n\,\xi_n}\,\widehat{u}(\xi',\xi_n) \right.$$

$$\times \left. \left(\int_{\mathbf{R}^{n-1}} e^{i\,x'\cdot\xi'}\,a\,(x',x_n,\xi',\xi_n)\,\varphi(x')\,dx' \right) d\xi_n \right) d\xi'.$$

Now the proof of Theorem 4.19 is complete. $\square$

Remark 4.24. It is easy to see that the function $G(x_n, \xi')$ is rapidly decreasing with respect to the variable ξ'.

4.3.2 Jump Formulas

If $u \in \mathcal{D}'(\Omega)$ has a sectional trace on $\partial\Omega$ of order zero, we can define its extension u^0 in $\mathcal{D}'(M)$ as follows: Choose functions $\theta \in C_0^\infty(W)$ and $\psi \in C_0^\infty(\Omega)$ such that $\theta + \psi = 1$ on $\overline{\Omega}$, and define u^0 by the formula

$$\langle u^0, \varphi \cdot \mu \rangle = \int_0^1 \langle u(t), (\theta\varphi)(\cdot, t) \cdot \omega \rangle \, dt + \langle u, \psi\varphi \cdot \mu \rangle, \quad \varphi \in C^\infty(M).$$

The distribution u^0 is an extension to M of u which is equal to zero in $M \setminus \overline{\Omega}$.

If $v \in \mathcal{D}'(\partial\Omega)$, we define a multiple layer

$$v \otimes D_t^j \delta_{\partial\Omega} \quad \text{for } j = 0, 1, \ldots,$$

by the formula

$$\left\langle v \otimes D_t^j \delta_{\partial\Omega}, \varphi \cdot \mu \right\rangle = (-1)^j \left\langle v, D_t^j \varphi(\cdot, 0) \cdot \omega \right\rangle \quad \text{for all } \varphi \in C^\infty(M).$$

It is clear that $v \otimes D_t^j \delta$ is a distribution on M with support in $\partial\Omega$.

Let P be a differential operator of order m with C^∞ coefficients on M. In a neighborhood of $\partial\Omega$, we can write uniquely $P = P(x, D_x)$ in the form

$$P(x, D_x) = \sum_{j=0}^m P_j(x, D_{x'})D_t^j, \quad x = (x', t),$$

where $P_j(x, D_{x'})$ is a differential operator of order $m - j$ acting along the surfaces parallel to the boundary $\partial\Omega$.

Then we have the following two formulas (1) and (2):

(1) If $u \in \mathcal{D}'(\Omega)$ has sectional traces on $\partial\Omega$ up to order j, then we have the formula

$$D_t^j(u^0) = (D_t^j u)^0 + \frac{1}{\sqrt{-1}} \sum_{k=0}^{j-1} \gamma_{j-k-1} u \otimes D_t^k \delta_{\partial\Omega}. \tag{4.14}$$

(2) If $u \in \mathcal{D}'(\Omega)$ has sectional traces on $\partial\Omega$ up to order m, then we have the jump formula

$$P\left(u^0\right) = (Pu)^0 + \frac{1}{\sqrt{-1}} \sum_{\ell+k+1\leq m} P_{\ell+k+1}(x, D_{x'})\gamma_\ell u \otimes D_t^k \delta_{\partial\Omega}. \tag{4.15}$$

4.4 Fourier Integral Operators

In this section, we present a brief description of basic concepts and results of the theory of Fourier integral operators.

4.4.1 Symbol Classes

First, we introduce symbol classes for Fourier integral operators:

Definition 4.25. *Let Ω be an open subset of $\mathbf{R}^n$. If $m \in \mathbf{R}$ and $0 \leq \delta < \rho \leq 1$, we let*

$$S^m_{\rho,\delta}(\Omega \times \mathbf{R}^N) = \text{the set of all functions } a(x,\theta) \in C^\infty(\Omega \times \mathbf{R}^N) \text{ with}$$

the property that, for any compact $K \subset \Omega$ and

any multi-indices α, β there exists a constant $C_{K,\alpha,\beta} > 0$

such that we have, for all $x \in K$ and $\theta \in \mathbf{R}^N$,

$$\left| \partial_\theta^\alpha \partial_x^\beta a(x,\theta) \right| \leq C_{K,\alpha,\beta} (1 + |\theta|)^{m - \rho|\alpha| + \delta|\beta|}.$$

If K is a compact subset of Ω and if j is a non-negative integer, we define a seminorm $p_{K,j,m}$ on $S^m_{\rho,\delta}(\Omega \times \mathbf{R}^N)$ by the formula

$$S^m_{\rho,\delta}(\Omega \times \mathbf{R}^N) \ni a \longmapsto p_{K,j,m}(a) = \sup_{\substack{x \in K, \theta \in \mathbf{R}^N \\ |\alpha| \leq j}} \frac{\left| \partial_\theta^\alpha \partial_x^\beta a(x,\theta) \right|}{(1 + |\theta|)^{m - \rho|\alpha| + \delta|\beta|}}.$$

We equip the space $S^m_{\rho,\delta}(\Omega \times \mathbf{R}^N)$ with the topology defined by the family $\{p_{K,j,m}\}$ of seminorms where K ranges over all compact subsets of Ω and $j = 0, 1, \ldots$. The space $S^m_{\rho,\delta}(\Omega \times \mathbf{R}^N)$ is a Fréchet space.

The elements of $S^m_{\rho,\delta}(\Omega \times \mathbf{R}^N)$ are called symbols *of order m. We drop the $\Omega \times \mathbf{R}^N$ and use $S^m_{\rho,\delta}$ when the context is clear.*

We give simple and typical examples of symbols:

Example 4.5. (1) A polynomial $p(x,\xi) = \sum_{|\alpha| \leq m} a_\alpha(x) \xi^\alpha$ of order m with coefficients in $C^\infty(\Omega)$ is in $S^m_{1,0}(\Omega \times \mathbf{R}^n)$.

(2) If $m \in \mathbf{R}$, the function

$$\Omega \times \mathbf{R}^n \ni (x,\xi) \longmapsto \left(1 + |\xi|^2\right)^{m/2}$$

is in $S^m_{1,0}(\Omega \times \mathbf{R}^n)$.

(3) A function $a(x,\theta) \in C^\infty(\Omega \times (\mathbf{R}^N \setminus \{0\}))$ is said to be *positively homogeneous* of degree m in θ if it satisfies the condition

$$a(x,t\theta) = t^m a(x,\theta) \quad \text{for all } t > 0 \text{ and } \theta \in \mathbf{R}^N \setminus \{0\}.$$

If $a(x,\theta)$ is positively homogeneous of degree m in θ and if $\varphi(\theta)$ is a smooth function such that $\varphi(\theta) = 0$ for $|\theta| \leq 1/2$ and $\varphi(\theta) = 1$ for $|\theta| \geq 1$, then the function $\varphi(\theta) a(x,\theta)$ is in $S^m_{1,0}(\Omega \times \mathbf{R}^N)$.

We set

$$S^{-\infty}(\Omega \times \mathbf{R}^N) = \bigcap_{m \in \mathbf{R}} S_{\rho,\delta}^m(\Omega \times \mathbf{R}^N).$$

The next theorem gives a meaning to a formal sum of symbols of decreasing order:

Theorem 4.26. *Let* $a_j(x, \theta) \in S_{\rho,\delta}^{m_j}(\Omega \times \mathbf{R}^N)$, $m_j \downarrow -\infty$, $j = 0, 1, \ldots$. *Then there exists a symbol* $a(x, \theta) \in S_{\rho,\delta}^{m_0}(\Omega \times \mathbf{R}^N)$, *unique modulo* $S^{-\infty}(\Omega \times \mathbf{R}^N)$, *such that we have, for all positive integer* k,

$$a(x, \theta) - \sum_{j=0}^{k-1} a_j(x, \theta) \in S_{\rho,\delta}^{m_k}(\Omega \times \mathbf{R}^N). \tag{4.16}$$

If formula (4.16) holds true, we write

$$a(x, \theta) \sim \sum_{j=0}^{\infty} a_j(x, \theta).$$

The formal sum $\sum_{j=0}^{\infty} a_j(x, \theta)$ is called an asymptotic expansion of $a(x, \theta)$.

Now we introduce the most important symbol class that naturally enters in the study of elliptic boundary value problems:

Definition 4.27. *A symbol* $a(x, \theta) \in S_{1,0}^m(\Omega \times \mathbf{R}^N)$ *is said to be* classical *if there exist smooth functions* $a_j(x, \theta)$, *positively homogeneous of degree* $m - j$ *in* θ *for* $|\theta| \geq 1$, *such that we have, for all positive integer* k,

$$a(x, \theta) - \sum_{j=0}^{k-1} a_j(x, \theta) \in S_{1,0}^{m-k}(\Omega \times \mathbf{R}^N).$$

The homogeneous function $a_0(x, \theta)$ *of degree* m *is called the* principal part *of* $a(x, \theta)$.

We let

$$S_{\mathrm{cl}}^m(\Omega \times \mathbf{R}^N) = \text{the set of all classical symbols of order } m.$$

For example, the symbols in Example 4.5 are all classical, and they have respectively as principal part the following functions:

(1) $p_m(x, \xi) = \sum_{|\alpha|=m} a_\alpha(x) \xi^\alpha$.
(2) $|\xi|^m$.
(3) $a(x, \theta)$.

A symbol $a(x, \theta)$ in $S_{\rho,\delta}^m(\Omega \times \mathbf{R}^N)$ is said to be *elliptic* of order m if, for any compact $K \subset \Omega$, there exists a positive constant C_K such that

$$|a(x, \theta)| \geq C_K(1 + |\theta|)^m \quad \text{for all } x \in K \text{ and } |\theta| \geq \frac{1}{C_K}.$$

There is a simple criterion for ellipticity in the case of classical symbols:

Theorem 4.28. *Let $a(x,\theta)$ be in $S_{cl}^m(\Omega \times \mathbf{R}^N)$ with principal part $a_0(x,\theta)$. Then $a(x,\theta)$ is* elliptic *if and only if it satisfies the condition*

$$a_0(x,\theta) \neq 0 \quad \text{for all } x \in \Omega \text{ and } |\theta| = 1.$$

4.4.2 Phase Functions

Secondly, we introduce phase functions for Fourier integral operators:

Definition 4.29. *Let Ω be an open subset of $\mathbf{R}^n$. A function*

$$\varphi(x,\theta) \in C^\infty \left(\Omega \times \left(\mathbf{R}^N \setminus \{0\}\right)\right)$$

is called a phase function *on $\Omega \times (\mathbf{R}^N \setminus \{0\})$ if it satisfies the following three conditions (a), (b) and (c):*

(a) $\varphi(x,\theta)$ is real-valued.
(b) $\varphi(x,\theta)$ is positively homogeneous of degree one in the variable θ.
(c) The differential $d\varphi(x,\theta)$ does not vanish on $\Omega \times (\mathbf{R}^N \setminus \{0\})$.

We give a typical example of phase functions:

Example 4.6. Let U be an open subset of $\mathbf{R}^p$ and $\Omega = U \times U$. The function

$$\varphi(x,y,\xi) = (x - y) \cdot \xi$$

is a phase function on the space $\Omega \times (\mathbf{R}^p \setminus \{0\})$ with $n = 2p$ and $N = p$.

The next lemma due to Peter Lax [75] will play a fundamental role in defining oscillatory integrals.

Lemma 4.30 (Lax). *If $\varphi(x,\theta)$ is a phase function on $\Omega \times (\mathbf{R}^N \setminus \{0\})$, then there exists a first-order differential operator*

$$L = \sum_{j=1}^N a_j(x,\theta)\frac{\partial}{\partial\theta_j} + \sum_{k=1}^n b_k(x,\theta)\frac{\partial}{\partial x_k} + c(x,\theta)$$

such that

$$L\left(e^{i\varphi}\right) = e^{i\varphi},$$

and that its coefficients $a_j(x,\theta)$, $b_k(x,\theta)$, $c(x,\theta)$ enjoy the following properties:

$$\begin{cases} a_j(x,\theta) \in S_{1,0}^0(\Omega \times \mathbf{R}^N), \\ b_k(x,\theta) \in S_{1,0}^{-1}(\Omega \times \mathbf{R}^N), \\ c(x,\theta) \in S_{1,0}^{-1}(\Omega \times \mathbf{R}^N). \end{cases}$$

Furthermore, the transpose L' of L has coefficients $a'_j(x,\theta)$, $b'_k(x,\theta)$, $c'(x,\theta)$ in the same symbol classes as $a_j(x,\theta)$, $b_k(x,\theta)$, $c(x,\theta)$, respectively.

For example, if $\varphi(x, y, \xi)$ is a phase function as in Example 4.6

$$\varphi(x, y, \xi) = (x - y) \cdot \xi \quad \text{for } (x, y) \in U \times U \text{ and } \xi \in (\mathbf{R}^p \setminus \{0\}),$$

then the operator L is given by the formula

$$L = \frac{1}{\sqrt{-1}} \frac{1 - \rho(\xi)}{2 + |x - y|^2} \left\{ \sum_{j=1}^{p} (x_j - y_j) \frac{\partial}{\partial \xi_j} + \sum_{k=1}^{p} \frac{\xi_j}{|\xi|^2} \frac{\partial}{\partial x_k} \right.$$

$$\left. + \sum_{k=1}^{p} \frac{-\xi_j}{|\xi|^2} \frac{\partial}{\partial y_k} \right\} + \rho(\xi),$$

where $\rho(\xi)$ is a function in $C_0^\infty(\mathbf{R}^p)$ such that $\rho(\xi) = 1$ for $|\xi| \leq 1$.

4.4.3 Oscillatory Integrals

If Ω is an open subset of $\mathbf{R}^n$, we let

$$S_{\rho,\delta}^\infty (\Omega \times \mathbf{R}^N) = \bigcup_{m \in \mathbf{R}} S_{\rho,\delta}^m (\Omega \times \mathbf{R}^N).$$

If $\varphi(x, \theta)$ is a phase function on $\Omega \times (\mathbf{R}^N \setminus \{0\})$, we wish to give a meaning to the integral

$$I_\varphi(aw) = \iint_{\Omega \times \mathbf{R}^N} e^{i\varphi(x,\theta)} a(x, \theta) u(x) \, dx \, d\theta, \quad u \in C_0^\infty(\Omega), \tag{4.17}$$

for each symbol $a(x, \theta) \in S_{\rho,\delta}^\infty (\Omega \times \mathbf{R}^N)$.

By Lemma 4.30, we can replace $e^{i\varphi}$ in formula (4.17) by $L(e^{i\varphi})$. Then a *formal* integration by parts gives us that

$$I_\varphi(au) = \iint_{\Omega \times \mathbf{R}^N} e^{i\varphi(x,\theta)} L'(a(x, \theta) w(x, y)) \, dx \, d\theta.$$

However, the properties of the coefficients of the transpose L' imply that L' maps $S_{\rho,\delta}^r$ continuously into $S_{\rho,\delta}^{r-\eta}$ for all $r \in \mathbf{R}$, where $\eta = \min(\rho, 1 - \delta)$. By continuing this process, we can reduce the growth of the integrand at infinity until it becomes integrable, and give a meaning to the integral (4.17) for each symbol $a(x, \theta) \in S_{\rho,\delta}^\infty (\Omega \times \mathbf{R}^n)$.

More precisely, we have the following theorem:

Theorem 4.31. *Let $0 \leq \delta < \rho \leq 1$. Then we have the following three assertions (i), (ii) and (iii):*

(i) The linear functional

$$S^{-\infty} (\Omega \times \mathbf{R}^N) \ni a \longmapsto I_\varphi(au) \in \mathbf{C}$$

extends uniquely to a linear functional ℓ on $S_{\rho,\delta}^{\infty}\left(\Omega \times \mathbf{R}^N\right)$ whose restriction to each $S_{\rho,\delta}^m\left(\Omega \times \mathbf{R}^N\right)$ is continuous. Furthermore, the restriction of the linear functional ℓ to $S_{\rho,\delta}^m\left(\Omega \times \mathbf{R}^N\right)$ is expressed as the formula

$$\ell(a) = \iint_{\Omega \times \mathbf{R}^N} e^{i\,\varphi(x,\theta)}\left(L'\right)^k\!\left(a(x,\theta)w(x,y)\right)dx\,d\theta,$$

where $k > (m+N)/\eta$ and $\eta = \min(\rho, 1-\delta)$.
(ii) For any fixed $a(x,\theta) \in S_{\rho,\delta}^m\left(\Omega \times \mathbf{R}^N\right)$, the mapping

$$C_0^{\infty}(\Omega) \ni u \longmapsto I_{\varphi}(au) = \ell(a) \in \mathbf{C} \tag{4.18}$$

is a distribution of order $\leq k$ for $k > (m+N)/\eta$ with $\eta = \min(\rho, 1-\delta)$.
(iii) Let $\chi(\theta)$ be a function in $C_0^{\infty}(\mathbf{R}^N)$ such that $\chi(\theta) = 1$ in a neighborhood of $\theta = 0$. Then we have the formula

$$\ell(a) = \lim_{j \to \infty} \iint_{\Omega \times \mathbf{R}^N} e^{i\,\varphi(x,\theta)} \chi\left(\frac{\theta}{j}\right) a(x,\theta)u(x)\,dx\,d\theta.$$

Part (iii) in Theorem 4.31 asserts that the linear functional $\ell(a)$ defined in part (i) is independent of the positive integer $k > (m+N)/\eta$ chosen, while part (i) asserts that $\ell(a)$ is independent of the function $\chi(\theta) \in C_0^{\infty}(\mathbf{R}^N)$ used in part (iii).

We call the linear functional ℓ on $S_{\rho,\delta}^{\infty}$ an *oscillatory integral*, but use the standard notation as in formula (4.17). The distribution (4.18) is called the *Fourier integral distribution* associated with the phase function $\varphi(x,\theta)$ and the amplitude $a(x,\theta)$, and will be denoted as follows:

$$\boxed{K(x) = \int_{\mathbf{R}^N} e^{i\,\varphi(x,\theta)} a(x,\theta)\,d\theta.}$$

If u is a distribution on Ω, then the *singular support* of u is the smallest closed subset of Ω outside of which u is smooth. The singular support of u is denoted by sing supp u.

The next theorem estimates the singular support of a Fourier integral distribution.

Theorem 4.32. *If $\varphi(x,\theta)$ is a phase function on $\Omega \times \left(\mathbf{R}^N \setminus \{0\}\right)$ and if $a(x,\theta)$ is in $S_{\rho,\delta}^{\infty}\left(\Omega \times \mathbf{R}^N\right)$, then the distribution*

$$K(x) = \int_{\mathbf{R}^N} e^{i\,\varphi(x,\theta)} a(x,\theta)\,d\theta \in \mathcal{D}'(\Omega)$$

satisfies the condition

$$\text{sing supp } K \subset \left\{x \in \Omega : d_\theta \varphi(x\theta) = 0 \text{ for some } \theta \in \mathbf{R}^N \setminus \{0\}\right\}.$$

4.4.4 Definitions and Basic Properties of Fourier Integral Operators

Let U and V be open subsets of $\mathbf{R}^p$ and $\mathbf{R}^q$, respectively. If $\varphi(x, y, \theta)$ is a phase function on $U \times V \times (\mathbf{R}^N \setminus \{0\})$ and if $a(x, y, \theta) \in S_{\rho,\delta}^{\infty}(U \times V \times \mathbf{R}^N)$, then there is associated a distribution $K \in \mathcal{D}'(U \times V)$ defined by the formula

$$K(x, y) = \int_{\mathbf{R}^N} e^{i\,\varphi(x,y,\theta)} a(x, y, \theta)\, d\theta.$$

By applying Theorem 4.32 to our situation, we obtain that

$$\text{sing supp}\, K$$
$$\subset \left\{ (x, y) \in U \times V : d_\theta \varphi(x, y, \theta) = 0 \text{ for some } \theta \in \mathbf{R}^N \setminus \{0\} \right\}.$$

The distribution $K \in \mathcal{D}'(U \times V)$ defines a continuous linear operator

$$A \colon C_0^\infty(V) \longrightarrow \mathcal{D}'(U)$$

by the formula

$$\langle Av, u \rangle = \langle K, u \otimes v \rangle \quad \text{for } u \in C_0^\infty(U) \text{ and } v \in C_0^\infty(V).$$

The operator A is called the *Fourier integral operator* associated with the phase function $\varphi(x, y, \theta)$ and the amplitude $a(x, y, \theta)$, and will be denoted as follows:

$$Av(x) = \iint_{V \times \mathbf{R}^N} e^{i\,\varphi(x,y,\theta)} a(x, y, \theta) v(y)\, dy\, d\theta \quad \text{for } v \in C_0^\infty(V).$$

The next theorem summarizes some basic mapping properties of the Fourier integral operator A (see Figure 4.13 below):

Theorem 4.33. *(i) If $d_{y,\theta}\varphi(x, y, \theta) \neq 0$ on $U \times V \times (\mathbf{R}^N \setminus \{0\})$, then the operator A maps $C_0^\infty(V)$ continuously into $C^\infty(U)$.*
   (ii) If $d_{x,\theta}\varphi(x, y, \theta) \neq 0$ on $U \times V \times (\mathbf{R}^N \setminus \{0\})$, then the operator A extends to a continuous linear operator on $\mathcal{E}'(V)$ into $\mathcal{D}'(U)$.
   (iii) If $d_{y,\theta}\varphi(x, y, \theta) \neq 0$ and $d_{x,\theta}\varphi(x, y, \theta) \neq 0$ on $U \times V \times (\mathbf{R}^N \setminus \{0\})$, then we have, for all $v \in \mathcal{E}'(V)$,

$$\text{sing supp}\, Av \subset$$
$$\left\{ x \in U : d_\theta \varphi(x, y, \theta) = 0 \text{ for some } y \in \text{sing supp}\, v \text{ and } \theta \in \mathbf{R}^N \setminus \{0\} \right\}.$$

4.5 Pseudo-Differential Operators

In this section, we present a brief description of the basic concepts and results of the L^p theory of pseudo-differential operators.

$$\begin{array}{ccc}
\mathcal{E}'(V) & \xrightarrow{\ \ A\ \ } & \mathcal{D}'(U) \\
\uparrow & & \uparrow \\
C_0^\infty(V) & \xrightarrow[\ \ A\ \ ]{} & C^\infty(U)
\end{array}$$

Fig. 4.13. The mapping properties of the Fourier integral operator A

4.5.1 Definitions and Basic Properties of Pseudo-Differential Operators

Let Ω_1 and Ω_2 be open subsets of $\mathbf{R}^{n_1}$ and $\mathbf{R}^{n_2}$, respectively. If $K(x_1, x_2)$ is a distribution in $\mathcal{D}'(\Omega_1 \times \Omega_2)$, we can define a continuous linear operator

$$A \in L\left(C_0^\infty(\Omega_2), \mathcal{D}'(\Omega_1)\right)$$

by the formula

$$\langle A\psi, \varphi \rangle = \langle K, \varphi \otimes \psi \rangle \quad \text{for all } \varphi \in C_0^\infty(\Omega_1) \text{ and } \psi \in C_0^\infty(\Omega_2).$$

We then write $A = \operatorname{Op}(K)$. Since the tensor space $C_0^\infty(\Omega_1) \otimes C_0^\infty(\Omega_2)$ is *sequentially dense* in $C_0^\infty(\Omega_1 \times \Omega_2)$, it follows that the mapping

$$\mathcal{D}'(\Omega_1 \times \Omega_2) \ni K \longmapsto \operatorname{Op}(K) \in L(C_0^\infty(\Omega_2), \mathcal{D}'(\Omega_1))$$

is injective. The next theorem asserts that it is also *surjective* ([26, Chapitre I, Théorème 4.4], [122, Theorem 5.36]):

Theorem 4.34 (the Schwartz kernel theorem). *If A is a continuous linear operator on $C_0^\infty(\Omega_2)$ into $\mathcal{D}'(\Omega_1)$, then there exists a unique distribution $K_A(x_1, x_2)$ in $\mathcal{D}'(\Omega_1 \times \Omega_2)$ such that $A = \operatorname{Op}(K)$.*

The distribution K_A is called the kernel *of A. Formally, we have the formula*

$$A\psi(x_1) = \int_{\Omega_2} K_A(x_1, x_2)\, \psi(x_2)\, dx_2 \quad \text{for all } \psi \in C_0^\infty(\Omega_2).$$

Now we give some important examples of distributions kernels (see Table 4.2 below and also Example 4.2):

Example 4.7. (a) Riesz potentials: $\Omega_1 = \Omega_2 = \mathbf{R}^n$ and $0 < \alpha < n$.

$$(-\Delta)^{-\alpha/2} u(x) = R_\alpha * u(x)$$
$$= \frac{\Gamma((n-\alpha)/2)}{2^\alpha \pi^{n/2} \Gamma(\alpha/2)} \int_{\mathbf{R}^n} \frac{1}{|x-y|^{n-\alpha}}\, u(y)\, dy \quad \text{for } u \in C_0^\infty(\mathbf{R}^n).$$

(b) Newtonian potentials: $\Omega_1 = \Omega_2 = \mathbf{R}^n$ for $n \geq 3$.

$$(-\Delta)^{-1} u(x) = N * u(x)$$
$$= \frac{\Gamma((n-2)/2)}{4\,\pi^{n/2}} \int_{\mathbf{R}^n} \frac{1}{|x-y|^{n-2}}\, u(y)\, dy \quad \text{for } u \in C_0^\infty(\mathbf{R}^n).$$

(c) Bessel potentials: $\Omega_1 = \Omega_2 = \mathbf{R}^n$ and $\alpha > 0$.

$$(I - \Delta)^{-\alpha/2} u(x)$$
$$= G_\alpha * u(x) = \int_{\mathbf{R}^n} G_\alpha(x-y)\, u(y)\, dy \quad \text{for } u \in C_0^\infty(\mathbf{R}^n).$$

(d) Riesz operators: $\Omega_1 = \Omega_2 = \mathbf{R}^n$ and $1 \leq j \leq n$.

$$Y_j u(x) = R_j * u(x)$$
$$= \sqrt{-1}\, \frac{\Gamma((n+1)/2)}{\pi^{(n+1)/2}} \,\text{v. p.} \int_{\mathbf{R}^n} \frac{x_j - y_j}{|x-y|^{n+1}}\, u(y)\, dy \quad \text{for } u \in C_0^\infty(\mathbf{R}^n).$$

(e) The Calderón–Zygmund integro-differential operator: $\Omega_1 = \Omega_2 = \mathbf{R}^n$.

$$(-\Delta)^{1/2} u(x) = \frac{1}{\sqrt{-1}} \sum_{j=1}^n Y_j \left(\frac{\partial u}{\partial x_j} \right)(x)$$
$$= \frac{\Gamma((n+1)/2)}{\pi^{(n+1)/2}} \sum_{j=1}^n \text{v. p.} \int_{\mathbf{R}^n} \frac{x_j - y_j}{|x-y|^{n+1}} \frac{\partial u}{\partial y_j}(y)\, dy \quad \text{for } u \in C_0^\infty(\mathbf{R}^n).$$

Now we are in a position to define pseudo-differential operators:

Definition 4.35. *Let Ω be an open subset of $\mathbf{R}^n$ and $m \in \mathbf{R}$. A pseudo-differential operator of order m on Ω is a Fourier integral operator of the form*

$$Au(x) = \iint_{\Omega \times \mathbf{R}^n} e^{i(x-y)\cdot\xi} a(x,y,\xi) u(y)\, dy d\xi \quad \text{for } u \in C_0^\infty(\Omega), \quad (4.19)$$

with some $a(x,y,\xi) \in S_{\rho,\delta}^m(\Omega \times \Omega \times \mathbf{R}^n)$. In other words, a pseudo-differential operator of order m is a Fourier integral operator associated with the phase function $\varphi(x,y,\xi) = (x-y)\cdot\xi$ and some amplitude $a(x,y,\xi) \in S_{\rho,\delta}^m(\Omega \times \Omega \times \mathbf{R}^n)$.

We let

$$L_{\rho,\delta}^m(\Omega) = \text{the set of all pseudo-differential operators of order } m \text{ on } \Omega.$$

The set $L_{\rho,\delta}^m(\Omega)$ is called the *Hörmander class*.

By applying Theorems 4.32 and 4.33 to our situation, we obtain the following three assertions (1), (2) and (3):

Operators	**Notation**	**Distribution kernels**
Riesz potential	$(-\Delta)^{-\alpha/2}$ $(0 < \alpha < n)$	$\frac{\Gamma((n-\alpha)/2)}{2^\alpha \pi^{n/2} \Gamma(\alpha/2)} \frac{1}{\|x\|^{n-\alpha}}$
Newtonian potential	$(-\Delta)^{-1}$ $(\alpha = 2)$	$\frac{\Gamma((n-2)/2)}{4\pi^{n/2}} \frac{1}{\|x\|^{n-2}}$
Bessel potential	$(I - \Delta)^{-\alpha/2}$ $(\alpha > 0)$	$G_\alpha(x)$
Riesz operator	Y_j $(1 \leq j \leq n)$	$i \frac{\Gamma((n+1)/2)}{\pi^{(n+1)/2}} \text{v. p.} \frac{x_j}{\|x\|^{n+1}}$

Table 4.2. Operators and their kernels in Example 4.7

$$\begin{array}{ccc} \mathcal{E}'(\Omega) & \xrightarrow{A} & \mathcal{D}'(\Omega) \\ \uparrow & & \uparrow \\ C_0^\infty(\Omega) & \xrightarrow[A]{} & C^\infty(\Omega) \end{array}$$

Fig. 4.14. The mapping properties of the pseudo-differential operator A

(1) A pseudo-differential operator A maps $C_0^\infty(\Omega)$ continuously into $C^\infty(\Omega)$ and extends to a continuous linear operator $A \colon \mathcal{E}'(\Omega) \to \mathcal{D}'(\Omega)$ (see Figure 4.14 above).

(2) The distribution kernel $K_A(x, y)$, defined by the formula

$$\boxed{K_A(x, y) = \frac{1}{(2\pi)^n} \int_{\mathbf{R}^n} e^{i(x-y)\cdot\xi} \, a(x, y, \xi) \, d\xi,}$$

of a pseudo-differential operator A satisfies the condition

$$\text{sing supp}\, K_A \subset \{(x, x) : x \in \Omega\},$$

that is, the kernel K_A is smooth off the diagonal $\{(x, x) : x \in \Omega\}$ in $\Omega \times \Omega$.

(3) sing supp $Au \subset$ sing supp u, $u \in \mathcal{E}'(\Omega)$. In other words, Au is smooth when-ever u is. This property is referred to as the *pseudo-local property*.

We set

$$L^{-\infty}(\Omega) = \bigcap_{m \in \mathbf{R}} L^m_{\rho,\delta}(\Omega).$$

The next theorem characterizes the class $L^{-\infty}(\Omega)$.

Theorem 4.36. *The following three conditions (i), (ii) and (iii) are equivalent:*

(i) $A \in L^{-\infty}(\Omega)$.
(ii) A is written in the form (4.19) with some $a \in S^{-\infty}(\Omega \times \Omega \times \mathbf{R}^n)$.
(iii) A is a regularizer, or equivalently, its distribution kernel $K_A(x,y)$ is in the space $C^\infty(\Omega \times \Omega)$.

The next definition plays an important role in the study of partial differential operators:

Definition 4.37. *A continuous linear operator*

$$A \colon C_0^\infty(\Omega) \longrightarrow \mathcal{D}'(\Omega)$$

is said to be properly supported *if the following two conditions (a) and (b) are satisfied (see Figures 4.15 and 4.16 below):*

(a) For any compact subset K of Ω, there exists a compact subset K' of Ω such that

$$\text{supp } v \subset K \implies \text{supp } Av \subset K'.$$

(b) For any compact subset K' of Ω, there exists a compact subset $K \supset K'$ of Ω such that

$$\text{supp } v \cap K = \emptyset \implies \text{supp } Av \cap K' = \emptyset.$$

If A is properly supported, then it maps $C^\infty(\Omega)$ continuously into itself, and extends to a continuous linear operator on $\mathcal{D}'(\Omega)$ into itself. The situation can be visualized as in Figure 4.17 below.

The next theorem states that every pseudo-differential operator can be written as the sum of a properly supported operator and a regularizer.

Theorem 4.38. *If $A \in L^m_{\rho,\delta}(\Omega)$, then we have the decomposition*

$$A = A_0 + R,$$

where $A_0 \in L^m_{\rho,\delta}(\Omega)$ is properly supported and $R \in L^{-\infty}(\Omega)$.

Proof. Take a function $\sigma \in C^\infty(\Omega \times \Omega)$ (see [41, Proposition (8.15)]) that satisfies the following two conditions (a) and (b):

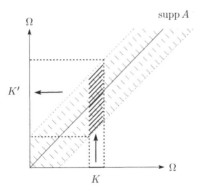

Fig. 4.15. Condition (a) in Definition 4.37

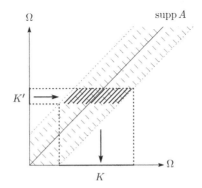

Fig. 4.16. Condition (b) in Definition 4.37

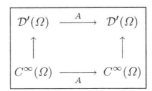

Fig. 4.17. The mapping properties of the properly supported, pseudo-differential operator A

(a) $\sigma(x,y) = 1$ in a neighborhood of the diagonal $\Delta_\Omega = \{(x,x) : x \in \Omega\}$ in $\Omega \times \Omega$;

(b) The restrictions to supp σ of the projections

$$p_i \colon \Omega \times \Omega \ni (x_1, x_2) \longmapsto x_i \in \Omega \quad \text{for } i = 1, 2$$

are proper mappings.

Then it is easy to see that the operators A_0 and R, defined respectively by the distribution kernels

$$K_{A_0}(x, y) = \sigma(x, y)\, K_A(x, y),$$
$$K_R(x, y) = (1 - \sigma(x, y))\, K_A(x, y),$$

are the desired ones, since the distribution kernel K_A is of class C^∞ off the diagonal Δ_Ω.

The proof of Theorem 4.38 is complete. □

If $p(x, \xi) \in S^m_{\rho,\delta}(\Omega \times \mathbf{R}^n)$, then the operator $p(x, D)$, defined by the formula

$$p(x, D)u(x) = \frac{1}{(2\pi)^n} \int_{\mathbf{R}^n} e^{i\,x\cdot\xi}\, p(x, \xi)\, \widehat{u}(\xi)\, d\xi \quad \text{for } u \in C_0^\infty(\Omega), \qquad (4.20)$$

is a pseudo-differential operator of order m on Ω, that is, $p(x, D) \in L^m_{\rho,\delta}(\Omega)$.

The next theorem asserts that every properly supported pseudo-differential operator can be reduced to the form (4.20).

Theorem 4.39. *If $A \in L^m_{\rho,\delta}(\Omega)$ is properly supported, then we have the assertion*

$$p(x, \xi) = e^{-i\,x\cdot\xi}\, A(e^{i\,x\cdot\xi}) \in S^m_{\rho,\delta}(\Omega \times \mathbf{R}^n),$$

and

$$A = p(x, D).$$

Furthermore, if $a(x, y, \xi) \in S^m_{\rho,\delta}(\Omega \times \Omega \times \mathbf{R}^n)$ is an amplitude for A, then we have the following asymptotic expansion:

$$p(x, \xi) \sim \sum_{\alpha \geq 0} \frac{1}{\alpha!} \partial_\xi^\alpha D_y^\alpha \left(a(x, y, \xi)\right)\bigg|_{y=x}.$$

The function $p(x, \xi)$ is called the complete symbol *of A.*

We can extend the notion of a complete symbol to the whole space $L^m_{\rho,\delta}(\Omega)$ in the following way: If $A \in L^m_{\rho,\delta}(\Omega)$, then we choose a properly supported operator $A_0 \in L^m_{\rho,\delta}(\Omega)$ such that $A - A_0 \in L^{-\infty}(\Omega)$, and define

$$\sigma(A) = \text{the equivalence class of the complete symbol of } A_0$$
$$\text{in } S^m_{\rho,\delta}(\Omega \times \mathbf{R}^n)/S^{-\infty}(\Omega \times \mathbf{R}^n).$$

In view of Theorems 4.38 and 4.39, it follows that $\sigma(A)$ does not depend on the operator A_0 chosen. The equivalence class $\sigma(A)$ is called the *complete symbol* of A. It is easy to see that the mapping

$$L^m_{\rho,\delta}(\Omega) \ni A \longmapsto \sigma(A) \in S^m_{\rho,\delta}(\Omega \times \mathbf{R}^n)/S^{-\infty}(\Omega \times \mathbf{R}^n)$$

induces an isomorphism

$$L^m_{\rho,\delta}(\Omega)/L^{-\infty}(\Omega) \longrightarrow S^m_{\rho,\delta}(\Omega \times \mathbf{R}^n)/S^{-\infty}(\Omega \times \mathbf{R}^n).$$

We shall often identify the complete symbol $\sigma(A)$ with a representative in the class $S^m_{\rho,\delta}(\Omega \times \mathbf{R}^n)$ for notational convenience, and call any member of $\sigma(A)$ a complete symbol of A.

Now we introduce the most important Hörmander class that enter naturally in connection with the classical symbol class:

Definition 4.40. *A pseudo-differential operator* $A \in L^m_{1,0}(\Omega)$ *is said to be* classical *if its complete symbol* $\sigma(A)$ *has a representative in the class* $S^m_{\mathrm{cl}}(\Omega \times \mathbf{R}^n)$.

We let

$$L^m_{\mathrm{cl}}(\Omega) = \text{the set of all classical pseudo-differential operators}$$
$$\text{of order } m \text{ on } \Omega.$$

Then the mapping

$$L^m_{\mathrm{cl}}(\Omega) \ni A \longmapsto \sigma(A) \in S^m_{\mathrm{cl}}(\Omega \times \mathbf{R}^n)/S^{-\infty}(\Omega \times \mathbf{R}^n)$$

induces an isomorphism

$$L^m_{\mathrm{cl}}(\Omega)/L^{-\infty}(\Omega) \longrightarrow S^m_{\mathrm{cl}}(\Omega \times \mathbf{R}^n)/S^{-\infty}(\Omega \times \mathbf{R}^n).$$

Also we have the assertion

$$L^{-\infty}(\Omega) = \bigcap_{m \in \mathbf{R}} L^m_{\mathrm{cl}}(\Omega).$$

If $A \in L^m_{\mathrm{cl}}(\Omega)$, then the principal part of $\sigma(A)$ has a canonical representative $\sigma_A(x,\xi) \in C^\infty(\Omega \times (\mathbf{R}^n \setminus \{0\}))$ which is positively homogeneous of degree m in the variable ξ. The function $\sigma_A(x,\xi)$ is called the *homogeneous principal symbol* of A.

The next two theorems assert that the class of pseudo-differential operators forms an algebra closed under the operations of composition of operators and taking the transpose or adjoint of an operator.

Theorem 4.41. *If* $A \in L^m_{\rho,\delta}(\Omega)$, *then its transpose* A' *and its adjoint* A^* *are both in* $L^m_{\rho,\delta}(\Omega)$, *and the complete symbols* $\sigma(A')$ *and* $\sigma(A^*)$ *have respectively the following asymptotic expansions:*

$$\sigma(A')(x,\xi) \sim \sum_{\alpha \geq 0} \frac{1}{\alpha!} \partial^\alpha_\xi D^\alpha_x \left(\sigma(A)(x,-\xi) \right),$$

$$\sigma(A^*)(x,\xi) \sim \sum_{\alpha \geq 0} \frac{1}{\alpha!} \partial^\alpha_\xi D^\alpha_x \left(\overline{\sigma(A)(x,\xi)} \right).$$

Theorem 4.42. *If $A \in L^{m'}_{\rho',\delta'}(\Omega)$ and $B \in L^{m''}_{\rho'',\delta''}(\Omega)$ where $0 \le \delta' < \rho'' \le 1$ and if one of them is properly supported, then the composition AB is in $L^{m'+m''}_{\rho,\delta}(\Omega)$ with $\rho = \min(\rho',\rho'')$, $\delta = \max(\delta',\delta'')$, and we have the following asymptotic expansion:*

$$\sigma(AB)(x,\xi) \sim \sum_{\alpha \ge 0} \frac{1}{\alpha!} \partial_\xi^\alpha \left(\sigma(A)(x,\xi) \right) \cdot D_x^\alpha \left(\sigma(B)(x,\xi) \right).$$

We introduce the special Hörmander class that are the "invertible" elements in the algebra of pseudo-differential operators:

Definition 4.43. *A pseudo-differential operator $A \in L^m_{\rho,\delta}(\Omega)$ is said to be elliptic of order m if its complete symbol $\sigma(A)$ is elliptic of order m. In view of Theorem 4.28, it follows that a classical pseudo-differential operator $A \in L^m_{cl}(\Omega)$ is elliptic if and only if its homogeneous principal symbol $\sigma_A(x,\xi)$ does not vanish on the space $\Omega \times (\mathbf{R}^n \setminus \{0\})$.*

In fact, we can prove the following theorem:

Theorem 4.44. *An operator $A \in L^m_{\rho,\delta}(\Omega)$ is elliptic if and only if there exists a properly supported operator $B \in L^{-m}_{\rho,\delta}(\Omega)$ such that:*

$$\begin{cases} AB \equiv I & mod\ L^{-\infty}(\Omega), \\ BA \equiv I & mod\ L^{-\infty}(\Omega). \end{cases}$$

Such an operator B is called a *parametrix* for A. In other words, a parametrix for A is a two-sided inverse of A modulo $L^{-\infty}(\Omega)$. We observe that a parametrix is unique modulo $L^{-\infty}(\Omega)$.

The next theorem proves the invariance of pseudo-differential operators under change of coordinates.

Theorem 4.45. *Let Ω_1 and Ω_2 be two open subsets of $\mathbf{R}^n$ and let $\chi \colon \Omega_1 \to \Omega_2$ be a C^∞ diffeomorphism. If $A \in L^m_{\rho,\delta}(\Omega_1)$, where $1 - \rho \le \delta < \rho \le 1$, then the mapping*

$$A_\chi \colon C_0^\infty(\Omega_2) \longrightarrow C^\infty(\Omega_2)$$
$$v \longmapsto A(v \circ \chi) \circ \chi^{-1}$$

is in $L^m_{\rho,\delta}(\Omega_2)$, and we have the asymptotic expansion

$$\sigma(A_\chi)(y,\eta) \sim \sum_{\alpha \ge 0} \frac{1}{\alpha!} \left(\partial_\xi^\alpha \sigma(A) \right)(x, {}^t\chi'(x) \cdot \eta) \cdot D_z^\alpha \left(e^{i\,r(x,z,\eta)} \right) \Big|_{z=x} \qquad (4.21)$$

with

$$r(x,z,\eta) = \langle \chi(z) - \chi(x) - \chi'(x) \cdot (z - x), \eta \rangle.$$

Here $x = \chi^{-1}(y)$, $\chi'(x)$ is the derivative of χ at x and ${}^t\chi'(x)$ is its transpose.

Remark 4.46. Formula (4.21) shows that

$$\sigma(A_\chi)(y, \eta) \equiv \sigma(A)\left(x, {}^t\chi'(x) \cdot \eta\right) \quad \mod S_{\rho,\delta}^{m-(\rho-\delta)}.$$

Note that the mapping

$$\Omega_2 \times \mathbf{R}^n \ni (y, \eta) \longmapsto \left(x, {}^t\chi'(x) \cdot \eta\right) \in \Omega_1 \times \mathbf{R}^n$$

is just a transition map of the cotangent bundle $T^*(\mathbf{R}^n)$. This implies that the principal symbol $\sigma_m(A)$ of $A \in L_{\rho,\delta}^m(\mathbf{R}^n)$ can be invariantly defined on $T^*(\mathbf{R}^n)$ when $1 - \rho \leq \delta < \rho \leq 1$.

The situation may be visualized as in Figure 4.18 below.

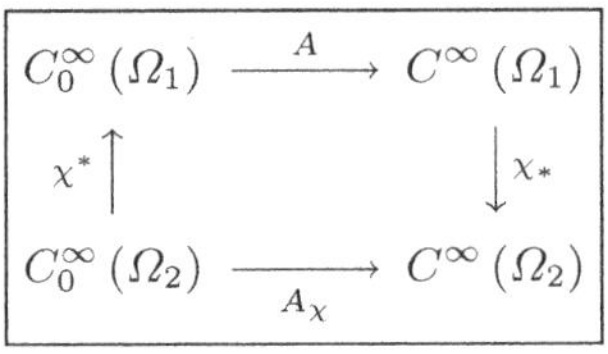

$$
\begin{array}{ccc}
C_0^\infty(\Omega_1) & \xrightarrow{\ A\ } & C^\infty(\Omega_1) \\[4pt]
{\chi^*}\Big\uparrow & & \Big\downarrow{\chi_*} \\[4pt]
C_0^\infty(\Omega_2) & \xrightarrow[\ A_\chi\ ]{} & C^\infty(\Omega_2)
\end{array}
$$

Fig. 4.18. The pseudo-differential operators A and A_χ in Theorem 4.45

Here $\chi^* v = v \circ \chi$ is the pull-back of v by χ and $\chi_* u = u \circ \chi^{-1}$ is the push-forward of u by χ, respectively.

4.5.2 L^p Boundedness of Pseudo-Differential Operators

In this subsection, we study the boundedness of pseudo-differential operators in the framework of L^p Sobolev spaces.

A differential operator of order m with smooth coefficients on Ω is continuous on $H_{\mathrm{loc}}^{s,p}(\Omega)$ (resp. $B_{\mathrm{loc}}^{s,p}(\Omega)$) into $H_{\mathrm{loc}}^{s-m,p}(\Omega)$ (resp. $B_{\mathrm{loc}}^{s-m,p}(\Omega)$) for all $s \in \mathbf{R}$. This result extends to pseudo-differential operators:

More precisely, we can obtain the following L^p *boundedness theorem* for properly supported, pseudo-differential operators due to Bourdaud [16]:

Theorem 4.47 (the Besov space boundedness theorem). *Every properly supported operator* $A \in L_{1,\delta}^m(\Omega)$ *with* $0 \leq \delta < 1$ *extends to continuous linear operators (see Figures 4.19 and 4.20 below)*

$$A \colon H_{\mathrm{loc}}^{s,p}(\Omega) \longrightarrow H_{\mathrm{loc}}^{s-m,p}(\Omega) \quad \textit{for all } s \in \mathbf{R} \textit{ and } 1 < p < \infty,$$

$$A \colon B_{\mathrm{loc}}^{s,p}(\Omega) \longrightarrow B_{\mathrm{loc}}^{s-m,p}(\Omega) \quad \textit{for all } s \in \mathbf{R} \textit{ and } 1 \leq p \leq \infty.$$

For a proof of Theorem 4.47, the reader might refer to Bourdaud [16, Theorem 1] (see also [122, Appendix A]).

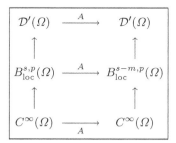

Fig. 4.19. The mapping properties of A in the localized Sobolev spaces $H_{\mathrm{loc}}^{s,p}(\Omega)$

$$
\begin{array}{ccc}
\mathcal{D}'(\Omega) & \xrightarrow{\ A\ } & \mathcal{D}'(\Omega) \\
\uparrow & & \uparrow \\
B_{\mathrm{loc}}^{s,p}(\Omega) & \xrightarrow{\ A\ } & B_{\mathrm{loc}}^{s-m,p}(\Omega) \\
\uparrow & & \uparrow \\
C^{\infty}(\Omega) & \xrightarrow[A]{} & C^{\infty}(\Omega)
\end{array}
$$

Fig. 4.20. The mapping properties of A in the localized Besov spaces $B_{\mathrm{loc}}^{s,p}(\Omega)$

4.5.3 Pseudo-Differential Operators on a Manifold

Now we define the concept of a pseudo-differential operator on a manifold, and transfer all the machinery of pseudo-differential operators to manifolds. Let M be an n-dimensional *compact* smooth manifold without boundary. Theorem 4.45 leads us to the following definition (see Figure 4.21 below):

Definition 4.48. *Let* $1 - \rho \le \delta < \rho \le 1$. *A continuous linear operator* $A\colon C^{\infty}(M) \to C^{\infty}(M)$ *is called a* pseudo-differential operator *of order* $m \in \mathbf{R}$ *if it satisfies the following two conditions (i) and (ii):*

(i) The distribution kernel $K_A(x,y)$ *of* A *is smooth off the diagonal* $\{(x,x) : x \in M\}$ *in* $M \times M$.
(ii) For any chart (U, χ) *on* M, *the mapping*

$$
A_\chi \colon C_0^{\infty}\left(\chi(U)\right) \longrightarrow C^{\infty}\left(\chi(U)\right)
$$
$$
v \longmapsto A\left(v \circ \chi\right) \circ \chi^{-1}
$$

belongs to the class $L_{\rho,\delta}^{m}\left(\chi(U)\right)$.

$$\begin{array}{ccc}
C_0^\infty(U) & \xrightarrow{\ A|_U\ } & C^\infty(U) \\[2mm]
{\scriptstyle \chi^*}\Big\uparrow & & \Big\downarrow{\scriptstyle \chi_*} \\[2mm]
C_0^\infty(\chi(U)) & \xrightarrow[\ A_\chi\ ]{} & C^\infty(\chi(U))
\end{array}$$

Fig. 4.21. The mapping $A_\chi = \chi_* (A|_U) \chi^*$ in Definition 4.48

We let

$$L_{\rho,\delta}^m(M)$$

= the set of all pseudo-differential operators of order m on M,

and set

$$L^{-\infty}(M) = \bigcap_{m \in \mathbf{R}} L_{\rho,\delta}^m(M).$$

Some results about pseudo-differential operators on $\mathbf{R}^n$ stated above are also true for pseudo-differential operators on M. In fact, pseudo-differential operators on M are defined to be locally pseudo-differential operators on $\mathbf{R}^n$.

For example, we have the following five assertions (1) through (5) (see Figures 4.22 and 4.23 below):

(1) A pseudo-differential operator A extends to a continuous linear operator $A \colon \mathcal{D}'(M) \to \mathcal{D}'(M)$.
(2) sing supp $Au \subset$ sing supp u, $u \in \mathcal{D}'(M)$.
(3) A continuous linear operator $A \colon C^\infty(M) \to \mathcal{D}'(M)$ is a —em regularizer if and only if it is in the class $L^{-\infty}(M)$.
(4) The class $L_{\rho,\delta}^m(M)$ is stable under the operations of composition of operators and taking the transpose or adjoint of an operator.
(5) (**The Besov space boundedness theorem**) A pseudo-differential operator $A \in L_{1,\delta}^m(M)$ with $0 \le \delta < 1$ extends to continuous linear operators (see Figures 4.22 and 4.23)

$$A \colon H^{s,p}(M) \longrightarrow H^{s-m,p}(M) \quad \text{for all } s \in \mathbf{R} \text{ and } 1 < p < \infty,$$
$$A \colon B^{s,p}(M) \longrightarrow B^{s-m,p}(M) \quad \text{for all } s \in \mathbf{R} \text{ and } 1 \le p \le \infty.$$

A pseudo-differential operator $A \in L_{1,0}^m(M)$ is said to be *classical* if, for any chart (U, χ) on M, the mapping $A_\chi \colon C_0^\infty(\chi(U)) \to C^\infty(\chi(U))$ belongs to the class $L_{\mathrm{cl}}^m(\chi(U))$.

We let

$$L_{\mathrm{cl}}^m(M) = \text{the set of all classical, pseudo-differential operators of}$$
$$\text{order } m \text{ on } M.$$

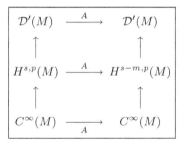

Fig. 4.22. The mapping properties of A in the Sobolev spaces $H^{s,p}(M)$

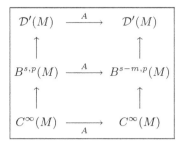

Fig. 4.23. The mapping properties of A in the Besov spaces $B^{s,p}(M)$

We observe that
$$L^{-\infty}(M) = \bigcap_{m \in \mathbf{R}} L^m_{\mathrm{cl}}(M).$$

Let $A \in L^m_{\mathrm{cl}}(M)$. If (U, χ) is a chart on M, there is associated a homo-geneous principal symbol $\sigma_{A_\chi} \in C^\infty (\chi(U) \times (\mathbf{R}^n \setminus \{0\}))$. In view of Remark 4.46, by smoothly patching together the functions σ_{A_χ} we can obtain a smooth function $\sigma_A(x, \xi)$ on $T^*(M) \setminus \{0\} = \{(x, \xi) \in T^*(M) : \xi \neq 0\}$, which is pos-itively homogeneous of degree m in the variable ξ. The function $\sigma_A(x, \xi)$ is called the *homogeneous principal symbol* of A.

A classical pseudo-differential operator $A \in L^m_{\mathrm{cl}}(M)$ is said to be *elliptic* of order m if its homogeneous principal symbol $\sigma_A(x, \xi)$ does not vanish on the bundle $T^*(M) \setminus \{0\}$ of non-zero cotangent vectors.

Then we have the following assertion (6):

(6) An operator $A \in L^m_{\mathrm{cl}}(M)$ is elliptic if and only if there exists a parametrix $B \in L^{-m}_{\mathrm{cl}}(M)$ for A:

$$\begin{cases} AB \equiv I \mod L^{-\infty}(M), \\ BA \equiv I \mod L^{-\infty}(M). \end{cases}$$

4.5.4 Hypoelliptic Pseudo-Differential Operators

Let Ω be an open subset of $\mathbf{R}^n$. A properly supported pseudo-differential operator A on Ω is said to be *hypoelliptic* if it satisfies the condition

$$\text{sing supp } u = \text{sing supp } Au \quad \text{for all } u \in \mathcal{D}'(\Omega).$$

For example, Theorem 4.44 asserts that elliptic operators are hypoelliptic. We remark that this notion may be transferred to manifolds.

The following criterion for hypoellipticity is due to Hörmander (cf. [59], Theorem 4.2]):

Theorem 4.49 (Hörmander). *Let* $A = p(x, D)$ *be a properly supported pseudo-differential operator in the Hörmander class* $L^m_{\rho,\delta}(\Omega)$ *with* $1 - \rho \leq \delta < \rho \leq 1$. *Assume that, for any compact* $K \subset \Omega$ *and any multi-indices* α, β *there exist constants* $C_{K,\alpha,\beta} > 0$, $C_K > 0$ *and* $\mu \in \mathbf{R}$ *such that we have, for all* $x \in K$ *and* $|\xi| \geq C_K$,

$$\left| D^\alpha_\xi D^\beta_x p(x,\xi) \right| \leq C_{K,\alpha,\beta} |p(x,\xi)| \left(1 + |\xi|\right)^{-\rho|\alpha| + \delta|\beta|}, \tag{4.22a}$$

$$|p(x,\xi)|^{-1} \leq C_K \left(1 + |\xi|\right)^\mu. \tag{4.22b}$$

Then there exists a parametrix $B \in L^\mu_{\rho,\delta}(\Omega)$ *for* A.

Remark 4.50. It should be emphasized that Theorem 4.49 extends to the Hörmander class $\mathbf{L}^m_{\rho,\delta}(\Omega, \mathbf{R}^n)$ of $n \times n$ *matrix-valued*, pseudo-differential operators on Ω (see [61, Chapter XXII, Theorem 22.1.3]; [119]).

4.5.5 Distribution Kernel of Pseudo-Differential Operators

In this subsection, following Coifman–Meyer [30, Chapitre IV, Proposition 1]) we state that the distribution kernel

$$s(x,y) = \frac{1}{(2\pi)^n} \int_{\mathbf{R}^n} e^{i(x-y)\cdot\xi} a(x,\xi)\, d\xi$$

of a pseudo-differential operator $S \in L^m_{1,0}(\mathbf{R}^n)$ with symbol $a(x,\theta)$ satisfies the estimate

$$|s(x,y)| \leq \frac{C}{|x-y|^{n+m}} \quad \text{for all } x,\, y \in \mathbf{R}^n \text{ and } x \neq y.$$

More precisely, we can obtain the following fundamental result (see [122, Section 7.8, Theorem 7.36]):

Theorem 4.51. *Let* $\sigma(x,\xi)$ *be a symbol in the class* $S^m_{1,0}(\mathbf{R}^n \times \mathbf{R}^n)$ *such that*

$$\left| \sigma^{(\alpha)}_{(\beta)}(x,\xi) \right| \leq C_{\alpha\beta} \left(1 + |\xi|\right)^{m-|\alpha|} \quad \text{for all } (x,\xi) \in \mathbf{R}^n \times \mathbf{R}^n. \tag{4.23}$$

We let

$$r(x, z) = \frac{1}{(2\pi)^n} \int_{\mathbf{R}^n} e^{iz \cdot \xi} \sigma(x, \xi)\, d\xi,$$

where the integral is taken in the sense of oscillatory integrals (see Theorem 4.31). Then the distribution $r(x, z)$ satisfies the condition

$$r(x, z) \in C^\infty \left(\mathbf{R}^n \times \mathbf{R}^n \setminus \{0\}\right),$$

and the estimate

$$|r(x, z)| \leq \frac{C}{|z|^{n+m}} \quad \text{for all } z \neq 0. \tag{4.24}$$

We can summarize the three operators $(-\Delta)^{-\alpha/2}$, $(-\Delta)^{-1}$ and $(I-\Delta)^{-\alpha/2}$ in Example 4.7 as in Table 4.3 below.

Notation	Distribution kernels	Symbols
$(-\Delta)^{-\alpha/2}$ $(0 < \alpha < n)$	$\dfrac{\Gamma((n-\alpha)/2)}{2^\alpha \pi^{n/2} \Gamma(\alpha/2)} \dfrac{1}{|x|^{n-\alpha}}$	$\dfrac{1}{|\xi|^\alpha}$
$(-\Delta)^{-1}$ $(\alpha = 2)$	$\dfrac{\Gamma((n-2)/2)}{4\pi^{n/2}} \dfrac{1}{|x|^{n-2}}$	$\dfrac{1}{|\xi|^2}$
$(I-\Delta)^{-\alpha/2}$ $(\alpha > 0)$	$G_\alpha(x)$	$\dfrac{1}{(1+|\xi|^2)^{\alpha/2}}$

Table 4.3. Riesz, Newtonian and Bessel potentials in Example 4.7

For the calculation of their distribution kernels via the *heat kernel*, the reader might refer to Section 4.7.

Moreover, we give an example of the density function of Lévy measure in the symmetric α-stable process for $0 < \alpha < 2$:

Example 4.8. Let $0 < \alpha < 2$. If $A_\alpha = -(-\Delta)^{\alpha/2} \in L^\alpha_{1,0}(\mathbf{R}^n)$ is a pseudo-differential operator with symbol $-|\xi|^\alpha$, then we find from Example 4.3 that its distribution kernel $a_\alpha(x, y)$ is equal to the following:

$$a_\alpha(x, y) = \frac{2^\alpha \, \Gamma((\alpha + n)/2)}{\pi^{n/2} \, |\Gamma(-\alpha/2)|} \, \text{v.p.} \, \frac{1}{|x - y|^{n+\alpha}}.$$

Hence we have, for every $f \in C_0^\infty(\mathbf{R}^n)$,

$$
\begin{aligned}
A_\alpha f(x) &= -(-\Delta)^{\alpha/2} f(x) = \frac{1}{(2\pi)^n} \int_{\mathbf{R}^n} e^{i\,x\cdot\xi} \left(-|\xi|^\alpha\right) \widehat{f}(\xi)\,d\xi \\
&= \frac{2^\alpha\,\Gamma((\alpha+n)/2)}{\pi^{n/2}\,|\Gamma(-\alpha/2)|} \int_{\mathbf{R}^n} \text{v.p.} \frac{1}{|x-y|^{n+\alpha}} f(y)\,dy \\
&= \frac{2^\alpha\,\Gamma((\alpha+n)/2)}{\pi^{n/2}\,|\Gamma(-\alpha/2)|} \lim_{\varepsilon\downarrow 0} \int_{|y|\geq\varepsilon} \frac{f(x+y)-f(x)}{|y|^{n+\alpha}}\,dy
\end{aligned}
$$

for $0 < \alpha < 1$,

and also

$$
\begin{aligned}
A_\alpha f(x) &= -(-\Delta)^{\alpha/2} f(x) = \frac{1}{(2\pi)^n} \int_{\mathbf{R}^n} e^{i\,x\cdot\xi} \left(-|\xi|^\alpha\right) \widehat{f}(\xi)\,d\xi \\
&= \frac{2^\alpha\,\Gamma((\alpha+n)/2)}{\pi^{n/2}\,|\Gamma(-\alpha/2)|} \int_{\mathbf{R}^n} \text{v.p.} \frac{1}{|x-y|^{n+\alpha}} f(y)\,dy \\
&= \frac{2^\alpha\,\Gamma((\alpha+n)/2)}{\pi^{n/2}\,|\Gamma(-\alpha/2)|} \lim_{\varepsilon\downarrow 0} \int_{|y|\geq\varepsilon} \frac{f(x+y)+f(x-y)-2f(x)}{|y|^{n+\alpha}}\,dy
\end{aligned}
$$

for $1 \leq \alpha < 2$.

We remark (see Table 4.4 below) that the distribution

$$
\nu_\alpha(y) := \frac{2^\alpha\,\Gamma((\alpha+n)/2)}{\pi^{n/2}\,|\Gamma(-\alpha/2)|} \,\text{v.p.} \frac{1}{|y|^{n+\alpha}}
$$

is the density function of *Lévy measure* in the symmetric α-stable process (see [63, Chapter 1]) and further that the operator A_α is the infinitesimal generator of the *probabilistic convolution semigroup* $\{e^{t\,A_\alpha}\}_{t\geq 0}$ (see [27, Theorem 1.4]). More precisely, we have the formula

$$
e^{t\,A_\alpha} f(x) = (P_t^\alpha * f)(x) = \int_{\mathbf{R}^n} P_t^\alpha(x-y)f(y)\,dy \quad \text{for all } f \in C_0^\infty(\mathbf{R}^n),
$$

where

$$
P_t^\alpha(x) = \frac{1}{(2\pi)^n} \int_{\mathbf{R}^n} e^{i\,x\cdot\xi} e^{-t\,|\xi|^\alpha}\,d\xi \quad \text{for } x \in \mathbf{R}^n \text{ and } t > 0.
$$

4.6 Elliptic Pseudo-differential Operators and their Indices

In this section, by using the Riesz–Schauder theory we prove some of the most important results about elliptic pseudo-differential operators on a manifold and their indices in the framework of Sobolev spaces. These results will be useful for the study of elliptic boundary value problems in Chapter 5.

Throughout this section, let M be an n-dimensional, *compact* smooth manifold without boundary.

Infinitesimal generator	$A_\alpha = -(-\Delta)^{\alpha/2}$
Distribution kernel	$\dfrac{2^\alpha}{\pi^{n/2}} \dfrac{\Gamma((\alpha+n)/2)}{|\Gamma(-\alpha/2)|}$ v.p. $\dfrac{1}{|x-y|^{n+\alpha}}$
Symbol	$-|\xi|^\alpha$

Table 4.4. An overview of symmetric α-stable processes for $0 < \alpha < 2$

4.6.1 Pseudo-Differential Operators on Sobolev Spaces

Let

$$H^s(M) = H^{s,2}(M) \quad (p = 2)$$

be the Sobolev space of order $s \in \mathbf{R}$ on M. Recall that

$$C^\infty(M) = \bigcap_{s \in \mathbf{R}} H^s(M),$$

$$\mathcal{D}'(M) = \bigcup_{s \in \mathbf{R}} H^s(M).$$

A linear operator $T \colon C^\infty(M) \to C^\infty(M)$ is said to be *of order* $m \in \mathbf{R}$ if it extends to a continuous linear operator on $H^s(M)$ into $H^{s-m}(M)$ for each $s \in \mathbf{R}$. For example, every pseudo-differential operator in $L^m(M)$ is of order m.

We say that $T \colon C^\infty(M) \to C^\infty(M)$ is of order $-\infty$ if it extends to a continuous linear operator on $H^s(M)$ into $C^\infty(M)$ for each $s \in \mathbf{R}$. This is equivalent to saying that T is a regularizer; hence we have the formula

$$L^{-\infty}(M) = \text{the set of all operators of order } -\infty. \qquad (4.25)$$

Let $T \colon H^s(M) \to H^t(M)$ be a linear operator with domain $\mathcal{D}(T)$ dense in $H^s(M)$. Each element v of $H^{-t}(M)$ defines a linear functional G on $\mathcal{D}(T)$ by the formula

$$G(u) = (Tu, v) \quad \text{for } u \in \mathcal{D}(T),$$

where $(\cdot, \cdot)$ on the right-hand side is the sesquilinear pairing of $H^t(M)$ and $H^{-t}(M)$. If this functional G is continuous everywhere on $\mathcal{D}(T)$, by applying [123, Theorem 2.6] we obtain that G can be extended uniquely to a continuous linear functional $\tilde{G}$ on $\overline{\mathcal{D}(T)} = H^s(M)$. Hence, there exists a unique element v^* of $H^{-s}(M)$ such that

$$\widetilde{G}(u) = (u, v^*) \quad \text{for } u \in H^s(M),$$

since the sesquilinear form $(\cdot, \cdot)$ on the product space $H^s(M) \times H^{-s}(M)$ permits us to identify the strong dual space of $H^s(M)$ with $H^{-s}(M)$. In particular, we have the formula

$$(Tu, v) = G(u) = (u, v^*) \quad \text{for all } u \in \mathcal{D}(T).$$

So we let

$\mathcal{D}(T^*) = $ the totality of those $v \in H^{-t}(M)$ such that the mapping
$u \mapsto (Tu, v)$ is continuous everywhere on $\mathcal{D}(T)$,

and define

$$T^* v = v^*.$$

Therefore, it follows that T^* is a linear operator from $H^{-t}(M)$ into $H^{-s}(M)$ with domain $\mathcal{D}(T^*)$ such that

$$(Tu, v) = (u, T^* v) \quad \text{for all } u \in \mathcal{D}(T) \text{ and } v \in \mathcal{D}(T^*). \tag{4.26}$$

The operator T^* is called the *adjoint* of T.

The *transpose* of T is a linear operator T' from $H^{-t}(M)$ into $H^{-s}(M)$ with domain $\mathcal{D}(T')$ such that

$\mathcal{D}(T') = $ the totality of those $v \in H^{-t}(M)$ such that the mapping
$u \mapsto \langle Tu, v \rangle$ is continuous everywhere on $\mathcal{D}(T)$,

and satisfies the formula

$$\langle Tu, v \rangle = \langle u, T'v \rangle \quad \text{for all } u \in \mathcal{D}(T) \text{ and } v \in \mathcal{D}(T'). \tag{4.27}$$

Here $\langle \cdot, \cdot \rangle$ on the left-hand (resp. right-hand) side is the bilinear pairing of $H^t(M)$ and $H^{-t}(M)$ (resp. $H^s(M)$ and $H^{-s}(M)$).

In view of formulas (4.26) and (4.27), it follows that

(a) $v \in \mathcal{D}(T') \iff \overline{v} \in \mathcal{D}(T^*)$,
(b) $T'v = \overline{T^* \overline{v}}$ for every $v \in \mathcal{D}(T')$. Here $\overline{}$ denotes complex conjugation.

Hence, we have the following two assertions (1) and (2):

$$\begin{cases} (1) \text{ The ranges } \mathcal{R}(T^*) \text{ and } \mathcal{R}(T') \text{ are isomorphic.} \\ (2) \text{ The null spaces } \mathcal{N}(T^*) \text{ and } \mathcal{N}(T') \text{ are isomorphic.} \end{cases} \tag{4.28}$$

Now let $A \in L^m(M)$. Then the operator $A: C^\infty(M) \to C^\infty(M)$ extends uniquely to a continuous linear operator

$$A_s: H^s(M) \longrightarrow H^{s-m}(M)$$

$$
\begin{array}{ccc}
\mathcal{D}'(M) & \xrightarrow{\ \overline{A}\ } & \mathcal{D}'(M) \\
\uparrow & & \uparrow \\
H^s(M) & \xrightarrow{\ A_s\ } & H^{s-m}(M) \\
\uparrow & & \uparrow \\
C^\infty(M) & \xrightarrow[\ A\ ]{} & C^\infty(M)
\end{array}
$$

Fig. 4.24. The mapping properties of the continuous operators A, A_s and $\overline{A}$

for all $s \in \mathbf{R}$, and hence to a continuous linear operator

$$\overline{A} \colon \mathcal{D}'(M) \longrightarrow \mathcal{D}'(M).$$

The situation can be visualized as in Figure 4.24 above.

The adjoint A^* of A is also in $L^m(M)$; hence the operator

$$A^* \colon C^\infty(M) \longrightarrow C^\infty(M)$$

extends uniquely to a continuous linear operator

$$A^*_s \colon H^s(M) \longrightarrow H^{s-m}(M)$$

for all $s \in \mathbf{R}$.

The next lemma states a fundamental relationship between the operators A_s and A^*_s:

Lemma 4.52. *If $A \in L^m(M)$, we have, for all $s \in \mathbf{R}$,*

$$
\begin{cases}
(A_s)^* = A^*_{-s+m}, \\
(A^*_{-s+m})^* = A_s.
\end{cases}
\tag{4.29}
$$

Proof. If $u \in \mathcal{D}(A_s) = H^s(M)$ and $v \in \mathcal{D}(A^*_{-s+m}) = H^{-s+m}(M)$, there exist sequences $\{u_j\}$ and $\{v_j\}$ in $C^\infty(M)$ such that

$$
\begin{cases}
u_j \longrightarrow u & \text{in } H^s(M) \text{ as } j \to \infty, \\
v_j \longrightarrow v & \text{in } H^{-s+m}(M) \text{ as } j \to \infty.
\end{cases}
$$

Then we have the assertions

$$
\begin{cases}
Au_j \longrightarrow A_s u & \text{in } H^{s-m}(M), \\
A^* v_j \longrightarrow A^*_{-s+m} v & \text{in } H^{-s}(M),
\end{cases}
$$

so that

$$(A_s u, v) = \lim_{j \to \infty} (A u_j, v_j) = \lim_{j \to \infty} (u_j, A^* v_j) = \left(u, A^*_{-s+m} v\right).$$

This proves the desired formulas (4.29).

The proof of Lemma 4.52 is complete. $\square$

4.6.2 The Index of an Elliptic Pseudo-Differential Operator

In this section, we study the operators A_s when A is a classical elliptic pseudo-differential operator.

The next theorem is an immediate consequence of Theorem 4.44:

Theorem 4.53 (the elliptic regularity theorem). *Let $A \in L_{\mathrm{cl}}^m(M)$ be elliptic. Then we have, for all $s \in \mathbf{R}$,*

$$u \in \mathcal{D}'(M), \ Au \in H^s(M) \implies u \in H^{s+m}(M).$$

In particular, we have the assertions

$$\mathcal{R}(A_s) \cap C^\infty(M) = \mathcal{R}(A), \tag{4.30}$$
$$N(A_s) = \mathcal{N}(A). \tag{4.31}$$

Here

$$\mathcal{R}(A_s) = \{Au : u \in H^s(M)\}, \quad \mathcal{R}(A) = \{Au : u \in C^\infty(M)\};$$
$$\mathcal{N}(A_s) = \{u \in H^s(M) : A_s u = 0\}, \quad \mathcal{N}(A) = \{u \in C^\infty(M) : Au = 0\}.$$

Here it is worth pointing out that assertion (4.31) is a generalization of the celebrated Weyl theorem, which states that harmonic functions on Euclidean space are smooth.

Now let X, Y be Banach spaces. A bounded (continuous) linear operator

$$T \colon X \longrightarrow Y$$

is called a *Fredholm operator* if it satisfies the following three conditions (i), (ii) and (iii):

(i) The null space $\mathcal{N}(T) = \{x \in X : Tx = 0\}$ of T has finite dimension, that is, $\dim \mathcal{N}(T) < \infty$.
(ii) The range $\mathcal{R}(T) = \{Tx : x \in X\}$ of T is closed in X.
(iii) The range $\mathcal{R}(T)$ has finite codimension in X, that is, $\operatorname{codim} \mathcal{R}(T) = \dim X/\mathcal{R}(T) < \infty$.

In this case the *index* $\operatorname{ind} T$ of T is defined by the formula

$$\operatorname{ind} T = \dim \mathcal{N}(T) - \operatorname{codim} \mathcal{R}(T).$$

The next theorem states that the operator $A_s \colon H^s(M) \to H^{s-m}(M)$ is a Fredholm operator for every $s \in \mathbf{R}$:

Theorem 4.54. *If $A \in L_{\mathrm{cl}}^m(M)$ is elliptic, the operator $A_s \colon H^s(M) \to H^{s-m}(M)$ is a Fredholm operator for every $s \in \mathbf{R}$.*

Proof. Take a parametrix $B \in L_{\mathrm{cl}}^{-m}(M)$ for A:

$$\begin{cases} BA = I + P, & P \in L^{-\infty}(M), \\ AB = I + Q, & Q \in L^{-\infty}(M). \end{cases}$$

Then we have, for all $s \in \mathbf{R}$,

$$\begin{cases} B_{s-m} \cdot A_s = I + P_s, \\ A_s \cdot B_{s-m} = I + Q_{s-m}. \end{cases}$$

Furthermore, in view of assertion (4.25) it follows from an application of the Rellich–Kondrachov theorem (Theorem 4.10) that the operators

$$P_s \colon H^s(M) \longrightarrow H^s(M)$$

and

$$Q_{s-m} \colon H^{s-m}(M) \longrightarrow H^{s-m}(M)$$

are both *compact*. Therefore, by applying [100, Theorem 7.2] (cf. [80, Theorem 2.24]) to our situation we obtain that A_s is a Fredholm operator.

The proof of Theorem 4.54 is complete. □

Corollary 4.55. *Let $A \in L_{\mathrm{cl}}^m(M)$ be elliptic. Then we have the following two assertions (i) and (ii):*

(i) The range $\mathcal{R}(A)$ of A is a closed linear subspace of $C^\infty(M)$.
(ii) The null space $\mathcal{N}(A)$ of A is a finite dimensional, closed linear subspace of $C^\infty(M)$.

Proof. (i) It follows from Theorem 4.54 that the range $\mathcal{R}(A_s)$ of A_s is closed in $H^{s-m}(M)$; hence it is closed in $C^\infty(M)$, since the injection

$$C^\infty(M) \longrightarrow H^{s-m}(M)$$

is continuous. In view of formula (4.30), this proves part (i).

(ii) Similarly, in view of formula (4.31) it follows from Theorem 4.54 that $\mathcal{N}(A)$ has finite dimension; so it is closed in each $H^s(M)$ and hence in $C^\infty(M)$.

The proof of Corollary 4.55 is complete. □

The next theorem asserts that the index

$$\mathrm{ind}\, A_s = \dim \mathcal{N}(A_s) - \mathrm{codim}\, \mathcal{R}(A_s)$$

is *independent* of $s \in \mathbf{R}$:

Theorem 4.56. *If $A \in L_{\mathrm{cl}}^m(M)$ is elliptic, then we have, for all $s \in \mathbf{R}$,*

$$\mathrm{ind}\, A_s = \dim \mathcal{N}(A) - \dim \mathcal{N}(A^*), \tag{4.32}$$

where

$$\mathcal{N}(A^*) = \{v \in C^\infty(M) : A^* v = 0\}.$$

Proof. Since the range $\mathcal{R}(A_s)$ is closed in $H^{s-m}(M)$, by applying the closed range theorem (see [100, Theorem 3.16], [147, p. 205, Theorem]) to our situation we obtain that

$$\operatorname{codim}\mathcal{R}(A_s) = \dim\mathcal{N}(A'_s).$$

However, in view of assertion (4.28) it follows that

$$\dim\mathcal{N}(A'_s) = \dim\mathcal{N}(A^*_s).$$

Furthermore, we have, by formulas (4.29) and (4.31),

$$\mathcal{N}(A^*_s) = \mathcal{N}(A^*_{-s+m}) = \mathcal{N}(A^*), \tag{4.33}$$

since $A^* \in L^m_{\mathrm{cl}}(M)$ is also elliptic (cf. Theorem 4.41).

Summing up, we obtain that

$$\operatorname{codim}\mathcal{R}(A_s) = \dim\mathcal{N}(A^*). \tag{4.34}$$

Therefore, the desired formula (4.32) follows from formulas (4.31) and (4.34). The proof of Theorem 4.56 is complete. □

We give another useful expression for $\operatorname{ind} A_s$. To do this, we need the following lemma:

Lemma 4.57. *Let $A \in L^m_{\mathrm{cl}}(M)$ be elliptic. Then the spaces $\mathcal{N}(A^*)$ and $\mathcal{R}(A)$ are orthogonal complements of each other in the space $C^\infty(M)$ relative to the inner product of $L^2(M)$ (see Figure 4.25 below):*

$$C^\infty(M) = \mathcal{N}(A^*) \oplus \mathcal{R}(A). \tag{4.35}$$

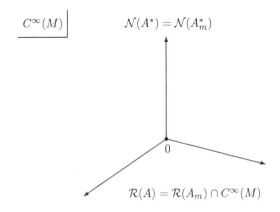

Fig. 4.25. The orthogonal decomposition (4.35) in the space $C^\infty(M)$

Proof. Since the range $\mathcal{R}(A_m)$ is closed in $L^2(M)$, it follows from an application of the closed range theorem (see [100, Theorem 3.16], [147, p. 205, Theorem]) that

$$L^2(M) = \mathcal{N}(A_m^*) \oplus \mathcal{R}(A_m). \tag{4.36}$$

However we have, by formulas (4.33) and (4.30),

$$\mathcal{N}(A_m^*) = \mathcal{N}(A^*), \tag{4.37a}$$
$$\mathcal{R}(A_m) \cap C^\infty(M) = \mathcal{R}(A). \tag{4.37b}$$

Therefore, the desired orthogonal decomposition (4.35) follows from assertions (4.37), by restricting the orthogonal decomposition (4.36) to the space $C^\infty(M)$.

The proof of Lemma 4.57 is complete. ☐

Now we can prove the following theorem:

Theorem 4.58. *If $A \in L_{\mathrm{cl}}^m(M)$ is elliptic, then we have, for all $s \in \mathbf{R}$,*

$$\operatorname{ind} A_s = \dim \mathcal{N}(A) - \operatorname{codim} \mathcal{R}(A). \tag{4.38}$$

Here

$$\operatorname{codim} \mathcal{R}(A) = \dim C^\infty(M)/\mathcal{R}(A).$$

Proof. The orthogonal decomposition (4.35) tells us that

$$\dim \mathcal{N}(A^*) = \operatorname{codim} \mathcal{R}(A).$$

Hence, the desired formula (4.38) follows from formula (4.32).

The proof of Theorem 4.58 is complete. ☐

We let

$$\operatorname{ind} A = \dim \mathcal{N}(A) - \dim \mathcal{N}(A^*) \tag{4.39}$$
$$= \dim \mathcal{N}(A) - \operatorname{codim} \mathcal{R}(A).$$

The next theorem states that the index of an elliptic pseudo-differential operator depends only on its principal symbol:

Theorem 4.59. *If A, $B \in L_{\mathrm{cl}}^m(M)$ are elliptic and if they have the same homogeneous principal symbol, then it follows that*

$$\operatorname{ind} A = \operatorname{ind} B. \tag{4.40}$$

Proof. Since the difference $A - B$ belongs to the class $L_{\mathrm{cl}}^{m-1}(M)$, it follows from Rellich's theorem that the operator

$$A_s - B_s \colon H^s(M) \longrightarrow H^{s-m}(M)$$

is compact. Hence, by applying [100, Theorem 7.8] (cf. [80, Theorem 2.26]) we obtain that

$$\text{ind}\, A_s = \text{ind}\, (B_s + (A_s - B_s)) = \text{ind}\, B_s.$$

In view of Theorem 4.58, this proves the desired formula (4.40).

The proof of Theorem 4.59 is complete.   □

As for the product of elliptic pseudo-differential operators, we have the following theorem:

Theorem 4.60. *If $A \in L^m_{\text{cl}}(M)$ and $B \in L^{m'}_{\text{cl}}(M)$ are elliptic, then we have the formula*

$$\text{ind}\, BA = \text{ind}\, B + \text{ind}\, A. \qquad (4.41)$$

Proof. We remark that we have, for each $s \in \mathbf{R}$,

$$(BA)_s = B_{s-m} \cdot A_s.$$

Hence, by applying [100, Theorem 7.3] (cf. [80, Theorem 2.21]) to our situation we obtain that

$$\text{ind}\, (BA)_s = \text{ind}\, B_{s-m} + \text{ind}\, A_s.$$

This proves the desired formula (4.41), since BA is an elliptic operator in $L^{m+m'}_{\text{cl}}(M)$ (cf. Theorem 4.42).

The proof of Theorem 4.60 is complete.   □

As for the adjoints, we have the following theorem:

Theorem 4.61. *If $A \in L^m_{\text{cl}}(M)$ is elliptic, then we have the formula*

$$\text{ind}\, A^* = -\text{ind}\, A. \qquad (4.42)$$

Indeed, it suffices to note that $A^{**} = A$.

We give some useful criteria for $\text{ind}\, A = 0$:

Theorem 4.62. *If $A \in L^m_{\text{cl}}(M)$ is elliptic and if A and A^* have the same homogeneous principal symbol, then it follows that*

$$\text{ind}\, A = 0. \qquad (4.43)$$

Proof. Theorem 4.59 tells us that $\text{ind}\, A = \text{ind}\, A^*$. However, in view of formula (4.42) this implies the desired formula (4.43).

The proof of Theorem 4.62 is complete.   □

Corollary 4.63. *If $A \in L^m_{\text{cl}}(M)$ is elliptic and if its homogeneous principal symbol is* real, *then we have the assertion*

$$\text{ind}\, A = 0.$$

Indeed, by Theorem 4.41 it follows that A and A^* have the same homogeneous real principal symbol. Therefore, Theorem 4.62 applies.

Theorem 4.64. *If $A \in L_{cl}^m(M)$ is elliptic and if $A^* = \lambda A$ for some $\lambda \in \mathbf{C}$, then we have the assertions*

$$|\lambda| = 1,$$

and

$$\text{ind } A = 0.$$

Proof. First, we remark that

$$A = A^{**} = (\lambda A)^* = \bar{\lambda} A^* = |\lambda|^2 A.$$

Hence we have $|\lambda| = 1$ and so

$$\begin{cases} \mathcal{N}(\lambda A) = \mathcal{N}(A), \\ \mathcal{R}(\lambda A) = \mathcal{R}(A). \end{cases}$$

Therefore, it follows from formula (4.42) that

$$\text{ind } A = \text{ind } \lambda A = \text{ind } A^* = -\text{ind } A.$$

This proves that ind $A = 0$.
The proof of Theorem 4.64 is complete.   □

The next theorem describes conditions under which an elliptic pseudo-differential operator is invertible on Sobolev spaces:

Theorem 4.65. *Let $A \in L_{cl}^m(M)$ be elliptic. Assume that*

$$\begin{cases} \text{ind } A = 0, \\ \mathcal{N}(A) = \{0\}. \end{cases}$$

Then we have the following three assertions (i), (ii) and (iii):

(i) The operator $A \colon C^\infty(M) \to C^\infty(M)$ is bijective.
(ii) The operator $A_s \colon H^s(M) \to H^{s-m}(M)$ is an isomorphism for each $s \in \mathbf{R}$.
(iii) The inverse A^{-1} of A belongs to the class $L_{cl}^{-m}(M)$.

Proof. (i) Since ind $A = 0$ and $\mathcal{N}(A) = \{0\}$, it follows from formula (4.39) that

$$\mathcal{N}(A^*) = \{0\}.$$

Hence the surjectivity of A follows from the orthogonal decomposition (4.35).
(ii) Since $\mathcal{N}(A_s) = \mathcal{N}(A) = \{0\}$ and ind $A_s = $ ind $A = 0$, it follows that the operator

$$A_s \colon H^s(M) \longrightarrow H^{s-m}(M)$$

is bijective for each $s \in \mathbf{R}$. Therefore, by applying the closed graph theorem (see [100, Theorem 4.10], [147, p. 79, Theorem 1]) to our situation we obtain that the inverse

$$A_s^{-1} \colon H^{s-m}(M) \longrightarrow H^s(M)$$

is continuous for each $s \in \mathbf{R}$.

(iii) Since we have the formula

$$A^{-1} = A_s^{-1}\big|_{C^\infty(M)}$$

and since each $A_s^{-1} \colon H^{s-m}(M) \to H^s(M)$ is continuous, it follows that the operator

$$A^{-1} \colon C^\infty(M) \longrightarrow C^\infty(M)$$

is continuous, and also it is of order $-m$. The situation can be visualized as in Figure 4.26 above.

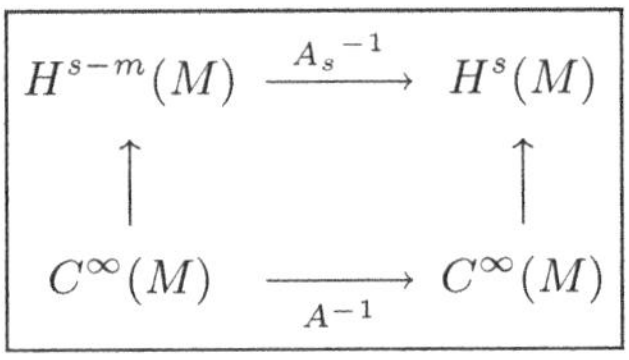

Fig. 4.26. The mapping properties of the inverses A^{-1} and A_s^{-1}

It remains to prove that $A^{-1} \in L_{\mathrm{cl}}^{-m}(M)$ takes a parametrix $B \in L_{\mathrm{cl}}^{-m}(M)$ for A:

$$\begin{cases} AB = I + P, & P \in L^{-\infty}(M), \\ BA = I + Q, & Q \in L^{-\infty}(M). \end{cases}$$

Then we have the formula

$$A^{-1} - B = (I - BA)\,A^{-1} = -Q \cdot A^{-1}.$$

However, in view of assertion (4.25) it follows that the operator

$$Q \cdot A^{-1} \in L^{-\infty}(M),$$

since it is of order $-\infty$. This proves that

$$A^{-1} = B - Q \cdot A^{-1} \in L_{\mathrm{cl}}^{-m}(M).$$

The proof of Theorem 4.65 is complete. □

The next theorem states that the Sobolev spaces $H^s(M)$ can be characterized in terms of elliptic pseudo-differential operators:

Theorem 4.66. *Let $A \in L_{\mathrm{cl}}^m(M)$ be elliptic with $m > 0$. Assume that*

$$\begin{cases} A = A^*, \\ \mathcal{N}(A) = \{0\}. \end{cases}$$

Then we have the following two assertions (i) and (ii):

(i) There exists a complete orthonormal system $\{\varphi_j\}$ of $L^2(M)$ consisting of eigenfunctions of A, and its corresponding eigenvalues $\{\lambda_j\}$ are real and $|\lambda_j| \to +\infty$ as $j \to \infty$.

(ii) A distribution $u \in \mathcal{D}'(M)$ belongs to $H^{mr}(M)$ for some integer r if and only if we have the condition

$$\sum_{j=1}^{\infty} \lambda_j^{2r} |(u, \varphi_j)|^2 < +\infty.$$

More precisely, the quantity

$$(u, v)_{mr} = \sum_{j=1}^{\infty} \lambda_j^{2r} (u, \varphi_j) \overline{(v, \varphi_j)} \tag{4.44}$$

is an admissible inner product for the Hilbert space $H^{mr}(M)$.

Proof. (i) Since $A = A^*$, it follows from Theorem 4.61 that ind $A = 0$. Hence, by applying Theorem 4.65 we obtain that the operator

$$A \colon C^{\infty}(M) \longrightarrow C^{\infty}(M)$$

is bijective, and its inverse A^{-1} is an elliptic operator in $L_{\mathrm{cl}}^{-m}(M)$.
We let

$$\widetilde{A^{-1}} = \text{the composition of } (A^{-1})_0 \colon L^2(M) \to H^m(M)$$
$$\text{and the injection: } H^m(M) \to L^2(M).$$

Then it follows from Rellich's theorem that the operator

$$\widetilde{A^{-1}} \colon L^2(M) \longrightarrow L^2(M)$$

is compact. Furthermore, since $C^{\infty}(M)$ is dense in $L^2(M)$ and $A = A^*$, we have the formula

$$\left(\widetilde{A^{-1}} u, v \right) = \left(u, \widetilde{A^{-1}} v \right) \quad \text{for } u, v \in L^2(M),$$

where $(\cdot, \cdot)$ is the inner product of $L^2(M)$. This implies that the operator $\widetilde{A^{-1}}$ is self-adjoint. Also, we have the assertion

$$\mathcal{N}\left(\widetilde{A^{-1}} \right) = \mathcal{N}\left((A^{-1})_0 \right) = \mathcal{N}(A^{-1}) = \{0\},$$

since $A^{-1} \in L_{\mathrm{cl}}^{-m}(M)$ is elliptic. Therefore, by applying the Hilbert–Schmidt theorem (see [123, Theorem 2.56]) to the self-adjoint operator $\widetilde{A^{-1}}$ we obtain that there exists a complete orthonormal system $\{\varphi_j\}$ of $L^2(M)$ consisting of eigenfunctions of $\widetilde{A^{-1}}$, and its corresponding eigenvalues $\{\mu_j\}$ are real and converges to zero as $j \to \infty$.

Since the eigenvalues μ_j are all non-zero, it follows that

$$\varphi_j = \frac{1}{\mu_j} \widetilde{A^{-1}} \varphi_j = \frac{1}{\mu_j} \left(A^{-1}\right)_0 \varphi_j \in H^m(M).$$

However, note that

$$\left(A^{-1}\right)_0 \big|_{H^m(M)} = \left(A^{-1}\right)_m$$

and that

$$\left(A^{-1}\right)_m : H^m(M) \longrightarrow H^{2m}(M).$$

Hence, we have the assertion

$$\varphi_j = \frac{1}{\mu_j} \left(A^{-1}\right)_m \varphi_j \in H^{2m}(M).$$

Continuing this way, we obtain that

$$\varphi_j \in \bigcap_{k \in \mathbf{N}} H^{km}(M) = C^\infty(M).$$

Therefore, we have the formula

$$A^{-1}\varphi_j = \mu_j \varphi_j,$$

and so

$$A\varphi_j = \lambda_j \varphi_j, \quad \lambda_j = \frac{1}{\mu_j},$$

with

$$|\lambda_j| = \frac{1}{|\mu_j|} \longrightarrow +\infty \quad \text{as } j \to \infty.$$

(ii) For each integer r, we let

$$A^r = \begin{cases} A^r & \text{if } r \geq 0, \\ \left(A^{-1}\right)^{|r|} & \text{if } r < 0, \end{cases}$$

where $A^0 = I$. Then it follows that A^r is an elliptic operator in $L_{\mathrm{cl}}^{mr}(M)$ and that

$$\mathcal{N}(A^r) = \{0\}.$$

Moreover we have, by Theorem 4.60,

$$\text{ind } A^r = \begin{cases} r \text{ ind } A = 0 & \text{if } r \geq 0, \\ |r| \text{ ind } (A^{-1}) = 0 & \text{if } r < 0. \end{cases}$$

Therefore, by applying Theorem 4.65 we obtain that the operator

$$(A^r)_{mr} : H^{mr}(M) \longrightarrow L^2(M)$$

is an isomorphism. Thus the quantity

$$(u, v)_{mr} = ((A^r)_{mr} u, (A^r)_{mr} v)$$

is an admissible inner product for the Hilbert space $H^{mr}(M)$.

Furthermore, since $\{\varphi_j\}$ is a complete orthonormal system of $L^2(M)$, we have, by Parseval's formula,

$$(u, v)_{mr} = \sum_{j=1}^{\infty} ((A^r)_{mr} u, \varphi_j) \overline{((A^r)_{mr} v, \varphi_j)}. \tag{4.45}$$

However, we have, by formula (4.29),

$$((A^r)_{mr})^* = ((A^r)^*)_0 = (A^r)_0,$$

since $A = A^*$. Hence, it follows that

$$((A^r)_{mr} u, \varphi_j) = (u, (A^r)_0 \varphi_j) = (u, A^r \varphi_j) \tag{4.46}$$
$$= \lambda_j^r (u, \varphi_j).$$

Therefore, the desired formula (4.44) follows from formulas (4.45) and (4.46). The proof of Theorem 4.66 is complete. □

As one of the important applications of Theorem 4.66, we can obtain the following theorem:

Theorem 4.67. *Let Δ be the Laplace–Beltrami operator on M and let $\{\chi_j\}$ be the orthonormal system of $L^2(M)$ consisting of eigenfunctions of $-\Delta$ and $\{\lambda_j\}$ its corresponding eigenvalues:*

$$-\Delta \chi_j = \lambda_j \chi_j, \quad \lambda_j \geq 0.$$

Then the functions χ_j span the Sobolev spaces $H^s(M)$ for each $s \in \mathbf{R}$. More precisely, the quantity

$$(u, v)_s = \sum_{j=1}^{\infty} (1 + \lambda_j)^s (u, \chi_j) \overline{(v, \chi_j)}$$

is an admissible inner product for the Hilbert space $H^s(M)$.

4.7 Functional Calculus for the Laplacian via the Heat Kernel

The heat kernel

$$K_t(x) = \frac{1}{(4\pi t)^{n/2}}\, e^{-\frac{|x|^2}{4t}} \quad \text{for } t > 0 \tag{4.47}$$

has many important and interesting applications in partial differential equations. Physically, the heat kernel $K_t(x)$ expresses a thermal distribution of position x at time t in a homogeneous isotropic medium $\mathbf{R}^n$ with unit coefficient of thermal diffusivity, given that the initial thermal distribution is the Dirac measure $\delta(x)$ (see Figure 4.27 above).

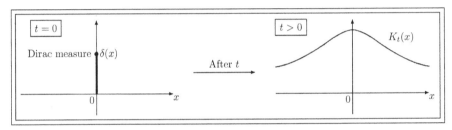

Fig. 4.27. An intuitive meaning of the heat kernel $K_t(x)$

In this section we derive Newtonian, Riesz and Bessel potentials in Example 4.7 and also the Poisson kernel for the Dirichlet boundary value problem, by calculating various convolution kernels for the Laplacian.

(I) First, it is easy to verify that the function

$$K_t * f(x) = \int_{\mathbf{R}^n} K_t(x-y)f(y)\,dy \quad \text{for } f \in \mathcal{S}(\mathbf{R}^n),$$

is a solution of the initial value problem for the heat equation

$$\begin{cases} \left(\dfrac{\partial}{\partial t} - \Delta\right) u(x,t) = 0 & \text{in } \mathbf{R}^n \times (0,\infty), \\ u(x,0) = f(x) & \text{on } \mathbf{R}^n. \end{cases}$$

Therefore, we have the **formal** representation formula

$$K_t * f = e^{t\Delta} f \quad \text{for } t > 0. \tag{4.48}$$

(II) Next, we consider the *Laplace transform*:

$$\mathcal{L}\phi(s) = \int_0^\infty e^{-st}\,\phi(t)\,dt,$$

where $\phi(s)$ is a function to be chosen later on.

If we make the **formal** change of variables

$$s := -\Delta,$$

then we obtain that

$$\mathcal{L}\phi(-\Delta)f = \int_0^\infty (e^{t\Delta}f)\phi(t)\,dt$$

$$= \int_0^\infty \left(\int_{\mathbf{R}^n} K_t(x-y)f(y)\,dy \right) \phi(t)\,dt$$

$$= \int_{\mathbf{R}^n} \left(\int_0^\infty K_t(x-y)\,\phi(t)\,dt \right) f(y)\,dy.$$

This implies that the *convolution kernel* of the operator

$$\mathcal{L}\phi(-\Delta)$$

is equal to the following:

$$\int_0^\infty K_t(x)\,\phi(t)\,dt.$$

(A) First, we let

$$\phi(t) := t^{\beta-1} \quad \text{for Re } \beta > 0.$$

Since we have the formula

$$\mathcal{L}\phi(s) = \int_0^\infty e^{-st}\,\phi(t)\,dt = \int_0^\infty e^{-st}\,t^{\beta-1}\,dt = \Gamma(\beta)\,s^{-\beta}, \tag{4.49}$$

it follows from formula (4.49) that the convolution kernel of the operator

$$(-\Delta)^{-\beta}$$

is equal to the following:

$$\frac{1}{\Gamma(\beta)} \int_0^\infty K_t(x)\,t^{\beta-1}\,dt. \tag{4.50}$$

However, we have the following claim:

Claim 4.68. *Let $0 < \text{Re } \beta < n/2$. We have, for $x \neq 0$,*

$$\frac{1}{\Gamma(\beta)} \int_0^\infty K_t(x)\,t^{\beta-1}\,dt = \frac{\Gamma(n/2-\beta)}{\Gamma(\beta)\,4^\beta\,\pi^{n/2}}\,\frac{1}{|x|^{n-2\beta}}. \tag{4.51}$$

Proof. By formula (4.47), we obtain that

$$\frac{1}{\Gamma(\beta)} \int_0^\infty \frac{1}{(4\pi t)^{n/2}}\,e^{-\frac{|x|^2}{4t}}\,t^{\beta-1}\,dt$$

$$= \frac{1}{\Gamma(\beta)} \frac{1}{(4\pi)^{n/2}} \int_0^\infty \tau^{n/2} \tau^{1-\beta} e^{-\frac{|x|^2}{4}\tau} \frac{d\tau}{\tau^2}$$

$$= \frac{1}{\Gamma(\beta)} \frac{1}{(4\pi)^{n/2}} \int_0^\infty \tau^{n/2-\beta-1} e^{-\frac{|x|^2}{4}\tau} d\tau$$

$$= \frac{1}{\Gamma(\beta)} \frac{1}{(4\pi)^{n/2}} \int_0^\infty e^{-\sigma} \left(\frac{4\sigma}{|x|^2} \right)^{n/2-\beta-1} \frac{4}{|x|^2} d\sigma$$

$$= \frac{1}{\Gamma(\beta)} \frac{1}{\pi^{n/2}} \frac{1}{4^\beta} \frac{1}{|x|^{n-2\beta}} \int_0^\infty e^{-\sigma} \sigma^{n/2-\beta-1} d\sigma$$

$$= \frac{\Gamma(n/2 - \beta)}{\Gamma(\beta) 4^\beta \pi^{n/2}} \frac{1}{|x|^{n-2\beta}}.$$

The proof of Claim 4.68 is complete. □

By combining formulas (4.50) and (4.51), we have proved the following theorem (see Table 4.3 with $\alpha := 2\,\mathrm{Re}\,\beta$):

Theorem 4.69. *Let* $0 < \mathrm{Re}\,\beta < n/2$. *The convolution kernel of the operator*

$$(-\Delta)^{-\beta}$$

is equal to the following:

$$\frac{\Gamma(n/2 - \beta)}{\Gamma(\beta) 4^\beta \pi^{n/2}} \frac{1}{|x|^{n-2\beta}} \quad \text{for } x \neq 0.$$

Example 4.9 (the Newtonian potential). If $n \geq 3$ and $\beta = 1$, then it follows from an application of Theorem 4.53 that the convolution kernel of the operator

$$(-\Delta)^{-1}$$

is equal to the following:

$$\frac{\Gamma(n/2 - 1)}{4\pi^{n/2}} \frac{1}{|x|^{n-2}}.$$

However, we remark that

$$\Gamma(n/2 - 1) = \frac{2}{n-2} \Gamma\left(\frac{n}{2}\right),$$

$$\omega_n = \frac{2\pi^{n/2}}{\Gamma(n/2)} \quad \text{(the surface area of the unit ball)},$$

so that

$$\frac{\Gamma(n/2 - 1)}{4\pi^{n/2}} = \frac{1}{(n-2)\omega_n}.$$

Hence it follows that the convolution kernel of the operator

$$(-\Delta)^{-1}$$

is equal to the following:

$$\frac{1}{(n-2)\omega_n} \frac{1}{|x|^{n-2}}.$$

This is the *Newtonian potential* (see Table 4.3).

Example 4.10 (the Riesz potential). If $\beta = \alpha/2$ and $0 < \operatorname{Re}\alpha < n$, then it follows from an application of Theorem 4.53 that the convolution kernel of the operator

$$(-\Delta)^{-\alpha/2}$$

is equal to the following:

$$\frac{\Gamma((n-\alpha)/2)}{\Gamma(\alpha/2)\,2^\alpha\,\pi^{n/2}} \frac{1}{|x|^{n-\alpha}}.$$

This is the *Riesz potential* of order α.

(B) Secondly, if we replace s by $s+1$ in formula (4.49) we obtain that

$$(s+1)^{-\beta} = \frac{1}{\Gamma(\beta)} \int_0^\infty e^{-(s+1)t}\, t^{\beta-1}\, dt \tag{4.52}$$

$$= \frac{1}{\Gamma(\beta)} \int_0^\infty e^{-st}\, e^{-t}\, t^{\beta-1}\, dt$$

$$= \frac{1}{\Gamma(\beta)} \mathcal{L}\left(e^{-s}\, s^{\beta-1}\right).$$

Therefore, by taking

$$s := -\Delta,\ \ \beta := \alpha/2 \ \text{for}\ \operatorname{Re}\alpha > 0,$$

we have, by formula (4.24),

Theorem 4.70 (the Bessel potential). *If* $\operatorname{Re}\alpha > 0$, *then the convolution kernel of the operator*

$$(I-\Delta)^{-\alpha/2}$$

is equal to the following:

$$\frac{1}{\Gamma(\alpha/2)} \frac{1}{(4\pi)^{n/2}} \int_0^\infty e^{-t-\frac{|x|^2}{4t}}\, t^{\frac{\alpha-n}{2}}\, \frac{dt}{t}.$$

This is the Bessel potential *of order α (see Table 4.3).*

If we let

$$G_\alpha(x) = \frac{1}{\Gamma(\alpha/2)} \frac{1}{(4\pi)^{n/2}} \int_0^\infty e^{-t-\frac{|x|^2}{4t}}\, t^{\frac{\alpha-n}{2}}\, \frac{dt}{t} \quad \text{for}\ \operatorname{Re}\alpha > 0,$$

then we have the following claim:

Claim 4.71. *For* $\operatorname{Re}\alpha > 0$, *we have the assertion*

$$G_\alpha \in L^1(\mathbf{R}^n).$$

Proof. We remark that

$$\frac{1}{(4\pi t)^{n/2}} \int_{\mathbf{R}^n} e^{-\frac{|x|^2}{4t}}\, dx = 1,$$

so that

$$
\begin{aligned}
\int_{\mathbf{R}^n} |G_\alpha(x)|\, dx &= \frac{1}{|\Gamma(\alpha/2)|} \int_0^\infty e^{-t}\, t^{(\operatorname{Re}\alpha)/2-1}\, dt \\
&= \frac{\Gamma(\operatorname{Re}\alpha/2)}{|\Gamma(\alpha/2)|}.
\end{aligned}
$$

The proof of Claim 4.71 is complete. □

Therefore, by Young's inequality it follows that

$$
\begin{aligned}
\left\| (I-\Delta)^{-\alpha} f \right\|_p &= \| G_\alpha * f \|_p \\
&\le \frac{\Gamma(\operatorname{Re}\alpha/2)}{|\Gamma(\alpha/2)|} \, \|f\|_p \quad \text{for } f \in L^p(\mathbf{R}^n).
\end{aligned}
$$

(C) Thirdly, if we let

$$\phi(s) := \frac{1}{\sqrt{\pi s}}\, e^{-\frac{\beta^2}{4s}} \quad \text{for } \beta > 0,$$

then we have the formula (see formula (6.8) in Claim 6.1)

$$e^{-\beta} = \int_0^\infty e^{-s}\, \frac{1}{\sqrt{\pi s}}\, e^{-\frac{\beta^2}{4s}}\, ds. \tag{4.53}$$

Indeed, by the *residue theorem* it follows that

$$\int_{-\infty}^\infty \frac{e^{i\beta\tau}}{1+\tau^2}\, d\tau = 2\pi i \operatorname{Res}\left[\frac{e^{i\beta\tau}}{1+\tau^2} \right]_{\tau=i} = \pi e^{-\beta}.$$

Hence we obtain the desired formula (4.53)

$$
\begin{aligned}
e^{-\beta} &= \frac{1}{\pi} \int_{-\infty}^\infty \frac{e^{i\beta\tau}}{1+\tau^2}\, d\tau \\
&= \frac{1}{\pi} \int_{-\infty}^\infty e^{i\beta\tau} \left(\int_0^\infty e^{-(1+\tau^2)s}\, ds \right) d\tau \\
&= \frac{1}{\pi} \int_0^\infty e^{-s} \left(\int_{-\infty}^\infty e^{i\beta\tau}\, e^{-s\tau^2}\, d\tau \right) ds
\end{aligned}
$$

$$= \frac{1}{\pi} \int_0^\infty e^{-s} \left(\sqrt{\frac{\pi}{s}} e^{-\frac{\beta^2}{4s}} \right) ds$$

$$= \int_0^\infty e^{-s} \frac{1}{\sqrt{\pi s}} e^{-\frac{\beta^2}{4s}} ds.$$

Now, by taking

$$\beta := t \sqrt{-\Delta},$$

in formula (4.53), we have the **formal** representation formula

$$e^{-t\sqrt{-\Delta}} f = \int_0^\infty e^{-s} \frac{1}{\sqrt{\pi s}} \left(e^{\frac{t^2}{4s}\Delta} f \right) ds$$

$$= \int_0^\infty \frac{e^{-s}}{\sqrt{\pi s}} \left(\int_{\mathbf{R}^n} K_{t^2/4s}(x-y) f(y) \, dy \right) ds$$

$$= \int_{\mathbf{R}^n} \left(\int_0^\infty \frac{e^{-s}}{\sqrt{\pi s}} K_{t^2/4s}(x-y) \, ds \right) f(y) \, dy.$$

This implies that the convolution kernel of the operator

$$e^{-t\sqrt{-\Delta}}$$

is equal to the following:

$$\int_0^\infty \frac{e^{-s}}{\sqrt{\pi s}} K_{t^2/4s}(x) \, ds. \tag{4.54}$$

However, we have the following claim:

Claim 4.72. *We have, for $t > 0$,*

$$\int_0^\infty \frac{e^{-s}}{\sqrt{\pi s}} K_{t^2/4s}(x) \, ds = \frac{\Gamma((n+1)/2)}{\pi^{(n+1)/2}} \frac{t}{(|x|^2 + t^2)^{(n+1)/2}}. \tag{4.55}$$

Proof. Indeed, by using formula (4.47) we have the desired formula (4.55)

$$\int_0^\infty \frac{e^{-s}}{\sqrt{\pi s}} K_{t^2/4s}(x), ds$$

$$= \int_0^\infty \frac{e^{-s}}{\sqrt{\pi s}} \frac{1}{(\pi t^2/s)^{n/2}} e^{-\frac{|x|^2 s}{t^2}} ds$$

$$= \frac{1}{\pi^{(n+1)/2}} \frac{1}{t^n} \int_0^\infty e^{-s(1+\frac{|x|^2}{t^2})} s^{(n-1)/2} ds$$

$$= \frac{1}{\pi^{(n+1)/2}} \frac{1}{t^n} \left(1 + \frac{|x|^2}{t^2} \right)^{-(n+1)/2} \int_0^\infty e^{-\sigma} \sigma^{(n+1)/2-1} d\sigma$$

$$= \frac{\Gamma((n+1)/2)}{\pi^{(n+1)/2}} \frac{t}{(|x|^2 + t^2)^{(n+1)/2}}.$$

The proof of Claim 4.72 is complete. □

By combining formulas (4.54) and (4.55), we have proved the following theorem:

Theorem 4.73 (the Poisson kernel). *The convolution kernel of the operator*

$$e^{-t\sqrt{-\Delta_x}}$$

is equal to the following:

$$\frac{\Gamma((n+1)/2)}{\pi^{(n+1)/2}} \frac{t}{(|x|^2 + t^2)^{(n+1)/2}}.$$

This is called the Poisson kernel.

More precisely, we can prove that the function

$$
\begin{aligned}
u(x,t) &= e^{-t\sqrt{-\Delta_x}} f(x) \\
&= \frac{\Gamma((n+1)/2)}{\pi^{(n+1)/2}} \int_{\mathbf{R}^n} \frac{t}{(|x-y|^2 + t^2)^{(n+1)/2}} f(y)\, dy
\end{aligned}
$$

is a solution of the following Dirichlet problem

$$
\begin{cases}
\left(\dfrac{\partial^2}{\partial t^2} + \Delta_x \right) u(x,t) = 0 & \text{in } \mathbf{R}_+^{n+1}, \\
u(x,0) = f(x) & \text{on } \partial\mathbf{R}_+^{n+1} = \mathbf{R}^n,
\end{cases}
$$

where

$$\mathbf{R}_+^{n+1} = \left\{ (x,t) \in \mathbf{R}^{n+1} : t > 0 \right\}.$$

4.8 Notes and Comments

Sections 4.1 through 4.3: The function spaces discussed here are adapted from Adams–Fournier [2], Bergh–Löfström [13], Calderón [21], Friedman [43], Taibleson [112] and Triebel [135].

Schwartz [102] and Gelfand–Shilov [49] are the classics for distribution theory. Our treatment in this book follows the expositions of Chazarain–Piriou [26], Hörmander [56] and Treves [134].

Section 4.3: This section is a modern version of trace and sectional trace theorems in terms of pseudo-differential operators.

Sections 4.4 and 4.5: For detailed studies of Fourier integral operators and pseudo-differential operators, the reader is referred to Chazarain–Piriou [26], Duistermaat [31], Duistermaat–Hörmander [32], Hörmander [60], [61], Kumano-go [73], Rempel–Schulze [93] and Taylor [133].

Section 4.6: This section is devoted to a modern Fredholm operator theory of elliptic pseudo-differential operators on a manifold, which is adapted from Palais [86, Chapter VII] and Taira [114, Section 8.7]. Gohberg–Kreĭn [51] is the classic for index theory of linear Fredholm operators in Banach spaces. See also McLean [80, Chapter 2] and Schechter [100, Chapter 7].

Section 4.7: The material is adapted from Folland [41, Chapter 4, Section B].

This chapter is an expanded and revised version of Chapter 3 of the second edition [121].

5

Boutet de Monvel Calculus

In this chapter we introduce the notion of *transmission property* due to Boutet de Monvel [17], [18] and [19], which is a condition about symbols in the normal direction at the boundary. Elliptic boundary value problems cannot be treated directly by pseudo-differential operator methods. It was Boutet de Monvel [19] who brought in the operator-algebraic aspect with his calculus in 1971. He constructed a relatively small "algebra", called the *Boutet de Monvel algebra*, which contains the boundary value problems for elliptic differential operators as well as their parametrices.

We take a close look at Boutet de Monvel's work.

Let Ω be a bounded, domain of Euclidean space $\mathbf{R}^n$ with smooth boundary $\partial\Omega$. Without loss of generality, we may assume that Ω is a relatively compact, open subset of an n-dimensional, compact smooth manifold $M = \widehat{\Omega}$ without boundary (see Figure 5.1 below). The manifold M is called the *double* of Ω (see [83]).

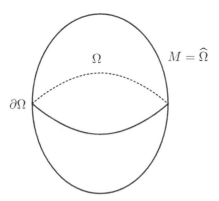

Fig. 5.1. The double $M = \widehat{\Omega}$ of Ω

© Springer Nature Switzerland AG 2020
K. Taira, *Boundary Value Problems and Markov Processes*, Lecture Notes in Mathematics 1499,
https://doi.org/10.1007/978-3-030-48788-1_5

If we study the Dirichlet problem in the domain Ω, it is natural to use the zero extension e^+u of functions u outside of the closure $\overline{\Omega} = \Omega \cup \partial\Omega$:

$$e^+u(x) = \begin{cases} u(x) & \text{for } x \in \overline{\Omega}, \\ 0 & \text{for } x \in M \setminus \overline{\Omega}. \end{cases}$$

The zero extension has a probabilistic interpretation. Namely, this corresponds to stopping the diffusion process with jumps into the double M at the first exit time of the closure $\overline{\Omega}$.

Boutet de Monvel [19] introduced a 2×2 matrix of operators

$$\mathcal{A} = \begin{pmatrix} P_\Omega + G & K \\ T & S \end{pmatrix} : \begin{array}{c} C^\infty(\overline{\Omega}) \\ \oplus \\ C^\infty(\partial\Omega) \end{array} \longrightarrow \begin{array}{c} C^\infty(\overline{\Omega}) \\ \oplus \\ C^\infty(\partial\Omega) \end{array}$$

Here:

(1) P is a pseudo-differential operator on the full manifold M and

$$P_\Omega u = r^+ \left(P(e^+u) \right) = P(e^+u)\big|_\Omega \quad \text{for all } u \in C^\infty(\overline{\Omega}),$$

where e^+u is the extension of u by zero to M and r^+v is the *restriction* $v|_\Omega$ to Ω of a distribution v on M. In view of the *pseudo-local property* of P, we find that the operator P_Ω can be visualized as follows:

$$\boxed{P_\Omega \colon C^\infty(\overline{\Omega}) \xrightarrow{e^+} \mathcal{D}'(M) \xrightarrow{P} \mathcal{D}'(M) \xrightarrow{r^+} C^\infty(\Omega).}$$

The crucial requirement is that the symbol of P has the *transmission property* in order that P_Ω maps $C^\infty(\overline{\Omega})$ into itself. In Section 5.1 we introduce three basic function spaces H, H^+ and H^- (Proposition 5.1). In Section 5.2 we illustrate how the transmission property of the symbol ensures that the associated operator preserves smoothness up to the boundary (Theorems 5.6 and 5.7).

(2) S is a pseudo-differential operator on $\partial\Omega$.

(3) The potential operator K and trace operator T are generalizations of the potentials and trace operators known from the classical theory of elliptic boundary value problems, respectively.

(4) The entry G, a *singular Green operator*, is an operator which is smoothing in the interior Ω while it acts like a pseudo-differential operator in directions tangential to the boundary $\partial\Omega$. As an example, we may take the difference of two solution operators to (invertible) classical boundary value problems with the same differential part in the interior but different boundary conditions. In Section 5.3 we give typical examples of a potential operator K, a trace operator T and a singular Green operator G (see Examples 5.5 through 5.12).

Boutet de Monvel [19] proved that these operator matrices form an algebra in the following sense (see [93, p. 175, Theorem 1 and p. 195, Theorem 2]): Given another element of the calculus, say,

$$\mathcal{A}' = \begin{pmatrix} P'_\Omega + G' & K' \\ T' & S' \end{pmatrix} : \begin{matrix} C^\infty(\overline{\Omega}) \\ \oplus \\ C^\infty(\partial\Omega) \end{matrix} \longrightarrow \begin{matrix} C^\infty(\overline{\Omega}) \\ \oplus \\ C^\infty(\partial\Omega) \end{matrix}$$

the composition $\mathcal{A}'\mathcal{A}$ is again an operator matrix of the type described above. It is worth pointing out here that the product $P'_\Omega P_\Omega$ does not coincide with $(P'P)_\Omega$; in fact, the difference

$$P'_\Omega P_\Omega - (P'P)_\Omega$$

turns out to be a singular Green operator (see [93, p. 100, Lemma 5]).

A general symbol class of the Boutet de Monvel calculus except for the directions tangential to $\partial\Omega$ may be summarized as in Table 5.1 below.

Operators	Pseudo-differential operators with transmission property	Singular Green operators	Potential operators	Trace operators
Notation	P_Ω	G	K	T
Symbol classes	$H = H^+ \oplus H^-$	$H^+ \otimes H^-$	H^+	$H^- = H^-_0 \oplus H'$

Table 5.1. A general symbol class of the Boutet de Monvel calculus

It should be emphasized that the Boutet de Monvel calculus is closely related to the classical Wiener–Hopf technique (see [145], [84]) that remains an extremely important tool for modern scientists, and the areas of application continue to broaden. The last Section 5.4 is devoted to a brief historical perspective of the Wiener–Hopf technique.

5.1 The Spaces H, H^+ and H^-

In this section we introduce three basic function spaces H, H^+ and H^- (Proposition 5.1). We let

$\mathcal{S}(\mathbf{R}) =$ the Schwartz space of all rapidly decreasing functions on $\mathbf{R}$.

The dual space is the space $\mathcal{S}'(\mathbf{R})$ of tempered distributions on $\mathbf{R}$. The Fourier transform

$$\mathcal{F} \colon \mathcal{S}\,(\mathbf{R}) \longrightarrow \mathcal{S}\,(\mathbf{R})$$

will in general be indicated by a hat: For $u \in \mathcal{S}\,(\mathbf{R})$, we let

$$\widehat{u}(\xi) = (\mathcal{F}u)(\xi) = \int_{\mathbf{R}} e^{-i\,x\cdot\xi} u(x)\, dx \quad \text{for } \xi \in \mathbf{R}.$$

Given a distribution u on $\mathbf{R}$, we write $r^{+}u$ for its restriction to the right half-axis $\mathbf{R}_{+}$:

$$r^{+}u = u|_{\mathbf{R}_{+}},$$

where

$$\mathbf{R}_{+} = \{x \in \mathbf{R} : x > 0\}.$$

Similarly, we write $r^{-}u$ for its restriction to the left half-axis $\mathbf{R}_{-}$:

$$r^{-}u = u|_{\mathbf{R}_{-}},$$

where

$$\mathbf{R}_{-} = \{x \in \mathbf{R} : x < 0\}.$$

We let

$$\mathcal{S}\,(\overline{\mathbf{R}_{+}}) = \{r^{+}u : u \in \mathcal{S}(\mathbf{R})\},$$
$$\mathcal{S}\,(\overline{\mathbf{R}_{-}}) = \{r^{-}u : u \in \mathcal{S}(\mathbf{R})\},$$

and

$$\mathcal{S}_{\theta}(\mathbf{R}) = \mathcal{S}(\overline{\mathbf{R}_{+}}) \oplus \mathcal{S}(\overline{\mathbf{R}_{+}}).$$

Given a function f on $\mathbf{R}_{+}$, we denote by $e^{+}f$ its extension by zero to a function on $\mathbf{R}$:

$$\left(e^{+}f\right)(x) = \begin{cases} f(x) & \text{if } x \in \mathbf{R}_{+}, \\ 0 & \text{if } x \in \mathbf{R} \setminus \mathbf{R}_{+}. \end{cases}$$

Similarly, given a function g on $\mathbf{R}_{-}$, we denote by $e^{-}g$ its extension by zero to a function on $\mathbf{R}$:

$$\left(e^{-}g\right)(x) = \begin{cases} g(x) & \text{if } x \in \mathbf{R}_{-}, \\ 0 & \text{if } x \in \mathbf{R}^{n} \setminus \mathbf{R}_{-}, \end{cases}$$

where

$$\mathbf{R}_{-} = \{x \in \mathbf{R} : x < 0\}.$$

We let

$$H^{+} := \mathcal{F}\left(\mathcal{S}(\overline{\mathbf{R}_{+}})\right) = \{(e^{+}u)^{\wedge} : u \in \mathcal{S}(\overline{\mathbf{R}_{+}})\},$$
$$H_{0}^{-} := \mathcal{F}\left(\mathcal{S}(\overline{\mathbf{R}_{-}})\right) = \{(e^{-}u)^{\wedge} : u \in \mathcal{S}(\overline{\mathbf{R}_{-}})\},$$

and

$$H_{0} = H^{+} \oplus H_{0}^{-}.$$

We remark that H^+ and H_0^- are spaces of smooth functions on $\mathbf{R}$ decaying to the first order near infinity. We consider H^+ and H_0^- in the image topology of $\mathcal{S}(\overline{\mathbf{R}_+})$ and $\mathcal{S}(\overline{\mathbf{R}_-})$, respectively. Then the Fourier transform

$$\mathcal{F}\colon \mathcal{S}_\theta(\mathbf{R}) = \mathcal{S}(\overline{\mathbf{R}_+}) \oplus \mathcal{S}(\overline{\mathbf{R}_+}) \longrightarrow H_0 = H^+ \oplus H_0^-$$

is an isomorphism of Fréchet spaces.

Now we define the continuous projections

$$\Pi^\pm := \mathcal{F} r^\pm \mathcal{F}^{-1}$$

as follows:

$$\Pi^+\colon H_0 \xrightarrow{\mathcal{F}^{-1}} \mathcal{S}_\theta(\mathbf{R}) \xrightarrow{r^+} \mathcal{S}(\overline{\mathbf{R}_+}) \xrightarrow{\mathcal{F}} H^+,$$

$$\Pi^-\colon H_0 \xrightarrow{\mathcal{F}^{-1}} \mathcal{S}_\theta(\mathbf{R}) \xrightarrow{r^-} \mathcal{S}(\overline{\mathbf{R}_-}) \xrightarrow{\mathcal{F}} H_0^-.$$

We denote by H' the space of all polynomials, and let

$$H^- = H_0^- \oplus H',$$
$$H = H^+ \oplus H_0^- \oplus H' = H^+ \oplus H^-.$$

We extend the projections $\Pi^\pm$ to the space $H = H_0 \oplus H'$ by setting

$$\begin{cases} \Pi^+ = 0 \quad \text{on } H', \\ \Pi^- = I \quad \text{on } H', \end{cases}$$

so that

$$\Pi^- = I - \Pi^+ \quad \text{on } H.$$

We give simple but important examples of elements of H^+ and H^-:

Example 5.1. Let λ be a positive number. For any non-negative integer k, the function

$$\frac{(\lambda - it)^k}{(\lambda + it)^{k+1}}$$

is the Fourier transform of the product of a *Laguerre polynomial* by an exponential:

$$\varphi_k(x) = \begin{cases} \left(\lambda - \dfrac{d}{dx}\right)^k \left(\dfrac{x^k}{k!} e^{-\lambda x}\right) & \text{if } x > 0, \\ 0 & \text{if } x < 0. \end{cases}$$

In a similar way, a function $f(\xi) \in H^-$ is the Fourier transform of the sum of a function $\varphi_k(-x)$, and of a linear combination of derivatives of the Dirac measure at the origin.

Part (a) of the following Proposition 5.1 is easily verified. Parts (b), (c) and (d) provide Paley–Wiener type characterizations of the spaces H^+, H_0^- and H_0, respectively (see Rempel–Schulze [93, Chapter 2, Subsection 2.1.1]):

Proposition 5.1. *(a) The spaces H^+, H_0^-, H^-, H_0 and H are algebras.*

(b) A function $h(t) \in C^\infty(\mathbf{R})$ belongs to H^+ if and only if it has an analytic extension $h(\zeta)$ to the lower open half-plane $\{\operatorname{Im}\zeta < 0\}$, continuous in the lower closed half-plane $\{\operatorname{Im}\zeta \leq 0\}$, together with an asymptotic expansion

$$h(\zeta) \sim \sum_{k=1}^\infty a_k \zeta^{-k}, \tag{5.1}$$

for $|\zeta| \to \infty$ in the lower closed half-plane $\{\operatorname{Im}\zeta \leq 0\}$, which can be differentiated formally (a complex analytic version).

A function $h(t) \in C^\infty(\mathbf{R})$ belongs to H^+ if and only if it has a unique expansion

$$h(t) = \sum_{k=0}^\infty \alpha_k \frac{(1-it)^k}{(1+it)^{k+1}} = \alpha_0 \frac{1}{1+it} + \alpha_1 \frac{1-it}{(1+it)^2} + \dots,$$

where the coefficients α_k form a rapidly decreasing sequence (a real analytic version).

(c) A function $h(t) \in C^\infty(\mathbf{R})$ belongs to H_0^- if and only if it has an analytic extension $h(\zeta)$ to the upper open half-plane $\{\operatorname{Im}\zeta > 0\}$, continuous in the upper closed half-plane $\{\operatorname{Im}\zeta \geq 0\}$, together with an asymptotic expansion

$$h(\zeta) \sim \sum_{k=1}^\infty a_k \zeta^{-k}, \tag{5.2}$$

for $|\zeta| \to \infty$ in the upper closed half-plane $\{\operatorname{Im}\zeta \geq 0\}$, which can be differentiated formally (a complex analytic version).

A function $h(t) \in C^\infty(\mathbf{R})$ belongs to H_0^- if and only if it has a unique expansion

$$h(t) = \sum_{k=-1}^{-\infty} \alpha_k \frac{(1+it)^{|k|-1}}{(1-it)^{|k|}} = \alpha_{-1} \frac{1}{1-it} + \alpha_{-2} \frac{1+it}{(1-it)^2} + \dots,$$

where the coefficients α_k form a rapidly decreasing sequence (a real analytic version).

(d) $h(t) \in C^\infty(\mathbf{R})$ belongs to $H_0 = H^+ \oplus H_0^-$ if and only if it has an expansion

$$h(\zeta) \sim \sum_{k=1}^\infty a_k \zeta^{-k},$$

for $|\zeta| \to \infty$ in $\mathbf{R}$, which can be differentiated formally (a complex analytic version).

A function $h(t) \in C^\infty(\mathbf{R})$ belongs to H_0 if and only if it has a unique expansion

$$h(t) = \sum_{k=-\infty}^{\infty} \alpha_k \frac{(1-it)^k}{(1+it)^{k+1}},$$

where the coefficients α_k form a rapidly decreasing sequence (a real analytic version).

Proposition 5.2. *Assume that $h(t)$ is analytic on the real line $\mathbf{R}$ and meromorphic at infinity. If $h(\zeta)$ is a meromorphic extension of h to the whole complex plane $\mathbf{C}$, then we let*

$$\int^+ h(t)\, dt = \int_\gamma h(\zeta)\, d\zeta, \tag{5.3}$$

where γ is a large circle in the upper open half-plane $\{\operatorname{Im}\zeta > 0\}$.
    Then we have the following two assertions (i) and (ii):

(i) If $f(t) \in H^- = H_0^- \oplus H'$, it follows that $\int^+ f(t)\, dt = 0$.
(ii) If $h(t) \in H^+$ is integrable on $\mathbf{R}$, we have the formula

$$\int^+ h(t)\, dt = \int_{-\infty}^\infty h(t)\, dt = 0.$$

Remark 5.3. Assume that $f(t)$ and $g(t)$ are analytic on the real line $\mathbf{R}$ and meromorphic at infinity. By Proposition 5.2, we have the formula

$$\int^+ f(t)g(t)\, dt = \begin{cases} \int^+ f(t) \cdot \Pi^- g(t)\, dt & \text{if } f \in H^+, \\ \int^+ f(t) \cdot \Pi^+ g(t)\, dt & \text{if } f \in H^-. \end{cases}$$

5.2 Transmission Property of Pseudo-Differential Operators

Given an arbitrary pseudo-differential operator A, it is in general not true that the operator

$$A_\Omega : u \longmapsto r^+ \left(A(e^+ u) \right) = A(e^+ u)\big|_\Omega$$

maps functions u which are smooth up to the boundary $\partial\Omega$ into functions $A_\Omega u$ with the same property. The crucial requirement here is that the symbol of A has the transmission property. On one hand, this restricts the class of boundary value problems in the calculus, on the other hand, however, it ensures that solutions to elliptic equations with smooth data are smooth; it therefore helps to avoid problems with singularities of solutions at the boundary.
    In this section, following Boutet de Monvel [17], [18] and [19] we introduce the notion of transmission property which is a condition about symbols in the normal direction at the boundary in order to ensure the stated regularity property (see Rempel–Schulze [93]).

If $x = (x_1, x_2, \ldots, x_n)$ is the variable in $\mathbf{R}^n$, we write

$$x = (x', x_n), \quad x' = (x_1, x_2, \ldots, x_{n-1}),$$

so $x' \in \mathbf{R}^{n-1}$ is the tangential component of x with dual variables

$$\xi' = (\xi_1, \xi_2, \ldots, \xi_{n-1}),$$

and $x_n \in \mathbf{R}$ is its normal component with dual variable ξ_n.

If $m \in \mathbf{R}$, we let

$$S_{1,0}^m \left(\overline{\mathbf{R}_+^n} \times \mathbf{R}^n \right) = \text{the space of symbols } a(x, \xi) \text{ in } S_{1,0}^m(\mathbf{R}_+^n \times \mathbf{R}^n)$$
$$\text{which have an extension in } S_{1,0}^m(\mathbf{R}^n \times \mathbf{R}^n).$$

Now we are in a position to define the transmission property of symbols in the class $S_{1,0}^m(\overline{\mathbf{R}_+^n} \times \mathbf{R}^n)$:

Definition 5.1. A symbol $a(x, \xi) \in S_{1,0}^m(\overline{\mathbf{R}_+^n} \times \mathbf{R}^n)$ is said to have the *transmission property* with respect to the boundary $\mathbf{R}^{n-1}$ if all its derivatives

$$\left(\frac{\partial}{\partial x_n} \right)^\gamma a(x', 0, \xi', \nu) \quad \text{for } \gamma \geq 0$$

admit an expansion of the form

$$a_{(\gamma)}(x', 0, \xi', \nu) \tag{5.4}$$
$$= \left(\frac{\partial}{\partial x_n} \right)^\gamma a(x', 0, \xi', \nu)$$
$$= \sum_{j=0}^m b_{\gamma j}(x', \xi') \left(\nu \langle \xi' \rangle^{-1} \right)^j + \sum_{k=-\infty}^\infty a_{\gamma k}(x', \xi') \frac{(1 - i\nu \langle \xi' \rangle^{-1})^k}{(1 + i\nu \langle \xi' \rangle^{-1})^{k+1}} \quad \text{for } \nu \in \mathbf{R},$$

where (cf. [93, p. 119, Proposition 3])

$$b_{\gamma j}(x', \xi') \in S_{1,0}^m \left(\mathbf{R}_{x'}^{n-1} \times \mathbf{R}_{\xi'}^{n-1} \right),$$

and

$$a_{\gamma k}(x', \xi') \in S_{1,0}^m \left(\mathbf{R}_{x'}^{n-1} \times \mathbf{R}_{\xi'}^{n-1} \right)$$

form a rapidly decreasing sequence with respect to k, and

$$\langle \xi' \rangle = \left(1 + |\xi'|^2 \right)^{1/2}.$$

It is easy to see that the expansion condition (5.4) is equivalent to the following:

$$\left(\frac{\partial}{\partial x_n} \right)^\gamma a(x', 0, \xi', \langle \xi' \rangle \nu) \in S_{1,0}^m \left(\mathbf{R}_{x'}^{n-1} \times \mathbf{R}_{\xi'}^{n-1} \right) \widehat{\otimes}_\pi H_\nu.$$

Here the space $S_{1,0}^m(\mathbf{R}^{n-1} \times \mathbf{R}^{n-1}) \widehat{\otimes}_\pi H_\nu$ is the completed π-topology (or projective topology) tensor product of the Fréchet spaces $S_{1,0}^m(\mathbf{R}^{n-1} \times \mathbf{R}^{n-1})$ and H_ν (see Schaefer [99, Chapter III, Section 6], Treves [134, Chapter 45]).

For classical pseudo-differential symbols of integer order, the transmission property can be expressed via *homogeneity properties* of the terms in the asymptotic expansion (see [17, Proposition (2.3.2)]):

Remark 5.4. If $a(x, \xi) \in S_{1,0}^m(\mathbf{R}^n \times \mathbf{R}^n)$ is a classical symbol of order $m \in \mathbf{Z}$ with an asymptotic expansion

$$a(x, \xi) \sim \sum_{j=0}^\infty a_{m-j}(x, \xi),$$

where $a_{m-j}(x, \xi) \in S_{1,0}^{m-j}(\mathbf{R}^n \times \mathbf{R}^n)$ is positively homogeneous of degree $m - j$ for $|\xi| \geq 1$, then it is easy to verify (see [93, p. 123, Proposition 1], [61, Lemma 18.2.14]) that $a(x, \xi)$ has the transmission property if and only if we have, for all multi-indices $\alpha = (\alpha', \alpha_n)$,

$$\left(\frac{\partial}{\partial x_n}\right)^{\alpha_n} \left(\frac{\partial}{\partial \xi'}\right)^{\alpha'} a_{m-j}(x', 0, 0, +1) \tag{5.5}$$

$$= (-1)^{m-j-|\alpha'|} \left(\frac{\partial}{\partial x_n}\right)^{\alpha_n} \left(\frac{\partial}{\partial \xi'}\right)^{\alpha'} a_{m-j}(x', 0, 0, -1).$$

It should be emphasized that this is no longer true for non-classical pseudo-differential symbols. For details, see the analysis by Grubb–Hörmander [53].

We give two typical examples of classical symbols having the transmission property:

Example 5.2. (1) The symbol $a(x, \xi)$ of the second-order, elliptic differential operator

$$-\Delta + 1 = -\sum_{i=1}^n \frac{\partial^2}{\partial x_i^2} + 1$$

is given by the formula $(m = 2)$

$$a(x, \xi) = |\xi|^2 + 1 = \langle \xi' \rangle^2 + \nu^2 \in S_{1,0}^2\left(\mathbf{R}_{x'}^{n-1} \times \mathbf{R}_{\xi'}^{n-1}\right) \widehat{\otimes}_\pi H_\nu$$

(2) The symbol $a(x, \xi)^{-1}$ is given by the formula $(m = -2)$

$$a(x, \xi)^{-1} = \frac{1}{|\xi|^2 + 1}$$

$$= \frac{1}{2 \langle \xi' \rangle} \frac{1}{\langle \xi' \rangle + i\nu} + \frac{1}{2 \langle \xi' \rangle} \frac{1}{\langle \xi' \rangle - i\nu} \in S_{1,0}^{-2}\left(\mathbf{R}_{x'}^{n-1} \times \mathbf{R}_{\xi'}^{n-1}\right) \widehat{\otimes}_\pi H_\nu.$$

We recall (Subsection 4.1.5) that the symbol

$$\frac{1}{|\xi|^2 + 1}$$

corresponds to the Bessel potential of order 2 (see Aronszajn–Smith [12])

$$G_2 * u(x) = \int_{\mathbf{R}^n} G_2(x - y)u(y)\,dy,$$

and further that the function $G_2(x)$ can be expressed in the form (see Stein [108, Chapter V, Section 3])

$$G_2(x) = \frac{1}{(4\pi)^{n/2}} \int_0^\infty e^{-t - \frac{|x|^2}{4t}} t^{\frac{2-n}{2}} \frac{dt}{t}$$

$$= \frac{1}{2^n \pi^{n/2}} K_{(n-2)/2}(|x|) \frac{1}{|x|^{(n-2)/2}},$$

where $K_{(n-2)/2}(z)$ is the modified Bessel function of the third kind (see Watson [142]).

We let

$$L_{1,0}^m(\overline{\mathbf{R}_+^n}) = \text{the space of pseudo-differential operators in } L_{1,0}^m(\mathbf{R}_+^n)$$

which can be extended to a pseudo-differential operator
in the class $L_{1,0}^m(\mathbf{R}^n)$.

Now we can define pseudo-differential operators in the class $L_{1,0}^m(\mathbf{R}^n)$ to have the transmission property:

Definition 5.2. A pseudo-differential operator $A \in L_{1,0}^m(\overline{\mathbf{R}_+^n})$ is said to have the *transmission property* with respect to the boundary $\mathbf{R}^{n-1}$ if any complete symbol of A has the transmission property with respect to the boundary $\mathbf{R}^{n-1}$.

As a first useful example, we formulate the pseudo-differential operators that have the transmission property in dimension one ([19, Theorem 2.7]):

Theorem 5.5. *Let $A(x, D)$ be a pseudo-differential operator defined in a neighborhood of the half-line $x \geq 0$. In order that the transmission property with respect to the origin holds true for A, it is necessary and sufficient that A admits a decomposition*

$$A = A_0 + A_1 + A_2.$$

Here:

(1) The symbol of A_0 vanishes to the infinite order at the origin $x = 0$.
(2) A_1 is a differential operator with smooth coefficients.
(3) The distribution kernel of A_2 is a function $f(x, y)$ which is smooth up to the diagonal for $x > y$, and also for $x < y$.

We give two simple examples A and B of pseudo-differential operators which have the transmission property (see Example 4.1):

Example 5.3. (i) The symbol $a(x, \xi)$ of $A \in L^m_{1,0}(\mathbf{R}^n)$ is given by the formula

$$a(x, \xi) = \langle \xi' \rangle^m \exp\left[-\frac{\xi_n^2}{\langle \xi' \rangle^2}\right] \in S^m_{1,0}\left(\mathbf{R}^{n-1}_{x'} \times \mathbf{R}^{n-1}_{\xi'}\right) \widehat{\otimes}_\pi H_{\xi_n}.$$

(ii) The symbol $b(x, \xi)$ of $B \in L^m_{1,0}(\mathbf{R}^n)$ is given by the formula

$$b(x, \xi)$$
$$= \langle \xi' \rangle^m \left(1 - \frac{2\xi_n^2}{\langle \xi' \rangle^2}\right) \exp\left[-\frac{\xi_n^2}{\langle \xi' \rangle^2}\right] \in S^m_{1,0}\left(\mathbf{R}^{n-1}_{x'} \times \mathbf{R}^{n-1}_{\xi'}\right) \widehat{\otimes}_\pi H_{\xi_n}.$$

Here
$$\xi = (\xi', \xi_n), \quad \xi' = (\xi_1, \xi_2, \dots, \xi_{n-1}), \quad \langle \xi' \rangle = \sqrt{1 + |\xi'|^2}.$$

Now we illustrate how the transmission property of the symbol ensures that the associated operator preserves smoothness up to the boundary. If A is a pseudo-differential operator in $L^m_{1,0}(\overline{\mathbf{R}^n_+})$, then we define a new operator

$$A_{\mathbf{R}^n_+} : C^\infty_{(0)}\left(\overline{\mathbf{R}^n_+}\right) \longrightarrow C^\infty\left(\overline{\mathbf{R}^n_+}\right)$$
$$u \longmapsto A(e^+ u)\big|_{\mathbf{R}^n_+},$$

where
$$C^\infty_{(0)}(\overline{\mathbf{R}^n_+}) = \left\{u\big|_{\overline{\mathbf{R}^n_+}} : u \in C^\infty_0(\mathbf{R}^n)\right\}.$$

The transmission property implies that if u is smooth up to the boundary, then so is $A_{\mathbf{R}^n_+} u$. More precisely, we have the following theorem (see [26, Chapitre V, Théorème 2.5]; [93, p. 137, Corollary 3 and p. 168, Theorem 8]):

Theorem 5.6. *Let $A \in L^m_{1,0}(\overline{\mathbf{R}^n_+})$. Then we have the following two assertions (i) and (ii) (see Figure 5.2 below):*

(i) *If A has the* transmission property *with respect to the boundary $\mathbf{R}^{n-1}$, then the operator $A_{\mathbf{R}^n_+}$ maps $C^\infty_{(0)}\left(\overline{\mathbf{R}^n_+}\right)$ continuously into $C^\infty\left(\overline{\mathbf{R}^n_+}\right)$.*

(ii) *If A has the* transmission property *with respect to the boundary $\mathbf{R}^{n-1}$, then the operator $A_{\mathbf{R}^n_+}$ maps $C^{k+\theta}_{\mathrm{comp}}\left(\overline{\mathbf{R}^n_+}\right)$ continuously into $C^{k-m+\theta}_{\mathrm{loc}}\left(\overline{\mathbf{R}^n_+}\right)$ for any integer $k \geq m$ and $0 < \theta < 1$.*
Here

$$C^{k+\theta}_{\mathrm{comp}}\left(\overline{\mathbf{R}^n_+}\right) = \left(C^{k+\theta}(\mathbf{R}^n) \cap C_0(\mathbf{R}^n)\right)\big|_{\mathbf{R}^n_+}$$
$$= \left\{u\big|_{\mathbf{R}^n_+} : u \in C^{k+\theta}\left(\mathbf{R}^n\right) \text{ with compact support in } \mathbf{R}^n\right\}$$

and

$$\begin{array}{ccc} C_{\text{comp}}^{k+\theta}\left(\overline{\mathbf{R}_+^n}\right) & \xrightarrow{A_{\mathbf{R}_+^n}} & C_{\text{loc}}^{k-m+\theta}\left(\overline{\mathbf{R}_+^n}\right) \\ \uparrow & & \uparrow \\ C_{(0)}^{\infty}\left(\overline{\mathbf{R}_+^n}\right) & \xrightarrow{\quad\quad} & C^{\infty}\left(\overline{\mathbf{R}_+^n}\right) \\ & A_{\mathbf{R}_+^n} & \end{array}$$

Fig. 5.2. The mapping properties of the operator $A_{\mathbf{R}_+^n}$

$$C_{\text{loc}}^{k-m+\theta}\left(\overline{\mathbf{R}_+^n}\right) = C_{\text{loc}}^{k-m+\theta}(\mathbf{R}^n)\big|_{\mathbf{R}_+^n}$$

$$= \left\{ u\big|_{\mathbf{R}_+^n} : \varphi\, u \in C^{k-m+\theta}\left(\mathbf{R}^n\right) \text{ for all } \varphi \in C_0^{\infty}\left(\mathbf{R}^n\right) \right\}.$$

Moreover, it should be noticed that the notion of transmission property is invariant under a change of coordinates which preserves the boundary. Hence the notion of transmission property can be transferred to a bounded domain Ω of Euclidean space $\mathbf{R}^n$ with smooth boundary $\partial\Omega$. Indeed, if Ω is a relatively compact open subset of the double $M = \widehat{\Omega}$ (see Figure 5.1), then the notion of transmission property can be extended to the class $L_{1,0}^m(M)$, upon using local coordinate systems flattening out the boundary $\partial\Omega$.

Then we have the following theorem (see [93, p. 139, Theorem 4 and p. 176, Theorem 1]):

Theorem 5.7. *Let $A \in L_{1,0}^m(M)$. Then we have the following two assertions (i) and (ii) (see Figure 5.3 below):*

(i) If A has the transmission property *with respect to the boundary $\partial\Omega$, then the operator*

$$A_\Omega : C^{\infty}(\overline{\Omega}) \longrightarrow C^{\infty}(\Omega)$$
$$u \longmapsto A(e^+ u)\big|_\Omega$$

maps $C^{\infty}(\overline{\Omega})$ continuously into itself, where $e^+ u$ is the zero extension of u to M by zero outside of $\overline{\Omega}$.

(ii) If A has the transmission property *with respect to the boundary $\partial\Omega$, then the operator A_Ω maps $C^{k+\theta}(\overline{\Omega})$ continuously into $C^{k-m+\theta}(\overline{\Omega})$ for any integer $k \geq m$ and $0 < \theta < 1$.*

Following Schrohe [101, Definition 2.3], we can introduce an equivalent definition of the transmission property for general symbols $a(x, y, \xi)$ in the class $S_{1,0}^m(\mathbf{R}^n \times \mathbf{R}^n \times \mathbf{R}^n)$. To do this, let H be the linear space of all complex-valued functions $f(t)$ on the real line $\mathbf{R}$ which are smooth and have a regular pole at infinity. More precisely, a function $f(t) \in C^{\infty}(\mathbf{R})$ belongs to H if and only if it has a unique expansion

$$\begin{array}{ccc}
C^{k+\theta}\left(\overline{\Omega}\right) & \xrightarrow{\;\;A_\Omega\;\;} & C^{k-m+\theta}\left(\overline{\Omega}\right) \\
\big\uparrow & & \big\uparrow \\
C^\infty\left(\overline{\Omega}\right) & \xrightarrow[\;\;A_\Omega\;\;]{} & C^\infty\left(\overline{\Omega}\right)
\end{array}$$

Fig. 5.3. The mapping properties of the operator A_Ω

$$f(t) = \sum_{s=1}^{N} \alpha_s t^s + \sum_{k=-\infty}^{\infty} \alpha_k \left(\frac{1-it}{1+it}\right)^k, \quad i = \sqrt{-1},$$

where the coefficients α_k form a rapidly decreasing sequence (see [93, Chapter 2, Section 2.1.1]).

Now we are in a position to define the transmission property of symbols in the class $S^m_{1,0}(\mathbf{R}^n \times \mathbf{R}^n \times \mathbf{R}^n)$:

Definition 5.3. A symbol

$$a(x, y, \xi) = a(x', x_n, y', y_n, \xi', \xi_n) \in S^m_{1,0}(\mathbf{R}^n_{x',x_n} \times \mathbf{R}^n_{y',y_n} \times \mathbf{R}^n_{\xi',\xi_n})$$

is said to have the *transmission property* at $\{x_n = y_n = 0\}$, provided that we have, for all non-negative integers k and ℓ,

$$\frac{\partial^{k+\ell} a}{\partial x_n^k \partial y_n^\ell}(x', 0, y', 0, \xi', \langle\xi'\rangle\xi_n) \in S^m_{1,0}\left(\mathbf{R}^{n-1}_{x'} \times \mathbf{R}^{n-1}_{y'} \times \mathbf{R}^{n-1}_{\xi'}\right) \widehat{\otimes}_\pi H_{\xi_n},$$

where

$$\langle\xi'\rangle = \left(1 + |\xi'|^2\right)^{1/2}.$$

The subscripts x', y', ξ' and ξ_n are used in order to indicate the variable for which we have the corresponding property.

5.3 Trace, Potential and Singular Green Operators on $\mathbf{R}^n_+$

In this section we give basic definitions and properties of classes of trace, potential and singular Green operators on the half-space $\mathbf{R}^n_+$. The presentation here is based on Rempel–Schulze [93, Chapter 2, Subsection 2.3.2].

Pseudo-differential trace operators T are a natural generalization of the usual differential trace operators from elliptic boundary value problems, while potential operators K can be described as the adjoints of trace operators T with respect to the L^2 inner products. Singular Green operators G are introduced in order to get algebra of matrices of operators of the form

$$\mathcal{A} = \begin{pmatrix} P_\Omega + G & K \\ T & S \end{pmatrix},$$

called the *Boutet de Monvel algebra*. In fact, a typical example of a singular Green is the composition $K \circ T$ of a trace operator T and a potential operator K (see Example 5.11 below).

A general symbol class of the operator $\mathcal{A}$ may be expressed schematically as follows:

$$\begin{pmatrix} P_\Omega + G & K \\ T & S \end{pmatrix}$$

$$\Longleftrightarrow \begin{pmatrix} S_{1,0}^* \widehat{\otimes}_\pi H_\nu + S_{1,0}^* \widehat{\otimes}_\pi H_\nu^+ \widehat{\otimes}_\pi H_{d,\tau}^- & S_{1,0}^* \widehat{\otimes}_\pi H_\nu^+ \\ S_{1,0}^* \widehat{\otimes}_\pi H_{d,\tau}^- & S_{1,0}^* \end{pmatrix}.$$

Example 5.4. We give a concrete example of the symbol class of the operator $\mathcal{A}$ (see Examples 5.5 through 5.12):

$$\begin{pmatrix} \dfrac{1}{\langle \xi' \rangle^2 + \nu^2} + \dfrac{1}{\langle \xi' \rangle + i\nu} \dfrac{1}{\langle \xi' \rangle - i\tau} \dfrac{1}{\langle \xi' \rangle + i\nu} & \dfrac{1}{\langle \xi' \rangle + i\nu} \\ \dfrac{1}{\langle \xi' \rangle - i\tau} & \dfrac{1}{\langle \xi' \rangle} \end{pmatrix}.$$

Here and in the following we use the notation:

$$\xi = (\xi', \nu) = (\xi_1, \xi_2, \ldots, \xi_{n-1}, \nu) \in \mathbf{R}^n,$$

$$\langle \xi' \rangle = \sqrt{1 + |\xi'|^2},$$

$$\xi = (\xi', \nu) = (\xi_1, \xi_2, \ldots, \xi_{n-1}, \nu) \in \mathbf{R}^n \quad \text{for potential operators } K,$$

$$\xi = (\xi', \tau) = (\xi_1, \xi_2, \ldots, \xi_{n-1}, \tau) \in \mathbf{R}^n \quad \text{for trace operators } T.$$

5.3.1 Potential Operators on $\mathbf{R}_+^n$

A function

$$k(x', y', \xi', \nu) \in C^\infty(\mathbf{R}^{n-1} \times \mathbf{R}^{n-1} \times \mathbf{R}^{n-1} \times \mathbf{R})$$

is called a *potential symbol* of order m if it satisfies the condition

$$k(x', y', \xi', \nu) = \sum_{j=0}^\infty k_j(x', y', \xi') \frac{(1 - i\nu \langle \xi' \rangle^{-1})^j}{(1 + i\nu \langle \xi' \rangle^{-1})^{j+1}}. \tag{5.6}$$

Here the symbols $k_j(x', y', \xi')$ form a rapidly decreasing sequence in the class $S_{1,0}^m(\mathbf{R}^{n-1} \times \mathbf{R}^{n-1} \times \mathbf{R}^{n-1})$. The condition (5.6) is equivalent to the following:

$$k(x', y', \xi', \langle\xi'\rangle \nu) \in S^m_{1,0}\left(\mathbf{R}^{n-1}_{x'} \times \mathbf{R}^{n-1}_{y'} \times \mathbf{R}^{n-1}_{\xi'}\right) \widehat{\otimes}_\pi H^+_\nu.$$

Then the *potential operator*

$$\boxed{K: C^\infty_0(\mathbf{R}^{n-1}) \longrightarrow C^\infty(\overline{\mathbf{R}^n_+})}$$

is defined as an oscillatory integral by the formula

$$(Kv)(x', x_n) \tag{5.7}$$
$$= \frac{1}{(2\pi)^{n-1}} \iint_{\mathbf{R}^{n-1}\times\mathbf{R}^{n-1}} e^{i(x'-y')\cdot\xi'} \, \check{k}(x', y', \xi', x_n)\, v(y')\, dy'\, d\xi'$$
for all $v \in C^\infty_0(\mathbf{R}^{n-1})$.

Here:

$$\check{k}(x', y', \xi', x_n) = \frac{1}{2\pi} \int_{\mathbf{R}} e^{i x_n \nu}\, k(x', y', \xi', \nu)\, d\nu.$$

If the potential symbol $k(x', y', \xi', \nu)$ does not depend on x' and y', then the oscillatory integral (5.7) reduces to the following:

$$(Kv)(x', x_n) = \frac{1}{(2\pi)^n} \iint_{\mathbf{R}^n} e^{i(x'\cdot\xi' + x_n \nu)}\, k(\xi', \nu)\, \widehat{v}(\xi')\, d\xi'\, d\nu$$
for all $v \in C^\infty_0(\mathbf{R}^{n-1})$.

The mapping properties of the potential operator K of order m can be visualized as in Figure 5.4 below.

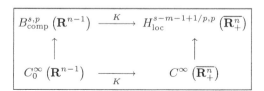

Fig. 5.4. The condition that $1 < p < \infty$ and $s \in \mathbf{R}$

Here

$$B^{s,p}_{\mathrm{comp}}(\mathbf{R}^{n-1}) = B^{s,p}(\mathbf{R}^{n-1}) \cap \mathcal{E}'(\mathbf{R}^{n-1})$$
$$= \left\{u \in B^{s,p}(\mathbf{R}^{n-1}) : \operatorname{supp} u \text{ is compact in } \mathbf{R}^{n-1}\right\}$$

and

$$H^{s-m-1+1/p,p}_{\mathrm{loc}}(\overline{\mathbf{R}^n_+}) = H^{s-m-1+1/p,p}_{\mathrm{loc}}(\mathbf{R}^n)\Big|_{\mathbf{R}^n_+}$$
$$= \left\{u|_{\mathbf{R}^n_+} : \varphi u \in H^{s-m-1+1/p,p}(\mathbf{R}^n) \text{ for all } \varphi \in C^\infty_0(\mathbf{R}^n)\right\}.$$

We give a typical example of potential operators:

Example 5.5 (Poisson operator). We let

$$k\left(x', \xi', \nu\right) = \frac{1}{i\nu + \langle \xi' \rangle}$$

$$= \frac{1}{\langle \xi' \rangle} \frac{1}{1 + i\nu \langle \xi' \rangle^{-1}} \in S_{1,0}^{-1}\left(\mathbf{R}_{x'}^{n-1} \times \mathbf{R}_{\xi'}^{n-1}\right) \widehat{\otimes}_\pi H_\nu^+, \quad m = -1.$$

Then we have, by formula (5.3) and the residue theorem,

$$\check{k}(x', \xi', x_n)$$

$$= \frac{1}{2\pi} \int_{\mathbf{R}} e^{i x_n \nu} \frac{1}{i\nu + \langle \xi' \rangle} \, d\nu = \frac{1}{2\pi} \int^+ e^{i x_n \nu} \frac{1}{i\nu + \langle \xi' \rangle} \, d\nu$$

$$= e^{-x_n \langle \xi' \rangle},$$

and the formula for the potential operator

$$(Kv)(x', x_n) = \frac{1}{(2\pi)^{n-1}} \int_{\mathbf{R}^{n-1}} e^{i x' \cdot \xi'} e^{-x_n \langle \xi' \rangle} \, \widehat{v}(\xi') \, d\xi'$$

for all $v \in C_0^\infty(\mathbf{R}^{n-1})$.

We remark that the function $u(x', x_n) = (Kv)(x', x_n)$ is a (unique) solution of the Dirichlet problem

$$\begin{cases} \Delta u(x', x_n) = 0 & \text{in } \mathbf{R}_+^n, \\ u(x', 0) = v(x') & \text{on } \mathbf{R}^{n-1} = \partial\mathbf{R}_+^n, \end{cases}$$

where

$$\Delta = \frac{\partial^2}{\partial x_1^2} + \frac{\partial^2}{\partial x_2^2} + \ldots + \frac{\partial^2}{\partial x_n^2}$$

is the usual Laplacian. Namely, the potential operator K is the *Poisson operator* or *Poisson kernel* for the Dirichlet problem (see formulas (5.1) and (5.2) in Section 5.1).

5.3.2 Trace Operators on $\mathbf{R}_+^n$

A function

$$t\left(x', y', \xi', \tau\right) \in C^\infty(\mathbf{R}^{n-1} \times \mathbf{R}^{n-1} \times \mathbf{R}^{n-1} \times \mathbf{R})$$

is called a *trace symbol* of order m and type d if it satisfies the condition

$$t\left(x', y', \xi', \tau\right) = \sum_{k=0}^{d-1} b_k(x', y', \xi') \left(\tau \langle \xi' \rangle^{-1}\right)^k \tag{5.8}$$

$$+ \sum_{j=0}^\infty t_j\left(x', y', \xi'\right) \frac{(1 + i\tau \langle \xi' \rangle^{-1})^j}{(1 - i\tau \langle \xi' \rangle^{-1})^{j+1}}.$$

Here
$$b_k\left(x', y', \xi'\right) \in S_{1,0}^m(\mathbf{R}^{n-1} \times \mathbf{R}^{n-1} \times \mathbf{R}^{n-1})$$
and the symbols $t_j\left(x', y', \xi'\right)$ form a rapidly decreasing sequence in the class $S_{1,0}^m(\mathbf{R}^{n-1} \times \mathbf{R}^{n-1} \times \mathbf{R}^{n-1})$. The condition (5.8) is equivalent to the following:

$$t\left(x', y', \xi', \langle\xi'\rangle\,\tau\right) \in S_{1,0}^m\left(\mathbf{R}_{x'}^{n-1} \times \mathbf{R}_{y'}^{n-1} \times \mathbf{R}_{\xi'}^{n-1}\right) \widehat{\otimes}_\pi H_{d,\tau}^-.$$

Then the *trace operator*

$$\boxed{T \colon C_{(0)}^\infty(\overline{\mathbf{R}_+^n}) \longrightarrow C^\infty(\mathbf{R}^{n-1})}$$

is defined as an oscillatory integral by the formula

$$(Tu)(x') \tag{5.9}$$
$$= \frac{1}{(2\pi)^{n-1}} \iint_{\mathbf{R}^{n-1} \times \mathbf{R}^{n-1}} e^{i(x'-y')\cdot\xi'}$$
$$\times \left[\frac{1}{2\pi}\int_{\mathbf{R}} t(x', y', \xi', \tau)\left(\int_{\mathbf{R}} e^{-i\,y_n\,\tau}\,(e^+u)(y', y_n)dy_n\right)d\tau\right] dy'\,d\xi'$$
for all $u \in C_{(0)}^\infty(\overline{\mathbf{R}_+^n})$.

Here e^+u is the extension of u to $\mathbf{R}^n$ by zero outside of $\overline{\mathbf{R}_+^n}$

$$(e^+u)(y', y_n) = \begin{cases} u(y', y_n) & \text{for } y_n \geq 0, \\ 0 & \text{for } y_n < 0. \end{cases}$$

If the trace symbol $t(x', y', \xi', \nu)$ does not depend on x' and y', then the oscillatory integral (5.9) reduces to the following (see Remark 5.3):

$$(Tu)(x') = \frac{1}{(2\pi)^{n-1}} \int_{\mathbf{R}^{n-1}} e^{i\,x'\cdot\xi'} \left(\frac{1}{2\pi}\int^+ t(\xi', \tau)\,\widehat{e^+u}(\xi', \tau)\,d\tau\right) d\xi'$$
for all $u \in C_{(0)}^\infty(\overline{\mathbf{R}_+^n})$.

We let

$$H_{\text{comp}}^{s,p}\left(\overline{\mathbf{R}_+^n}\right) := (H^{s,p}(\mathbf{R}^n) \cap \mathcal{E}'(\mathbf{R}^n))|_{\mathbf{R}_+^n}$$
$$= \left\{u|_{\mathbf{R}_+^n} : u \in H^{s,p}(\mathbf{R}^n) \text{ with compact support in } \mathbf{R}^n\right\}$$

and

$$B_{\text{loc}}^{s-m-1/p,p}\left(\mathbf{R}^{n-1}\right)$$
$$:= \left\{u \in \mathcal{D}'(\mathbf{R}^{n-1}) : \psi u \in B^{s-m-1/p,p}(\mathbf{R}^{n-1}) \text{ for all } \psi \in C_0^\infty(\mathbf{R}^{n-1})\right\}.$$

Then the mapping properties of the trace operator T of order m and type d can be visualized as in Figure 5.5 below.

We give three typical examples of trace operators:

$$
\begin{array}{ccc}
H^{s,p}_{\text{comp}}\left(\overline{\mathbf{R}^n_+}\right) & \xrightarrow{\quad T \quad} & B^{s-m-1/p,p}_{\text{loc}}\left(\mathbf{R}^{n-1}\right) \\[2mm]
\uparrow & & \uparrow \\[2mm]
C^\infty_{(0)}\left(\overline{\mathbf{R}^n_+}\right) & \xrightarrow{\quad T \quad} & C^\infty\left(\mathbf{R}^{n-1}\right)
\end{array}
$$

Fig. 5.5. The condition that $1 < p < \infty$, $s - 1/p \notin \mathbf{Z}$ and $s > d - 1 + 1/p$

Example 5.6 (Dirichlet condition). We let

$$
t_0(x', \xi', \tau) = 1 \in S^0_{1,0}\left(\mathbf{R}^{n-1}_{x'} \times \mathbf{R}^{n-1}_{\xi'}\right) \widehat{\otimes}_\pi H^-_{1,\tau}, \quad m = 0, \ d = 1.
$$

Then we have the formula

$$
\frac{1}{2\pi} \int_{\mathbf{R}} 1 \left(\int_{\mathbf{R}} e^{-i\,y_n\,\tau}\,(e^+ u)(y', y_n)\, dy_n \right) d\tau
$$
$$
= \frac{1}{2\pi} \int_{\mathbf{R}} e^{i\,0\,\tau} \left(\int_{\mathbf{R}} e^{-i\,y_n\,\tau}\,(e^+ u)(y', y_n)\, dy_n \right) d\tau
$$
$$
= (e^+ u)(y', 0) = u(x', 0) \quad \text{for all } u \in C^\infty_{(0)}(\overline{\mathbf{R}^n_+}).
$$

Therefore, we have the formula for the trace operator

$$
(T_0 u)(x')
$$
$$
= \frac{1}{(2\pi)^{n-1}} \iint_{\mathbf{R}^{n-1} \times \mathbf{R}^{n-1}} e^{i(x'-y')\cdot\xi'}
$$
$$
\times \left[\frac{1}{2\pi} \int_{\mathbf{R}} e^{i\,0\,\tau} \left(\int_{\mathbf{R}} e^{-i\,y_n\,\tau}\,(e^+ u)(y', y_n)\, dy_n \right) d\tau \right] dy'\, d\xi'
$$
$$
= \frac{1}{(2\pi)^{n-1}} \iint_{\mathbf{R}^{n-1} \times \mathbf{R}^{n-1}} e^{i(x'-y')\cdot\xi'}\, u(y', 0)\, dy'\, d\xi'
$$
$$
= u(x', 0)
$$
$$
= (\gamma_0 u)(x') \quad \text{for all } u \in C^\infty_{(0)}(\overline{\mathbf{R}^n_+}).
$$

This proves that the trace operator $T_0 = \gamma_0$ is the Dirichlet boundary condition (see Section 5.3).

Example 5.7 (Neumann condition). We let

$$
t_1(x', \xi', \tau) = i\tau
$$
$$
= i\,\langle \xi' \rangle \left(\langle \xi' \rangle^{-1} \right) \in S^1_{1,0}\left(\mathbf{R}^{n-1}_{x'} \times \mathbf{R}^{n-1}_{\xi'}\right) \widehat{\otimes}_\pi H^-_{2,\tau}, \quad m = 1, \ d = 2.
$$

Then we have the formula

$$\frac{1}{2\pi}\left(\int_{\mathbf{R}}(i\tau)\int_{\mathbf{R}}e^{-iy_n\,\tau}\,(e^+u)(y',y_n)\,dy_n\right)d\tau$$

$$=\int_{\mathbf{R}}e^{-y_n\langle\xi'\rangle}\,(e^+u)(y',y_n)\,dy_n$$

$$=\frac{\partial}{\partial x_n}\left[\frac{1}{2\pi}\int_{\mathbf{R}}e^{i\tau\,x_n}\left(\int_{\mathbf{R}}e^{-iy_n\,\tau}\,(e^+u)(y',y_n)\,dy_n\right)d\tau\right]\Bigg|_{x_n=0}$$

$$=\frac{\partial}{\partial x_n}\left((e^+u)(y',x_n)\right)\Bigg|_{x_n=0}$$

$$=\frac{\partial u}{\partial x_n}(y',0)\quad\text{for all }u\in C_{(0)}^\infty(\overline{\mathbf{R}_+^n}).$$

Therefore, we have the formula for the trace operator

$$(T_1u)(x')=\frac{1}{(2\pi)^{n-1}}\iint_{\mathbf{R}^{n-1}\times\mathbf{R}^{n-1}}e^{i(x'-y')\cdot\xi'}$$

$$\times\left[\frac{1}{2\pi}\int_{\mathbf{R}}(i\tau)\left(\int_{\mathbf{R}}e^{-iy_n\,\tau}\,(e^+u)(y',y_n)\,dy_n\right)d\tau\right]dy'\,d\xi'$$

$$=\frac{1}{(2\pi)^{n-1}}\iint_{\mathbf{R}^{n-1}\times\mathbf{R}^{n-1}}e^{i(x'-y')\cdot\xi'}\,\frac{\partial u}{\partial x_n}(y',0)\,dy'\,d\xi'$$

$$=\frac{\partial u}{\partial x_n}(x',0)$$

$$=(\gamma_1u)(x')\quad\text{for all }u\in C_{(0)}^\infty(\overline{\mathbf{R}_+^n}).$$

This proves that the trace operator $T_1=\gamma_1$ is the Neumann boundary condition (see Section 5.3).

Example 5.8 (Trace symbol and operator). We let

$$t_2(x',\xi',\tau)=\frac{1}{\langle\xi'\rangle-i\tau}=\frac{1}{\langle\xi'\rangle}\frac{1}{1-i\tau\,\langle\xi'\rangle^{-1}}$$

$$\in S_{1,0}^{-1}\left(\mathbf{R}_{x'}^{n-1}\times\mathbf{R}_{\xi'}^{n-1}\right)\widehat{\otimes}_\pi H_{0,\tau}^-,\quad m=-1,\ d=0.$$

Then we have the formula for symbols

$$\frac{1}{2\pi}\int_{\mathbf{R}}e^{-iy_n\,\tau}\,\frac{1}{\langle\xi'\rangle-i\tau}\,d\tau=\frac{1}{2\pi}\overline{\int^+e^{iy_n\,\tau}\,\frac{1}{i\tau+\langle\xi'\rangle}\,d\tau}\tag{5.10}$$

$$=e^{-y_n\langle\xi'\rangle}.$$

This proves that

$$\frac{1}{2\pi}\int_{\mathbf{R}}\frac{1}{\langle\xi'\rangle-i\tau}\int_{\mathbf{R}}e^{-iy_n\,\tau}\,(e^+u)(y',y_n)\,d\tau\,dy_n$$

$$=\int_{\mathbf{R}}e^{-y_n\langle\xi'\rangle}\,(e^+u)(y',y_n)\,dy_n=\int_0^\infty e^{-y_n\langle\xi'\rangle}\,u(y',y_n)\,dy_n.$$

Therefore, we have the formula for the trace operator

$$
\begin{aligned}
& (T_2 u)\,(x') \\
& = \frac{1}{(2\pi)^{n-1}} \iint_{\mathbf{R}^{n-1} \times \mathbf{R}^{n-1}} e^{i(x'-y')\cdot\xi'} \left(\int_0^\infty e^{-y_n \langle \xi' \rangle} u(y', y_n)\, dy_n \right) dy'\, d\xi' \\
& = \frac{1}{(2\pi)^{n-1}} \int_{\mathbf{R}^{n-1}} e^{i\,x'\cdot\xi'} \left(\int_0^\infty e^{-y_n \langle \xi' \rangle} \tilde{u}(\xi', y_n)\, dy_n \right) d\xi' \\
& \quad \text{for all } u \in C^\infty_{(0)}(\overline{\mathbf{R}^n_+}),
\end{aligned}
$$

and so

$$
\boxed{T_2 \colon C^\infty_{(0)}(\overline{\mathbf{R}^n_+}) \longrightarrow C^\infty(\mathbf{R}^{n-1}).}
$$

5.3.3 Singular Green Operators on $\mathbf{R}^n_+$

A function

$$
g\,(x', y', \xi', \nu, \tau) \in C^\infty(\mathbf{R}^{n-1} \times \mathbf{R}^{n-1} \times \mathbf{R}^{n-1} \times \mathbf{R} \times \mathbf{R})
$$

is called a *singular Green symbol* of order m and type d if it satisfies the condition

$$
\begin{aligned}
& g\,(x', y', \xi', \nu, \tau) && (5.11) \\
& = \sum_{\ell=0}^{d-1} \left(\sum_{j=0}^{\infty} c_{j\ell}\,(x', y', \xi') \frac{(1 - i\nu \langle \xi' \rangle^{-1})^j}{(1 + i\nu \langle \xi' \rangle^{-1})^{j+1}} \right) \left(\tau \langle \xi' \rangle^{-1} \right)^\ell \\
& \quad + \sum_{j=0}^{\infty} \sum_{\ell=0}^{\infty} b_{j\ell}(x', y', \xi') \frac{(1 - i\nu \langle \xi' \rangle^{-1})^j}{(1 + i\nu \langle \xi' \rangle^{-1})^{j+1}} \frac{(1 + i\tau \langle \xi' \rangle^{-1})^\ell}{(1 - i\tau \langle \xi' \rangle^{-1})^{\ell+1}}.
\end{aligned}
$$

Here the symbols $c_{j\ell}(x', y', \xi')$ form a rapidly decreasing sequence in the class $S^m_{1,0}(\mathbf{R}^{n-1} \times \mathbf{R}^{n-1} \times \mathbf{R}^{n-1})$ with respect to j, for each $\ell = 0, 1, 2, \ldots, d-1$, and the symbols $b_{j\ell}(x', y', \xi')$ form a rapidly decreasing double sequence in the class $S^m_{1,0}(\mathbf{R}^{n-1} \times \mathbf{R}^{n-1} \times \mathbf{R}^{n-1})$. The condition (5.11) is equivalent to the following:

$$
g(x', y', \xi', \langle \xi' \rangle \nu, \langle \xi' \rangle \tau) \in S^m_{1,0}\left(\mathbf{R}^{n-1}_{x'} \times \mathbf{R}^{n-1}_{y'} \times \mathbf{R}^{n-1}_{\xi'} \right) \widehat{\otimes}_\pi H^+_\nu \widehat{\otimes}_\pi H^-_{d,\tau}.
$$

Then the *singular Green operator*

$$
\boxed{G \colon C^\infty_{(0)}(\overline{\mathbf{R}^n_+}) \longrightarrow C^\infty(\overline{\mathbf{R}^n_+})}
$$

is defined as an oscillatory integral by the formula

$$
(Gu)(x', x_n) \tag{5.12}
$$

$$= \frac{1}{(2\pi)^{n-1}} \iint_{\mathbf{R}^{n-1} \times \mathbf{R}^{n-1}} e^{i(x'-y')\cdot\xi'}$$

$$\times \left[\frac{1}{2\pi} \int_{\mathbf{R}} \breve{g}(x',y',\xi',x_n,\tau) \left(\int_{\mathbf{R}} e^{-i y_n \tau} (e^+ u)(y',y_n) dy_n \right) d\tau \right] dy' \, d\xi'$$

for all $u \in C^\infty_{(0)}(\overline{\mathbf{R}^n_+})$.

Here

$$\breve{g}(x',y',\xi',x_n,\tau) = \frac{1}{2\pi} \int_{\mathbf{R}} e^{i x_n \nu} g(x',y',\xi',\nu,\tau) \, d\nu.$$

If the singular Green symbol $g(x',y',\xi',\nu,\tau)$ does not depend on x' and y', then the oscillatory integral (5.12) reduces to the following (see Remark 5.3):

$$(Gu)(x',x_n)$$
$$= \frac{1}{(2\pi)^n} \int_{\mathbf{R}^n} e^{i(x'\cdot\xi'+x_n \nu)} \left(\frac{1}{2\pi} \int^+ g(\xi',\nu,\tau) \widehat{e^+u}(\xi',\tau) d\tau \right) d\xi' \, d\nu$$

for all $u \in C^\infty_{(0)}(\overline{\mathbf{R}^n_+})$.

The mapping properties of the singular Green operator G of order $m-1$ can be visualized in Figure 5.6 below.

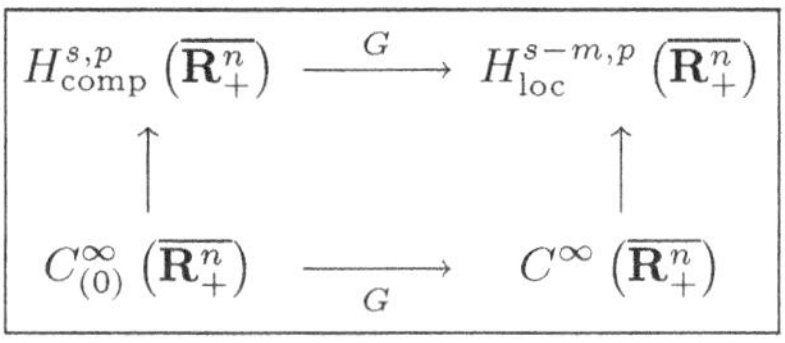

Fig. 5.6. The condition that $1 < p < \infty$ and $s - 1/p \notin \mathbf{Z}$

We give three typical examples of singular Green operators:

Example 5.9 (Composition of Poisson operator and Dirichlet condition). We let

$$g_0(x',y',\xi',\nu,\tau) = k(x',\xi',\nu) \cdot t_0(y',\xi',\tau)$$
$$= \frac{1}{\langle\xi'\rangle + i\nu} \cdot 1 \in S^{-1}_{1,0}\left(\mathbf{R}^{n-1}_{x'} \times \mathbf{R}^{n-1}_{\xi'}\right) \widehat{\otimes}_\pi H^+_\nu \widehat{\otimes}_\pi H^-_{1,\tau},$$

where (see Examples 5.5 and 5.6)

$$k(x',\xi',\nu) = \frac{1}{\langle\xi'\rangle + i\nu} = \frac{1}{\langle\xi'\rangle} \frac{1}{1 + i\nu \langle\xi'\rangle^{-1}} \in S^{-1}_{1,0}\left(\mathbf{R}^{n-1}_{x'} \times \mathbf{R}^{n-1}_{\xi'}\right) \widehat{\otimes}_\pi H^+_\nu,$$

$$t_0(y', \xi', \tau) = 1 \in S^0_{1,0}\left(\mathbf{R}^{n-1}_{x'} \times \mathbf{R}^{n-1}_{\xi'}\right) \widehat{\otimes}_\pi H^-_{1,\tau}.$$

Then we have, by Examples 5.5 and 5.6,

$$(G_0 u)(x', x_n) = (K(T_0 u))(x', x_n)$$
$$= \frac{1}{(2\pi)^{n-1}} \int_{\mathbf{R}^{n-1}} e^{i\,x'\cdot\xi'} e^{-x_n\langle\xi'\rangle} \widetilde{u}(\xi', 0)\, d\xi' \qquad \text{for all } u \in C^\infty_{(0)}(\overline{\mathbf{R}^n_+}),$$

and so

$$\boxed{G_0 = K \circ T_0 \colon C^\infty_{(0)}(\overline{\mathbf{R}^n_+}) \longrightarrow C^\infty(\overline{\mathbf{R}^n_+}).}$$

Here

$$\widetilde{u}(\xi', 0) = \int_{\mathbf{R}^{n-1}} e^{-i\,y'\cdot\xi'} u(y', 0)\, dy'$$

is the partial Fourier transform of the function $u(y', 0)$ with respect to the variable y'.

Example 5.10 (Composition of Poisson operator and Neumann condition). We let

$$g_1(x', y', \xi', \nu, \tau) = k(x', \xi', \nu) \cdot t_1(y', \xi', \tau)$$
$$= \frac{1}{\langle\xi'\rangle + i\nu} \cdot i\tau \in S^0_{1,0}\left(\mathbf{R}^{n-1}_{x'} \times \mathbf{R}^{n-1}_{\xi'}\right) \widehat{\otimes}_\pi H^+_\nu \widehat{\otimes}_\pi H^-_{2,\tau},$$

where (see Examples 5.5 and 5.6)

$$k(x', \xi', \nu) = \frac{1}{\langle\xi'\rangle + i\nu} = \frac{1}{\langle\xi'\rangle}\, \frac{1}{1 + i\nu\langle\xi'\rangle^{-1}} S^{-1}_{1,0}\left(\mathbf{R}^{n-1}_{x'} \times \mathbf{R}^{n-1}_{\xi'}\right) \widehat{\otimes}_\pi H^+_\nu,$$
$$t_1(y', \xi', \tau) = i\tau = i\langle\xi'\rangle\left(\tau\langle\xi'\rangle^{-1}\right) \in S^1_{1,0}\left(\mathbf{R}^{n-1}_{x'} \times \mathbf{R}^{n-1}_{\xi'}\right) \widehat{\otimes}_\pi H^-_{2,\tau}.$$

Then we have, by Examples 5.5 and 5.7,

$$(G_1 u)(x', x_n) = (K(T_1 u))(x', x_n)$$
$$= \frac{1}{(2\pi)^{n-1}} \int_{\mathbf{R}^{n-1}} e^{i\,x'\cdot\xi'} e^{-x_n\langle\xi'\rangle} \frac{\partial\widetilde{u}}{\partial x_n}(\xi', 0)\, d\xi' \quad \text{for all } u \in C^\infty_{(0)}(\overline{\mathbf{R}^n_+}),$$

and so

$$\boxed{G_1 = K \circ T_1 \colon C^\infty_{(0)}(\overline{\mathbf{R}^n_+}) \longrightarrow C^\infty(\overline{\mathbf{R}^n_+}).}$$

Here

$$\frac{\partial\widetilde{u}}{\partial x_n}(\xi', 0 = \int_{\mathbf{R}^{n-1}} e^{-i\,y'\cdot\xi'} \frac{\partial u}{\partial x_n}(y', 0)\, dy'$$

is the partial Fourier transform of the function $\partial u/\partial x_n(y', 0)$ with respect to the variable y'.

Example 5.11 (Singular Green symbol and operator). We let

$$g_2(x', y', \xi', \nu, \tau) = k(x', \xi', \nu) \cdot t_2(y', \xi', \tau)$$

$$= \frac{1}{\langle \xi' \rangle + i\nu} \frac{1}{\langle \xi' \rangle - i\tau} \in S_{1,0}^{-2} \left(\mathbf{R}_{x'}^{n-1} \times \mathbf{R}_{\xi'}^{n-1} \right) \widehat{\otimes}_\pi H_\nu^+ \widehat{\otimes}_\pi H_{0,\tau}^-,$$

where (see Example 5.5 and 5.8)

$$k(x', \xi', \nu) = \frac{1}{\langle \xi' \rangle + i\nu} = \frac{1}{\langle \xi' \rangle} \frac{1}{1 + i\nu \langle \xi' \rangle^{-1}} \in S_{1,0}^{-1} \left(\mathbf{R}_{x'}^{n-1} \times \mathbf{R}_{\xi'}^{n-1} \right) \widehat{\otimes}_\pi H_\nu^+,$$

$$t_2(y', \xi', \tau) = \frac{1}{\langle \xi' \rangle - i\tau} = \frac{1}{\langle \xi' \rangle} \frac{1}{1 - i\tau \langle \xi' \rangle^{-1}} \in S_{1,0}^{-1} \left(\mathbf{R}_{x'}^{n-1} \times \mathbf{R}_{\xi'}^{n-1} \right) \widehat{\otimes}_\pi H_{0,\tau}^-.$$

Then we have the formula for symbols

$$\check{g}_2(x', y', \xi', x_n, \tau) = \frac{1}{2\pi} \int_{\mathbf{R}} e^{i x_n \nu} \frac{1}{i\nu + \langle \xi' \rangle} d\nu \cdot \frac{1}{\langle \xi' \rangle - i\tau}$$

$$= \frac{1}{2\pi} \int^+ e^{i x_n \nu} \frac{1}{i\nu + \langle \xi' \rangle} d\nu \cdot \frac{1}{\langle \xi' \rangle - i\tau}$$

$$= \frac{1}{\langle \xi' \rangle - i\tau} e^{-x_n \langle \xi' \rangle},$$

and also, by formula (5.10),

$$\frac{1}{2\pi} \int_{\mathbf{R}} \check{g}_2(x', y', \xi', x_n, \tau) \left(\int_{\mathbf{R}} e^{-i y_n \tau} (e^+ u)(y', y_n) \, dy_n \right) d\tau$$

$$= e^{-x_n \langle \xi' \rangle} \int_{\mathbf{R}} \frac{1}{2\pi} \frac{1}{\langle \xi' \rangle - i\tau} \left(\int_{\mathbf{R}} e^{-i y_n \tau} (e^+ u)(y', y_n) \, dy_n \right) d\tau$$

$$= e^{-x_n \langle \xi' \rangle} \int_{\mathbf{R}} \left(\frac{1}{2\pi} \int_{\mathbf{R}} e^{-i y_n \tau} \frac{1}{\langle \xi' \rangle - i\tau} d\tau \right) (e^+ u)(y', y_n) \, dy_n$$

$$= e^{-x_n \langle \xi' \rangle} \int_{\mathbf{R}} e^{-y_n \langle \xi' \rangle} (e^+ u)(y', y_n) \, dy_n$$

$$= e^{-x_n \langle \xi' \rangle} \int_0^\infty e^{-y_n \langle \xi' \rangle} u(y', y_n) \, dy_n.$$

Therefore, we have the formula for the singular Green operator

$$(G_2 u)(x', x_n) = (K(T_2 u))(x', x_n)$$

$$= \frac{1}{(2\pi)^{n-1}} \iint_{\mathbf{R}^{n-1} \times \mathbf{R}^{n-1}} e^{i(x'-y') \cdot \xi'} e^{-x_n \langle \xi' \rangle}$$

$$\times \left(\int_0^\infty e^{-y_n \langle \xi' \rangle} u(y', y_n) \, dy_n \right) dy' \, d\xi'$$

$$= \frac{1}{(2\pi)^{n-1}} \int_{\mathbf{R}^{n-1}} e^{i x' \cdot \xi'} e^{-x_n \langle \xi' \rangle} \left(\int_0^\infty e^{-y_n \langle \xi' \rangle} \tilde{u}(\xi', y_n) \, dy_n \right) d\xi'$$

for all $u \in C_{(0)}^\infty(\overline{\mathbf{R}_+^n})$,

and so

$$\boxed{G_2 = K \circ T_2 \colon C_{(0)}^\infty(\overline{\mathbf{R}_+^n}) \longrightarrow C^\infty(\overline{\mathbf{R}_+^n}).}$$

5.3.4 Boundary Operators on $\mathbf{R}^{n-1}$

Finally, we give two typical examples of boundary operators:

Example 5.12 (Composition of Neumann condition and Poisson operator). We let

$$t_1(y', \xi', \tau) = i\tau \in S_{1,0}^1\left(\mathbf{R}_{x'}^{n-1} \times \mathbf{R}_{\xi'}^{n-1}\right) \widehat{\otimes}_\pi H_{2,\tau}^-,$$

$$k(x', \xi', \nu) = \frac{1}{\langle \xi' \rangle + i\nu} = \frac{1}{\langle \xi' \rangle} \frac{1}{1 + i\nu \langle \xi' \rangle^{-1}} \in S_{1,0}^{-1}\left(\mathbf{R}_{x'}^{n-1} \times \mathbf{R}_{\xi'}^{n-1}\right) \widehat{\otimes}_\pi H_\nu^+.$$

Then we have the composition formula for symbols (see [19, formula (1.13), 7])

$$\frac{1}{2\pi} \int^+ \frac{i\nu}{\langle \xi' \rangle + i\nu} \, d\nu = -\langle \xi' \rangle \in S_{1,0}^1(\mathbf{R}^{n-1} \times \mathbf{R}^{n-1}).$$

Therefore, we have the composition formula for the operators T_1 and K

$$((T_1 \circ K) v)(x') = (T_1(Kv))(x')$$

$$= \frac{1}{(2\pi)^{n-1}} \int_{\mathbf{R}^{n-1}} e^{i x' \cdot \xi'} \left(-\langle \xi' \rangle \right) \widehat{v}(\xi') \, d\xi' \quad \text{for all } v \in C_0^\infty(\mathbf{R}^{n-1}),$$

and so

$$\boxed{T_1 \circ K \colon C_0^\infty(\mathbf{R}^{n-1}) \longrightarrow C^\infty(\mathbf{R}^{n-1}).}$$

This operator $T_1 \circ K$ is called the *Dirichlet-to-Neumann operator* (see Section 6.4 in Chapter 6).

Example 5.13 (Composition of potential and trace operators). We let

$$k(x', \xi', \nu) = \frac{1}{\langle \xi' \rangle + i\nu} = \frac{1}{\langle \xi' \rangle} \frac{1}{1 + i\nu \langle \xi' \rangle^{-1}} \in S_{1,0}^{-1}\left(\mathbf{R}_{x'}^{n-1} \times \mathbf{R}_{\xi'}^{n-1}\right) \widehat{\otimes}_\pi H_\nu^+,$$

and

$$t_3(x', \xi', \tau) = \frac{q(x', \xi')}{\langle \xi' \rangle - i\tau}$$

$$= \frac{q(x', \xi')}{\langle \xi' \rangle} \frac{1}{1 - i\tau \langle \xi' \rangle^{-1}} \in S_{1,0}^k\left(\mathbf{R}_{x'}^{n-1} \times \mathbf{R}_{\xi'}^{n-1}\right) \widehat{\otimes}_\pi H_{0,\tau}^-, \quad d = 0,$$

where

$$q(x', \xi') \in S_{1,0}^{k+1}(\mathbf{R}^{n-1} \times \mathbf{R}^{n-1}).$$

Then we have the composition formula for symbols (see [19, formula (1.13), 7])

$$\frac{1}{2\pi} \int_{\mathbf{R}} \frac{1}{\langle \xi' \rangle + i\nu} \frac{q(x', \xi')}{\langle \xi' \rangle - i\nu} \, d\nu$$

$$= q(x', \xi') \frac{1}{2\pi} \int^{+} \frac{1}{\langle \xi' \rangle + i\nu} \frac{1}{\langle \xi' \rangle - i\nu} \, d\nu$$

$$= \frac{q(x', \xi')}{2 \langle \xi' \rangle} \frac{1}{2\pi} \int^{+} \left(\frac{1}{\langle \xi' \rangle + i\nu} + \frac{1}{\langle \xi' \rangle - i\nu} \right) d\nu$$

$$= \frac{q(x', \xi')}{2 \langle \xi' \rangle} \in S_{1,0}^{k}(\mathbf{R}^{n-1} \times \mathbf{R}^{n-1}).$$

Therefore, we have the composition formula for the operators T_3 and K

$$((T_3 \circ K) \, v) \, (x') = (T_3 \, (Kv)) \, (x')$$

$$= \frac{1}{(2\pi)^{n-1}} \int_{\mathbf{R}^{n-1}} e^{i x' \cdot \xi'} \frac{q(x' \, \xi')}{2 \langle \xi' \rangle} \, \widehat{v}(\xi') \, d\xi' \quad \text{for all } v \in C_0^{\infty}(\mathbf{R}^{n-1}),$$

and so

$$\boxed{T_3 \circ K \colon C_0^{\infty}(\mathbf{R}^{n-1}) \longrightarrow C^{\infty}(\mathbf{R}^{n-1}).}$$

5.4 Historical Perspective of the Wiener–Hopf Technique

It should be emphasized that the Boutet de Monvel calculus is closely related to the classical Wiener–Hopf technique (see [145], [84]) that remains a source of inspiration to Mathematicians, Physicists and Engineers working in many diverse fields, and the areas of application continue to broaden.

The Wiener–Hopf technique was first propounded as a means to solve, for a given function $f(x)$, an integral equation of the form

$$\int_0^\infty k(x - y) f(y) \, dy = g(x) \quad \text{for } 0 < x < \infty. \tag{5.13}$$

Here:

(i) $k(x - y)$ is a known difference kernel.
(ii) $g(x)$ is a specified function defined over $\mathbf{R}_+ = (0, \infty)$.

Full details can be found in the textbook by Noble [84].

The method proceeds by extending the domain of the integral equation (5.13) to negative real values of x, that is,

$$\int_0^\infty k(x - y) f(y) \, dy = \begin{cases} g(x) & \text{for } 0 < x < \infty, \\ h(x) & \text{for } -\infty < x < 0, \end{cases} \tag{5.14}$$

where $h(x)$ is unknown. Then the Fourier transform of equation (5.14) yields the typical Wiener–Hopf equation

$$G_+(\alpha) + H_-(\alpha) = F_+(\alpha)\,K(\alpha). \tag{5.15}$$

Here:

(a) $F_+(\alpha)$ is the half-range Fourier transform defined over the positive real axis $\mathbf{R}_+ = (0, \infty)$ of the unknown function $f(x)$.

$$F_+(\alpha) = \int_0^\infty e^{-i\alpha x}\, f(x)\, dx.$$

(b) $H_-(\alpha)$ is the half-range Fourier transform defined over the negative real axis $\mathbf{R}_- = (-\infty, 0)$ of the unknown function $h(x)$.

$$H_-(\alpha) = \int_{-\infty}^0 e^{-i\alpha x}\, h(x)\, dx.$$

(c) $G_+(\alpha)$ is the half-range Fourier transform defined over the positive real axis $\mathbf{R}_+ = (0, \infty)$ of the known function $g(x)$.

$$G_+(\alpha) = \int_0^\infty e^{-i\alpha x}\, g(x)\, dx.$$

(d) $K(\alpha)$ is the full-range Fourier transform defined over the real axis $\mathbf{R} = (-\infty, \infty)$ of the known function $k(x)$.

$$K(\alpha) = \int_{-\infty}^0 e^{-i\alpha x}\, k(x)\, dx.$$

We remark that the product form of the right-hand side of equation (5.15) is due to the fact that the original integral operator is of *convolution* type. The subscripts $+$ and $-$ indicate that respective functions are analytic in upper and lower half regions of the complex α-plane.

The Wiener–Hopf procedure hinges on finding a product-factorization for the Fourier transformed kernel in the form

$$K(\alpha) = K_+(\alpha)\,K_-(\alpha). \tag{5.16}$$

For example, the functions $K_\pm(\alpha)$ are explicitly given by the formulas:

$$K_-(\alpha) = \exp\left[-\frac{1}{2\pi i}\int_{-\infty}^\infty \frac{\ln K(z)}{z-\alpha}\, dz\right] \quad \text{for } \operatorname{Im}\alpha < 0,$$

$$K_+(\alpha) = \exp\left[\frac{1}{2\pi i}\int_{-\infty}^\infty \frac{\ln K(z)}{z-\alpha}\, dz\right] \quad \text{for } \operatorname{Im}\alpha > 0.$$

By the factorization (5.16), we can rewrite the equation (5.15) as follows:

$$\frac{G_+(\alpha)}{K_-(\alpha)} + \frac{H_-(\alpha)}{K_-(\alpha)} = F_+(\alpha)\,K_+(\alpha). \tag{5.17}$$

It should be noticed that the factors on the right-hand side are zero-free in their indicated half planes of analyticity.

Moreover, we consider a sum-factorization of the form

$$\frac{G_+(\alpha)}{K_-(\alpha)} = L_+(\alpha) + L_-(\alpha). \tag{5.18}$$

For example, the functions $L_\pm(\alpha)$ are explicitly given by the formulas:

$$L_-(\alpha) = -\frac{1}{2\pi i}\int_{-\infty}^{\infty}\frac{G_+(z)}{K_-(z)}\frac{1}{z-\alpha}\,dz \quad \text{for } \operatorname{Im}\alpha < 0,$$

$$L_+(\alpha) = \frac{1}{2\pi i}\int_{-\infty}^{\infty}\frac{G_+(z)}{K_-(z)}\frac{1}{z-\alpha}\,dz \quad \text{for } \operatorname{Im}\alpha > 0.$$

By using the factorization (5.18), we can express the equation (5.17) in the form

$$L_-(\alpha) + \frac{H_-(\alpha)}{K_-(\alpha)} = F_+(\alpha)\,K_+(\alpha) - L_+(\alpha). \tag{5.19}$$

The left-hand side of the equation (5.19) is analytic in the lower half-plane, while the right-hand side of the equation (5.19) is analytic in the overlapping upper half region. Arguments involving analytic continuation enable both sides of the equation (5.19) to be equated to an entire function $E(\alpha)$. Moreover, physical constraints on the behavior of $f(x)$, $g(x)$ and $k(x)$ as $x \to 0$, and their Fourier transformed quantities in the equation (5.19) as $|\alpha| \to \infty$ allows us to specify the function $E(\alpha)$:

$$E(\alpha) = L_-(\alpha) + \frac{H_-(\alpha)}{K_-(\alpha)} = F_+(\alpha)\,K_+(\alpha) - L_+(\alpha).$$

In this way, the functions $H_-(\alpha)$ and $F_+(\alpha)$ are (uniquely) determined respectively as follows:

$$H_-(\alpha) = (E(\alpha) - L_-(\alpha))\,K_-(\alpha),$$

$$F_+(\alpha) = \frac{E(\alpha) + L_+(\alpha)}{K_+(\alpha)}.$$

Finally, by using the Fourier inversion formula we can determine the unknown function

$$f(x) = \mathcal{F}^*\left(\frac{E(\alpha) + L_+(\alpha)}{K_+(\alpha)}\right) = \frac{1}{2\pi}\int_{-\infty}^{\infty} e^{i x \alpha}\,\frac{E(\alpha) + L_+(\alpha)}{K_+(\alpha)}\,d\alpha.$$

5.5 Notes and Comments

The material of this chapter is adapted from Boutet de Monvel [17], [18], [19], Noble [84], Rempel–Schulze [93], Schrohe [101] and also Taira [122, Chapter 7, Section 7.7].

Section 5.3: This section is devoted to several important examples of Boutet de Monvel calculus with the special emphasis on elliptic boundary value problems in terms of pseudo-differential operators. These examples are taken from Taira [124] and [126].

6

L^p Theory of Elliptic Boundary Value Problems

In this chapter we consider the non-homogeneous general Robin problem

$$\begin{cases} Au = f & \text{in } \Omega, \\ B\gamma u = a(x')\,\dfrac{\partial u}{\partial \nu} + b(x')u\Big|_{\partial\Omega} = \varphi & \text{on } \partial\Omega \end{cases} \tag{6.1}$$

under the following two conditions (H.1) and (H.2) (corresponding to conditions (A) and (B) with $\mu = a$ and $\gamma = -b$):

(H.1) $a(x') \geq 0$ and $b(x') \geq 0$ on $\partial\Omega$.
(H.2) $a(x') + b(x') > 0$ on $\partial\Omega$.

Here $\nu = -\mathbf{n}$ is the unit *outward* normal to the boundary $\partial\Omega$ (see Figure 6.1 below).

The general Robin boundary operator $B\gamma$ is defined as follows:

$$B\gamma = a(x')\,\gamma_1 + b(x')\,\gamma_0, \tag{6.2}$$

where the trace map $\gamma = (\gamma_0, \gamma_1)$ is given by the formulas

$$\begin{cases} \gamma_0 u = u|_{\partial\Omega}, \\ \gamma_1 u = \dfrac{\partial u}{\partial \nu}\big|_{\partial\Omega} = -\dfrac{\partial u}{\partial \mathbf{n}}\big|_{\partial\Omega}. \end{cases}$$

Section 6.1 is devoted to the study of the classical surface and volume potentials arising in boundary value problems for elliptic differential operators, in terms of pseudo-differential operators, This calculus of pseudo-differential operators is applied to elliptic boundary value problems in Part III.

In Section 6.2 we consider the Dirichlet problem in the framework of Sobolev spaces of L^p type. This is a generalization of the classical potential approach to the Dirichlet problem.

© Springer Nature Switzerland AG 2020
K. Taira, *Boundary Value Problems and Markov Processes*, Lecture Notes in Mathematics 1499,
https://doi.org/10.1007/978-3-030-48788-1_6

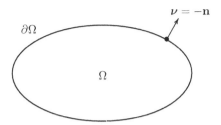

Fig. 6.1. The bounded domain Ω and the *outward normal* ν to the boundary $\partial\Omega$

In Section 6.3 we formulate elliptic boundary value problems in the framework of Sobolev spaces of L^p type. The pseudo-differential operator approach to elliptic boundary value problems can be traced back to the pioneering work of Calderón [22] in early 1960s ([57], [105]).

In Section 6.4, by using the Poisson operator $\mathcal{P}$ and the Green operator $\mathcal{G}_N$ for the Neumann problem we show that the general Robin problem (6.1) can be reduced to the study of a pseudo-differential operator

$$T = B\gamma\mathcal{P} = a(x')\,\gamma_1\mathcal{P} + b(x')\,\gamma_0\mathcal{P} = -a(x')\,\Pi + b(x')$$

on the boundary, where Π is the *Dirichlet-to-Neumann operator* given by the formula

$$\Pi\varphi := -\gamma_1\left(\mathcal{P}\varphi\right) = \left.\frac{\partial}{\partial\mathbf{n}}\left(\mathcal{P}\varphi\right)\right|_{\partial\Omega} \qquad \text{for all } \varphi \in C^\infty(\partial\Omega). \tag{6.3}$$

The virtue of this reduction to the boundary is that there is no difficulty in taking adjoints or transposes after restricting the attention to the boundary, whereas boundary value problems in general do not have adjoints or transposes. This allows us to discuss the *existence* theory more easily.

In Section 6.5 we study the non-homogeneous general Robin problem (6.1) and the non-homogeneous Neumann problem

$$\begin{cases} Av = g & \text{in } \Omega, \\ \left.\dfrac{\partial v}{\partial\mathbf{n}}\right|_{\partial D} = \psi & \text{on } \partial\Omega \end{cases} \tag{6.4}$$

in the framework of L^p Sobolev spaces from the viewpoint of the Boutet de Monvel calculus. Then we derive an index formula of Agranovič–Dynin type for the Neumann problem (6.4) and the general Robin problem (6.1) in the framework of L^p Sobolev spaces (Theorem 6.26).

In Section 6.6 we study an intimate relationship between the Dirichlet-to-Neumann operator Π and the reflecting diffusion in a bounded, domain Ω of Euclidean space $\mathbf{R}^N$ with smooth boundary ∂D (Theorem 6.27). This section is a *probabilistic approach* to pseudo-differential operators.

In Section 6.7, following Mizohata [82, Chapter 3] and Wloka [146] we prove that all the sufficiently large eigenvalues of the Dirichlet problem for the differential operator A lie in the *parabolic* type region (see assertion (6.74)).

Hence, by considering $A - \lambda_1$ and $A^* - \lambda_1$ we may assume that the fundamental condition

$$\mathcal{N}_0 (A - \lambda_1) = \mathcal{N}_0 (A^* - \lambda_1) = \{0\} \tag{SC}$$

is satisfied for certain positive number λ_1.

6.1 Classical Potentials and Pseudo-Differential Operators

The purpose of this section is to describe, in terms of pseudo-differential operators, the classical surface and volume potentials arising in boundary value problems for elliptic differential operators. This calculus of pseudo-differential operators will be applied to elliptic boundary value problems in Chapters 12, 13 and 14.

6.1.1 Single and Double Layer Potentials

Let Δ be the usual Laplacian

$$\Delta = \frac{\partial^2}{\partial x_1^2} + \frac{\partial^2}{\partial x_2^2} + \ldots + \frac{\partial^2}{\partial x_n^2}.$$

Then the Newtonian potential $N(x - y)$ is given by the formula

$$
\begin{aligned}
(-\Delta)^{-1} f(x) &= N * f(x) \\
&= \frac{\Gamma((n-2)/2)}{4\pi^{n/2}} \int_{\mathbf{R}^n} \frac{1}{|x - y|^{n-2}} f(y)\, dy \\
&= \frac{1}{(n-2)\omega_n} \int_{\mathbf{R}^n} \frac{1}{|x - y|^{n-2}} f(y)\, dy \quad \text{for all } f \in C_0^\infty(\mathbf{R}^n),
\end{aligned}
$$

where

$$\omega_n = \frac{2\pi^{n/2}}{\Gamma(n/2)}$$

is the surface area of the unit sphere. In the case $n = 3$, we have the formula

$$u(x) = \frac{1}{4\pi} \int_{\mathbf{R}^3} \frac{f(y)}{|x - y|}\, dy.$$

Up to an appropriate constant of proportionality, the Newtonian potential $N(x - y)$ is the gravitational potential at position x due to a unit point mass at position y, and so the function $u(x)$ is the gravitational potential due to a continuous mass distribution with density $f(x)$. In terms of electrostatics, the function $u(x)$ describes the electrostatic potential due to a charge distribution with density $f(x)$.

We define a *single layer potential* with density φ by the formula

$$N * (\varphi(x') \otimes \delta(x_n))$$
$$= \frac{\Gamma((n-2)/2)}{4\pi^{n/2}} \int_{\mathbf{R}^{n-1}} \frac{\varphi(y')}{(|x'-y'|^2 + x_n^2)^{(n-2)/2}} \, dy'$$
$$= \frac{1}{(n-2)\omega_n} \int_{\mathbf{R}^{n-1}} \frac{\varphi(y')}{(|x'-y'|^2 + x_n^2)^{(n-2)/2}} \, dy' \quad \text{for all } \varphi \in C_0^\infty(\mathbf{R}^{n-1}).$$

In the case $n = 3$, the function $N * (\varphi \otimes \delta)$ is related to the distribution of electric charge on a conductor Ω. In equilibrium, mutual repulsion causes all the charge to reside on the surface $\partial\Omega$ of the conducting body with density φ, and $\partial\Omega$ is an equipotential surface.

We define a *double layer potential* with density ψ by the formula

$$N * (\psi(x') \otimes \delta'(x_n)) = \frac{1}{\omega_n} \int_{\mathbf{R}^{n-1}} \frac{x_n \, \psi(y')}{(|x'-y'|^2 + x_n^2)^{(n-2)/2}} \, dy'$$
$$\text{for all } \psi \in C_0^\infty(\mathbf{R}^{n-1}).$$

In the case $n = 3$, the function $N * (\psi \otimes \delta')$ is the potential induced by a distribution of dipoles on $\mathbf{R}^2$ with density $\psi(y')$, the axes of the dipoles being normal to $\mathbf{R}^2$.

On the other hand, it is easy to verify that if φ is bounded and continuous on $\mathbf{R}^{n-1}$, then the function

$$u(x', x_n) = \frac{2}{\omega_n} \int_{\mathbf{R}^{n-1}} \frac{x_n}{(|x'-y'|^2 + x_n^2)^{n/2}} \varphi(y') \, dy' \tag{6.5}$$

is well-defined for $(x', x_n) \in \mathbf{R}_+^n$, and is a (unique) solution of the Dirichlet problem

$$\begin{cases} \Delta u = 0 & \text{in } \mathbf{R}_+^n, \\ \gamma_0 u = u|_{\mathbf{R}^{n-1}} = \varphi & \text{on } \mathbf{R}^{n-1}. \end{cases}$$

Formula (6.5) is called the *Poisson integral formula* for the solution of the Dirichlet problem.

Furthermore, by using the Fourier transform we can express formula (6.5) for $\varphi \in \mathcal{S}(\mathbf{R}^{n-1})$ as follows:

$$u(x', x_n) = \frac{1}{(2\pi)^{n-1}} \int_{\mathbf{R}^{n-1}} e^{i x' \cdot \xi'} e^{-x_n |\xi'|} \widehat{\varphi}(\xi') \, d\xi'. \tag{6.6}$$

To do this, we need the following elementary formulas:

Claim 6.1. *(1) For any $a > 0$, we have the formula*

$$\int_{-\infty}^{\infty} e^{i\alpha x} e^{-a x^2} \, dx = \sqrt{\frac{\pi}{a}} \, e^{-\frac{\alpha^2}{4a}}. \tag{6.7}$$

(2) For any $\beta > 0$, we have the formula

$$e^{-\beta} = \frac{1}{\sqrt{\pi}} \int_0^\infty \frac{e^{-s}}{\sqrt{s}}\, e^{-\frac{\beta^2}{4s}}\, ds. \tag{6.8}$$

Proof. We only prove formula (6.8). To do so, we remark that the function

$$\mathbf{C} \ni z \longmapsto f(z) = \frac{e^{i\beta z}}{1 + z^2}$$

has a pole at $z = i$ in the closed half-plane $\{z \in \mathbf{C} : \operatorname{Im} z \geq 0\}$, and further that its residue is given by the formula

$$\operatorname{Res}\left[f(z)\right]_{z=i} = \lim_{z \to i}(z - i)f(z) = \frac{e^{-\beta}}{2i} = -\frac{i}{2}e^{-\beta}.$$

Hence we have, by the residue theorem,

$$\int_\Gamma f(z)\, dz = 2\pi i \left(-\frac{i}{2}e^{-\beta}\right) = \pi\, e^{-\beta}. \tag{6.9}$$

Here Γ is a path consisting of the semicircle and the segment as in Figure 6.2 below.

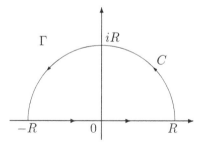

Fig. 6.2. The integral path Γ consisting of the semicircle C completed by the segment $[-R, R]$

Then we can rewrite formula (6.9) as follows:

$$\pi\, e^{-\beta} = \int_\Gamma f(z)\, dz = \int_{C_R} f(z)\, dz + \int_{-R}^R f(x)\, dx := I + II. \tag{6.10}$$

However, since we have, for all $z = x + iy \in C$,

$$\left|e^{i\beta z}\right| = \left|e^{i\beta x}e^{-\beta y}\right| = \left|e^{-\beta y}\right| \leq 1,$$

we can estimate the first term I as follows:

$$\left|\int_{C_R} f(z)\, dz\right| = \left|\int_0^\pi \frac{e^{i\beta R e^{i\theta}}}{1 + R^2 e^{2i\theta}}\, R\, i\, e^{i\theta}\, d\theta\right|$$

$$\leq \int_0^\pi \frac{R}{R^2 - 1}\, d\theta = \frac{\pi R}{R^2 - 1} \longrightarrow 0 \quad \text{as } R \to \infty.$$

Therefore, by using Fubini's theorem we obtain from formula (6.7) with $\alpha := \beta$ and $a := s$ that

$$e^{-\beta} = \frac{1}{\pi}\lim_{R\to\infty}\int_{-R}^R f(x)\,dx = \frac{1}{\pi}\int_{-\infty}^\infty \frac{e^{i\beta x}}{1+x^2}\,dx$$

$$= \frac{1}{\pi}\int_{-\infty}^\infty e^{i\beta x}\left(\int_0^\infty e^{-(1+x^2)s}\,ds\right) dx$$

$$= \frac{1}{\pi}\int_0^\infty e^{-s}\left(\int_{-\infty}^\infty e^{i\beta x}\,e^{-sx^2}\,dx\right) ds$$

$$= \frac{1}{\sqrt{\pi}}\int_0^\infty \frac{e^{-s}}{\sqrt{s}}\,e^{-\frac{\beta^2}{4s}}\,ds.$$

This proves the desired formula (6.8).

The proof of Claim 6.1 is complete. □

Therefore, it follows from an application of Fubini's theorem and Claim 6.1 with $\beta := x_n|\xi'|$ that

$$\frac{1}{(2\pi)^{n-1}}\int_{\mathbf{R}^{n-1}} e^{i\,x'\cdot\xi'}e^{-x_n|\xi'|}\,\widehat{\varphi}(\xi')\,d\xi'$$

$$= \int_{\mathbf{R}^{n-1}}\varphi(y')\left(\frac{1}{(2\pi)^{n-1}}\int_{\mathbf{R}^{n-1}} e^{i(x'-y')\cdot\xi'}\,e^{-x_n|\xi'|}\,d\xi'\right) dy'$$

$$= \int_{\mathbf{R}^{n-1}}\varphi(y')\left(\frac{1}{\pi^{n/2}}\frac{1}{x_n^{n-1}}\int_0^\infty e^{-s\left(1+|x'-y'|^2/x_n^2\right)}s^{n/2-1}\,ds\right) dy'$$

$$= \int_{\mathbf{R}^{n-1}}\varphi(y')\left(\frac{1}{\pi^{n/2}}\frac{1}{x_n^{n-1}}\int_0^\infty e^{-s}\,s^{n/2-1}\,ds\,\frac{x_n^n}{(|x'-y'|^2+x_n^2)^{n/2}}\right) dy'$$

$$= \frac{\Gamma(n/2)}{\pi^{n/2}}\int_{\mathbf{R}^{n-1}}\frac{x_n}{(|x'-y'|^2+x_n^2)^{n/2}}\varphi(y')\,dy'$$

$$= \frac{2}{\omega_n}\int_{\mathbf{R}^{n-1}}\frac{x_n}{(|x'-y'|^2+x_n^2)^{n/2}}\varphi(y')\,dy'.$$

This proves the desired formula (6.6). □

6.1.2 The Green Representation Formula

First, we have the jump formula (4.15) for the minus Laplacian $-\Delta$ in the half-space $\mathbf{R}_+^n$:

$$(-\Delta)\left(u^0\right) = (-\Delta u)^0 - \frac{\partial u}{\partial x_n}(x',0)\otimes\delta(x_n) - u(x',0)\otimes\delta'(x_n).$$

By applying the Newtonian potential $N(x-y)$ to the both sides (see Subsection 6.1.1), we obtain that

$$
\begin{aligned}
u^0 &= (N * (-\Delta)) \left(u^0\right) \\
&= N * \left((-\Delta u)^0\right) - N * (\gamma_1 u \otimes \delta(x_n)) - N * (\gamma_0 u \otimes \delta'(x_n)) \\
&= -\int_{\mathbf{R}^n} N(x-y) \, \Delta u(y) \, dy - \int_{\mathbf{R}^{n-1}} N(x'-y', x_n) \frac{\partial u}{\partial y_n}(y', 0) \, dy' \\
&\quad + \int_{\mathbf{R}^{n-1}} \frac{\partial N}{\partial y_n}(x'-y', x_n) \, u(y', 0) \, dy.
\end{aligned}
$$

Therefore, we arrive at the classical *Green representation formula*:

$$
\begin{aligned}
u(x) &= \frac{1}{(2-n)\omega_n} \int_{\mathbf{R}^n_+} \frac{1}{|x-y|^{n-2}} \Delta u(y) \, dy \\
&\quad + \frac{1}{(2-n)\omega_n} \int_{\mathbf{R}^{n-1}} \frac{1}{(|x'-y'|^2 + x_n^2)^{(n-2)/2}} \frac{\partial u}{\partial y_n}(y', 0) \, dy' \\
&\quad + \frac{1}{\omega_n} \int_{\mathbf{R}^{n-1}} \frac{x_n}{(|x'-y'|^2 + x_n^2)^{n/2}} u(y', 0) \, dy' \quad \text{for } x \in \mathbf{R}^n_+.
\end{aligned}
$$

We remark that the first term is the Newtonian potential and the second and third terms are the single and double layer potentials, respectively.

6.1.3 Surface and Volume Potentials

We give a formal description of a background. Let Ω be a bounded domain in Euclidean space $\mathbf{R}^n$ with smooth boundary $\partial\Omega$. Its closure $\overline{\Omega}$ is an n-dimensional, compact smooth manifold with boundary. We may assume that $\overline{\Omega}$ is the closure of a relatively compact, open subset Ω of an n-dimensional compact smooth manifold M without boundary in which Ω has a smooth boundary $\partial\Omega$ (see Figure 4.1). The manifold M is called the *double* of Ω.

Let P be a differential operator of order m with smooth coefficients on M. Then we have the *jump formula* (4.15)

$$
P\left(u^0\right) = (Pu)^0 + \widetilde{P}\gamma u \quad \text{for all } u \in C^\infty(\overline{\Omega}), \tag{6.11}
$$

where u^0 is the extension of u to M by zero outside Ω, and $\widetilde{P}\gamma u$ is a distribution on M with support in $\partial\Omega$. If P admits an "inverse" Q, then the function u may be expressed as follows:

$$
u = Q\left((Pu)^0\right)\Big|_\Omega + Q\left(\widetilde{P}\gamma u\right)\Big|_\Omega.
$$

The first term on the right-hand side is a volume potential and the second term is a surface potential with m "layers". For example, if P is the usual Laplacian Δ and if $\Omega = \mathbf{R}^n_+$, then the first term is the classical Newtonian potential and the second term is the familiar combination of single and double layer potentials (see Subsection 6.1.2).

(I) First, we state a theorem which covers *surface potentials* (see [17, Théorème (1.3.5)], [26, Chapitre V, Théorème 2.4]):

Theorem 6.2. *Let* $A \in L_{\mathrm{cl}}^m(M)$ *be properly supported. Assume that*

$$\text{Every term in the complete symbol } \sum_{j=0}^{\infty} a_j(x, \xi) \text{ of } A \qquad (6.12)$$

$$\text{is a rational function of } \xi$$

Then we have the following three assertions (i), (ii) and (iii):

(i) The operator

$$H : v \longmapsto A(v \otimes \delta)|_{\Omega}$$

is continuous on $C^{\infty}(\partial\Omega)$ *into* $C^{\infty}(\overline{\Omega})$. *If* $v \in \mathcal{D}'(\partial\Omega)$, *the distribution* Hv *has sectional traces on* $\partial\Omega$ *of any order.*

(ii) The operator

$$S : C^{\infty}(\partial\Omega) \longrightarrow C^{\infty}(\partial\Omega)$$

$$v \longmapsto Hv|_{\partial\Omega}$$

belongs to the class $L_{\mathrm{cl}}^{m+1}(\partial\Omega)$. *Furthermore, its homogeneous principal symbol is given by the formula*

$$(x', \xi') \longmapsto \frac{1}{2\pi} \int_{\Gamma} a_0(x', 0, \xi', \xi_n) \, d\xi_n \qquad (6.13)$$

where $a_0(x', x_n, \xi', \xi_n) \in C^{\infty}(T^*(M) \setminus \{0\})$ *is the homogeneous principal symbol of* A, *and* Γ *is a circle in the plane* $\{\xi_n \in \mathbf{C} : \operatorname{Im} \xi_n > 0\}$ *that encloses the poles* ξ_n *of* $a_0(x', 0, \xi', \xi_n)$ *there (see Figure 6.3 below).*

(iii) If $1 < p < \infty$, *then the operator* H *extends to a continuous linear operator*

$$H : B^{s,p}(\partial\Omega) \longrightarrow H^{s-m-1+1/p,p}(\Omega)$$

for all $s \in \mathbf{R}$ *(see Figure 6.4 below).*

Proof. We only prove the case where $p = 2$, by following [26, Chapitre V, Théorème 2.4].

By using local coordinate systems flattening out the boundary $\partial\Omega$, together with a partition of unity, we may assume that

$$\Omega = \mathbf{R}_+^n,$$

$$u \in \mathcal{E}'(\mathbf{R}^n).$$

The proof of Theorem 6.2 is divided into three parts.

Proof of Assertion (i): As in the proof of Theorem 4.19, we have, for all $\Phi(x) \in C_0^{\infty}(\mathbf{R}^n)$,

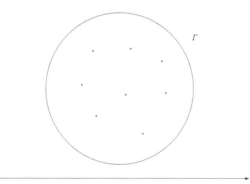

Fig. 6.3. The circle Γ in the plane $\{\operatorname{Im}\xi_n > 0\}$ that encloses the poles ξ_n of $a_0(x',0,\xi',\xi_n)$

$$
\begin{array}{ccc}
B^{s,p}(\partial\Omega) & \xrightarrow{\ H\ } & H^{s-m-1+1/p,p}(\partial\Omega) \\[4pt]
\uparrow & & \uparrow \\[4pt]
C^\infty(\partial\Omega) & \xrightarrow{\ \ \ \ } & C^\infty(\overline{\Omega}) \\
& H &
\end{array}
$$

Fig. 6.4. The mapping properties of the surface potential H for all $s \in \mathbf{R}$ and $1 < p < \infty$

$$
\begin{aligned}
\left\langle \widetilde{A}u, \Phi \right\rangle &= \frac{1}{(2\pi)^n} \int_{\mathbf{R}^n} \widehat{u}(\xi) \left(\int_{\mathbf{R}^n} e^{i\,x\cdot\xi}\,\widetilde{a}(x,\xi)\Phi(x)\,dx \right) d\xi \\
&= \frac{1}{(2\pi)^n} \int_{|\xi|\geq 1} \widehat{u}(\xi) \left(\int_{\mathbf{R}^n} e^{i\,x\cdot\xi} a(x,\xi)\Phi(x)\,dx \right) d\xi \\
&\quad + \frac{1}{(2\pi)^n} \int_{|\xi|\leq 1} \widehat{u}(\xi) \left(\int_{\mathbf{R}^n} e^{i\,x\cdot\xi}\chi(\xi)a(x,\xi)\Phi(x)\,dx \right) d\xi \\
&= \langle Bu, \Phi \rangle + \langle Ru, \Phi \rangle .
\end{aligned}
$$

However, we recall that

$$
Ru \in C^\infty(\mathbf{R}^n).
$$

Hence we are reduced to the study of the term

$$
Bu := \widetilde{A}u - Ru,
$$

where

$$
\langle Bu, \Phi \rangle = \frac{1}{(2\pi)^n} \int_{|\xi|\geq 1} \widehat{u}(\xi) F(\xi)\,d\xi,
$$

$$
F(\xi) = \int_{\mathbf{R}^n} e^{i\,x\cdot\xi}\, a(x,\xi)\,\Phi(x)\,dx.
$$

By taking

$$u(x', x_n) := v(x') \otimes \delta(x_n) \quad \text{with } v \in C_0^\infty(\mathbf{R}^{n-1}),$$
$$\Phi(x', x_n) := \varphi(x') \otimes \psi(x_n) \quad \text{with } \varphi \in C_0^\infty(\mathbf{R}^{n-1}) \text{ and } \psi \in C_0^\infty(\mathbf{R}_+),$$

we have, by Claim 4.23,

$$\langle B(v(x') \otimes \delta(x_n)), \varphi(x') \otimes \psi(x_n) \rangle \tag{6.14}$$
$$= \frac{1}{(2\pi)^n} \int_0^\infty \psi(x_n)$$
$$\times \left(\int_{\mathbf{R}^{n-1}} \left(\int_{\Gamma_{\xi'}} \int_{\mathbf{R}^{n-1}} e^{i\,x\cdot\xi}\, a(x,\xi)\, \varphi(x')\, \widehat{v}(\xi')\, d\xi_n\, dx' \right) dx' \right) dx_n$$
$$= \frac{1}{(2\pi)^n} \int_0^\infty \psi(x_n) \left(\int_{\mathbf{R}^{n-1}} \left(\int_{\mathbf{R}^{n-1}} e^{i\,x'\cdot\xi'}\, \widehat{v}(\xi') \right. \right.$$
$$\times \left. \left. \left(\int_{\Gamma_{\xi'}} e^{i\,x_n\,\xi_n}\, a(x', x_n, \xi', \xi_n)\, d\xi_n \right) dx' \right) \varphi(x')\, dx' \right) dx_n$$
$$= \left\langle \int_{\mathbf{R}^{n-1}} e^{i\,x'\cdot\xi'}\, k(x', x_n, \xi')\, \widehat{v}(\xi')\, d\xi', \varphi(x') \otimes \psi(x_n) \right\rangle$$
$$\text{for all } \varphi \in C_0^\infty(\mathbf{R}^{n-1}) \text{ and } \psi \in C_0^\infty(\mathbf{R}_+),$$

where

$$k(x', x_n, \xi') = \int_{\Gamma_{\xi'}} e^{i\,x_n\,\xi_n}\, a(x', x_n, \xi', \xi_n)\, d\xi_n.$$

Therefore, we obtain from formula (6.14) that

$$B(v(x') \otimes \delta(x_n))|_{\mathbf{R}_+^n}$$
$$= \int_{\mathbf{R}^{n-1}} e^{i\,x'\cdot\xi'} k(x, \xi')\, \widehat{v}(\xi')\, d\xi' \quad \text{for all } v \in C_0^\infty(\mathbf{R}^{n-1}).$$

The next claim proves that the mapping

$$C_0^\infty(\mathbf{R}^{n-1}) \ni v \longmapsto B(v(x') \otimes \delta(x_n))|_{\mathbf{R}_+^n} \in C^\infty(\overline{\mathbf{R}_+^n})$$

is continuous:

Claim 6.3. *For every integer $\ell \geq 0$, there exists a constant $C_\ell > 0$ such that*

$$\left| D_n^\ell k(x', x_n, \xi') \right| \leq C_\ell (1 + |\xi'|)^{\ell + \mu + 1} \tag{6.15}$$
$$\text{for all } x_n \geq 0 \text{ and } |\xi'| \geq 1.$$

Proof. First, we have, for all $x_n \geq 0$, $|\xi'| \geq 1$ and $\operatorname{Im} \xi_n \geq 0$,

$$\bullet \ \left| e^{i\,x_n\,\xi_n} \right| = e^{-x_n \operatorname{Im} \xi_n} \leq 1,$$

- $\Gamma_{\xi'} = |\xi'|\,\Gamma_1$.

Moreover, we remark that

$$D_n\left(e^{i\,x_n\,\xi_n}\,a(x',x_n,\xi',\xi_n)\right)$$
$$= \left(\xi_n\,a(x',x_n,\xi',\xi_n) + D_n a(x',x_n,\xi',\xi_n)\right)e^{i\,x_n\,\xi_n}.$$

Hence we have, for every integer $\ell \geq 0$,

$$D_n^\ell k(x',x_n,\xi') = \int_{\Gamma_{\xi'}} e^{i\,x_n\,\xi_n}\left(D_n + \xi_n\right)^\ell a(x',x_n,\xi',\xi_n)\,d\xi_n. \tag{6.16}$$

Therefore, the desired estimate (6.15) follows from formula (6.16), if we make use of the inequality

$$|\xi_n| \leq c_0\,(1 + |\xi'|) \quad \text{for all } \xi_n \in \Gamma_{\xi'} \text{ and } |\xi'| \geq 1.$$

The proof of Claim 6.3 is complete. □

Proof of Assertion (ii): The proof is divided into three steps.
Step (1): First, we have the following claim:

Claim 6.4. *Let $A' \in L_{\mathrm{cl}}^{\mu'}(\mathbf{R}^n)$ be properly supported. If $\mu' < -1$, then we have the assertion*

$$A'\left(v(x') \otimes \delta(x_n)\right) \in C(\mathbf{R}^n) \quad \text{for all } v \in C_0^\infty(\mathbf{R}^{n-1}).$$

Proof. If $a'(x,\xi)$ is a symbol of A', we have the formula

$$A'\left(v(x') \otimes \delta(x_n)\right) = \frac{1}{(2\pi)^n}\int_{\mathbf{R}^n} e^{i\,x\cdot\xi} a'(x,\xi)\,\widehat{v}(\xi')\,d\xi.$$

However, we have, for all $x,\,y \in \mathbf{R}^n$,

$$\left|e^{i\,x\cdot\xi}a'(x,\xi)\,\widehat{v}(\xi') - e^{i\,y\cdot\xi}a'(y,\xi)\,\widehat{v}(\xi')\right|$$
$$\leq C\left(1 + |\xi'|^2 + \xi_n^2\right)^{\mu'/2}|\widehat{v}(\xi')|,$$

where $C > 0$ is a constant. On the other hand, since $\mu' < -1$, it follows that

$$\left(1 + |\xi'|^2 + \xi_n^2\right)^{\mu'/2} \leq \left(1 + \xi_n^2\right)^{\mu'/2} \quad \text{for all } \xi = (\xi',\xi_n) \in \mathbf{R}^n.$$

Hence we have, for all $x,\,y \in \mathbf{R}^n$,

$$\left|e^{i\,x\cdot\xi}a'(x,\xi)\,\widehat{v}(\xi') - e^{i\,y\cdot\xi}a'(y,\xi)\,\widehat{v}(\xi')\right|$$
$$\leq C\left(1 + \xi_n^2\right)^{\mu'/2}|\widehat{v}(\xi')| \in L^1(\mathbf{R}^n).$$

By applying Lebesgue's dominated convergence theorem, we find that

$$A'\left(v(x') \otimes \delta(x_n)\right) \in C(\mathbf{R}^n).$$

The proof of Claim 6.4 is complete. □

Step (2): By Claim 6.4, it follows that

$$A'\left(v \otimes \delta\right)(x', 0) = \frac{1}{(2\pi)^n} \int_{\mathbf{R}^n} e^{i\,x' \cdot \xi'} a'(x', 0, \xi)\, \widehat{v}(\xi')\, d\xi \in C(\mathbf{R}^{n-1}).$$

Moreover, since $\mu' < -1$, we can write the formula $A'\left(v \otimes \delta\right)(x', 0)$ in the form

$$A'\left(v \otimes \delta\right)(x', 0)$$
$$= \frac{1}{(2\pi)^{n-1}} \int_{\mathbf{R}^{n-1}} e^{i\,x' \cdot \xi'} \left(\frac{1}{2\pi} \int_{-\infty}^{\infty} a'(x', 0, \xi', \xi_n)\, d\xi_n\right) \widehat{v}(\xi')\, d\xi'.$$

Indeed, it suffices to note that

$$|a'(x', 0, \xi', \xi_n)| \leq C\left(1 + |\xi'|^2 + \xi_n^2\right)^{\mu'/2}$$
$$\leq C\left(1 + \xi_n^2\right)^{\mu'/2} \quad \text{for all } \xi_n \in \mathbf{R}.$$

More precisely, we have the following claim:

Claim 6.5. *The operator*

$$C_0^\infty(\mathbf{R}^{n-1}) \ni v \longmapsto A'\left(v \otimes \delta\right)(x', 0)$$

is a pseudo-differential operator on $\mathbf{R}^{n-1}$ with symbol

$$a''(x', \xi') := \frac{1}{2\pi} \int_{-\infty}^{\infty} a'(x', 0, \xi', \xi_n)\, d\xi_n \in S^{\mu'+1}\left(\mathbf{R}^{n-1} \times \mathbf{R}^{n-1}\right).$$

Proof. For each compact set $K' \subset \mathbf{R}^{n-1}$ and multi-indices α', β', there exists a constant $C_{\alpha',\beta',K'} > 0$ such that

$$\left|D_{x'}^{\alpha'} D_{\xi'}^{\beta'} a'(x', 0, \xi', \xi_n)\right| \leq C_{\alpha',\beta',K'} \left(1 + |\xi'|^2 + \xi_n^2\right)^{\mu'/2 - |\beta'|/2}$$
$$\text{for all } x' \in K' \text{ and } \xi = (\xi', \xi_n) \in \mathbf{R}^n.$$

Hence we have the inequality

$$\left|D_{x'}^{\alpha'} D_{\xi'}^{\beta'} a''(x', \xi')\right|$$
$$= \left|\frac{1}{2\pi} \int_{-\infty}^{\infty} D_{x'}^{\alpha'} D_{\xi'}^{\beta'} a'(x', 0, \xi', \xi_n)\, d\xi_n\right|$$
$$\leq \frac{C_{\alpha',\beta',K'}}{2\pi} \int_{-\infty}^{\infty} \left(1 + |\xi'|^2 + \xi_n^2\right)^{\mu'/2 - |\beta'|/2}\, d\xi_n$$
$$= \frac{C_{\alpha',\beta',K'}}{2\pi} \left(1 + |\xi'|^2\right)^{\mu'/2 - |\beta'|/2 + 1/2} \int_{-\infty}^{\infty} \left(1 + \eta_n^2\right)^{\mu'/2 - |\beta'|/2}\, d\eta_n$$
$$\leq \frac{C_{\alpha',\beta',K'}}{2\pi} \left(1 + |\xi'|^2\right)^{(\mu'+1)/2 - |\beta'|/2} \int_{-\infty}^{\infty} \left(1 + \eta_n^2\right)^{\mu'/2}\, d\eta_n$$
$$:= C_{\alpha',\beta',K'}' \left(1 + |\xi'|^2\right)^{(\mu'+1)/2 - |\beta'|/2} \quad \text{for all } x' \in K' \text{ and } \xi' \in \mathbf{R}^{n-1}.$$

The proof of Claim 6.5 is complete. $\quad\square$

Summing up, we have proved that if $A' \in L_{\mathrm{cl}}^{\mu'}(\mathbf{R}^n)$ with $\mu' < -1$, there exists a pseudo-differential operator $A'' \in L_{\mathrm{cl}}^{\mu'+1}(\mathbf{R}^{n-1})$ such that

$$A'(v \otimes \delta)(x', 0) = A''v(x') \quad \text{for all } v \in C_0^\infty(\mathbf{R}^{n-1}).$$

Step (3): In this way, we are reduced to the study of the operator

$$B(v \otimes \delta)(x', 0)$$
$$= \int_{\mathbf{R}^{n-1}} e^{i x' \cdot \xi'} k(x', 0, \xi') \widehat{v}(\xi') \, d\xi' \quad \text{for all } v \in C_0^\infty(\mathbf{R}^{n-1}),$$

where

$$k(x', 0, \xi') = \int_{\Gamma_{\xi'}} a(x', 0, \xi', \xi_n) \, d\xi_n.$$

However, we can prove the following claim:

Claim 6.6. *The function* $k(x', 0, \xi')$ *is positively homogeneous of degree* $\mu + 1$ *with respect to the variable* ξ', *for* $|\xi'| \geq 1$. *In particular, we have the assertion*

$$k(x', 0, \xi') \in S_{\mathrm{cl}}^{\mu+1}\left(\mathbf{R}^{n-1} \times \mathbf{R}^{n-1}\right).$$

Proof. Since we have the formula

$$\Gamma_{\xi'} = |\xi'| \, \Gamma_1 \quad \text{for all } |\xi'| \geq 1,$$

it follows that

$$k(x', 0, \xi') = \int_{\Gamma_{\xi'}} a(x', 0, \xi', \xi_n) \, d\xi_n = \int_{|\xi'| \, \Gamma_1} a(x', 0, \xi', \xi_n) |\xi'| \, d\left(\frac{\xi_n}{|\xi'|}\right)$$
$$= \int_{\Gamma_1} |\xi'|^\mu a\left(x', 0, \frac{\xi'}{|\xi'|}, \eta_n\right) |\xi'| \, d\eta_n$$
$$= |\xi'|^{\mu+1} \int_{\Gamma_1} a\left(x', 0, \frac{\xi'}{|\xi'|}, \eta_n\right) d\eta_n$$
$$= |\xi'|^{\mu+1} k\left(x', 0, \frac{\xi'}{|\xi'|}\right) \quad \text{for all } |\xi'| \geq 1.$$

The proof of Claim 6.6 is complete. □

Proof of Assertion (iii): The proof is divided into six steps.

Step (1): For every function $\varphi \in C_0^\infty(\mathbf{R}^n)$, we have only to show that the mapping

$$\varphi K \colon H^s(\mathbf{R}^{n-1}) \longrightarrow H^{s-\mu-1/2}(\overline{\mathbf{R}_+^n}) \tag{6.17}$$

is continuous for all $s \in \mathbf{R}$.

Indeed, since A is properly supported, we can find a function $\psi \in C_0^\infty(\mathbf{R}^{n-1})$ such that

$$\varphi K v = \varphi K(\psi v) \quad \text{for all } v \in \mathcal{D}'(\mathbf{R}^{n-1}).$$

Then we have the assertion

$$v_j \longrightarrow 0 \quad \text{in } H^s_{\text{loc}}(\mathbf{R}^{n-1})$$
$$\Longrightarrow \psi v_j \longrightarrow 0 \quad \text{in } H^s(\mathbf{R}^{n-1})$$
$$\Longrightarrow \varphi K v_j = \varphi K(\psi v_j) \longrightarrow 0 \quad \text{in } H^{s-\mu-1/2}(\overline{\mathbf{R}^n_+}).$$

This proves that

$$K v_j \longrightarrow 0 \quad \text{in } H^{s-\mu-1/2}_{\text{loc}}(\overline{\mathbf{R}^n_+}).$$

Step (2): Now we assume that

$$A' \in L^{\mu'}_{\text{cl}}(\mathbf{R}^n) \quad \text{for } \mu' < \mu,$$

and that

$$s - \mu + \mu' < 0.$$

Then we have, for all $\varphi \in C^\infty_0(\mathbf{R}^n)$ and $v \in C^\infty_0(\mathbf{R}^{n-1})$,

$$\left\| \varphi A' \left(v(x') \otimes \delta(x_n) \right) \big|_\Omega \right\|_{s-\mu-1/2} \tag{6.18}$$
$$\leq C \left\| v(x') \otimes \delta(x_n) \right\|_{s-\mu-1/2+\mu'} \leq C' \left\| v \right\|_{s-(\mu-\mu')}$$
$$\leq C'' \left\| v \right\|_s .$$

Indeed, it suffices to note that

$$\left\| v(x') \otimes \delta(x_n) \right\|^2_{s-\mu-1/2+\mu'}$$
$$= \frac{1}{(2\pi)^n} \int_{\mathbf{R}^n} \left(1 + |\xi'|^2 + |\xi_n|^2 \right)^{s-\mu-1/2+\mu'} |\widehat{v}(\xi')|^2 \, d\xi' \, d\xi_n$$
$$= \frac{1}{(2\pi)^n} \int_{-\infty}^{\infty} \left(1 + |\eta_n|^2 \right)^{s-\mu-1/2+\mu'} d\eta_n$$
$$\times \int_{\mathbf{R}^{n-1}} \left(1 + |\xi'|^2 \right)^{s-\mu+\mu'} |\widehat{v}(\xi')|^2 \, d\xi'$$
$$= C_{s,\mu,\mu'} \left\| v \right\|^2_{s-(\mu-\mu')} \quad \text{for } 2(s - \mu + \mu' - 1/2) < -1.$$

However, we have the decomposition

$$\varphi A (v \otimes \delta) = \varphi B (v \otimes \delta) + \varphi R (v \otimes \delta),$$

where

$$A = B + R, \quad B \in L^\mu_{\text{cl}}(\mathbf{R}^n), \ R \in L^{-\infty}(\mathbf{R}^n).$$

By applying inequality (6.18) to the operator R, we obtain that

$$\left\| \varphi R (v \otimes \delta) \big|_\Omega \right\|_{s-\mu-1/2} \leq C_s \left\| v \right\|_s .$$

Therefore, we are reduced to the proof of the following inequality:

$$\|\varphi\, B\,(v \otimes \delta)|_{\Omega}\|_{s-\mu-1/2} \leq C_s\, \|v\|_s \quad \text{for all } v \in C_0^\infty(\mathbf{R}^{n-1}). \qquad (6.19)$$

From now on, we may assume that

- $B\,(v(x') \otimes \delta(x_n))|_{\mathbf{R}^n_+} = \dfrac{1}{(2\pi)^{n-1}} \displaystyle\int_{\mathbf{R}^{n-1}} e^{i\,x'\cdot\xi'}\, k(x,\xi')\, \widehat{v}(\xi')\, d\xi',$

- $k(x,\xi') = \displaystyle\int_{\Gamma_{\xi'}} e^{i\,x_n\,\xi_n}\, a(x,\xi',\xi_n)\, d\xi_n,$

where the support of $a(x,\xi)$ (and hence that of $k(x,\xi')$) with respect to x is compact in $\overline{\mathbf{R}^n_+}$, depending on $\varphi \in C_0^\infty(\mathbf{R}^n)$.

Step (3): First, we prove the following lemma:

Lemma 6.7. *For any multi-indices α and β, there exists a constant $C_{\alpha,\beta} > 0$ such that*

$$\left|x^\beta\, D_x^\alpha k(x,\xi')\right| \leq C_{\alpha,\beta}\,(1+|\xi'|)^{\mu+1+\alpha_n-\beta_n} \qquad (6.20)$$
$$\text{for all } x = (x',x_n) \in \overline{\mathbf{R}^n_+} \text{ and } \xi' \in \mathbf{R}^{n-1}.$$

Here the support of $k(x,\xi')$ with respect to x is compact in $\overline{\mathbf{R}^n_+}$.

Proof. A general term $x^\beta\, D_x^\alpha k(x,\xi')$ is expressed as follows:

$$x'^{\beta'} \int_{\Gamma_{\xi'}} x_n^{\beta_n}\, e^{i\,x_n\,\xi_n}\, \xi_n^p\, D_{x'}^{\alpha'} a(x,\xi',\xi_n)\, d\xi_n \quad \text{for } 0 \leq p \leq \alpha_n.$$

However, we remark that

$$\left|x_n^{\beta_n}\, e^{i\,x_n\,\xi_n}\right| = (x_n\,|\xi'|)^{\beta_n}\, e^{-(x_n\,|\xi'|)\,\mathrm{Im}\,\zeta_n} \left(\frac{1}{|\xi'|}\right)^{\beta_n}$$
$$\text{for all } x_n \geq 0 \text{ and } \zeta_n \in \Gamma_1,$$

and further that

$$d\xi_n = |\xi'|\, d\zeta_n \quad \text{for } \xi_n \in \Gamma_{\xi'} \text{ and } \zeta_n \in \Gamma_1.$$

Therefore, we have the desired assertion

$$\left|x^\beta\, D_x^\alpha k(x,\xi')\right| = O\left(|\xi'|^{-\beta_n+\alpha_n+\mu+1}\right),$$

since the support of $a(x,\xi)$ is compact with respect to x.

The proof of Lemma 6.7 is complete. $\square$

Step (4) We extend the function $k(x, \xi')$ to the whole space $\mathbf{R}^n$ with respect to x in the following way: By virtue of Seeley [103] (see [123, Lemma 4.22]), we can find a function $w(t) \in \mathcal{S}(\mathbf{R})$ such that

$$\operatorname{supp} w \subset [1, \infty),$$

$$\int_0^\infty t^n w(t)\, dt = (-1)^n \quad \text{for each non-negative integer } n.$$

Moreover, we take a function $\phi(t) \in C_0^\infty(\mathbf{R})$ such that

$$\operatorname{supp} \phi \subset [-2, 2],$$
$$\phi(t) = 1 \quad \text{on the interval } [-1, 1].$$

If we let

$$k(x', x_n, \xi') := \begin{cases} k(x', x_n, \xi') & \text{for } x_n \geq 0, \\ \int_0^\infty w(s)\, \phi(-x_n s)\, k(x', -s\, x_n, \xi')\, ds & \text{for } x_n < 0, \end{cases}$$

then we have the following three assertions (a), (b) and (c):

(a) $k(x', x_n, \xi')$ is smooth with respect to the variable x_n, in the interval $(-\infty, 0)$.
(b) The support of $k(x, \xi')$ with respect to x is compact in the whole space $\mathbf{R}^n$.
(c) $k(x, \xi')$ satisfies the same estimate (6.20) for all $x = (x', x_n) \in \mathbf{R}^n$ and $\xi' \in \mathbf{R}^{n-1}$.

Step (5): Moreover, we can prove the following lemma:

Lemma 6.8. *Let*

$$\widehat{k}(\zeta, \xi') = \int_{\mathbf{R}^n} e^{-i\, x \cdot \zeta}\, k(x, \xi')\, dx \quad \text{for } \zeta \in \mathbf{R}^n \text{ and } \xi' \in \mathbf{R}^{n-1}.$$

For any integers $q \geq n$ and $r \geq 1$, there exists a constant $C = C_{q,r} > 0$ such that

$$\left| \widehat{k}(\zeta, \xi') \right| \leq C\, (1 + |\xi'|)^\mu\, (1 + |\zeta'|)^{-r} \left(1 + \frac{|\zeta_n|}{1 + |\xi'|} \right)^{-q} \tag{6.21}$$

$$\text{for all } \zeta \in \mathbf{R}^n \text{ and } \xi' \in \mathbf{R}^{n-1}.$$

Proof. For any multi-index α, it follows from an application of Lemma 6.7 that there exists a constant $C_{\alpha,q} > 0$ such that

$$(1 + |x'|)^q\, (1 + |x_n|\, (1 + |\xi'|))^{-2}\, |D_x^\alpha k(x, \xi')| \tag{6.22}$$

$$\leq C_{\alpha,q}\, (1 + |\xi'|)^{\mu + 1 + \alpha_n} \quad \text{for all } x = (x', x_n) \in \overline{\mathbf{R}_+^n} \text{ and } \xi' \in \mathbf{R}^{n-1}.$$

We recall that the support of $k(x, \xi')$ with respect to x is compact in the whole space $\mathbf{R}^n$.

Hence, by integration by parts it follows from inequality (6.22) that

$$
\left| \zeta^\alpha \widehat{k}(\zeta, \xi') \right| = \left| \int_{\mathbf{R}^n} D_x^\alpha \left(e^{-ix\cdot\zeta} \right) k(x, \xi') \right| \tag{6.23}
$$
$$
= \left| \int_{\mathbf{R}^n} e^{-ix\cdot\zeta} D_x^\alpha \left(k(x, \xi') \right) \right|
$$
$$
\leq C_1 \left(1 + |\xi'| \right)^{\mu+1+\alpha_n} \int_{-\infty}^{\infty} \frac{1}{\left(1 + |x_n| \left(1 + |\xi'| \right) \right)^2} dx_n
$$
$$
\times \int_{\mathbf{R}^{n-1}} \frac{1}{\left(1 + |x'| \right)^q} dx'
$$

for all $\zeta \in \mathbf{R}^n$ and $\xi' \in \mathbf{R}^{n-1}$.

However, we have the formula

$$
\int_{-\infty}^{\infty} \frac{dx_n}{\left(1 + |x_n| \left(1 + |\xi'| \right) \right)^2} = \int_{-\infty}^{\infty} \frac{dy_n}{\left(1 + |y_n| \right)^2} \left(1 + |\xi'| \right)^{-1}
$$
$$
= \frac{2}{1 + |\xi'|}.
$$

Hence we have, by inequality (6.23),

$$
\left| \zeta^\alpha \widehat{k}(\zeta, \xi') \right| \leq C_2 \left(1 + |\xi'| \right)^{\mu+\alpha_n}.
$$

Therefore, by taking

$$
\alpha = (\alpha', q), \quad |\alpha'| = r,
$$

we obtain the desired inequality

$$
\left(1 + |\zeta'| \right)^r \left(1 + \frac{|\zeta_n|}{1 + |\xi'|} \right)^q \left| \widehat{k}(\zeta, \xi') \right| \leq C_3 \left(1 + |\xi'| \right)^\mu.
$$

The proof of Lemma 6.8 is complete. $\square$

Step (6): If we let

$$
U(x) := \frac{1}{(2\pi)^n} \int_{\mathbf{R}^{n-1}} e^{ix'\cdot\xi'} k(x, \xi') \widehat{v}(\xi') d\xi' \quad \text{for } x \in \mathbf{R}^n,
$$

then we have the formula

$$
U|_{\mathbf{R}_+^n} = B \left(v \otimes \delta \right)|_{\mathbf{R}_+^n}.
$$

Hence it suffices to show that there exists a constant $C_s > 0$ such that

$$\|U\|_{s-\mu-1/2} \leq C_s \|v\|_s, \quad v \in C_0^\infty(\mathbf{R}^{n-1}),$$

or equivalently

$$|\langle U, \varphi \rangle| \leq C_s \|v\|_s \|\varphi\|_{-s+\mu+1/2} \quad \text{for all } \varphi \in C_0^\infty(\mathbf{R}^n). \tag{6.24}$$

Here recall that

$$\langle U, \varphi \rangle = \frac{1}{(2\pi)^n} \int_{\mathbf{R}^n} \widehat{U}(\eta) \, \widehat{\varphi}(-\eta) \, d\eta$$

is the pairing between the spaces $H^{s-\mu-1/2}(\mathbf{R}^n)$ and $H^{-s+\mu+1/2}(\mathbf{R}^n)$.

Substep (6-a): First, we calculate the Fourier transform $\widehat{U}(\eta)$:

$$\widehat{U}(\eta) = \int_{\mathbf{R}^n} e^{-i\,x\cdot\eta}\, U(x)\, dx$$

$$= \frac{1}{(2\pi)^n} \int_{\mathbf{R}^n} e^{-i\,x\cdot\eta} \left(\int_{\mathbf{R}^{n-1}} e^{i\,x'\cdot\xi'}\, k(x,\xi')\, \widehat{v}(\xi')\, d\xi' \right) dx.$$

However, we remark that the support of $k(x,\xi')$ is compact with respect to x and that $\widehat{v} \in \mathcal{S}(\mathbf{R}^{n-1})$. Hence, by using Fubini's theorem we have the formula

$$\widehat{U}(\eta) = \frac{1}{(2\pi)^n} \int_{\mathbf{R}^{n-1}} \left(e^{-i\,x'\cdot\eta'-ix_n\eta_n}\, k(x,\xi')\, dx \right) \widehat{v}(\xi')\, d\xi'$$

$$= \frac{1}{(2\pi)^n} \int_{\mathbf{R}^{n-1}} \widehat{k}(\eta - (\xi',0), \xi')\, \widehat{v}(\xi')\, d\xi'$$

$$= \frac{1}{(2\pi)^n} \int_{\mathbf{R}^{n-1}} \widehat{k}(\eta' - \xi', \eta_n, \xi')\, \widehat{v}(\xi')\, d\xi'.$$

Therefore, we have the formula

$$\langle U, \varphi \rangle$$

$$= \frac{1}{(2\pi)^n} \int_{\mathbf{R}^n} \widehat{U}(\eta)\, \widehat{\varphi}(-\eta)\, d\eta$$

$$= \frac{1}{(2\pi)^n} \left(\int_{\mathbf{R}^{n-1}} \widehat{k}(\eta - \xi', \eta_n, \xi')\, \widehat{v}(\xi')\, d\xi' \right) \widehat{\varphi}(-\eta)\, d\eta$$

$$= \frac{1}{(2\pi)^n} \left(1 + |\eta|^2 \right)^{(-s+\mu+1/2)/2} \widehat{\varphi}(-\eta)$$

$$\times \left(\frac{1}{(2\pi)^{n-1}} \int_{\mathbf{R}^{n-1}} \widehat{k}(\eta - \xi', \eta_n, \xi') \left(1 + |\eta|^2 \right)^{(s-\mu-1/2)/2} \widehat{v}(\xi')\, d\xi' \right) d\eta.$$

If we let

$$V(\eta) := \int_{\mathbf{R}^{n-1}} \left| \widehat{k}(\eta' - \xi', \eta_n, \xi') \right| \left(1 + |\eta| \right)^{s-\mu-1/2} |\widehat{v}(\xi')|\, d\xi',$$

by applying Schwarz's inequality we can find a constant $C > 0$ such that

$$|\langle U, \varphi \rangle| \le C \, \|\varphi\|_{-s+\mu+1/2} \, \|V\|_0 \, . \tag{6.25}$$

We are reduced to the study of the function $V(\eta)$.

Substep (6-b): The estimate of the norm $\|V\|_0$. By Lemma 6.8, there exists a constant $C = C_{q,r} > 0$ such that

$$\left| \widehat{k}(\eta' - \xi', \eta_n, \xi') \right| \le C \, (1 + |\xi'|)^\mu \, (1 + |\eta' - \xi'|)^{-r} \left(1 + \frac{|\eta_n|}{1 + |\xi'|} \right)^{-q}.$$

Hence we have the inequality

$$V(\eta) \le C_{q,r} \int_{\mathbf{R}^{n-1}} (1 + |\xi'|)^\mu \, (1 + |\eta' - \xi'|)^{-r} \left(1 + \frac{|\eta_n|}{1 + |\xi'|} \right)^{-q}$$
$$\times (1 + |\eta|)^{s-\mu-1/2} \, |\widehat{v}(\xi')| \, d\xi'$$
$$= C_{q,r} \int_{\mathbf{R}^{n-1}} (1 + |\xi'|)^s \, |\widehat{v}(\xi')| \, \chi(\eta, \xi') d\xi',$$

where

$$\chi(\eta, \xi') := (1 + |\xi'|)^{\mu-s} \, (1 + |\eta|)^{s-\mu-1/2} \, (1 + |\eta' - \xi'|)^{-r}$$
$$\times \left(1 + \frac{|\eta_n|}{1 + |\xi'|} \right)^{-q} \quad \text{for } \eta = (\eta', \eta_n) \in \mathbf{R}^n \text{ and } \xi' \in \mathbf{R}^{n-1}.$$

If we let

$$W(\eta) := \int_{\mathbf{R}^{n-1}} \chi(\eta, \xi') \, (1 + |\xi'|)^s \, |\widehat{v}(\xi')| \, d\xi' \quad \text{for } \eta \in \mathbf{R}^n,$$

then we have the inequality

$$V(\eta) \le C_{q,r} \, W(\eta) \quad \text{for all } \eta \in \mathbf{R}^n,$$

and so

$$\|V\|_0 \le C_{q,r} \, \|W\|_0 \, . \tag{6.26}$$

In this way, we are reduced to the study of the function $W(\eta)$.

Substep (6-c): The estimate of the norm $\|W\|_0$. To do this, we make the change of variables

$$\eta' = \zeta', \quad \eta_n = (1 + |\zeta'|) \, \zeta_n.$$

Then we have the formula

$$\int_{\mathbf{R}^n} |W(\eta', \eta_n)|^2 \, d\eta' \, d\eta_n = \int_{\mathbf{R}^n} |W(\zeta', (1 + |\zeta'|) \, \zeta_n)|^2 \, (1 + |\zeta'|) \, d\zeta' \, d\zeta_n.$$

However, we have the following:

(1) $1 + |\eta| = (1 + |\zeta'|) \, (1 + |\zeta_n|)$.

(2) $1 + |\xi'| \le 1 + |\zeta'| + |\xi' - \zeta'| \le (1 + |\zeta'|)(1 + |\xi' - \zeta'|)$, so that

$$\frac{|\zeta_n|}{1 + |\zeta' - \xi'|} \le \frac{(1 + |\zeta'|)|\eta_n|}{1 + |\xi'|} = \frac{|\eta_n|}{1 + |\xi'|}.$$

On the other hand, by applying Peetre's inequality (see [26, p. 79, Lemme 2.6.2])

$$\left(1 + |\xi'|^2\right)^{\sigma}$$
$$\le 2^{|\sigma|}\left(1 + |\xi' - \zeta'|^2\right)^{|\sigma|}\left(1 + |\zeta'|^2\right)^{\sigma} \quad \text{for all } \xi', \zeta' \in \mathbf{R}^{n-1} \text{ and } \sigma \in \mathbf{R},$$

we have the inequality

$$(1 + |\xi'|)^{\mu - s} \le 2^{|\mu - s|/2}(1 + |\xi' - \zeta'|)^{|\mu - s|}(1 + |\zeta'|)^{\mu - s}.$$

Summing up, we obtain that

$$\chi(\zeta', (1 + |\zeta'|)\zeta_n, \xi')$$
$$= (1 + |\xi'|)^{\mu - s}(1 + |\zeta'|)^{s - \mu - 1/2}(1 + |\zeta_n|)^{s - \mu - 1/2}(1 + |\zeta' - \xi'|)^{-r}$$
$$\times \left(1 + \frac{|\zeta_n|}{1 + |\zeta' - \xi'|}\right)^{-q}$$
$$\le 2^{|\mu - s|/2}(1 + |\xi' - \zeta'|)^{|\mu - s|}(1 + |\zeta'|)^{\mu - s}(1 + |\zeta'|)^{s - \mu - 1/2}$$
$$\times (1 + |\zeta_n|)^{s - \mu - 1/2}(1 + |\zeta' - \xi'|)^{-r}\left(1 + \frac{|\zeta_n|}{1 + |\zeta' - \xi'|}\right)^{-q}$$
$$= 2^{|\mu - s|/2}(1 + |\zeta_n|)^{s - \mu - 1/2}(1 + |\xi' - \zeta'|)^{|\mu - s| - r}$$
$$\times (1 + |\zeta'|)^{-1/2}\left(1 + \frac{|\zeta_n|}{1 + |\zeta' - \xi'|}\right)^{-q}.$$

Therefore, we have the inequality

$$\int_{\mathbf{R}^n} |W(\eta)|^2 \, d\eta \tag{6.27}$$
$$= \int_{\mathbf{R}^n} |W(\zeta', (1 + |\zeta'|)\zeta_n)|^2 (1 + |\zeta'|) \, d\zeta' \, d\zeta_n$$
$$= \int_{\mathbf{R}^n} \left[\int_{\mathbf{R}^{n-1}} \chi(\zeta', (1 + |\zeta'|)\zeta_n, \xi')(1 + |\xi'|)^s |\widehat{v}(\xi')| \, d\xi'\right]^2$$
$$\times (1 + |\zeta'|) \, d\zeta' \, d\zeta_n$$
$$\le 2^{|\mu - s|} \int_{\mathbf{R}^n} \left[\int_{\mathbf{R}^{n-1}} (1 + |\zeta_n|)^{s - \mu - 1/2}(1 + |\zeta' - \xi'|)^{|\mu - s| - r}\right.$$
$$\times (1 + |\zeta'|)^{-1/2}\left(1 + \frac{|\zeta_n|}{1 + |\zeta' - \xi'|}\right)^{-q}(1 + |\xi'|)^s |\widehat{v}(\xi')| \, d\xi'\Bigg]^2$$

$$\times \left(1 + |\zeta'|\right) d\zeta' \, d\zeta_n$$

$$\leq C \int_{\mathbf{R}^n} \left(1 + |\xi_n|^2\right)^{s-\mu-1/2}$$

$$\times \left[\int_{\mathbf{R}^{n-1}} \left(1 + |\zeta' - \xi'|\right)^{|\mu-s|-r} \left(1 + \frac{|\zeta_n|}{1 + |\zeta' - \xi'|}\right)^{-q}\right.$$

$$\left. \times \left(1 + |\xi'|\right)^s |\widehat{v}(\xi')| \, d\xi'\right]^2 d\zeta' d\zeta_n$$

$$= C \int_{\mathbf{R}^n} \left(1 + |\xi_n|^2\right)^{s-\mu-1/2}$$

$$\times \left[\int_{\mathbf{R}^{n-1}} \left(1 + |\zeta' - \xi'|\right)^{|\mu-s|-r+q} \left(1 + |\zeta' - \xi'| + |\zeta_n|\right)^{-q}\right.$$

$$\left. \times \left(1 + |\xi'|\right)^s |\widehat{v}(\xi')| \, d\xi'\right]^2 d\zeta' d\zeta_n.$$

If we let

- $Y(\zeta', \zeta_n) := \int_{\mathbf{R}^{n-1}} Z(\zeta' - \xi', \zeta_n) \left(1 + |\xi'|\right)^s |\widehat{v}(\xi')| \, d\xi',$

- $Z(\zeta', \zeta_n) := \left(1 + |\zeta'|\right)^{|\mu-s|-r+q} \left(1 + |\zeta| + |\zeta_n|\right)^{-q},$

then we can rewrite inequality (6.27) as follows:

$$\|W\|_0^2 \leq C \int_{\mathbf{R}} \left(1 + |\xi_n|^2\right)^{s-\mu-1/2} \|Y(\cdot, \zeta_n)\|_0^2 \, d\zeta_n. \tag{6.28}$$

However, by the Hausdorff–Young inequality it follows that

$$\|Y(\cdot, \zeta_n)\|_0 \leq \left(\int_{\mathbf{R}^{n-1}} |Z(\zeta', \zeta_n)| \, d\zeta'\right) \left(\int_{\mathbf{R}^{n-1}} \left(1 + |\xi'|\right)^{2s} |\widehat{v}(\xi')|^2 \, d\xi'\right)^{1/2}$$

$$\leq C \left(\int_{\mathbf{R}^{n-1}} |Z(\zeta', \zeta_n)| \, d\zeta'\right) \|v\|_s.$$

Moreover, if we choose q and r so large that

$$q > n - 1,$$
$$r \geq |\mu - s| + q,$$

then we have the inequality

$$\int_{\mathbf{R}^{n-1}} |Z(\zeta', \zeta_n)| \, d\zeta' = \int_{\mathbf{R}^{n-1}} \frac{\left(1 + |\zeta'|\right)^{|\mu-s|-r+q}}{\left(1 + |\zeta'| + |\zeta_n|\right)^q} d\zeta'$$

$$\leq \int_{\mathbf{R}^{n-1}} \frac{1}{\left(1 + |\zeta'| + |\zeta_n|\right)^q} d\zeta'$$

$$= (1 + |\xi_n|)^{-q+n-1} \int_{\mathbf{R}^{n-1}} \frac{1}{(1+|\eta'|)^q} d\eta'$$
$$= C (1 + |\xi_n|)^{-q+n-1} \quad \text{for } -q < -(n-1).$$

Hence we have the inequality

$$\|Y(\cdot, \xi_n)\|_0^2 \leq C \left(\int_{\mathbf{R}^{n-1}} |Z(\zeta', \zeta_n)| d\zeta' \right)^2 \|v\|_s^2 \tag{6.29}$$
$$\leq C \|v\|_s^2 \left(1 + |\xi_n|^2\right)^{-q+n-1}.$$

Therefore, by inequalities (6.28) and (6.29) it follows that

$$\|W\|_0 \leq C' \|v\|_s \left(\int_{\mathbf{R}} \left(1 + |\xi_n|^2\right)^{s-\mu-1/2-q+n-1} d\xi_n \right)^{1/2} \tag{6.30}$$
$$\leq C'' \|v\|_s \quad \text{for } 2(s - \mu - 1/2 - q + n - 1) < -1.$$

By combining (6.18), (6.19) and (6.30), we obtain that

$$|\langle U, \varphi \rangle| \leq C \|\varphi\|_{-s+\mu+1/2} \|V\|_0$$
$$\leq C \|\varphi\|_{-s+\mu+1/2} \|W\|_0$$
$$\leq C \|\varphi\|_{-s+\mu+1/2} \|v\|_s \quad \text{for all } \varphi \in C_0^\infty(\mathbf{R}^n).$$

This proves the desired inequality (6.24).

Now the proof of Theorem 6.2 is complete. $\square$

Remark 6.9. In view of Theorem 4.49, it follows that condition (6.12) is invariant under change of coordinates. Furthermore, it is easy to see that every parametrix for an elliptic differential operator satisfies condition (6.12).

(II) Secondly, the next theorem covers *volume potentials* (see [17, Théorème (2.2.2)], [18, Théorème 2.9], [26, Chapitre V, Théorème 2.5] for $p := 2$):

Theorem 6.10. *Let $A \in L_{\mathrm{cl}}^m(M)$ be as in Theorem 6.2. Then we have the following two assertions (i) and (ii):*

(i) The operator

$$A_\Omega : f \longmapsto A(f^0)|_\Omega$$

is continuous on $C^\infty(\overline{\Omega})$ into itself.

(ii) If $1 < p < \infty$, then the operator A_Ω extends to a continuous linear operator

$$A_\Omega : H^{s,p}(\Omega) \longrightarrow H^{s-m,p}(\Omega)$$

for $s > -1 + 1/p$ and $1 < p < \infty$ (see Figure 6.5 below).

$$\begin{array}{ccc} H^{s,p}(\Omega) & \xrightarrow{\ A_\Omega\ } & H^{s-m,p}(\Omega) \\[4pt] \uparrow & & \uparrow \\[4pt] C^\infty(\overline{\Omega}) & \xrightarrow[\ A_\Omega\ ]{} & C^\infty(\overline{\Omega}) \end{array}$$

Fig. 6.5. The mapping properties of the volume potential A_Ω

Remark 6.11. The operator A_Ω can be visualized as follows:

$$A_\Omega\colon C^\infty(\overline{\Omega}) \longrightarrow \mathcal{D}'(M) \xrightarrow{\ A\ } \mathcal{D}'(M) \longrightarrow C^\infty(\Omega),$$

where the first arrow is the zero extension to M and the last one is the restriction to Ω. Part (i) of Theorem 6.10 asserts that the volume potential A_Ω preserves smoothness up to the boundary $\partial\Omega$.

Proof. We only prove part (i), by using the Fourier transform ([26, Chapitre V, Théorème 2.5]).
    (i) Let

$$E\colon C^\infty(\overline{\Omega}) \longrightarrow C^\infty(\mathbf{R}^n)$$

be Seeley's extension operator (see Theorem 4.11). If $f \in C^\infty(\overline{\Omega})$, we let

$$u(x) = \begin{cases} Ef(x) & \text{for } x \in \Omega', \\ 0 & \text{for } x \in \overline{\Omega}, \end{cases}$$

where Ω' is the exterior domain of Ω

$$\Omega' = \mathbf{R}^n \setminus \overline{\Omega}.$$

Then we have the formulas

$$f^0 = Ef - u,$$

and

$$A\left(f^0\right)\big|_\Omega = A(Ef)\big|_\Omega - Au\big|_\Omega,$$

with $A(Ef) \in C_0^\infty(\mathbf{R}^n)$. Hence it suffices to show that the mapping

$$C^\infty(\overline{\Omega'}) \ni g \longmapsto (Au)\big|_\Omega \in C^\infty(\overline{\Omega})$$

is continuous.
    Just as in the proof of Theorem 4.19 (formula (4.8)), we may assume that

$$Bu(x) = \frac{1}{(2\pi)^{n-1}} \int_{\mathbf{R}^{n-1}} e^{i\,x'\cdot\xi'} \left(\int_{\Gamma_{\xi'}} e^{ix_n\xi_n}\, a\left(x, \xi', \xi_n\right) \widehat{u}\left(\xi', \xi_n\right) d\xi_n \right) d\xi',$$

where

$$\widehat{u}(\xi', \xi_n) = \int_{\mathbf{R}^n} \widetilde{f}(x', x_n)\, dx$$

$$= \int_{\mathbf{R}^{n-1}} e^{-i\,x'\cdot\xi'} \left(\int_{-\infty}^0 e^{-i\,x_n\,\xi_n}\, \widetilde{f}(x', x_n)\, dx_n \right) dx'.$$

However, we remark that the function

$$\int_{-\infty}^0 e^{-i\,x_n\,\xi_n}\, \widetilde{f}(x', x_n)\, dx_n$$

admits an analytic extension into the upper complex half-plane $\operatorname{Im}\xi_n > 0$ and continuous in $\operatorname{Im}\xi_n \geq 0$. Hence we find that the Fourier transform $\widehat{u}(\xi', \xi_n)$ is rapidly decreasing with respect to the variable $\xi' \in \mathbf{R}^{n-1}$, for $\operatorname{Im}\xi_n \geq 0$.

In this way, we obtain that

$$Bu|_{\mathbf{R}^n_+} \in C^\infty(\overline{\mathbf{R}^n_+}).$$

The proof of Theorem 6.10 is complete. $\square$

6.2 Dirichlet Problem

In this section we shall consider the Dirichlet problem in the framework of Sobolev spaces of L^p type. This is a generalization of the classical potential approach to the Dirichlet problem in terms of pseudo-differential operators.

Let Ω be a bounded domain of Euclidean space $\mathbf{R}^n$ with smooth boundary $\partial\Omega$. Its closure $\overline{\Omega} = \Omega \cup \partial\Omega$ is an n-dimensional, compact smooth manifold with boundary. We may assume that $\overline{\Omega}$ is the closure of a relatively compact open subset Ω of an n-dimensional, compact smooth manifold M without boundary in which Ω has a smooth boundary $\partial\Omega$ ([1], [83]). This manifold $M = \widehat{\Omega}$ is the *double* of Ω (see Figure 4.1).

We let

$$Au = \sum_{i,j=1}^n a^{ij}(x)\frac{\partial^2 u}{\partial x_i \partial x_j} + \sum_{i=1}^n b^i(x)\frac{\partial u}{\partial x_i} + c(x)u \tag{6.31}$$

be a second-order, *uniformly elliptic* differential operator with real coefficients on the double $M = \widehat{\Omega}$ of Ω such that:

(1) The $a^{ij}(x)$ are the components of a C^∞ symmetric contravariant tensor of type $\binom{2}{0}$ on M and there exists a constant $a_0 > 0$ such that

$$\sum_{i,j=1}^n a^{ij}(x)\xi_i\xi_j \geq a_0|\xi|^2 \quad \text{on } T^*(M),$$

where $T^*(M)$ is the cotangent bundle of M.

(2) $b^i \in C^\infty(M)$ for $1 \leq i \leq n$.
(3) $c \in C^\infty(M)$ and $c(x) \leq 0$ in Ω.

Following Seeley [105] and [106], we let

$$\mathcal{N}_0(A) := \left\{ u \in C^\infty(M) : \operatorname{supp} u \subset \overline{\Omega}, \; Au = 0 \text{ in } \Omega \right\}.$$

It is known (see [105, Theorem 7]) that $\mathcal{N}_0(A)$ is finite dimensional. It should be emphasized that all norms on the finite-dimensional space $\mathcal{N}_0(A)$ are *equivalent*.

We remark (see [3, pp. 274–277] and [82, Theorem 3.20]) that all the sufficiently large eigenvalues of the Dirichlet problem for the differential operator A and its formal adjoint A^* lie in the *parabolic* type region, as will be discussed in Section 6.7 (see assertion (6.74)). Hence, by considering $A - \lambda_1$ and $A^* - \lambda_1$ for certain positive number λ_1 we may assume that

$$\boxed{\mathcal{N}_0(A) = \mathcal{N}_0(A^*) = \{0\}.} \tag{SC}$$

The next theorem states the existence of a *volume potential* for A, which plays the same role for A as the Newtonian potential plays for the Laplacian (cf. [105, Theorem 1], [106, p. 239, Theorem 2] and [114, Theorem 8.2.1]; [122, Theorem 7.29]):

Theorem 6.12 (Seeley). *Assume that condition (SC) is satisfied. Then we have the following two assertions (i) and (ii):*

(i) The operator $A \colon C^\infty(M) \to C^\infty(M)$ is bijective, and its inverse C is a classical, elliptic pseudo-differential operator of order -2 on M.
(ii) The operators A and C extend respectively to isomorphisms

$$
\begin{array}{ccc}
H^{s,p}(M) & \xrightarrow{\;\;A\;\;} & H^{s-2,p}(M) \\
\| & & \| \\
H^{s,p}(M) & \xleftarrow[\;\;C\;\;]{} & H^{s-2,p}(M)
\end{array}
$$

for all $s \in \mathbf{R}$, which are still inverses of each other.

Proof. First, by condition (SC) we can apply [106, Theorem 2] to find a classical, pseudo-differential operator C_1 of order -2 on M such that

$$(AC_1)\, f = A\,(C_1 f) = f \quad \text{for all } f \in C^\infty(M).$$

Similarly, by applying [106, Theorem 2] to the adjoint A^* we can find a classical, pseudo-differential operator C_2 of order -2 on M such that

$$(A^*C_2)\, g = A^*\,(C_2 g) = g \quad \text{for all } g \in C^\infty(M).$$

Hence, by passing to the adjoint we have the formula

$$C_2{}^* f = C_2{}^* (A C_1 f) = (C_2{}^* A)(C_1 f) = (A^* C_2)^* (C_1 f)$$
$$= C_1 f \quad \text{for all } g \in C^\infty(M),$$

and so

$$(C_1 A) g = C_1 (A g) = C_2{}^* (A g) = ((A^* C_2))^* g = g \quad \text{for all } g \in C^\infty(M).$$

Summing up, we have proved that $\mathcal{C} = C_1 = C_2{}^*$ is the inverse of A. The proof of Theorem 6.12 is complete. $\square$

Next we construct a *surface potential* for A, which is a generalization of the classical Poisson kernel for the Laplacian.

We let

$$K v = \gamma_0 \left(\mathcal{C} \left(v \otimes \delta_{\partial\Omega} \right) \right) = \mathcal{C} \left(v \otimes \delta_{\partial\Omega} \right) \big|_{\partial\Omega} \quad \text{for } v \in C^\infty(\partial\Omega).$$

Here $v \otimes \delta_{\partial\Omega} = v(x') \otimes \delta(t)$ is a distribution on M defined by the formula

$$\langle v \otimes \delta_{\partial\Omega}, \varphi \cdot \mu \rangle = \langle v, \varphi(\cdot, 0) \cdot \omega \rangle \quad \text{for all } \varphi(x', t) \in C^\infty(M),$$

where μ is a strictly positive density on M and ω is a strictly positive density on $\partial\Omega = \{t = 0\}$, respectively.

Then we have the following theorem (cf. [114, Theorem 8.2.2]):

Theorem 6.13. *Assume that condition (SC) is satisfied. Then we have the following two assertions (i) and (ii):*

(i) *The operator K is a classical elliptic pseudo-differential operator of order -1 on $\partial\Omega$.*

(ii) *The operator $K \colon C^\infty(\partial\Omega) \to C^\infty(\partial\Omega)$ is bijective, and its inverse L is a classical elliptic pseudo-differential operator of first order on $\partial\Omega$. Furthermore, the operators K and L extend respectively to isomorphisms*

$$
\begin{array}{ccc}
B^{\sigma,p}(\partial\Omega) & \xrightarrow{\ K\ } & B^{\sigma+1,p}(\partial\Omega) \\
\Big\| & & \Big\| \\
B^{\sigma,p}(\partial\Omega) & \xleftarrow[\ L\ ]{} & B^{\sigma+1,p}(\partial\Omega)
\end{array}
$$

for all $\sigma \in \mathbf{R}$ and $1 < p < \infty$, which are still inverses of each other.

Now we let

$$\mathcal{P}\varphi := \mathcal{C} \left((L\varphi) \otimes \delta_{\partial\Omega} \right) \big|_\Omega \quad \text{for } \varphi \in C^\infty(\partial\Omega).$$

Then the operator $\mathcal{P}$ maps $C^\infty(\partial\Omega)$ continuously into $C^\infty(\overline{\Omega})$, and extends to a continuous linear operator

$$\mathcal{P} \colon B^{s-1/p,p}(\partial\Omega) \longrightarrow H^{s,p}(\Omega)$$

for all $s \in \mathbf{R}$. Furthermore, we have, for all $\varphi \in B^{s-1/p,p}(\partial\Omega)$,

$$\begin{cases} A\mathcal{P}\varphi = A\mathcal{C}\left((L\varphi) \otimes \delta_{\partial\Omega}\right)|_\Omega = \left((L\varphi) \otimes \delta_{\partial\Omega}\right)|_\Omega = 0 & \text{in } \Omega, \\ \gamma_0\left(\mathcal{P}\varphi\right) = \mathcal{P}\varphi|_{\partial\Omega} = \gamma_0\left(\mathcal{C}\left((L\varphi) \otimes \delta_{\partial\Omega}\right)\right) = K\left(L\varphi\right) = \varphi & \text{on } \partial\Omega. \end{cases}$$

The operator $\mathcal{P}$ is called the *Poisson operator* or *Poisson kernel*.
We let

$$\mathcal{N}\left(A, s, p\right) = \{u \in H^{s,p}(\Omega) : Au = 0 \text{ in } \Omega\}, \quad s \in \mathbf{R}.$$

Since the injection $H^{s,p}(\Omega) \to \mathcal{D}'(\Omega)$ is continuous, it follows that the null space $\mathcal{N}\left(A, s, p\right)$ is a closed subspace of $H^{s,p}(\Omega)$; hence it is a Banach space.

Then we have the following fundamental result (cf. [105, Theorems 5 and 6]):

Theorem 6.14 (Seeley). *The Poisson operator $\mathcal{P}$ maps the Besov space $B^{s-1/p,p}(\partial\Omega)$ isomorphically onto the null space $\mathcal{N}\left(A, s, p\right)$ for all $s \in \mathbf{R}$ and $1 < p < \infty$. More precisely, the spaces $\mathcal{N}\left(A, s, p\right)$ and $B^{s-1/p,p}(\partial\Omega)$ are isomorphic for all $\sigma \in \mathbf{R}$ and $1 < p < \infty$ in such a way that*

$$\begin{array}{ccc} B^{s-1/p,p}(\partial\Omega) & \xrightarrow{\ \mathcal{P}\ } & \mathcal{N}\left(A, s, p\right) \\ \| & & \| \\ B^{s-1/p,p}(\partial\Omega) & \xleftarrow[\gamma_0]{} & \mathcal{N}\left(A, s, p\right). \end{array}$$

By combining Theorems 6.12 and 6.14, we can obtain the following existence and uniqueness theorem for the Dirichlet problem (cf. [4], [50], [77], [133], [136], [146]):

Theorem 6.15 (the Dirichlet case). *Let $1 < p < \infty$ and $s > -2 + 1/p$. If condition (SC) is satisfied, then the Dirichlet problem*

$$\begin{cases} Au = f & \text{in } \Omega, \\ \gamma_0 u = u|_{\partial\Omega} = \varphi & \text{on } \partial\Omega \end{cases} \tag{6.32}$$

has a unique solution $u(x)$ in $H^{s,p}(\Omega)$ for any $f \in H^{s-2,p}(\Omega)$ and any $\varphi \in B^{s-1/p,p}(\partial\Omega)$.

Indeed, it suffices to note that the unique solution u of the Dirichlet problem (6.32) is given by the following formula:

$$u = \mathcal{C}(Ef)|_\Omega + \mathcal{P}\left(\varphi - (\mathcal{C}Ef)|_{\partial\Omega}\right) \quad \text{in } \Omega. \tag{6.33}$$

Here $E \colon H^{s,p}(\Omega) \to H^{s,p}(M)$ is the Seeley extension operator (see Theorem 4.11).

Furthermore, in the Neumann problem for the differential operator A and its formal adjoint A^*, we have *parabolic* condensation of eigenvalues along the negative real axis, as discussed in Agmon [3, pp. 276–277]. Hence, we replace A and A^* by $A - \lambda_1$ and $A^* - \lambda_1$ for certain positive number λ_1, respectively, we may assume that condition (SC) is satisfied.

Therefore, we can prove the following existence and uniqueness theorem for the Neumann problem (cf. [4], [50], [77], [133], [136], [146]):

Theorem 6.16 (the Neumann case). *Let $1 < p < \infty$ and $s > -1 + 1/p$. If condition (SC) is satisfied, then the Neumann problem*

$$\begin{cases} Av = g & \text{in } \Omega, \\ \dfrac{\partial v}{\partial \mathbf{n}}\bigg|_{\partial \Omega} = \psi & \text{on } \partial \Omega \end{cases} \tag{6.4}$$

has a unique solution $v(x)$ in $H^{s,p}(\Omega)$ for any $g \in H^{s-2,p}(\Omega)$ and any $\psi \in B^{s-1-1/p,p}(\partial \Omega)$. Here $\mathbf{n}$ is the unit inward normal to the boundary $\partial \Omega$.

By Theorem 6.16, we can introduce a linear operator

$$\mathcal{G}_N \colon H^{s-2,p}(\Omega) \longrightarrow H^{s,p}(\Omega)$$

as follows: For any $g \in H^{s-2,p}(\Omega)$, the function $v = \mathcal{G}_N g \in H^{s,p}(\Omega)$ is the unique solution of the homogeneous Neumann problem

$$\begin{cases} Av = g & \text{in } \Omega, \\ \gamma_1 v = \dfrac{\partial v}{\partial \boldsymbol{\nu}}\bigg|_{\partial \Omega} = -\dfrac{\partial v}{\partial \mathbf{n}}\bigg|_{\partial \Omega} = 0 & \text{on } \partial \Omega. \end{cases} \tag{6.34}$$

Here recall that $\boldsymbol{\nu} = -\mathbf{n}$ is the unit *outward* normal to the boundary $\partial \Omega$ (see Figure 1.2).

The operator $\mathcal{G}_N$ is called the *Green operator* for the Neumann problem.

6.3 Formulation of the Boundary Value Problem

In this section we formulate elliptic boundary value problems in the framework of L^p Sobolev spaces. If $u \in H^{2,p}(\Omega) = W^{2,p}(\Omega)$, we can define its traces $\gamma_0 u$ and $\gamma_1 u$ respectively by the formulas (see Theorems 4.17 and 4.18)

$$\begin{cases} \gamma_0 u = u|_{\partial \Omega}, \\ \gamma_1 u = \dfrac{\partial u}{\partial \boldsymbol{\nu}}\bigg|_{\partial \Omega} = -\dfrac{\partial u}{\partial \mathbf{n}}\bigg|_{\partial \Omega}. \end{cases}$$

Then we have the following theorem (cf. [2], [107]):

Theorem 6.17 (the trace theorem). *The trace map*

$$\gamma = (\gamma_0, \gamma_1) : H^{2,p}(\Omega) \longrightarrow B^{2-1/p,p}(\partial\Omega) \oplus B^{1-1/p,p}(\partial\Omega)$$

is continuous and surjective for all $1 < p < \infty$.

We define a first-order, boundary condition

$$B\gamma u \tag{6.2}$$

$$:= a(x') \frac{\partial u}{\partial \boldsymbol{\nu}} + b(x')u\bigg|_{\partial\Omega} = a(x')\gamma_1 u + b(x')\gamma_0 u \quad \text{for } u \in H^{2,p}(\Omega).$$

Here:

(1) $a \in C^\infty(\partial\Omega)$ and $a(x') \geq 0$ on $\partial\Omega$.
(2) $b \in C^\infty(\partial\Omega)$ and $b(x') \geq 0$ on $\partial\Omega$.

Then we have the following proposition:

Proposition 6.18. *The boundary operator*

$$B\gamma : H^{2,p}(\Omega) \longrightarrow B^{1-1/p,p}(\partial\Omega)$$

is continuous for all $1 < p < \infty$.

Now we can formulate our boundary value problem for $(A, B\gamma)$ as follows: Given functions $f \in L^p(\Omega)$ and $\varphi \in B^{2-1/p,p}(\partial\Omega)$, find a function $u \in H^{2,p}(\Omega)$ such that

$$\begin{cases} Au = f & \text{in } \Omega, \\ B\gamma u = \varphi & \text{on } \partial\Omega. \end{cases} \tag{6.1}$$

6.4 Special Reduction to the Boundary

In this section, by using the operators $\mathcal{P}$ and $\mathcal{G}_N$ we shall show that the boundary value problem (6.1) can be reduced to the study of a pseudo-differential operator on the boundary. The virtue of this reduction is that there is no difficulty in taking adjoints or transposes after restricting the attention to the boundary, whereas boundary value problems in general do not have adjoints or transposes. This allows us to discuss the *existence* theory more easily. Here it should be emphasized that our reduction approach would break down if we use the Dirichlet problem (Theorem 6.15) as usual, instead of the Neumann problem (Theorem 6.16).

First, we remark that every function $u(x)$ in $H^{2,p}(\Omega)$ can be written in the following form:

$$u(x) = v(x) + w(x), \tag{6.35}$$

where

$$\begin{cases} v = \mathcal{G}_N(Au) \in H^{2,p}(\Omega), \\ w = u - v \in \mathcal{N}(A, 2, p) = \left\{ z \in H^{2,p}(\Omega) : Az = 0 \text{ in } \Omega \right\}. \end{cases}$$

Since the operator $\mathcal{G}_N \colon L^p(\Omega) \to H^{2,p}(\Omega)$ is continuous, it follows that the decomposition (6.35) is continuous; more precisely, we have the inequalities

- $\|v\|_{2,p} \le C \|Au\|_p \le C \|u\|_{2,p}$;
- $\|w\|_{2,p} \le \|u\|_{2,p} + \|v\|_{2,p} \le C \|u\|_{2,p}$.

Here the letter C denotes a generic positive constant.

Now we assume that $u \in H^{2,p}(\Omega)$ is a solution of the boundary value problem

$$\begin{cases} Au = f & \text{in } \Omega, \\ B\gamma u = \varphi & \text{on } \partial\Omega. \end{cases} \tag{6.1}$$

Then, by virtue of the decomposition (6.35) of $u(x)$ this is equivalent to saying that $w = u - v \in H^{2,p}(\Omega)$ is a solution of the boundary value problem

$$\begin{cases} Aw = 0 & \text{in } \Omega, \\ B\gamma w = \varphi - B\gamma v = \varphi - b(x')\gamma_0(\mathcal{G}_N f) & \text{on } \partial\Omega, \end{cases} \tag{6.36}$$

since we have, by formulas (6.2) and (6.34),

$$\begin{aligned} B\gamma v &= a(x')\gamma_1(\mathcal{G}_N f) + b(x')\gamma_0(\mathcal{G}_N f) \\ &= b(x')\gamma_0(\mathcal{G}_N f) \quad \text{on } \partial\Omega. \end{aligned}$$

However, by Theorem 6.14 it follows that $\mathcal{N}(A, 2, p)$ and $B^{2-1/p,p}(\partial\Omega)$ are isomorphic in such a way that:

$$\begin{array}{ccc} \mathcal{N}(A, 2, p) & \xrightarrow{\ \gamma_0\ } & B^{2-1/p,p}(\partial\Omega) \\ \| & & \| \\ \mathcal{N}(A, 2, p) & \xleftarrow[\ \mathcal{P}\ ]{} & B^{2-1/p,p}(\partial\Omega). \end{array}$$

Therefore, we find that $w \in H^{2,p}(\Omega)$ is a solution of problem (6.14) if and only if $\psi(x') \in B^{2-1/p,p}(\partial\Omega)$ is a solution of the equation

$$B\gamma(\mathcal{P}\psi) = \varphi - b(x')\gamma_0(\mathcal{G}_N f) \quad \text{on } \partial\Omega. \tag{6.37}$$

Here $\psi = \gamma_0 w$, or equivalently, $w = \mathcal{P}\psi$. We remark that equation (6.37) is a generalization of the classical *Fredholm integral equation*.

Summing up, we obtain the following proposition:

Proposition 6.19. *Let* $1 < p < \infty$. *Assume that condition (SC) is satisfied. For given functions* $f \in L^p(\Omega)$ *and* $\varphi \in B^{2-1/p,p}(\partial\Omega)$, *there exists a solution* $u \in H^{2,p}(\Omega)$ *of the boundary value problem* (6.1) *if and only if there exists a solution* $\psi(x') \in B^{2-1/p,p}(\partial\Omega)$ *of equation* (6.37). *Furthermore, the solutions* $u(x)$ *and* $\psi(x')$ *are related as follows:*

$$u = \mathcal{G}_N f + \mathcal{P}\psi,$$
$$B\gamma(\mathcal{P}\psi) = \varphi - b(x')\gamma_0(\mathcal{G}_N f) \quad \text{on } \partial\Omega.$$

Remark 6.20. The advantage of equation (6.37) is that the terms in the right-hand side have the same regularity

$$\varphi - b(x')\gamma_0(\mathcal{G}_N f) \in B^{2-1/p,p}(\partial\Omega).$$

We let

$$T\colon C^\infty(\partial\Omega) \longrightarrow C^\infty(\partial\Omega)$$
$$\varphi \longmapsto B\gamma(\mathcal{P}\varphi).$$

Then we have, by formula (6.2),

$$T = B\gamma\mathcal{P} = -a(x')\Pi + b(x'),$$

where Π is the Dirichlet-to-Neumann operator defined as follows (cf. [138, p. 134, formula (4.13)]:

$$\Pi\varphi = \frac{\partial}{\partial\mathbf{n}}(\mathcal{P}\varphi)\Big|_{\partial\Omega} = -\frac{\partial}{\partial\boldsymbol{\nu}}(\mathcal{P}\varphi)\Big|_{\partial\Omega}$$
$$= -\gamma_1(\mathcal{P}\varphi) \quad \text{for all } \varphi \in C^\infty(\partial\Omega).$$

It is known (cf. [26], [57], [61], [73], [93], [105], [133]) that the operator Π is a classical, *elliptic* pseudo-differential operator of first order on $\partial\Omega$; hence the operator $T = a(x')\Pi + b(x')$ is a classical pseudo-differential operator of first order on the boundary $\partial\Omega$.

Consequently, Proposition 6.19 asserts that the boundary value problem (6.1) can be reduced to the study of the first-order pseudo-differential operator T on the boundary $\partial\Omega$. We shall formulate this fact more precisely in terms of functional analysis.

First, we remark that the operator $T\colon C^\infty(\partial\Omega) \to C^\infty(\partial\Omega)$ extends to a continuous linear operator

$$T\colon B^{s,p}(\partial\Omega) \longrightarrow B^{s-1,p}(\partial\Omega), \quad s \in \mathbf{R}.$$

Then we have the formula

$$T\varphi = B\gamma(\mathcal{P}\varphi) \quad \text{for all } \varphi \in B^{2-1/p,p}(\partial\Omega),$$

since the Poisson operator

$$\mathcal{P}: B^{2-1/p,p}(\partial\Omega) \longrightarrow N(A, 2, p)$$

and the boundary operator

$$B\gamma: H^{2,p}(\Omega) \longrightarrow B^{1-1/p,p}(\partial\Omega)$$

are both continuous.

We associate with the boundary value problem (6.1) a linear operator

$$\boxed{\mathcal{A} = (A, B\gamma) : H^{2,p}(\Omega) \longrightarrow L^p(\Omega) \oplus B^{2-1/p,p}(\partial\Omega)}$$

as follows.

(a) The domain $\mathcal{D}(\mathcal{A})$ of $\mathcal{A}$ is the space

$$\mathcal{D}(\mathcal{A}) = \left\{ u \in H^{2,p}(\Omega) = W^{2,p}(\Omega) : B\gamma u \in B^{2-1/p,p}(\partial\Omega) \right\}.$$

(b) $\mathcal{A}u = \{Au, B\gamma u\}$ for every $u \in \mathcal{D}(\mathcal{A})$.

Note that the space $B^{2-1/p,p}(\partial\Omega)$ is a right boundary space associated with the Dirichlet condition: $a(x') \equiv 0$ and $b(x') \equiv 1$ on $\partial\Omega$.

Since the operators

$$A: H^{2,p}(\Omega) \longrightarrow L^p(\Omega)$$

and

$$B\gamma: H^{2,p}(\Omega) \to B^{1-1/p,p}(\partial\Omega)$$

are both continuous, it follows that $\mathcal{A}$ is a closed operator. Furthermore, the operator $\mathcal{A}$ is densely defined, since the domain $\mathcal{D}(\mathcal{A})$ contains the space $C^\infty(\overline{\Omega})$. The situation can be visualized in Figure 6.6 below.

$$
\begin{array}{ccc}
H^{2,p}(\Omega) & \xrightarrow{\ (A,\,B\gamma)\ } & L^p(\Omega) \oplus B^{1-1/p,p}(\partial\Omega) \\
\uparrow & & \uparrow \\
\mathcal{D}(\mathcal{A}) & \xrightarrow{\ \ A\ \ } & L^p(\Omega) \oplus B^{2-1/p,p}(\partial\Omega) \\
\uparrow & & \uparrow \\
C^\infty(\overline{\Omega}) & \xrightarrow[\ (A,\,B\gamma)\ ]{} & C^\infty(\overline{\Omega}) \oplus C^\infty(\partial\Omega)
\end{array}
$$

Fig. 6.6. The mapping property of the operator $\mathcal{A}$

Similarly, we associate with equation (6.37) a linear operator

$$\boxed{\mathcal{T} = B\gamma\mathcal{P}: B^{2-1/p,p}(\partial\Omega) \longrightarrow B^{2-1/p,p}(\partial\Omega)}$$

as follows.

(α) The domain $\mathcal{D}(\mathcal{T})$ is the space

$$\mathcal{D}(\mathcal{T}) = \left\{ \varphi \in B^{2-1/p,p}(\partial\Omega) : T\varphi \in B^{2-1/p,p}(\partial\Omega) \right\}.$$

(β) $\mathcal{T}\varphi = T\varphi = B\gamma\,(\mathcal{P}\varphi)$ for every $\varphi \in \mathcal{D}(\mathcal{T})$.

Then the operator $\mathcal{T}$ is a densely defined, closed operator, since the operator

$$T \colon B^{2-1/p,p}(\partial\Omega) \longrightarrow B^{1-1/p,p}(\partial\Omega)$$

is continuous and since the domain $\mathcal{D}(\mathcal{T})$ contains the space $C^\infty(\partial\Omega)$. The situation can be visualized in Figure 6.7 below.

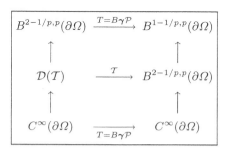

Fig. 6.7. The mapping property of the operator $\mathcal{T}$

The next theorem states that $\mathcal{A}$ has regularity property if and only if $\mathcal{T}$ has.

Theorem 6.21 (Regularity). *Let $1 < p < \infty$. The following two conditions* (6.38) *and* (6.39) *are equivalent:*

- $u \in L^p(\Omega)$, $Au \in L^p(\Omega)$ and $B\gamma u \in B^{2-1/p,p}(\partial\Omega)$ (6.38)
  $\implies u \in H^{2,p}(\Omega)$.
- $\varphi \in B^{-1/p,p}(\partial\Omega)$ and $T\varphi \in B^{2-1/p,p}(\partial\Omega)$ (6.39)
  $\implies \varphi \in B^{2-1/p,p}(\partial\Omega)$.

Proof. (i) First, we show that assertion (6.38) implies assertion (6.39). To do this, assume that

$$\varphi \in B^{-1/p,p}(\partial\Omega) \quad \text{and} \quad T\varphi \in B^{2-1/p,p}(\partial\Omega).$$

Then, by letting $u = \mathcal{P}\varphi$ we obtain that

$$u \in L^p(\Omega), \quad Au = 0 \quad \text{and} \quad B\gamma u = T\varphi \in B^{2-1/p,p}(\partial\Omega).$$

Hence it follows from condition (6.38) that

$$u \in H^{2,p}(\Omega),$$

so that, by Theorem 6.17,

$$\varphi = \gamma_0 u \in B^{2-1/p,p}(\partial\Omega).$$

(ii) Conversely, we show that assertion (6.17) implies estimate (6.16). To do this, assume that

$$u \in L^p(\Omega), \quad Au \in L^p(\Omega) \quad \text{and} \quad B\gamma u \in B^{2-1/p,p}(\partial\Omega).$$

Then the distribution $u(x)$ can be decomposed as follows:

$$u(x) = v(x) + w(x),$$

where

$$\begin{cases} v = \mathcal{G}_N(Au) \in H^{2,p}(\Omega), \\ w = u - v \in \mathcal{N}(A,0,p) = \{z \in L^p(\Omega) : Az = 0 \text{ in } \Omega\}. \end{cases}$$

Moreover, Theorem 6.14 asserts that the distribution $w(x)$ can be written in the form

$$w = \mathcal{P}\varphi, \quad \varphi = \gamma_0 w \in B^{-1/p,p}(\partial\Omega).$$

Hence we have, by Theorem 6.17,

$$T\varphi = B\gamma(\mathcal{P}\varphi) = B\gamma w = B\gamma u - B\gamma v = B\gamma u - b(x')\gamma_0 v \in B^{2-1/p,p}(\partial\Omega),$$

since $\gamma_1 v = 0$. Thus it follows from condition (6.39) that

$$\varphi \in B^{2-1/p,p}(\partial\Omega),$$

so that again, by Theorem 6.14,

$$w = \mathcal{P}\varphi \in H^{2,p}(\Omega).$$

This proves that

$$u = v + w \in H^{2,p}(\Omega).$$

The proof of Theorem 6.21 is complete. □

The next theorem states that *a priori* estimates for $\mathcal{A}$ are entirely equivalent to corresponding *a priori* estimates for $\mathcal{T}$.

Theorem 6.22 (Estimates). *Let* $1 < p < \infty$. *The following two estimates* (6.40) *and* (6.41) *are equivalent:*

$$\|u\|_{2,p} \leq C \left(\|Au\|_p + |B\gamma u|_{2-1/p,p} + \|u\|_p \right) \text{ for all } u \in \mathcal{D}(\mathcal{A}). \tag{6.40}$$

$$|\varphi|_{2-1/p,p} \leq C \left(|T\varphi|_{2-1/p,p} + |\varphi|_{-1/p,p} \right) \text{ for all } \varphi \in \mathcal{D}(\mathcal{T}). \tag{6.41}$$

Here and in the following the letter C *denotes a generic positive constant.*

Proof. (i) First, we show that estimate (6.40) implies estimate (6.41).

By taking $u = \mathcal{P}\varphi$ with $\varphi \in \mathcal{D}(\mathcal{T})$ in estimate (6.40), we obtain that

$$\|\mathcal{P}\varphi\|_{2,p} \leq C \left(|\mathcal{T}\varphi|_{2-1/p,p} + \|\mathcal{P}\varphi\|_p \right). \tag{6.42}$$

However, Theorem 6.14 asserts that the Poisson operator $\mathcal{P}$ maps the Besov space $B^{s-1/p,p}(\partial\Omega)$ isomorphically onto the null space $\mathcal{N}(A, s, p)$ for all $s \in \mathbf{R}$. Thus the desired estimate (6.41) follows from estimate (6.42).

(ii) Conversely, we show that estimate (6.41) implies estimate (6.40).

To do this, we express a function $u \in \mathcal{D}(\mathcal{A})$ in the form

$$u(x) = v(x) + w(x),$$

where

$$\begin{cases} v = \mathcal{G}_N(Au) \in H^{2,p}(\Omega), \\ w = u - v \in \mathcal{N}(A, 2, p) = \{z \in H^{2,p}(\Omega) : Az = 0 \text{ in } \Omega\}. \end{cases}$$

Then we have, by Theorem 6.16 with $s := 2$,

$$\|v\|_{2,p} = \|\mathcal{G}_N(Au)\|_{2,p} \leq C\|Au\|_p. \tag{6.43}$$

Furthermore, by applying estimate (6.41) to the distribution $\gamma_0 w$ we obtain that

$$\begin{aligned} |\gamma_0 w|_{2-1/p,p} &\leq C \left(|\mathcal{T}(\gamma_0 w)|_{2-1/p,p} + |\gamma_0 w|_{-1/p,p} \right) \\ &= C \left(|B\gamma w|_{2-1/p,p} + |\gamma_0 w|_{-1/p,p} \right) \\ &\leq C \left(|B\gamma u|_{2-1/p,p} + |B\gamma v|_{2-1/p,p} + |\gamma_0 w|_{-1/p,p} \right). \end{aligned}$$

In view of Theorem 6.14, this proves that

$$\begin{aligned} \|w\|_{2,p} &\leq C \left(|B\gamma u|_{2-1/p,p} + |B\gamma v|_{2-1/p,p} + \|w\|_p \right) \tag{6.44} \\ &\leq C \left(|B\gamma u|_{2-1/p,p} + |B\gamma v|_{2-1/p,p} + \|u\|_p + \|v\|_p \right) \\ &\leq C \left(|B\gamma u|_{2-1/p,p} + |B\gamma v|_{2-1/p,p} + \|u\|_p + \|v\|_{2,p} \right). \end{aligned}$$

However, since $\gamma_1 v = 0$, it follows from an application of Theorem 6.17 that

$$|B\gamma v|_{2-1/p,p} = |b(x')\gamma_0 v|_{2-1/p,p} \leq C\|v\|_{2,p}. \tag{6.45}$$

Thus, by carrying estimates (6.43) and (6.45) into estimate (6.44) we obtain that

$$\|w\|_{2,p} \leq C \left(\|Au\|_p + |B\gamma u|_{2-1/p,p} + \|u\|_p \right). \tag{6.46}$$

Therefore, the desired estimate (6.40) follows from estimates (6.43) and (6.46), since $u(x) = v(x) + w(x)$.

The proof of Theorem 6.22 is complete. $\square$

6.5 Boundary Value Problems via Boutet de Monvel Calculus

In this section we consider a second-order, *uniformly elliptic differential operator* A with real coefficients on the double $M = \widehat{\Omega}$ of Ω such that

$$Au(x) = \sum_{i,j=1}^{n} a^{ij}(x)\frac{\partial^2 u}{\partial x_i \partial x_j}(x) + \sum_{i=1}^{n} b^i(x)\frac{\partial u}{\partial x_i}(x) + c(x)u(x), \qquad (6.31)$$

and a first-order, boundary condition B such that

$$B\gamma u(x') = a(x')\frac{\partial u}{\partial \nu}(x') + b(x')u(x') \quad \text{on } \partial\Omega. \qquad (6.2)$$

Assume that the following three conditions (SC), (H.1) and (H.2) are satisfied:

(SC) $\mathcal{N}_0(A) = \mathcal{N}_0(A^*) = \{0\}$.
(H.1) $a(x') \geq 0$ and $b(x') \geq 0$ on $\partial\Omega$.
(H.2) $a(x') + b(x') > 0$ on $\partial\Omega$.

Now we can formulate the non-homogeneous Robin problem: Given functions f defined in Ω and φ defined on $\partial\Omega$, respectively, find a function u in Ω such that

$$\begin{cases} Au = f & \text{in } \Omega, \\ B\gamma u = \varphi & \text{on } \partial\Omega. \end{cases} \qquad (6.1)$$

In this section we study the general Robin problem (6.1) in the framework of L^p Sobolev spaces from the viewpoint of the Boutet de Monvel calculus, and prove index formulas of Agranovič–Dynin type that relate the indices of two elliptic boundary value problems using the index of a pseudo-differential operator on the boundary (see [5], [67]). In this section, we derive an index formula of Agranovič–Dynin type for the Neumann problem (6.4) and the general Robin problem (6.1) in the framework of L^p Sobolev spaces (Theorem 6.26).

6.5.1 Boundary Space $B_\star^{s-1-1/p,p}(\partial\Omega)$

First, we introduce a subspace of the Besov space $B^{s-1-1/p,p}(\partial\Omega)$ under the conditions (H.1) and (H.2), which is a suitable tool to investigate the general Robin boundary operator $B\gamma$ defined by formula (6.2). We let

$$B_\star^{s-1-1/p,p}(\partial\Omega) := a(x')\,B^{s-1-1/p,p}(\partial\Omega) + b(x')\,B^{s-1/p,p}(\partial\Omega) \qquad (6.47)$$

$$= \left\{ \varphi = a(x')\varphi_1 + b(x')\varphi_2 : \varphi_1 \in B^{s-1-1/p,p}(\partial\Omega),\ \varphi_2 \in B^{s-1/p,p}(\partial\Omega) \right\},$$

and define a norm

$$|\varphi|_{s-1-1/p,p}^* \qquad (6.48)$$

$$= \inf \left\{ |\varphi_1|_{s-1-1/p,p} + |\varphi_2|_{s-1/p,p} : \varphi = a(x')\varphi_1 + b(x')\varphi_2 \right\},$$

where $|\cdot|_{t,p}$ is the norm of the Besov space $B^{t,p}(\partial\Omega)$. Then we have the assertions

$$B_\star^{s-1-1/p,p}(\partial\Omega) = a(x')\, B^{s-1-1/p,p}(\partial\Omega) + b(x')\, B^{s-1/p,p}(\partial\Omega)$$

$$= \begin{cases} B^{s-1/p,p}(\partial\Omega) & \text{if } a(x') \equiv 0 \text{ and } b(x') > 0 \text{ on } \partial\Omega, \\ B^{s-1-1/p,p}(\partial\Omega) & \text{if } a(x') > 0 \text{ on } \partial\Omega. \end{cases}$$

Indeed, it suffices to note that if $a(x') > 0$ on $\partial\Omega$, then any function $\varphi \in B^{s-1-1/p,p}(\partial\Omega)$ can be written in the form

$$\varphi = a(x')\frac{\varphi}{a(x')} + b(x') \cdot 0, \quad \frac{\varphi}{a(x')} \in B^{s-1-1/p,p}(\partial\Omega).$$

Moreover, we have the following lemma (see [123, Lemma 6.8]):

Lemma 6.23. *Let $1 < p < \infty$ and $s > 1 + 1/p$. Assume that the conditions (H.1) and (H.2) are satisfied. Then we have the two assertions:*

(i) The space $B_\star^{s-1-1/p,p}(\partial\Omega)$ is a Banach space with respect to the norm $|\cdot|_{s-1-1/p,p}^*$.

(ii) For general $a(x')$, we have the continuous injections

$$B^{s-1/p,p}(\partial\Omega) \subset B_\star^{s-1-1/p,p}(\partial\Omega) \subset B^{s-1-1/p,p}(\partial\Omega).$$

If $u \in H^{s,p}(\Omega)$, we can define its traces $\gamma_0 u$ and $\gamma_1 u$ respectively by the formulas

$$\begin{cases} \gamma_0 u = u|_{\partial\Omega}, \\ \gamma_1 u = \dfrac{\partial u}{\partial\nu}\bigg|_{\partial\Omega}, \end{cases}$$

and let

$$\gamma u = \{\gamma_0 u, \gamma_1 u\}.$$

By applying Theorems 4.17 and 4.18 with $j := 0$ and $j := 1$, we obtain the following theorem:

Theorem 6.24 (the trace theorem). *Let $1 < p < \infty$. Then the trace map*

$$\gamma = (\gamma_0, \gamma_1) : H^{s,p}(\Omega) \longrightarrow B^{s-1/p,p}(\partial\Omega) \times B^{s-1-1/p,p}(\partial\Omega)$$

is continuous and surjective *for every $s > 1 + 1/p$.*

Furthermore, we can prove the following trace theorem for the general Robin boundary operator

$$B\gamma u = a(x')\,\gamma_1 u + b(x')\,\gamma_0 u. \tag{6.2}$$

Proposition 6.25. *Let* $1 < p < \infty$. *Then the general Robin boundary operator*

$$B\gamma \colon H^{s,p}(\Omega) \longrightarrow B^{s-1-1/p,p}_\star(\partial\Omega)$$

is continuous and surjective *for every* $s > 1 + 1/p$.

Proof. We have only to prove the surjectivity of the operator $B\gamma$.

By virtue of Theorem 6.24, we obtain that, for any given function

$$\varphi = a(x')\varphi_1 + b(x')\varphi_2 \in B^{s-1-1/p,p}_\star(\partial\Omega),$$

we can find a function $u \in H^{s,p}(\Omega)$ such that

$$\begin{cases} \gamma_0 u = \varphi_2 \in B^{s-1/p,p}_\star(\partial\Omega), \\ \gamma_1 u = \varphi_1 \in B^{s-1-1/p,p}_\star(\partial\Omega). \end{cases}$$

Then we have the formula

$$B\gamma u = a(x')\gamma_1 u + b(x')\gamma_0 u = a(x')\varphi_1 + b(x')\varphi_2$$
$$= \varphi \in B^{s-1-1/p,p}_\star(\partial\Omega).$$

This proves the surjectivity of $B\gamma \colon H^{s,p}(\Omega) \to B^{s-1-1/p,p}_\star(\partial\Omega)$.

The proof of Proposition 6.25 is complete. □

6.5.2 Index Formula of Agranovič–Dynin Type

Index formulas of Agranovič–Dynin type are rules that relate the indices of two elliptic boundary value problems using the index of a pseudo-differential operator on the boundary (see [5], [67]). In this subsection, we assume that condition (SC) is satisfied, and derive an index formula of Agranovič–Dynin type for the Neumann problem and the Robin problem in the framework of L^p Sobolev spaces (Theorem 6.26).

(I) First, we consider the non-homogeneous Robin problem

$$\begin{cases} Au = f & \text{in } \Omega, \\ B\gamma u = a(x')\,\gamma_1 u + b(x')\,\gamma_0 u = \varphi & \text{on } \partial\Omega. \end{cases} \tag{6.1}$$

We express the Robin problem (6.1), in terms of the Boutet de Monvel calculus, as a mapping of the matrix form

$$\mathcal{B} = \begin{pmatrix} A & 0 \\ B\gamma & 0 \end{pmatrix}.$$

We remark that the mapping

$$\mathcal{B} = \begin{pmatrix} A & 0 \\ B\gamma & 0 \end{pmatrix} : \begin{matrix} H^{s+2,p}(\Omega) \\ \oplus \\ B^{s+1-1/p,p}(\partial\Omega) \end{matrix} \longrightarrow \begin{matrix} H^{s,p}(\Omega) \\ \oplus \\ B^{s+1-1/p,p}_{\star}(\partial\Omega) \end{matrix} \tag{6.49}$$

is continuous for all $s > -1 + 1/p$ with $1 < p < \infty$, and its principal symbol $\sigma(\mathcal{B})$ is homotopic to the following:

$$\begin{pmatrix} -\left(\langle \xi' \rangle^2 + \nu^2\right) & 0 \\ -a(x')\, i\tau + b(x') & 0 \end{pmatrix}. \tag{6.50}$$

Here and in the following we use the notation

$$\xi = (\xi', \nu) = (\xi_1, \xi_2, \ldots, \xi_{n-1}, \nu) \in \mathbf{R}^n,$$

$$\langle \xi' \rangle = \sqrt{1 + |\xi'|^2},$$

$$\xi = (\xi', \nu) = (\xi_1, \xi_2, \ldots, \xi_{n-1}, \nu) \in \mathbf{R}^n \quad \text{for potential operators,}$$

$$\xi = (\xi', \tau) = (\xi_1, \xi_2, \ldots, \xi_{n-1}, \tau) \in \mathbf{R}^n \quad \text{for trace operators.}$$

(II) Secondly, we consider the non-homogeneous Neumann problem

$$\begin{cases} Av = g & \text{in } \Omega, \\ \gamma_1 v = \dfrac{\partial v}{\partial \nu}\bigg|_{\partial\Omega} = \psi & \text{on } \partial\Omega, \end{cases} \tag{6.51}$$

and express this Neumann problem, in terms of the Boutet de Monvel calculus, as a mapping of the matrix form

$$\mathcal{A} = \begin{pmatrix} A & 0 \\ \gamma_1 & 0 \end{pmatrix}.$$

We remark that the mapping

$$\mathcal{A} = \begin{pmatrix} A & 0 \\ \gamma_1 & 0 \end{pmatrix} : \begin{matrix} H^{s+2,p}(\Omega) \\ \oplus \\ B^{s+1-1/p,p}(\partial\Omega) \end{matrix} \longrightarrow \begin{matrix} H^{s,p}(\Omega) \\ \oplus \\ B^{s+1-1/p,p}(\partial\Omega) \end{matrix} \tag{6.52}$$

is continuous for all $s > -1 + 1/p$ with $1 < p < \infty$, and its principal symbol $\sigma(\mathcal{A})$ is homotopic to the following:

$$\begin{pmatrix} -\left(\langle \xi' \rangle^2 + \nu^2\right) & 0 \\ -i\tau & 0 \end{pmatrix}. \tag{6.53}$$

(III) Thirdly, if we introduce the *Fredholm boundary operator* T by the formula

$$T: C^\infty(\partial\Omega) \longrightarrow C^\infty(\partial\Omega)$$
$$\varphi \longmapsto B\gamma(\mathcal{P}\varphi),$$

then we have the formula

$$T = -a(x')\Pi + b(x'). \tag{6.54}$$

Here it should be noticed that the Fredholm boundary operator

$$\boxed{T = -a(x')\Pi + b(x'): B^{s+2-1/p,p}(\partial\Omega) \longrightarrow B^{s+1-1/p,p}_\star(\partial\Omega)}$$

is continuous for all $s \in \mathbf{R}$ and $1 < p < \infty$.

(IV) Now we can prove an index formula of Agranovič–Dynin type for the Neumann problem (6.4) and the degenerate Robin problem (6.1) in the framework of L^p Sobolev spaces:

Theorem 6.26. *Assume that the conditions (SC), (H.1) and (H.2) are satisfied. The next index formula relates the indices of the operators $\mathcal{B}$ and $\mathcal{A}$ using the index of the Fredholm boundary operator T:*

$$\boxed{\operatorname{ind}\mathcal{B} - \operatorname{ind}\mathcal{A} = \operatorname{ind}T.} \tag{6.55}$$

Proof. The proof is divided into four steps.

Step 1: By applying Theorem 6.2 with

$$a(x') \equiv 1, \quad b(x') \equiv 0,$$

we obtain that the homogeneous Neumann problem

$$\begin{cases} Av = f & \text{in } \Omega, \\ \gamma_1 v = \dfrac{\partial v}{\partial \boldsymbol{\nu}}\bigg|_{\partial\Omega} = 0 & \text{on } \partial\Omega \end{cases} \tag{6.34}$$

has a unique solution $v \in H^{s+2,p}(\Omega)$ for every function $f \in H^{s,p}(\Omega)$ (see [4], [50], [136]). If we let

$$v := \mathcal{G}_N f,$$

then it follows that the *Green operator*

$$\mathcal{G}_N: H^{s,p}(\Omega) \longrightarrow H^{s+2,p}(\Omega)$$

is continuous for all $s > -1 + 1/p$ with $1 < p < \infty$.

Moreover, by applying Proposition 5.2 and formula (5.13) we find that the *Dirichlet-to-Neumann operator* Π, defined by the formula

$$\Pi\varphi := \frac{\partial}{\partial \mathbf{n}} (\mathcal{P}\varphi)\Big|_{\partial\Omega} = -\frac{\partial}{\partial\nu} (\mathcal{P}\varphi)\Big|_{\partial\Omega}$$
$$= -\gamma_1 (\mathcal{P}\varphi) \quad \text{for all } \varphi \in C^\infty(\partial\Omega),$$

maps $B^{s+2-1/p,p}(\partial\Omega)$ *isomorphically* onto $B^{s+1-1/p,p}(\partial\Omega)$ for all $s \in \mathbf{R}$ and $1 < p < \infty$.

Now we consider an operator $\mathcal{D}$ of the matrix form

$$\mathcal{D} = \begin{pmatrix} \mathcal{G}_N & -\mathcal{P}\,\Pi^{-1} \\ \gamma_0\mathcal{G}_N & 0 \end{pmatrix} : \begin{array}{c} H^{s,p}(\Omega) \\ \oplus \\ B^{s+1-1/p,p}(\partial\Omega) \end{array} \to \begin{array}{c} H^{s+1,p}(\Omega) \\ \oplus \\ B^{s+1-1/p,p}(\partial\Omega) \end{array} \tag{6.56}$$

and its principal symbol $\sigma(\mathcal{D})$ is homotopic to the following (see Examples 5.2, 5.5 and 5.8):

$$\begin{pmatrix} \dfrac{1}{\langle\xi'\rangle^2 + \nu^2} + \dfrac{1}{2\langle\xi'\rangle} \dfrac{1}{\langle\xi'\rangle + i\nu} \dfrac{1}{\langle\xi'\rangle - i\tau} & \dfrac{1}{\langle\xi'\rangle + i\nu} \dfrac{1}{\langle\xi'\rangle} \\[2mm] \dfrac{1}{\langle\xi'\rangle} \dfrac{1}{\langle\xi'\rangle - i\tau} & 0 \end{pmatrix}. \tag{6.57}$$

Step 2: By combining two formulas (6.52) and (6.56), we find that

$$\mathcal{A}\mathcal{D} = \begin{pmatrix} A & 0 \\ \gamma_1 & 0 \end{pmatrix} \begin{pmatrix} \mathcal{G}_N & -\mathcal{P}\,\Pi^{-1} \\ \gamma_0\mathcal{G}_N & 0 \end{pmatrix} \tag{6.58}$$
$$= \begin{pmatrix} A\mathcal{G}_N & -(A\mathcal{P})\,\Pi^{-1} \\ \gamma_1\mathcal{G}_N & -(\gamma_1\mathcal{P})\,\Pi^{-1} \end{pmatrix} = \begin{pmatrix} I & 0 \\ 0 & \Pi\,\Pi^{-1} \end{pmatrix}$$
$$= \begin{pmatrix} I & 0 \\ 0 & I \end{pmatrix}.$$

Therefore, the matrix form operator $\mathcal{D}$ is the *right-inverse* of the matrix form operator $\mathcal{A}$, that is,

$$\mathcal{A}\mathcal{D} = \begin{pmatrix} I & 0 \\ 0 & I \end{pmatrix} : \begin{array}{c} H^{s,p}(\Omega) \\ \oplus \\ B^{s+1-1/p,p}(\partial\Omega) \end{array} \longrightarrow \begin{array}{c} H^{s,p}(\Omega) \\ \oplus \\ B^{s+1-1/p,p}(\partial\Omega) \end{array} \tag{6.59}$$

Therefore, we obtain from formulas (6.58) and (6.59) that

$$\text{ind}\,\mathcal{A} + \text{ind}\,\mathcal{D} = \text{ind} \begin{pmatrix} I & 0 \\ 0 & I \end{pmatrix} = 0. \tag{6.60}$$

Step 3: Furthermore, we obtain from formulas (6.49) and (6.56) that the composition of the matrix form operators $\mathcal{B}$ and $\mathcal{D}$ is equal to the following:

$$\mathcal{BD} = \begin{pmatrix} A & 0 \\ B\gamma & 0 \end{pmatrix} \begin{pmatrix} \mathcal{G}_N & -\mathcal{P}\,\Pi^{-1} \\ \gamma_0\mathcal{G}_N & 0 \end{pmatrix} = \begin{pmatrix} A\mathcal{G}_N & -(A\mathcal{P})\,\Pi^{-1} \\ B\gamma\mathcal{G}_N & -(B\gamma\mathcal{P})\,\Pi^{-1} \end{pmatrix} \tag{6.61}$$

$$= \begin{pmatrix} I & 0 \\ b(x')\,(\gamma_0\mathcal{G}_N) & (-a(x')\,\Pi + b(x'))\,\Pi^{-1} \end{pmatrix}$$

$$= \begin{pmatrix} I & 0 \\ b(x')\,(\gamma_0\mathcal{G}_N) & -a(x') + b(x')\,\Pi^{-1} \end{pmatrix},$$

since we have the formulas

$$B\gamma\mathcal{G}_N = a(x')\,(\gamma_1\mathcal{G}_N) + b(x')\,(\gamma_0\mathcal{G}_N) = b(x')\,(\gamma_0\mathcal{G}_N), \tag{6.62a}$$

$$B\gamma\mathcal{P} = a(x')\,(\gamma_1\mathcal{P}) + b(x')\,(\gamma_0\mathcal{P}) = -a(x')\,\Pi + b(x'). \tag{6.62b}$$

Moreover, we remark that the mapping

$$\mathcal{BD}: \begin{array}{c} H^{s,p}(\Omega) \\ \oplus \\ B^{s+1-1/p,p}(\partial\Omega) \end{array} \longrightarrow \begin{array}{c} H^{s,p}(\Omega) \\ \oplus \\ B_*^{s+1-1/p,p}(\partial\Omega). \end{array}$$

is continuous for all $s > -1 + 1/p$ with $1 < p < \infty$, Indeed, it suffices to note the following two assertions:

- $b(x')\gamma_0(\mathcal{G}_N f) \in B_*^{s+2-1/p,p}(\partial\Omega) \subset B_*^{s+1-1/p,p}(\partial\Omega)$ for all $f \in H^{s,p}(\Omega)$,
- $-a(x')\psi + b(x')\Pi^{-1}\psi \in B_*^{s+1-1/p,p}(\partial\Omega)$ for all $\psi \in B^{s+1-1/p,p}(\partial\Omega)$.

Step 4: Since the Dirichlet-to-Neumann operator

$$\Pi: B^{s+2-1/p,p}(\partial\Omega) \longrightarrow B^{s+1-1/p,p}(\partial\Omega)$$

is an isomorphism for all $s \in \mathbf{R}$ and $1 < p < \infty$, we find from formulas (6.61) and (6.62) that

$$\operatorname{ind}(\mathcal{B}\,\mathcal{D}) = \operatorname{ind} \begin{pmatrix} I & 0 \\ b(x')\gamma_0\mathcal{G}_N & -a(x') + b(x')\,\Pi^{-1} \end{pmatrix} \tag{6.63}$$

$$= \operatorname{ind}\left(-a(x') + b(x')\,\Pi^{-1}\right) = \operatorname{ind}\left(-a(x')\Pi + b(x')\right)$$

$$= \operatorname{ind} T.$$

Therefore, by combining formulas (6.60) and (6.63) we obtain the desired index formula

$$\operatorname{ind}\mathcal{B} - \operatorname{ind}\mathcal{A} = \operatorname{ind}\mathcal{B} + \operatorname{ind}\mathcal{D} = \operatorname{ind}(\mathcal{B}\,\mathcal{D})$$

$$= \operatorname{ind} T.$$

The proof of Theorem 6.26 is complete. □

6.6 Dirichlet-to-Neumann Operator and Reflecting Diffusion

In this section we study an intimate relationship between the Dirichlet-to-Neumann operator and the reflecting diffusion in a bounded, domain D of Euclidean space $\mathbf{R}^N$ with smooth boundary ∂D (Theorems 6.27 and 6.28 and Remark 6.29).

We consider the following Dirichlet problem for the Laplacian Δ: For given function φ defined on ∂D, find a function u in D such that

$$\begin{cases} \Delta u = 0 & \text{in } D, \\ \gamma_0 u = u|_{\partial D} = \varphi & \text{on } \partial D. \end{cases} \tag{6.64}$$

The existence and uniqueness theorem for the Dirichlet problem (6.42) is well established in the framework of Hölder spaces and Sobolev spaces (see [3], [4], [50], [77]). We denote its unique solution u as follows:

$$u = H_0 \varphi.$$

The operator H_0 is called the *harmonic operator* or *Poisson operator* for the Laplacian Δ.

In this section we state that the pseudo-differential operator Π_0, defined by the formula

$$\Pi_0 \varphi := -\gamma_1 (H_0 \varphi) = \frac{\partial}{\partial \mathbf{n}} (H_0 \varphi) \Big|_{\partial D} = -\sqrt{-\Lambda'}\, \varphi \quad \text{for all } \varphi \in C^\infty(\partial D),$$

is the generator of the Markov process on the boundary which is the *trace* of trajectories w of the *reflecting diffusion* (see Section 3.5)

$$\mathcal{X} = \left(x_t, \zeta, \mathcal{B}_t, \mathcal{B}, P_x, x \in \overline{D} \cup \{\partial\} \right).$$

Here Λ' is the Laplace–Beltrami operator on the boundary ∂D. The pseudo-differential operator Π_0 is called the *Dirichlet-to-Neumann operator* (cf. formula (6.3)).

More precisely, Sato and Ueno [98] constructed the following Markov process on the state space $\partial D \cup \{\partial\}$ (see [98, Section 9]):

Theorem 6.27 (Sato–Ueno). *(1) Let H^* be the* trace

$$w^* \colon [0, +\infty] \longrightarrow \partial D \cup \{\partial\}$$

on ∂D of trajectories w of the reflecting diffusion $\mathcal{X}$.
(2) A random variable

$$\zeta^* \colon H^* \longrightarrow [0, +\infty],$$

called the lifetime, *defined by the formula*

$$\zeta^*(w^*) = \tau^{-1}(\zeta(w), w) \quad \text{for } w^* = w|_{\partial D} \in H^* \text{ with } w \in W.$$

Here $\zeta(w)$ is the lifetime of the reflecting diffusion $\mathcal{X}$ and $\tau^{-1}(t, w)$ is the right-continuous inverse of the local time $\tau(t, w)$ (see Section 3.6).
   We recall that

$$P_x(\{w \in W : \lim_{t \to +\infty} \tau(t, w) = +\infty\}) = 1 \quad \text{if } x \in \overline{D}.$$

(3) Let

$$x_t^*(w^*) = x_{\tau^{-1}(t,w)}(w) = w\left(\tau^{-1}(t, w)\right)$$

for $w^* = w|_{\partial D} \in H^*$ with $w \in W$ (see Figure 6.8 below).
   We remark that

$$x_t^*(w^*) = w\left(\tau^{-1}(t, w)\right) \begin{cases} \in \partial D & \text{for } 0 \le t < \zeta^*(w^*), \\ = \partial & \text{for } \zeta^*(w^*) \le t \le +\infty, \end{cases}$$

and that $x_t^*(w^*)$ is right-continuous with left limits for $t \in [0, \tau(+\infty, w)]$.

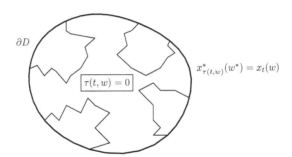

Fig. 6.8. The local time $\tau(t, w)$ and the trajectories $x_t^*(w^*)$

(4) Let $\mathcal{B}_t^* = \mathcal{B}_{\tau(t)}$ and $\mathcal{B}^* = \mathcal{B}_{\tau(\infty)}$, respectively.
(5) For each $t \in [0, +\infty]$, a pathwise shift mapping

$$\theta_t^* : H^* \longrightarrow H^*$$

defined by the formula

$$\theta_t^* w^*(s) = x_{\tau(t+s,w)}(w) = w\left(\tau(t + s, w)\right)$$

for all $w^*(t) = w|_{\partial D} \in H^*$ with coordinates $x_u(w) = w(u)$.
   (6) The system of measures $\{P_{x'}^*, x' \in \partial D \cup \{\partial\}\}$ on $(H^*, \mathcal{B}^*)$ such that

$$P_{x'}^*(B) = P_{x'}\left(\{w \in W : x_{\tau(t,w)}(w) \text{ belongs to } B \text{ as a function of } t\}\right)$$

for each $B \in \mathcal{B}^* = \mathcal{B}_{\tau(\infty)}$.

The Markov process

$$\mathcal{X}^* = (x_t^*, H^*, \mathcal{B}_t^*, \mathcal{B}^*, P_{x'}^*, x' \in \partial D \cup \{\partial\})$$

is called the *Markov process* on ∂D obtained from the reflecting diffusion

$$\mathcal{X} = (x_t, W, \mathcal{B}_t, \mathcal{B}, P_x, x \in \overline{D} \cup \{\partial\})$$

through *time change* by the local time $\tau(t)$ on the boundary (see Figure 6.9 below).

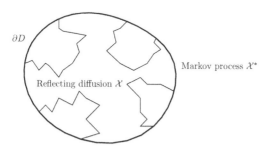

Fig. 6.9. The Markov processes $\mathcal{X}$ and $\mathcal{X}^*$

The next theorem asserts that the Dirichlet-to-Neumann operator

$$\Pi_0 = -\sqrt{-\Lambda'}$$

with *principal symbol*

$$-|\xi'|$$

is the generator of the Markov process $\mathcal{X}^*$ (see Volkonskiĭ [139], Sato–Ueno [98, Theorem 9.1], Ueno [137]):

Theorem 6.28 (Sato–Ueno). *(i) The Markov process $\mathcal{X}^*$ is conservative. Namely, we have the assertion*

$$P_{x'}^* \left(\{w^* \in H^* : \zeta^*(w^*) = +\infty\}\right) = 1 \quad \text{for each } x' \in \partial D.$$

(ii) A transition semigroup of linear operators

$$T_t^* \varphi(x') := E_{x'}^* \left(\varphi(x_t^*)\right) = E_{x'} \left(\varphi \left(x_{\tau^{-1}(t)}\right) \chi_{[0,\tau(\infty))}\right)$$

forms a Feller semigroup *on the state space ∂D, and its infinitesimal generator is equal to the* minimal closed extension $\overline{\Pi}_0$ *of the Dirichlet-to-Neumann operator $\Pi_0 = -\sqrt{-\Lambda'}$ with the domain $C^1(\partial D)$ in the Banach space $C(\partial D)$. More precisely, the infinitesimal generator $\overline{\Pi}_0$ is characterized as follows:*

(a) The domain of definition $\mathcal{D}\left(\overline{\Pi}_0\right)$ *is the space*

$$\mathcal{D}\left(\overline{\Pi}_0\right) = \{\varphi \in C(\partial D) : \Pi_0\varphi \in C(\partial D)\}.$$

(b) $\overline{\Pi}_0\varphi = \Pi_0\varphi$ *for every* $u \in \mathcal{D}\left(\overline{\Pi}_0\right)$.

We recall that $\Pi_0\varphi$ *is taken in the sense of* distributions.

Remark 6.29. By Example 4.8 with $n := N - 1$ and $\alpha := 1$, we find that the principal part of the *distribution kernel* of the Dirichlet-to-Neumann operator $\Pi_0 = -\sqrt{-\Lambda'}$ is equal to the following:

$$\frac{\Gamma(N/2)}{\pi^{N/2}} \frac{1}{|x' - y'|^N}$$

where $|x' - y'|$ is the geodesic distance between x' and y' with respect to the Riemannian metric of ∂D induced by the natural metric of $\mathbf{R}^N$ (see [46] and [122, Section 10.7]).

Theorems 6.27 and 6.28 and Remark 6.29 can be visualized as in Table 6.1 below.

Infinitesimal generator	$\Pi_0 = -\gamma_1 H_0$	Dirichlet-to-Neumann operator
Distribution kernel	$\frac{\Gamma(N/2)}{\pi^{N/2}}$ v.p. $\dfrac{1}{\|x' - y'\|^N}$	geodesic distance between x' and y'
Principal symbol	$-\|\xi'\|$	minus the length of ξ'

Table 6.1. An overview of the reflecting diffusion process

For a given parameter $\alpha > 0$, we study the following homogeneous Neumann boundary value problem:

$$\begin{cases} (\Delta - \alpha)\, u = f & \text{in } D, \\ \gamma_1 u = -\left.\dfrac{\partial u}{\partial \mathbf{n}}\right|_{\partial D} = 0 & \text{on } \partial D. \end{cases}$$

To do so, we take a probability measure $P(\cdot)$ on the interval $[0, +\infty]$ with density $\alpha\, e^{-\alpha t}$. Let $\overline{P}_x$ be the product measure of P_x and P on the product space $\Omega = W \times [0, +\infty]$. We define a sample path $\overline{x}_t(\omega)$ for $\omega = (w, T) \in \Omega$ and $t \in [0, +\infty]$ by the formula

$$\overline{x}_t(\omega) = \overline{x}_t(w, T) = \begin{cases} x_{\tau^{-1}(t,w)}(w) & \text{if } \tau^{-1}(t, w) < T, \\ \partial & \text{if } \tau^{-1}(t, w) \geq T, \end{cases}$$

and also by its life time $\overline{\zeta}(\omega)$ by the formula

$$\overline{\zeta}(\omega) = \inf \{t \geq 0 : \overline{x}_t(\omega) = \partial\}.$$

Then we have the following assertion:

$$\overline{P}_x\big(\{\overline{x}_t(\omega) \text{ is right-continuous with left limits as a function}$$
$$\text{of } t \in [0, \overline{\zeta}(\omega))\}\big) = 1 \text{ for each } x \in \overline{D}.$$

Moreover, we define a measure $P_{x'}^\sharp$ on the space $(H^*, \mathcal{B}^*)$ as follows:

$$P_{x'}^\sharp(B) = \overline{P}_{x'}\left(\{\overline{x}_t(\omega) \text{ belongs to } B \text{ as a function of } t\}\right)$$
$$\text{for each } B \in \mathcal{B}^* = \mathcal{B}_{\tau(\infty)}.$$

The Markov process

$$\mathcal{X}^\sharp = (x_t^*, H^*, \mathcal{B}_t^*, \mathcal{B}^*, P_{x'}^\sharp, x' \in \partial D \cup \{\partial\})$$

is called the Markov process on ∂D obtained from the reflecting diffusion

$$\mathcal{X} = (x_t, W, \mathcal{B}_t, \mathcal{B}, P_x, x \in \overline{D} \cup \{\partial\})$$

through *time change* by $\tau(t)$ and *killing rate* α.

A transition semigroup of linear operators

$$T_t^\sharp \varphi(x') := E_{x'}^\sharp \left(\varphi(x_t^*)\right) = E_{x'}\left(\varphi\left(x_{\tau^{-1}(t)}\right) e^{-\alpha \tau^{-1}(t)} \chi_{[0, \tau(\infty))}\right)$$

forms a *Feller semigroup* on the state space ∂D, and its infinitesimal generator is equal to the *minimal closed extension* $\overline{\Pi}_\alpha$ of the Dirichlet-to-Neumann operator, defined by the formula

$$\Pi_\alpha \varphi = -\gamma_1 (H_\alpha \varphi) = \frac{\partial}{\partial \mathbf{n}} (H_\alpha \varphi)\bigg|_{\partial D} \quad \text{for all } \varphi \in C^\infty(\partial D),$$

in the Banach space $C(\partial D)$, where H_α is the harmonic operator (or Poisson operator) for the operator $\Delta - \alpha$, that is, the function $u = H_\alpha \varphi$ is the unique solution of the Dirichlet problem

$$\begin{cases} (\Delta - \alpha) u = 0 & \text{in } D, \\ \gamma_0 u = u|_{\partial D} = \varphi & \text{on } \partial D. \end{cases}$$

More precisely, the infinitesimal generator $\overline{\Pi}_\alpha$ is characterized as follows:

(a) The domain of definition $\mathcal{D}\left(\overline{\Pi}_\alpha\right)$ is the space

$$\mathcal{D}\left(\overline{\Pi}_\alpha\right) = \{\varphi \in C(\partial D) : \Pi_\alpha \varphi \in C(\partial D)\}.$$

(b) $\overline{\Pi}_\alpha \varphi = \Pi_\alpha \varphi$ for every $u \in \mathcal{D}\left(\overline{\Pi}_\alpha\right)$.

Here $\Pi_\alpha \varphi$ is taken in the sense of *distributions*.

6.7 Spectral Analysis of the Dirichlet Eigenvalue Problem

Let Ω be a bounded domain in Euclidean space $\mathbf{R}^n$ with smooth boundary $\partial\Omega$. Its closure $\overline{\Omega} = \Omega \cup \partial\Omega$ is an n-dimensional, compact smooth manifold with boundary. We let

$$Au = \sum_{i,j=1}^{n} a^{ij}(x)\frac{\partial^2 u}{\partial x_i \partial x_j} + \sum_{i=1}^{n} b^i(x)\frac{\partial u}{\partial x_i} + c(x)u$$

be a second-order, elliptic differential operator with real coefficients such that:

(1) $a^{ij} \in C^\theta(\overline{\Omega})$ with $0 < \theta < 1$, $a^{ij}(x) = a^{ji}(x)$ for all $x \in \overline{\Omega}$ and $1 \le i,j \le n$, and there exists a constant $a_0 > 0$ such that

$$\sum_{i,j=1}^{n} a^{ij}(x)\xi_i\xi_j \ge a_0|\xi|^2 \quad \text{for all } x \in \overline{\Omega} \text{ and } \xi \in \mathbf{R}^n. \tag{6.65}$$

(2) $b^i \in C(\overline{\Omega})$ for $1 \le i \le n$.
(3) $c \in C(\overline{\Omega})$ and $c(x) \le 0$ in Ω.

In this section we study the following Dirichlet eigenvalue problem: Given a complex parameter λ, find a function $u(x)$ in Ω such that

$$\begin{cases} -Au = \lambda u & \text{in } \Omega, \\ u = 0 & \text{on } \partial\Omega, \end{cases} \tag{$*$}$$

where λ is a complex *spectral parameter*.

Following Mizohate [82, Chapter 3] and Wloka [146, Section 13], we introduce a closed linear operator

$$\boxed{\mathfrak{A}: L^2(\Omega) \longrightarrow L^2(\Omega)}$$

associated with the Dirichlet problem $(*)$ as follows:

(1) The domain $D(\mathfrak{A})$ of definition is the space (see [80, Theorem 3.40])

$$\mathcal{D}(\mathfrak{A}) = H_0^1(\Omega) \cap H^2(\Omega) = \left\{u \in H^2(\Omega) : \gamma_0 u = 0 \text{ on } \partial\Omega\right\}.$$

(2) $Au = \mathcal{A}u$ for every $u \in \mathcal{D}(\mathfrak{A})$.

We remark that the Dirichlet eigenvalue problem $(*)$ is equivalent to the eigenvalue problem of the closed operator $-\mathfrak{A}$:

$$- \mathfrak{A}u = \lambda u. \tag{$**$}$$

The purpose of this section is to prove that all the sufficiently large eigenvalues of the closed operator $-\mathfrak{A}$ lie in the *parabolic* type region (see assertion (6.52) and Figure 6.10 below).

6.7.1 Unique Solvability of the Dirichlet Problem

For all u, $v \in H_0^1(\Omega)$, we introduce a *Dirichlet form* $\mathcal{A}_0[u, v]$ by the formula

$$\mathcal{A}_0[u, v] = \sum_{i,j=1}^{n} \int_{\Omega} a_{ij}(x) \frac{\partial u}{\partial x_i} \cdot \overline{\frac{\partial v}{\partial x_j}} \, dx. \tag{6.66}$$

We remark that if u, $v \in C_0^\infty(\Omega)$, then we have, by integration by parts,

$$(Au + \lambda u, v)_{L^2(\Omega)} \tag{6.67}$$

$$= -\mathcal{A}_0[u, v] + \left(\sum_{i=1}^{n} \left(b^i(x) - \sum_{j=1}^{n} \frac{\partial a^{ij}}{\partial x_j} \right) \frac{\partial u}{\partial x_i} + c(x)u + \lambda u, v \right)_{L^2(\Omega)}$$

$$= -\mathcal{A}_0[u, v] + \left(\sum_{i=1}^{n} \widetilde{b}^i(x) \frac{\partial u}{\partial x_i} + c(x)u + \lambda u, v \right)_{L^2(\Omega)},$$

where

$$\widetilde{b}^i(x) := b^i(x) - \sum_{j=1}^{n} \frac{\partial a_{ij}}{\partial x_j} \quad \text{for } 1 \leq i \leq n.$$

Then we can prove the following existence and uniqueness theorem for the Dirichlet problem $(*)$:

Theorem 6.30. *Let $\lambda \in \mathbf{C}$. Assume that there exists a constant $C(\lambda) > 0$ such that the inequality*

$$\left| (Au + \lambda u, u)_{L^2(\Omega)} \right| \geq C(\lambda) \|u\|_{L^2(\Omega)}^2 \tag{6.68}$$

holds true for all $u \in H_0^1(\Omega) \cap H^2(\Omega)$. Then the Dirichlet problem $(*)$ has a unique solution

$$u \in H_0^1(\Omega) \cap H^2(\Omega)$$

for any function $f \in L^2(\Omega)$. This implies that λ belongs to the resolvent set $\rho(-\mathfrak{A})$ of $-\mathfrak{A}$ (see equation $(**)$)

$$\lambda \in \rho(-\mathfrak{A}) = -\rho(\mathfrak{A}).$$

Proof. The proof is divided into three steps.

Step (1): First, we decompose the differential operator A as follows:

$$Au = \sum_{i,j=1}^{n} a_{ij}(x)\frac{\partial^2 u}{\partial x_i \partial x_j} + \sum_{i=1}^{n} b^i(x)\frac{\partial u}{\partial x_i} + c(x)u$$

$$:= A_0 u + B_0 u + c(x)u$$

$$= \sum_{i,j=1}^{n} \frac{\partial}{\partial x_j}\left(a_{ij}(x)\frac{\partial u}{\partial x_i}\right) + \sum_{i=1}^{n}\left(b^i(x) - \sum_{j=1}^{n}\frac{\partial a_{ij}}{\partial x_j}\right)\frac{\partial u}{\partial x_i} + c(x)u,$$

where

$$A_0 u = \sum_{i,j=1}^{n} \frac{\partial}{\partial x_j}\left(a_{ij}(x)\frac{\partial u}{\partial x_i}\right),$$

$$B_0 u = \sum_{i=1}^{n}\left(b^i(x) - \sum_{j=1}^{n}\frac{\partial a_{ij}}{\partial x_j}\right)\frac{\partial u}{\partial x_i} = \sum_{i=1}^{n}\widetilde{b}^i(x)\frac{\partial u}{\partial x_i}.$$

We remark that A_0 is a *formally self-adjoint*, differential operator and further that B_0 is a *first-order* differential operator. Hence, it follows from an application of the Rellich–Kondrachov theorem (Theorem 4.10) that the operator

$$B_0 + c(x)\colon H^2(\Omega) \longrightarrow H^1(\Omega) \longrightarrow L^2(\Omega)$$

is *compact*.

However, we know from Gohberg–Kreĭn [51, Theorem 2.6] and Schechter [100, Theorem 5.10] (see also Theorem 4.59) that the index is *stable* under compact perturbations.

Therefore, we have the index formula

$$\operatorname{ind}\mathfrak{A} = \operatorname{ind}\mathfrak{A}_0. \tag{6.69}$$

Step (2): On the other hand, there is a *homotopy* between the formally self-adjoint operator A_0 and Laplacian Δ in the second-order differential operators. For example, we can take

$$A_t = (1-t)A_0 + t\Delta \quad \text{for } 0 \le t \le 1.$$

It is known (see [86, p. 187, Theorem 3], [146, Section 13, Theorem 13.2]) that the index of the Dirichlet problem is *homotopy invariant* in the class of second-order, formally self-adjoint differential operators:

$$\operatorname{ind}\mathfrak{A}_0 = \operatorname{ind}\mathfrak{A}_1. \tag{6.70}$$

Therefore, we are reduced to the study of the Dirichlet problem for the Laplacian:

$$\begin{cases} -\Delta u = \lambda u & \text{in } \Omega, \\ u = 0 & \text{on } \partial\Omega. \end{cases}$$

Step (3) However, we know from [146, Section 13, Theorem 13.5] (see also Corollary 4.63) that the index of the Dirichlet problem for the Laplacian is equal to *zero* in the space $H_0^1(\Omega) \cap H^2(\Omega)$:

$$\text{ind } \mathfrak{A}_1 = 0. \tag{6.71}$$

Hence we have, by formulas (6.47), (6.48) and (6.49),

$$\text{ind } \mathfrak{A} = \text{ind } \mathfrak{A}_0 = \text{ind } \mathfrak{A}_1 = 0. \tag{6.72}$$

In this way, we find that the index of the original Dirichlet eigenvalue problem (∗) is also equal to *zero* in the space $H_0^1(\Omega) \cap H^2(\Omega)$.

Therefore, if the fundamental inequality (6.68) holds true for some complex number λ, then we obtain from formula (6.72) that the Dirichlet eigenvalue problem (∗) has a unique solution u in the space $H_0^1(\Omega) \cap H^2(\Omega)$ for any function $f \in L^2(\Omega)$.

The proof of Theorem 6.30 is complete. $\square$

6.7.2 A Characterization of the Resolvent set of the Dirichlet Problem

In this subsection we characterize the complex parameters λ for which the fundamental inequality (6.68) holds true.

The next theorem asserts that the resolvent set $\rho(-\mathfrak{A})$ of $-\mathcal{A}$ contains the outside of a *parabola* (see Figure 6.10 below):

Theorem 6.31. *Let*

$$\lambda = \mu + i\nu, \quad i = \sqrt{-1}.$$

We can find two constants $p > 0$ and $q > 0$ such that if the condition

$$(|\nu| - p)^2 \geq 4p(\mu + q) \tag{6.73}$$

is satisfied, then the desired inequality (6.46) holds true.

Therefore, the Dirichlet eigenvalue problem (∗) has a unique solution $u \in H_0^1(\Omega) \cap H^2(\Omega)$ for any function $f \in L^2(\Omega)$. In particular, we have the assertion

$$\rho(-\mathfrak{A}) = -\rho(\mathfrak{A}) \supset \left\{ \lambda = \mu + i\nu : (|\nu| - p)^2 \geq 4p(\mu + q) \right\}. \tag{6.74}$$

Proof. We consider the case where

$$\mu > 0,$$

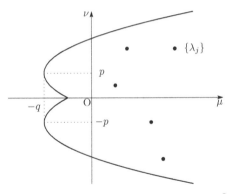

Fig. 6.10. The set $\rho\left(-\mathfrak{A}\right) = -\rho\left(\mathfrak{A}\right)$ contains $\left\{\left(\left|\nu\right| - p\right)^2 \geq 4p\left(\mu + q\right)\right\}$

and prove that if the condition (6.73) holds true, then there exists a constant $C(\lambda) > 0$ such that the fundamental inequality (6.68) holds true.

The proof is divided into three steps.

Step (1): First, we have, by inequality (6.65),

$$-\left(A_0 u, u\right)_{L^2(\Omega)} = \mathcal{A}_0[u, u] = \sum_{i,j=1}^{n} \int_{\Omega} a_{ij}(x) \frac{\partial u}{\partial x_i} \overline{\frac{\partial u}{\partial x_j}}\, dx \qquad (6.75)$$

$$\geq a_0 \sum_{i=1}^{n} \int_{\Omega} \left|\frac{\partial u}{\partial x_i}(x)\right|^2\, dx$$

$$= a_0 \sum_{i=1}^{n} \left\|\frac{\partial u}{\partial x_i}\right\|_{L^2(\Omega)}^2.$$

For any $\varepsilon > 0$, we have, by Schwarz's inequality and inequality (6.75),

$$\left|\sum_{i=1}^{n} \left(\tilde{b}^i(x)\frac{\partial u}{\partial x_i}, u\right)_{L^2(\Omega)}\right| \leq \sum_{i=1}^{n} \left|\left(\tilde{b}^i(x)\frac{\partial u}{\partial x_i}, u\right)_{L^2(\Omega)}\right| \qquad (6.76)$$

$$\leq \sum_{i=1}^{n} \max_{x \in \overline{\Omega}} \left|\tilde{b}^i(x)\right| \left\|\frac{\partial u}{\partial x_i}\right\|_{L^2(\Omega)} \|u\|_{L^2(\Omega)}$$

$$\leq C_0 \sum_{i=1}^{n} \left\|\frac{\partial u}{\partial x_i}\right\|_{L^2(\Omega)} \|u\|_{L^2(\Omega)}$$

$$\leq C_0 \sqrt{n} \sqrt{\sum_{i=1}^{n} \left\|\frac{\partial u}{\partial x_i}\right\|_{L^2(\Omega)}^2} \|u\|_{L^2(\Omega)}$$

$$\leq \frac{C_1}{\sqrt{a_0}} \left(\mathcal{A}_0[u, u]\right)^{1/2} \|u\|_{L^2(\Omega)}$$

$$\leq \varepsilon\, \mathcal{A}_0[u, u] + \frac{C_2}{\varepsilon} \|u\|_{L^2(\Omega)}^2 \qquad \text{for all } \varepsilon > 0,$$

with

$$C_0 = \max_{1 \le i \le n} \left\{ \max_{x \in \overline{\Omega}} \left| \widetilde{b}^i(x) \right| \right\}, \tag{6.77a}$$

$$C_1 = \sqrt{n}\, C_0, \tag{6.77b}$$

$$C_2 = \frac{C_1^{\,2}}{4a_0} = \frac{n}{4a_0}\, C_0^2. \tag{6.77c}$$

Here we have used the elementary inequality

$$2ab \le \varepsilon\, a^2 + \frac{1}{\varepsilon}\, b^2 \quad \text{for all } \varepsilon > 0.$$

On the other hand, we have, by formula (6.67),

$$\mathrm{Re}\,(Au + \lambda u, u)_{L^2(\Omega)} = -\mathcal{A}_0[u, u] + \mu\, \|u\|_{L^2(\Omega)}^2 + K_1, \tag{6.78a}$$

$$\mathrm{Im}\,(Au + \lambda u, u)_{L^2(\Omega)} = \nu\, \|u\|_{L^2(\Omega)}^2 + K_2. \tag{6.78b}$$

Here it is easy to see the following two formulas:

- $K_1 := \mathrm{Re}\left(\sum_{i=1}^n \widetilde{b}^i(x) \dfrac{\partial u}{\partial x_i} + c(x)u, u \right)_{L^2(\Omega)}$

  $= \left(\left(-\dfrac{1}{2}\left(\sum_{i=1}^n \dfrac{\partial \widetilde{b}^i}{\partial x_i} \right) + c(x) \right) u, u \right)_{L^2(\Omega)},$

- $K_2 := \mathrm{Im}\left(\sum_{i=1}^n \widetilde{b}^i(x) \dfrac{\partial u}{\partial x_i}, u \right)_{L^2(\Omega)}.$

Indeed, it suffices to note that we have, by integration by parts,

$$\left(\widetilde{b}^i(x)\frac{\partial u}{\partial x_i}, u \right)_{L^2(\Omega)} = \left(\frac{\partial u}{\partial x_i}, \widetilde{b}^i(x)\, u \right)_{L^2(\Omega)}$$

$$= -\left(u, \frac{\partial \widetilde{b}^i}{\partial x_i}\, u \right)_{L^2(\Omega)} - \left(u, \widetilde{b}^i(x)\frac{\partial u}{\partial x_i} \right)_{L^2(\Omega)}$$

$$= -\left(\frac{\partial \widetilde{b}^i}{\partial x_i}\, u, u \right)_{L^2(\Omega)} - \overline{\left(\widetilde{b}^i(x)\frac{\partial u}{\partial x_i}, u \right)_{L^2(\Omega)}},$$

since the $\widetilde{b}^i(x)$ are real-valued. Hence we have the formula

$$\mathrm{Re}\left(\widetilde{b}^i(x)\frac{\partial u}{\partial x_i}, u \right)_{L^2(\Omega)}$$

$$= \frac{1}{2}\left\{ \left(\widetilde{b}^i(x)\frac{\partial u}{\partial x_i}\, u, u \right)_{L^2(\Omega)} + \overline{\left(\widetilde{b}^i(x)\frac{\partial u}{\partial x_i}, u \right)_{L^2(\Omega)}} \right\}$$

$$= -\frac{1}{2}\left(\frac{\partial \widetilde{b}^i}{\partial x_i}u, u\right)_{L^2(\Omega)}.$$

Moreover, by using inequality (6.76) and Schwarz's inequality, we can estimate the terms K_2 and K_1 as follows:

$$\bullet \ |K_2| = \left|\mathrm{Im}\left(\sum_{i=1}^n \widetilde{b}^i(x)\frac{\partial u}{\partial x_i}, u\right)_{L^2(\Omega)}\right| \tag{6.79a}$$

$$\leq \left|\sum_{i=1}^n \left(\widetilde{b}^i(x)\frac{\partial u}{\partial x_i}, u\right)_{L^2(\Omega)}\right|$$

$$\leq \varepsilon\, \mathcal{A}_0[u, u] + \frac{C_2}{\varepsilon}\|u\|^2_{L^2(\Omega)} \quad \text{for all } \varepsilon > 0,$$

and

$$\bullet \ |K_1| \leq C_3\|u\|^2_{L^2(\Omega)}, \tag{6.79b}$$

where

$$C_3 = \max_{x \in \overline{\Omega}}\left|-\frac{1}{2}\sum_{i=1}^n\left(\frac{\partial \widetilde{b}^i}{\partial x_i}(x)\right) + c(x)\right|. \tag{6.80}$$

Step (2): Now, by using formulas (6.78) we can rewrite the term

$$(Au + \lambda u, u)_{L^2(\Omega)}$$

in the from

$$(Au + \lambda u, u)_{L^2(\Omega)}$$
$$= \mathrm{Re}\,(Au + \lambda u, u)_{L^2(\Omega)} + i\,\mathrm{Im}\,(Au + \lambda u, u)_{L^2(\Omega)}$$
$$= \left(-\mathcal{A}_0[u, u] + \mu\|u\|^2_{L^2(\Omega)} + K_1\right) + i\left(\nu\|u\|^2_{L^2(\Omega)} + K_2\right).$$

However, we remark that if $z = \alpha + i\beta$, then we have the inequality

$$|z| = \sqrt{\alpha^2 + \beta^2} \geq (1 - t)\,|\alpha| + t\,|\beta| \quad \text{for all } 0 \leq t \leq 1. \tag{6.81}$$

Therefore, we obtain from inequalities (6.79) and (6.81) that

$$\left|(Au + \lambda u, u)_{L^2(\Omega)}\right| \tag{6.82}$$

$$\geq (1 - t)\left|-\mathcal{A}_0[u, u] + \mu\|u\|^2_{L^2(\Omega)} + K_1\right| + t\left|\nu\|u\|^2_{L^2(\Omega)} + K_2\right|$$

$$\geq (1 - t)\left(\mathcal{A}_0[u, u] - \mu\|u\|^2_{L^2(\Omega)} - |K_1|\right) + t\left(|\nu|\|u\|^2_{L^2(\Omega)} - |K_2|\right)$$

$$\geq (1 - t)\left(\mathcal{A}_0[u, u] - \mu\|u\|^2_{L^2(\Omega)} - C_3\|u\|^2_{L^2(\Omega)}\right)$$

$$+ t \left(|\nu| \, \|u\|_{L^2(\Omega)}^2 - \varepsilon \, \mathcal{A}_0[u, u] - \frac{C_2}{\varepsilon} \|u\|_{L^2(\Omega)}^2 \right)$$

$$\geq (1 - t - \varepsilon) \, \mathcal{A}_0[u, u]$$

$$+ \left\{ t \left(|\nu| - \frac{C_2}{\varepsilon} \right) - (1 - t)(\mu + C_3) \right\} \|u\|_{L^2(\Omega)}^2.$$

Step (3): In order to obtain the desired inequality (6.68), it suffices to prove the following two inequalities (6.83) and (6.84):

$$1 - t - \varepsilon > 0 \tag{6.83}$$

$$t \left(|\nu| - \frac{C_2}{\varepsilon} \right) - (1 - t)(\mu + C_3) > 0 \tag{6.84}$$

If we take

$$\varepsilon = \frac{1 - t}{2},$$

then it follows that

$$g(t) := t \left(|\nu| - \frac{2C_2}{1 - t} \right) - (1 - t)(\mu + C_3)$$

$$= t(\mu + |\nu| + C_3) - \frac{2C_2 \, t}{1 - t} - \mu - C_3.$$

Since we have the formula

$$g'(t) = \mu + |\nu| + C_3 - \frac{2C_2}{(1 - t)^2},$$

the function $g(t)$ takes its *positive maximum* at the point

$$t_0 = 1 - \sqrt{\frac{2C_2}{\mu + |\nu| + C_3}},$$

under the condition

$$\mu + |\nu| + C_3 > 0. \tag{6.85}$$

Then we have the formula

$$g(t_0) = |\nu| - 2\sqrt{2C_2 (\mu + |\nu| + C_3)} + 2C_2.$$

Therefore, there exists a constant $C(\lambda) > 0$ in inequality (6.82) if we choose the complex number $\lambda = \mu + i\nu$ so large that

$$g(t_0) = |\nu| - 2\sqrt{2C_2 (\mu + |\nu| + C_3)} + 2C_2 > 0.$$

Summing up, we have proved that if the inequality

$$(|\nu| - p)^2 \geq 4p(\mu + q) \tag{6.73}$$

holds true for the constants $p = 2C_2$ and $q = C_3$, then the two inequalities (6.83) and (6.84) hold true.

The proof of Theorem 6.31 is complete. $\square$

Remark 6.32. If the operator A is of the *divergence form*, that is, if the conditions

$$\widetilde{b}^i(x) = b^i(x) - \sum_{j=1}^{n} \frac{\partial a_{ij}}{\partial x_j} \equiv 0 \quad \text{in } \Omega \text{ for } 1 \leq i \leq n$$

hold true, then we find from formulas (6.77), (6.80) and condition (6.85) that assertion (6.74) is reduced to the following:

$$\rho(-\mathfrak{A}) = -\rho(\mathfrak{A}) \supset \{\lambda = \mu + i\nu : \mu > -C_3, \ \nu = 0\},$$

where

$$C_3 = \max_{x \in \overline{\Omega}} |c(x)|.$$

6.8 Notes and Comments

Agmon–Douglis–Nirenberg [4] and Lions–Magenes [77] are the classics for the general theory for higher order elliptic boundary value problems. See also Agmon [3], Agranovich–Vishik [6] and Peetre [89].

For more thorough treatments of this subject, the reader might refer to Chazarain–Piriou [26], Hörmander [57], [61], Kumano-go [73], Seeley [105], [106], Taylor [133] and also Taira [122, Chapter 7].

Section 6.1: This section is an accessible and careful introduction to a modern version of the classical potential theory in terms of pseudo-differential operators. The material is taken from Boutet de Monvel [17], [18] and Chazarain–Piriou [26, Chapitre V, Section 2]. Theorems 6.2 and 6.10 for $p = 2$ correspond to Boutet de Monvel [17, Théorèmes (1.3.5) and (2.2.2)], respectively. See also Boutet de Monvel [18, Théorèmes 1.16 and 2.9].

Section 6.5: This section is devoted to the study of elliptic boundary value problems via the Boutet de Monvel calculus. The material is taken from Taira [126]. Moreover, this section is based on the talk entitled *Probabilistic approach to pseudo-differential operators* delivered at Special Seminar, Leibniz Universität Hannover, Germany, on October 16, 2019.

Section 6.6: This section is inspired by a probabilistic approach due to Sato–Ueno [98], Ueno [137] and Watanabe [141]. This section is also based on the above talk at Leibniz Universität Hannover.

Section 6.7: The material is taken from Agmon [3, Sections 14 and 15], Mizohate [82, Chapter 3] and Wloka [146, Section 13]. The proof of Theorem 6.31 may be new.

This chapter is an expanded and revised version of Chapter 4 of the second edition [121].

Analytic Semigroups in L^p Sobolev Spaces

7

Proof of Theorem 1.2

This chapter is devoted to the proof of Theorem 1.2. The idea of our proof is stated as follows. First, we reduce the study of the boundary value problem $(*)_\lambda$ to that of a first-order pseudo-differential operator $T(\lambda) = L\mathcal{P}(\lambda)$ on the boundary ∂D, just as in Section 6.4. Then we prove that conditions (A) and (B) are sufficient for the validity of the *a priori* estimate

$$\|u\|_{2,p} \leq C(\lambda) \left(\|f\|_p + |\varphi|_{2-1/p,p} + \|u\|_p \right). \tag{1.7}$$

More precisely, we construct a *parametrix* $S(\lambda)$ for $T(\lambda)$ in the Hörmander class $L^0_{1,1/2}(\partial D)$ (Lemma 7.2), and apply the Besov-space boundedness theorem (Theorem 4.47) to $S(\lambda)$ to obtain the desired estimate (1.7) (Lemma 7.1).

The proof of Theorem 1.2 can be flowcharted as in Table 7.1 below.

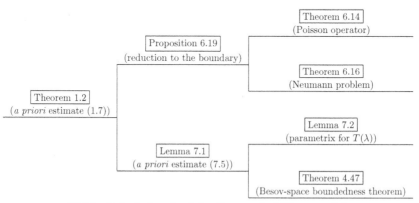

Table 7.1. A flowchart for the proof of Theorem 1.2

© Springer Nature Switzerland AG 2020
K. Taira, *Boundary Value Problems and Markov Processes*, Lecture Notes in Mathematics 1499,
https://doi.org/10.1007/978-3-030-48788-1_7

7.1 Boundary Value Problem with Spectral Parameter

Let D be a bounded domain of Euclidean space $\mathbf{R}^N$ with smooth boundary ∂D. Its closure $\overline{D} = D \cup \partial D$ is an N-dimensional, compact smooth manifold with boundary. We may assume that $\overline{D}$ is the closure of a relatively compact open subset D of an N-dimensional, compact smooth manifold $\widehat{D}$ without boundary in which D has a smooth boundary ∂D. This manifold $\widehat{D}$ is the *double* of D (see Figure 7.1 below).

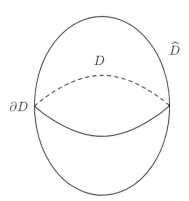

Fig. 7.1. The double $\widehat{D}$ of D

We let

$$A = \sum_{i,j=1}^{N} a^{ij}(x)\frac{\partial^2}{\partial x_i \partial x_j} + \sum_{i=1}^{N} b^i(x)\frac{\partial}{\partial x_i} + c(x)$$

be a second-order, *elliptic* differential operator with real coefficients such that:

(1) $a^{ij} \in C^\infty(\widehat{D})$ and $a^{ij}(x) = a^{ji}(x)$ for all $x \in \widehat{D}$, $1 \le i,j \le N$, and there exists a positive constant a_0 such that

$$\sum_{i,j=1}^{N} a^{ij}(x)\xi_i\xi_j \ge a_0|\xi|^2 \quad \text{on } T^*(\widehat{D}),$$

where $T^*(\widehat{D})$ is the cotangent bundle of the double $\widehat{D}$.

(2) $b^i \in C^\infty(\widehat{D})$ for $1 \le i \le N$.

(3) $c \in C^\infty(\widehat{D})$ and $c(x) \le 0$ in D.

In this chapter we consider the elliptic boundary value problem $(*)_\lambda$ with a spectral parameter

$$\begin{cases} (A - \lambda)u = f & \text{in } D, \\ Lu = \mu(x')\frac{\partial u}{\partial \mathbf{n}} + \gamma(x')u = \varphi & \text{on } \partial D. \end{cases} \tag{$*)_\lambda$}$$

Here we recall that:

(4) λ is a *complex* parameter.

(5) $\mu(x')$ and $\gamma(x')$ are real-valued, smooth functions on the boundary ∂D.

(6) $\mathbf{n} = (n_1, n_2, \ldots, n_N)$ is the unit interior normal to the boundary ∂D (see Figure 1.1).

The purpose of this chapter is to prove Theorem 1.2. More precisely, we prove the *a priori* estimate (1.7) provided that the following two conditions (A) and (B) are satisfied:

(A) $\mu(x') \geq 0$ and $\gamma(x') \leq 0$ on ∂D.

(B) $\mu(x') - \gamma(x') = \mu(x') + |\gamma(x')| > 0$ on ∂D.

7.2 Proof of the *A Priori* Estimate (1.7)

The proof of the *a priori* estimate (1.7) is divided into three steps.

Step I: It suffices to show that estimate (1.7) holds true for some $\lambda_0 > 0$, since we have, for all $\lambda \in \mathbf{C}$,

$$(A - \lambda_0)\, u = (A - \lambda)\, u + (\lambda - \lambda_0)\, u.$$

We take a positive constant λ_0 so large that the function $c(x) - \lambda_0$ satisfies the condition

$$c(x) - \lambda_0 < 0 \quad \text{on the double } \widehat{D} \text{ of } D. \tag{7.1}$$

This condition (7.1) implies that condition (SC) is satisfied for the operator $A - \lambda_0$:

$$\boxed{\mathcal{N}_0\, (A - \lambda_1) = \mathcal{N}_0\, (A^* - \lambda_1) = \{0\}.} \tag{SC}$$

Therefore, by applying Theorems 6.15 and 6.14 to the operator $A - \lambda_0$ we can obtain the following two fundamental results (a) and (b):

(a) The Dirichlet problem

$$\begin{cases} (A - \lambda_0)\, w = 0 & \text{in } D, \\ w = \varphi & \text{on } \partial D \end{cases} \tag{$D)_{\lambda_0}$}$$

has a unique solution $w(x) \in H^{t,p}(D)$ for any function $\varphi \in B^{t-1/p,p}(\partial D)$ with $t \in \mathbf{R}$.

(b) The Poisson operator

$$\mathcal{P}(\lambda_0) \colon B^{t-1/p,p}(\partial D) \longrightarrow H^{t,p}(D),$$

defined by $w = \mathcal{P}(\lambda_0)\varphi$, is an isomorphism of the space $B^{t-1/p,p}(\partial D)$ onto the null space

$$\mathcal{N}\, (A - \lambda_0, t, p) = \left\{ u \in H^{t,p}(D) : (A - \lambda_0)u = 0 \text{ in } D \right\}$$

for all $t \in \mathbf{R}$; and its inverse is the trace operator γ_0 on the boundary ∂D.

We let

$$T(\lambda_0) \colon C^\infty(\partial D) \longrightarrow C^\infty(\partial D)$$
$$\varphi \longmapsto LP(\lambda_0)\varphi.$$

Then we have the formula

$$T(\lambda_0) := LP(\lambda_0) = \mu(x')\,\Pi(\lambda_0) + \gamma(x'),$$

where $\Pi(\lambda_0)$ is the Dirichlet-to-Neumann operator defined by the formula

$$\Pi(\lambda_0)\varphi = \frac{\partial}{\partial \mathbf{n}}\,(P(\lambda_0)\varphi)\Big|_{\partial D}.$$

It is known that the operator $\Pi(\lambda_0)$ is a classical pseudo-differential operator of first order on the boundary ∂D and that its complete symbol is given by the following formula (cf. [122, Section 10.7]):

$$\big(p_1(x',\xi') + \sqrt{-1}\,q_1(x',\xi')\big) + \big(p_0(x',\xi') + \sqrt{-1}\,q_0(x',\xi')\big)$$
$$+ \text{ terms of order} \le -1 \text{ depending on } \lambda_0,$$

where

$$p_1(x',\xi') < 0 \tag{7.2}$$

on the bundle $T^*(\partial D) \setminus \{0\}$ of non-zero cotangent vectors.

For example, if A is the usual Laplacian

$$\Delta = \frac{\partial^2}{\partial x_1^2} + \frac{\partial^2}{\partial x_2^2} + \cdots + \frac{\partial^2}{\partial x_N^2},$$

then we have the formula

$$p_1(x',\xi') = \text{minus the length } |\xi'| \text{ of } \xi' \text{ with respect to the Riemannian}$$
$$\text{metric of } \partial D \text{ induced by the natural metric of } \mathbf{R}^N.$$

Therefore, we obtain that the operator

$$T(\lambda_0) = \mu(x')\,\Pi(\lambda_0) + \gamma(x')$$

is a classical pseudo-differential operator of first order on the boundary ∂D and further that its complete symbol $t(x',\xi';\lambda_0)$ is given by the following formula:

$$t(x',\xi';\lambda_0) = \mu(x')\,\big(p_1(x',\xi') + \sqrt{-1}\,q_1(x',\xi')\big) \tag{7.3}$$
$$+ \big([\gamma(x') + \mu(x')p_0(x',\xi')] + \sqrt{-1}\,\mu(x')q_0(x',\xi')\big)$$
$$+ \text{ terms of order} \le -1 \text{ depending on } \lambda_0.$$

Then, by arguing just as in Section 6.4 we can prove that the question of the validity of *a priori* estimates and the question of regularity for solutions of problem $(*)_\lambda$ for $\lambda = \lambda_0$ are reduced to the corresponding questions for the operator $T(\lambda_0)$ (see Theorems 6.21 and 6.22).

Step II: Therefore, in order to prove estimate (1.7) for $\lambda = \lambda_0$ it suffices to show the following lemma:

Lemma 7.1. *Assume that conditions (A) and (B) are satisfied. Then we have, for all $s \in \mathbf{R}$,*

$$\varphi \in \mathcal{D}'(\partial D),\ T(\lambda_0)\varphi \in B^{s,p}(\partial D) \implies \varphi \in B^{s,p}(\partial D). \tag{7.4}$$

Furthermore, for any $t < s$, there exists a positive constant $C_{s,t}$ such that

$$|\varphi|_{s,p} \leq C_{s,t}\left(|T(\lambda_0)\varphi|_{s,p} + |\varphi|_{t,p}\right). \tag{7.5}$$

Proof. (a) The proof of Lemma 7.1 is based on the following lemma (cf. [68, Theorem 3.1]):

Lemma 7.2. *Assume that conditions (A) and (B) are satisfied. Then, for each point x' of ∂D, we can find a neighborhood $U(x')$ of x' such that:*
For any compact $K \subset U(x')$ and any multi-indices α, β, there exist positive constants $C_{K,\alpha,\beta}$ and C_K such that we have, for all $x' \in K$ and all $|\xi'| \geq C_K$,

$$\left|D_{\xi'}^\alpha D_{x'}^\beta t(x', \xi'; \lambda_0)\right| \leq C_{K,\alpha,\beta}\, |t(x', \xi'; \lambda_0)|\, (1 + |\xi|)^{-|\alpha| + (1/2)|\beta|}, \tag{7.6a}$$

$$|t(x', \xi'; \lambda_0)|^{-1} \leq C_K. \tag{7.6b}$$

Granting Lemma 7.2 for the moment, we shall prove Lemma 7.1.

(b) First, we cover ∂D by a finite number of local charts

$$\{(U_j, \chi_j)\}_{j=1}^m$$

in each of which inequalities (7.6a) and (7.6b)) hold true. Since the operator $T(\lambda_0)$ satisfies conditions (4.22a) and (4.22b) of Theorem 4.49 with $\mu := 0$, $\rho := 1$ and $\delta = 1/2$, it follows from an application of the same theorem that there exists a *parametrix* $S(\lambda_0)$ in the Hörmander class $L^0_{1,1/2}(U_j)$ for $T(\lambda_0)$:

$$\begin{cases} T(\lambda_0)S(\lambda_0) \equiv I & \mathrm{mod}\ L^{-\infty}(U_j), \\ S(\lambda_0)T(\lambda_0) \equiv I & \mathrm{mod}\ L^{-\infty}(U_j). \end{cases}$$

Let $\{\varphi_j\}_{j=1}^m$ be a partition of unity subordinate to the covering $\{U_j\}_{j=1}^m$, and choose a function $\psi_j(x') \in C_0^\infty(U_j)$ such that $\psi_j(x') = 1$ on $\mathrm{supp}\,\varphi_j$, so that

$$\varphi_j(x')\psi_j(x') = \varphi_j(x').$$

Now we may assume that $\varphi \in B^{t,p}(\partial D)$ for some $t < s$ and that $T(\lambda_0)\varphi \in B^{s,p}(\partial D)$. We remark that the operator $T(\lambda_0)$ can be written in the following form:

$$T(\lambda_0) = \sum_{j=1}^{m} \varphi_j T(\lambda_0)\psi_j + \sum_{j=1}^{m} \varphi_j T(\lambda_0)(1 - \psi_j).$$

However, by applying Theorems 4.42 and 4.36 to our situation we obtain that the second terms $\varphi_j T(\lambda_0)(1 - \psi_j)$ are in $L^{-\infty}(\partial D)$. Indeed, it suffices to note that

$$\varphi_j(x')(1 - \psi_j(x')) = \varphi_j(x') - \varphi_j(x') = 0.$$

Hence we are reduced to the study of the first terms $\varphi_j T(\lambda_0)\psi_j$. This implies that we have only to prove the following *local version* of assertions (7.4) and (7.5):

$$\psi_j\varphi \in B^{t,p}(U_j), \; T(\lambda_0)\psi_j\varphi \in B^{s,p}(U_j) \implies \psi_j\varphi \in B^{s,p}(U_j). \tag{7.7}$$

$$|\psi_j\varphi|_{s,p} \leq C'_{s,t}\left(|T(\lambda_0)\psi_j\varphi|^2_{s,p} + |\psi_j\varphi|^2_{t,p}\right). \tag{7.8}$$

However, by applying the Besov-space boundedness theorem (Theorem 4.47) to our situation we find that the parametrix

$$S(\lambda_0)\colon B^{\sigma,p}_{\text{loc}}(U_j) \longrightarrow B^{\sigma,p}_{\text{loc}}(U_j)$$

is continuous for all $\sigma \in \mathbf{R}$. This proves the desired assertions (7.7) and (7.8), since we have the assertion (see Theorem 4.36)

$$\psi_j\varphi \equiv S(\lambda_0)(T(\lambda_0)\psi_j\varphi) \quad \text{mod } C^{-\infty}(U_j).$$

Lemma 7.1 is proved, apart from the proof of Lemma 7.2.   □

Step III: Proof of Lemma 7.2
The proof of Lemma 7.2 is divided into five steps.
Step III-1: First, we verify condition (7.6b):
By assertions (7.3) and (7.2), we can find positive constants c_0 and c_1 such that we have, for all sufficiently large $|\xi'|$,

$$|t(x', \xi'; \lambda_0)| \geq \mu(x')|p_1(x', \xi') + p_0(x', \xi')| - \gamma(x')$$

$$\geq \begin{cases} c_0\,\mu(x')|\xi'| - \frac{1}{2}\gamma(x') & \text{if } 0 \leq \mu(x') \leq c_1, \\ \frac{c_0}{2}\,\mu(x')|\xi'| - \gamma(x') & \text{if } c_1 \leq \mu(x') \leq 1, \end{cases}$$

since $\gamma(x') < 0$ on $M = \{x' \in \partial D : \mu(x') = 0\}$. Hence we have, for all sufficiently large $|\xi'|$,

$$|t(x', \xi'; \lambda_0)| \geq C\left(\mu(x')|\xi'| + 1\right). \tag{7.9}$$

Here in the following the letter C denotes a generic positive constant.
Inequality (7.9) implies the desired condition 7.6b:

$$|t(x', \xi'; \lambda_0)| \geq C. \tag{7.10}$$

Step III-2: Next we verify condition (7.6a) for $|\alpha| = 1$ and $|\beta| = 0$: Since we have, for all sufficiently large $|\xi'|$,

$$\left| D_{\xi'}^\alpha t(x', \xi'; \lambda_0) \right| \leq C \left(\mu(x') + |\xi'|^{-1} \right),$$

it follows from inequality (7.9) that

$$\left| D_{\xi'}^\alpha t(x', \xi'; \lambda_0) \right| \leq C \left(1 + |\xi'| \right)^{-1} \left(\mu(x')|\xi'| + 1 \right)$$
$$\leq C \left(1 + |\xi'| \right)^{-1} |t(x', \xi'; \lambda_0)|.$$

This inequality proves the desired condition (7.6a) for $|\alpha| = 1$ and $|\beta| = 0$.

Step III-3: We verify condition (7.6a) for $|\beta| = 1$ and $|\alpha| = 0$: To do this, we need the following elementary lemma on non-negative functions:

Lemma 7.3. *Let $f(x)$ be a non-negative, C^2 function on $\mathbf{R}$ such that we have, for some positive constant c,*

$$\sup_{x \in \mathbf{R}} |f''(x)| \leq c. \tag{7.11}$$

Then we have the inequality

$$|f'(x)| \leq \sqrt{2c} \sqrt{f(x)} \quad \text{on } \mathbf{R}. \tag{7.12}$$

Proof. In view of Taylor's formula, it follows that

$$0 \leq f(y) = f(x) + f'(x)(y - x) + \frac{f''(\xi)}{2}(y - x)^2,$$

where ξ is between x and y. Thus, by letting $z = x - y$ we obtain from estimate (7.11) that

$$0 \leq f(x) + f'(x)z + \frac{f''(\xi)}{2}z^2$$
$$\leq f(x) + f'(x)z + \frac{c}{2}z^2 \quad \text{for all } z \in \mathbf{R}.$$

This implies the desired inequality (7.12). $\quad\square$

Step III-4: Since we have, for all sufficiently large $|\xi'|$,

$$\left| D_{x'}^\beta t(x', \xi'; \lambda_0) \right| \leq C \left(\left| D_{x'}^\beta \mu(x') \right| \cdot |\xi'| + \mu(x')|\xi'| + 1 \right),$$

it follows from an application of Lemma 7.3 and inequalities (7.9) and (7.10) that

$$\left| D_{x'}^\beta t(x', \xi'; \lambda_0) \right| \leq C \left[\left(\sqrt{\mu(x')} \, |\xi'| + 1 \right) + \left(\mu(x')|\xi'| + 1 \right) \right]$$

$$\leq C \left[|\xi'|^{1/2} \left(\mu(x')|\xi'| + 1 \right)^{1/2} + \left(\mu(x')|\xi'| + 1 \right) \right]$$

$$\leq C \, |t(x', \xi'; \lambda_0)| \left(|\xi'|^{1/2} \, |t(x', \xi'; \lambda_0)|^{-1/2} + 1 \right)$$

$$\leq C \, |t(x', \xi'; \lambda_0)| \, (1 + |\xi'|)^{1/2} .$$

This inequality proves the desired condition (7.6a) for $|\beta| = 1$ and $|\alpha| = 0$.

Step III-5: Similarly, we can verify condition (7.6a) for the general case: $|\alpha| + |\beta| = k$ for $k \in \mathbf{N}$.

Now the proof of Lemma 7.1 and hence that of Theorem 1.2 is complete.
□

7.3 Notes and Comments

This chapter is a revised and expanded version of Chapter 5 of the second edition [121].

8

A Priori Estimates

This Chapter 8 and the next Chapter 9 are devoted to the proof of Theorem
1.4. In this chapter we study the operator A_p, and prove *a priori* estimates for
the operator $A_p - \lambda I$ (Theorem 8.3) which will play a fundamental role in the
next chapter. In the proof we make good use of Agmon's method (Proposition
8.4). This is a technique of treating a spectral parameter λ as a second-order,
elliptic differential operator of an extra variable and relating the old problem
to a new problem with the additional variable.

8.1 *A Priori* Estimates via Agmon's Method

Recall that the operator A_p is a unbounded linear operator from $L^p(D)$ into
itself defined as follows:

(a) The domain of definition $\mathcal{D}(A_p)$ of A_p is the space

$$\mathcal{D}(A_p) \tag{1.8}$$
$$= \left\{ u \in H^{2,p}(D) = W^{2,p}(D) : Lu = \mu(x')\frac{\partial u}{\partial \mathbf{n}} + \gamma(x')u = 0 \text{ on } \partial D \right\}.$$

(b) $A_p u = Au$ for every $u \in \mathcal{D}(A_p)$.

We remark that the operator A_p is densely defined, since the domain $\mathcal{D}(A_p)$
contains the space $C_0^\infty(D)$.

First, we have the following lemma:

Lemma 8.1. *Assume that the following two conditions (A) and (B) are satisfied:*

(A) $\mu(x') \geq 0$ and $\gamma(x') \leq 0$ on ∂D.
(B) $\mu(x') - \gamma(x') = \mu(x') + |\gamma(x')| > 0$ on ∂D.

© Springer Nature Switzerland AG 2020
K. Taira, *Boundary Value Problems and Markov Processes*, Lecture Notes in Mathematics 1499,
https://doi.org/10.1007/978-3-030-48788-1_8

Then we have the a priori *estimate*

$$\|u\|_{2,p} \leq C \left(\|Au\|_p + \|u\|_p \right) \quad \textit{for all } u \in \mathcal{D}(A_p). \tag{8.1}$$

Proof. The *a priori* estimate (8.1) follows immediately from estimate (1.7) of Theorem 1.2 with $\varphi := 0$. $\square$

Corollary 8.2. *The operator A_p is a closed operator.*

Proof. Let $\{u_j\}$ be an arbitrary sequence in the domain $\mathcal{D}(A_p)$ such that:

$$\begin{cases} u_j \longrightarrow u & \text{in } L^p(D), \\ Au_j \longrightarrow v & \text{in } L^p(D). \end{cases}$$

Then, by applying estimate (8.1) to the sequence $\{u_j\}$ we find that $\{u_j\}$ is a Cauchy sequence in the space $H^{2,p}(D)$. This proves that

$$u \in H^{2,p}(D),$$

and that

$$u_j \longrightarrow u \quad \text{in } H^{2,p}(D).$$

Hence we have the formula

$$Au = \lim_{j \to \infty} Au_j = v \quad \text{in } L^p(D),$$

and also, by Proposition 6.25 with $B := L$ and $s := 2$,

$$Lu = \lim_{j \to \infty} Lu_j = 0 \quad \text{in } B^{1-1/p,p}(\partial D).$$

Summing up, we have proved that $u \in \mathcal{D}(A_p)$ and $A_p u = v$.

The proof of Corollary 8.2 is complete. $\square$

The next theorem is an essential step in the proof of Theorem 1.4:

Theorem 8.3. *Assume that conditions (A) and (B) are satisfied. Then, for every $-\pi < \theta < \pi$ there exists a positive constant $R(\theta)$ depending on θ such that if $\lambda = r^2 e^{i\theta}$ and $|\lambda| = r^2 \geq R(\theta)$, we have, for all $u \in H^{2,p}(D)$ satisfying $Lu = 0$ on ∂D (i. e., $u \in \mathcal{D}(A_p)$),*

$$|u|_{2,p} + |\lambda|^{1/2} \cdot |u|_{1,p} + |\lambda| \cdot \|u\|_p \leq C(\theta) \, \|(A - \lambda)\, u\|_p, \tag{8.2}$$

with a positive constant $C(\theta)$ depending on θ. Here $|\cdot|_{j,p}$, $j = 1, 2$, is the seminorm on the space $H^{2,p}(D)$ defined by the formula

$$|u|_{j,p} = \left(\int_D \sum_{|\alpha|=j} |D^\alpha u(x)|^p \, dx \right)^{1/p}.$$

Proof. The proof of Theorem 8.3 is divided into two steps.

Step I: We shall make use of a method essentially due to Agmon (see [3], [45], [77], [113]).

We introduce an auxiliary variable y of the unit circle

$$S = \mathbf{R}/2\pi\,\mathbf{Z},$$

and replace the complex parameter λ by the second-order differential operator

$$-e^{i\theta}\frac{\partial^2}{\partial y^2}\quad\text{for }-\pi < \theta < \pi.$$

Namely, we replace the operator $A - \lambda$ by the operator

$$\widetilde{\Lambda}(\theta) := A + e^{i\theta}\frac{\partial^2}{\partial y^2}\quad\text{for }-\pi < \theta < \pi,$$

on the product domain $D \times S$ (see Figure 8.1 below), and consider instead of the problem with spectral parameter

$$\begin{cases}(A - \lambda)\,u = f & \text{in } D, \\ Lu = \mu(x')\frac{\partial u}{\partial \mathbf{n}} + \gamma(x')u = 0 & \text{on } \partial D\end{cases}\tag{$*$)$_\lambda$}$$

the following boundary value problem:

$$\begin{cases}\widetilde{\Lambda}(\theta)\widetilde{u} = \left(A + e^{i\theta}\frac{\partial^2}{\partial y^2}\right)\widetilde{u} = \widetilde{f} & \text{in } D \times S, \\ L\widetilde{u} = \mu(x')\frac{\partial \widetilde{u}}{\partial \mathbf{n}} + \gamma(x')\,\widetilde{u} = 0 & \text{on } \partial D \times S.\end{cases}\tag{8.3}$$

We remark that the operator $\widetilde{\Lambda}(\theta)$ is *elliptic* for $-\pi < \theta < \pi$.

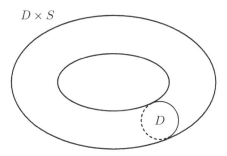

Fig. 8.1. The product domain $D \times S$

Then we have the following result, analogous to Lemma 8.1 (see [127, Theorem 9.1]):

Proposition 8.4. *Assume that conditions (A) and (B) are satisfied. Then we have, for all $\widetilde{u} \in H^{2,p}(D \times S)$ satisfying the boundary condition $L\widetilde{u} = 0$ on $\partial D \times S$,*

$$\|\widetilde{u}\|_{2,p} \leq \widetilde{C}(\theta) \left(\left\|\widetilde{\Lambda}(\theta)\widetilde{u}\right\|_p + \|\widetilde{u}\|_p \right), \tag{8.4}$$

with a positive constant $\widetilde{C}(\theta)$ depending on θ.

Proof. We reduce the study of the boundary value problem (8.3) to that of a pseudo-differential operator on the boundary $\partial D \times S$, just as in problem $(*)_\lambda$. We can prove (see [127, Theorem 10.1]) that Theorems 6.15 and 6.14 remain valid for the differential operator

$$\widetilde{\Lambda}(\theta) = A + e^{i\theta}\frac{\partial^2}{\partial y^2} \quad \text{for } -\pi < \theta < \pi.$$

($\widetilde{a}$) The Dirichlet problem

$$\begin{cases} \widetilde{\Lambda}(\theta)\widetilde{w} = 0 & \text{in } D \times S, \\ \widetilde{w} = \widetilde{\varphi} & \text{on } \partial D \times S \end{cases} \tag{$\widetilde{D}$}$$

has a unique solution $\widetilde{w}(x,y) \in H^{t,p}(D \times S)$ for any function $\widetilde{\varphi}(x',y) \in B^{t-1/p,p}(\partial D \times S)$ with $t \in \mathbf{R}$.

($\widetilde{b}$) The Poisson operator

$$\widetilde{\mathcal{P}}(\theta) \colon B^{t-1/p,p}(\partial D \times S) \longrightarrow H^{t,p}(D \times S),$$

defined by $\widetilde{w} = \widetilde{\mathcal{P}}(\theta)\widetilde{\varphi}$, is an isomorphism of the space $B^{t-1/p,p}(\partial D \times S)$ onto the null space

$$\mathcal{N}(\widetilde{\Lambda}(\theta),t,p) = \left\{ \widetilde{u} \in H^{t,p}(D \times S) : \widetilde{\Lambda}(\theta)\widetilde{u} = 0 \text{ in } D \times S \right\}$$

for all $t \in \mathbf{R}$; and its inverse is the trace operator on the boundary $\partial D \times S$.

We let

$$\widetilde{T}(\theta) \colon C^\infty(\partial D \times S) \longrightarrow C^\infty(\partial D \times S)$$
$$\widetilde{\varphi} \longmapsto L\widetilde{\mathcal{P}}(\theta)\widetilde{\varphi}.$$

Then the operator $\widetilde{T}(\theta)$ can be decomposed as follows:

$$\widetilde{T}(\theta) := L\widetilde{\mathcal{P}}(\theta) = \mu(x')\widetilde{\Pi}(\theta) + \gamma(x'), \tag{8.5}$$

where $\widetilde{\Pi}(\theta)$ is the Dirichlet-to-Neumann operator defined by the formula

$$\widetilde{\Pi}(\theta)\widetilde{\varphi} = \frac{\partial}{\partial \mathbf{n}}\left(\widetilde{\mathcal{P}}(\theta)\widetilde{\varphi}\right)\bigg|_{\partial D \times S} \qquad \text{for } \widetilde{\varphi} \in C^{\infty}(\partial D \times S).$$

It is known that the operator $\widetilde{\Pi}(\theta)$ is a classical pseudo-differential operator of first order on the boundary $\partial D \times S$ and further that its complete symbol is given by the following formula (cf. [122, Section 10.7]):

$$\left(\widetilde{p}_1(x', \xi', y, \eta; \theta) + \sqrt{-1}\,\widetilde{q}_1(x', \xi', y, \eta; \theta)\right)$$
$$+ \left(\widetilde{p}_0(x', \xi', y, \eta; \theta) + \sqrt{-1}\,\widetilde{q}_0(x', \xi', y, \eta; \theta)\right) + \text{terms of order} \leq -1,$$

where

$$\widetilde{p}_1(x', \xi', y, \eta; \theta) < 0$$
on the bundle $T^*(\partial D \times S) \setminus \{0\}$ of non-zero cotangent vectors,
for $-\pi < \theta < \pi.$ \hfill (8.6)

For example, if A is the usual Laplacian

$$\Delta = \frac{\partial^2}{\partial x_1^2} + \frac{\partial^2}{\partial x_2^2} + \ldots + \frac{\partial^2}{\partial x_N^2},$$

then we have the formula

$$\widetilde{p}_1(x', \xi', y, \eta; \theta)$$
$$= -\left[\frac{\left[\left(|\xi'|^2 + \cos\theta \cdot \eta^2\right)^2 + \sin^2\theta \cdot \eta^4\right]^{1/2} + \left(|\xi'|^2 + \cos\theta \cdot \eta^2\right)}{2}\right]^{1/2}.$$

Therefore, we obtain that the operator

$$\widetilde{T}(\theta) = \mu(x')\,\widetilde{\Pi}(\theta) + \gamma(x')$$

is a classical pseudo-differential operator of first order on the boundary $\partial D \times S$ and further that its complete symbol is given by the following formula:

$$\mu(x')\left(\widetilde{p}_1(x', \xi', y, \eta; \theta) + \sqrt{-1}\,\widetilde{q}_1(x', \xi', y, \eta; \theta)\right) \hfill (8.7)$$
$$+ \left(\left[\gamma(x') + \mu(x')\widetilde{p}_0(x', \xi', y, \eta; \theta)\right] + \sqrt{-1}\,\mu(x')\widetilde{q}_0(x', \xi', y, \eta; \theta)\right)$$
$$+ \text{terms of order} \leq -1.$$

Then, by virtue of assertions (8.7) and (8.6) we can verify that the operator $\widetilde{T}(\theta)$ satisfies conditions (4.22a) and (4.22b) of Theorem 4.49 with $\mu := 0$, $\rho := 1$ and $\delta := 1/2$, just as in the proof of Lemma 7.2. Hence we obtain the following lemma, analogous to Lemma 7.1 (see [127, Proposition 12.3]):

Lemma 8.5. *Assume that conditions (A) and (B) are satisfied. Then we have, for all* $s \in \mathbf{R}$,

$$\widetilde{\varphi} \in \mathcal{D}'(\partial D \times S), \ \widetilde{T}(\theta)\widetilde{\varphi} \in B^{s,p}(\partial D \times S) \implies \widetilde{\varphi} \in B^{s,p}(\partial D \times S).$$

Furthermore, for any $t < s$, *there exists a positive constant* $\widetilde{C}_{s,t}$ *such that*

$$|\widetilde{\varphi}|_{s,p} \leq \widetilde{C}_{s,t} \left(|\widetilde{T}(\theta)\widetilde{\varphi}|_{s,p} + |\widetilde{\varphi}|_{t,p} \right). \tag{8.8}$$

The desired estimate (8.4) follows from estimate (8.8) with $s := 2 - 1/p$ and $t := -1/p$, just as in the proof of Theorem 6.22.

The proof of Proposition 8.4 is complete.   □

Step II: Now let $u(x)$ be an arbitrary function in the domain $\mathcal{D}(A_p)$:

$$u \in H^{2,p}(D) \text{ and } Lu = 0 \text{ on } \partial D.$$

We choose a function $\zeta(y)$ in $C^\infty(S)$ such that

$$\begin{cases} 0 \leq \zeta(y) \leq 1 & \text{on } S, \\ \operatorname{supp}\zeta \subset \left[\frac{\pi}{3}, \frac{5\pi}{3}\right], \\ \zeta(y) = 1 & \text{for } \frac{\pi}{2} \leq y \leq \frac{3\pi}{2}, \end{cases}$$

and let

$$\widetilde{v}_\eta(x, y) = u(x) \otimes \zeta(y) e^{i\eta y} \quad \text{for } x \in D, \ y \in S \text{ and } \eta \geq 0.$$

Then we have the assertions

- $\widetilde{v}_\eta \in H^{2,p}(D \times S)$,
- $\widetilde{\Lambda}(\theta)\widetilde{v}_\eta = \left(A + e^{i\theta} \dfrac{\partial^2}{\partial y^2} \right) \widetilde{v}_\eta$

  $= (A - \eta^2 e^{i\theta})u \otimes \zeta(y)e^{i\eta y} + 2(i\eta)e^{i\theta}u \otimes \zeta'(y) e^{i\eta y} + e^{i\theta}u \otimes \zeta''(y) e^{i\eta y},$

and also
$$L\widetilde{v}_\eta(x', y) = (Lu(x')) \otimes \zeta(y) e^{i\eta y} = 0 \quad \text{on } \partial D \times S.$$

Thus, by applying inequality (8.4) to the functions

$$\widetilde{v}_\eta(x, y) = u(x) \otimes \zeta(y) e^{i\eta y} \quad \text{for } x \in D, \ y \in S \text{ and } \eta \geq 0,$$

we obtain that

$$\left\| u \otimes \zeta\, e^{i\eta y} \right\|_{2,p} \leq \widetilde{C}(\theta) \left(\left\| \widetilde{\Lambda}(\theta)\left(u \otimes \zeta\, e^{i\eta y} \right) \right\|_p + \left\| u \otimes \zeta\, e^{i\eta y} \right\|_p \right). \tag{8.9}$$

We can estimate each term of inequality (8.9) as follows:

- $\left\| u \otimes \zeta e^{i\eta y} \right\|_p = \left(\int_{D \times S} |u(x)|^p |\zeta(y)|^p \, dx \, dy \right)^{1/p} = \|\zeta\|_p \cdot \|u\|_p.$ (8.10)

- $\left\| \widetilde{\Lambda}(\theta) \left(u \otimes \zeta \, e^{i\eta y} \right) \right\|_p$ (8.11)

$\leq \left\| (A - \eta^2 e^{i\theta}) u \otimes \zeta e^{i\eta y} \right\|_p + 2\eta \left\| u \otimes \zeta' e^{i\eta y} \right\|_p + \left\| u \otimes \zeta'' e^{i\eta y} \right\|_p$

$\leq \|\zeta\|_p \cdot \left\| (A - \eta^2 e^{i\theta}) u \right\|_p + \left(2\eta \|\zeta'\|_p + \|\zeta''\|_p \right) \|u\|_p.$

- $\left\| u \otimes \zeta e^{i\eta y} \right\|_{2,p}^p$ (8.12)

$= \sum_{|\alpha| \leq 2} \int_{D \times S} \left| D_{x,y}^\alpha \left(u(x) \otimes \zeta(y) e^{i\eta y} \right) \right|^p \, dx \, dy$

$\geq \sum_{|\alpha| \leq 2} \int_D \int_{\pi/2}^{3\pi/2} \left| D_{x,y}^\alpha \left(u(x) \otimes e^{i\eta y} \right) \right|^p \, dx \, dy$

$= \sum_{k+|\beta| \leq 2} \int_D \int_{\pi/2}^{3\pi/2} \left| \eta^k D^\beta u(x) \right|^p \, dx \, dy$

$\geq \pi \left(\sum_{|\beta|=2} \int_D |D^\beta u(x)|^p \, dx + \eta^p \sum_{|\beta|=1} \int_D |D^\beta u(x)|^p \, dx + \eta^{2p} \int_D |u(x)|^p \, dx \right)$

$= \pi \left(|u|_{2,p}^p + \eta^p |u|_{1,p}^p + \eta^{2p} \|u\|_p^p \right).$

Therefore, by carrying these three inequalities (8.10), (8.11) and (8.12) into inequality (8.9) we obtain that

$$|u|_{2,p} + \eta |u|_{1,p} + \eta^2 \|u\|_p \leq \widetilde{C}'(\theta) \left(\left\| (A - \eta^2 e^{i\theta}) u \right\|_p + \eta \|u\|_p \right),$$

with a positive constant $\widetilde{C}'(\theta)$ independent of η.

If η is so large that

$$\eta \geq 2 \widetilde{C}'(\theta),$$

then we can eliminate the last term on the right-hand side to obtain that

$$|u|_{2,p} + \eta |u|_{1,p} + \eta^2 \|u\|_p \leq 2 \widetilde{C}'(\theta) \left\| (A - \eta^2 e^{i\theta}) u \right\|_p.$$

This proves the desired inequality (8.2)) if we take

$$\lambda := \eta^2 e^{i\theta},$$
$$R(\theta) := 4 \widetilde{C}'(\theta)^2,$$
$$C(\theta) := 2 \widetilde{C}'(\theta).$$

The proof of Theorem 8.3 is now complete. $\square$

8.2 Notes and Comments

This chapter is a revised and expanded version of Chapter 6 of the second edition [121]. For the detailed proof of Proposition 8.4, the reader is referred to Taira [127].

9

Proof of Theorem 1.4

In this chapter we prove Theorem 1.4 (Theorems 9.1 and 9.11). Once again we make use of Agmon's method in the proof of Theorems 9.1 and 9.11. In particular, Agmon's method plays an important role in the proof of the *surjectivity* of the operator $A_p - \lambda I$ (Proposition 9.2).

The proof of Theorem 1.4 can be flowcharted as in Table 9.1 below.

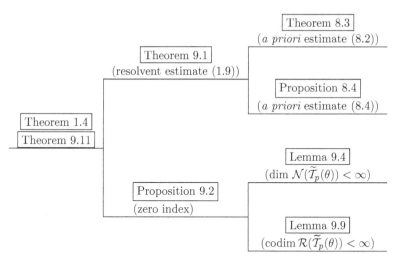

Table 9.1. A flowchart for the proof of Theorem 1.4

9.1 Proof of Theorem 1.4, Part (i)

First, we prove part (i) of Theorem 1.4:

Theorem 9.1. *Assume that the following two conditions (A) and (B) are satisfied:*

© Springer Nature Switzerland AG 2020
K. Taira, *Boundary Value Problems and Markov Processes*, Lecture Notes in Mathematics 1499,
https://doi.org/10.1007/978-3-030-48788-1_9

(A) $\mu(x') \geq 0$ and $\gamma(x') \leq 0$ on ∂D.

(B) $\mu(x') - \gamma(x') = \mu(x') + |\gamma(x')| > 0$ on ∂D.

Then, for every $0 < \varepsilon < \pi/2$ there exists a positive constant $r_p(\varepsilon)$ such that the resolvent set of A_p contains the set

$$\Sigma_p(\varepsilon) = \left\{ \lambda = r^2 e^{i\theta} : r \geq r_p(\varepsilon), -\pi + \varepsilon \leq \theta \leq \pi - \varepsilon \right\},$$

and that the resolvent $(A_p - \lambda I)^{-1}$ satisfies the estimate

$$\left\| (A_p - \lambda I)^{-1} \right\| \leq \frac{c_p(\varepsilon)}{|\lambda|} \quad \text{for all } \lambda \in \Sigma_p(\varepsilon), \tag{1.9}$$

where $c_p(\varepsilon)$ is a positive constant depending on ε.

Proof. The proof of Theorem 9.1 is divided into three steps.

Step I: By estimate (8.2), it follows that if $\lambda = r^2 e^{i\theta}$ for $-\pi < \theta < \pi$ and if $|\lambda| = r^2 \geq R(\theta)$, then we have, for all $u \in \mathcal{D}(A_p)$,

$$|u|_{2,p} + |\lambda|^{1/2} \cdot |u|_{1,p} + |\lambda| \cdot \|u\|_p \leq C(\theta) \left\| (A_p - \lambda I) u \right\|_p.$$

However, we find from the proof of Theorem 8.3 that the constants $R(\theta)$ and $C(\theta)$ depend *continuously* on $\theta \in (-\pi, \pi)$, so that they may be chosen uniformly in $\theta \in [-\pi + \varepsilon, \pi + \varepsilon]$, for every $\varepsilon > 0$. This proves the existence of the constants $r_p(\varepsilon)$ and $c_p(\varepsilon)$, that is, we have, for all $\lambda = r^2 e^{i\theta}$ satisfying the conditions $r \geq r_p(\varepsilon)$ and $-\pi + \varepsilon \leq \theta \leq \pi + \varepsilon$,

$$|u|_{2,p} + |\lambda|^{1/2} \cdot |u|_{1,p} + |\lambda| \cdot \|u\|_p \leq c_p(\varepsilon) \left\| (A_p - \lambda I) u \right\|_p. \tag{9.1}$$

By estimate (9.1), we find that the operator $A_p - \lambda I$ is injective and its range $\mathcal{R}(A_p - \lambda I)$ is closed in the space $L^p(D)$, for all $\lambda \in \Sigma_p(\varepsilon)$.

Step II: We show that the operator $A_p - \lambda I$ is surjective for all $\lambda \in \Sigma_p(\varepsilon)$, that is,

$$\mathcal{R}(A_p - \lambda I) = L^p(D) \quad \text{for all } \lambda \in \Sigma_p(\varepsilon). \tag{9.2}$$

To do this, it suffices to show that the operator $A_p - \lambda I$ is a Fredholm operator with

$$\text{ind}\,(A_p - \lambda I) = 0 \quad \text{for all } \lambda \in \Sigma_p(\varepsilon), \tag{9.3}$$

since $A_p - \lambda I$ is injective for all $\lambda \in \Sigma_p(\varepsilon)$.

Here we recall that a densely defined, closed linear operator T with domain $\mathcal{D}(T)$ from a Banach space X into itself is called a *Fredholm (closed) operator* if it satisfies the following three conditions (i), (ii) and (iii):

(i) The null space $\mathcal{N}(T) = \{x \in \mathcal{D}(T) : Tx = 0\}$ of T has finite dimension, that is, $\dim \mathcal{N}(T) < \infty$.

(ii) The range $\mathcal{R}(T) = \{Tx : x \in \mathcal{D}(T)\}$ of T is closed in X.

(iii) The range $\mathcal{R}(T)$ has finite codimension in X, that is, $\text{codim}\,\mathcal{R}(T) = \dim X/\mathcal{R}(T) < \infty$.

In this case the *index* ind T of T is defined by the formula

$$\operatorname{ind} T = \dim \mathcal{N}(T) - \operatorname{codim} \mathcal{R}(T).$$

Step II-1: We reduce the study of the operator $A_p - \lambda I$ for $\lambda \in \Sigma_p(\varepsilon)$ to that of a pseudo-differential operator on the boundary, just as in the proof of Theorem 1.2.

Let $T(\lambda)$ be a classical pseudo-differential operator of first order on the boundary ∂D defined as follows:

$$T(\lambda) = L\mathcal{P}(\lambda) = \mu(x')\,\Pi(\lambda) + \gamma(x') \quad \text{for } \lambda \in \Sigma_p(\varepsilon), \tag{9.4}$$

where $\Pi(\lambda)$ is the Dirichlet-to-Neumann operator defined by the formula

$$\Pi(\lambda) : C^\infty(\partial D) \longrightarrow C^\infty(\partial D)$$

$$\varphi \longmapsto \frac{\partial}{\partial \mathbf{n}}\left(\mathcal{P}(\lambda)\varphi\right)\Big|_{\partial D}.$$

Since the operator $T(\lambda) : C^\infty(\partial D) \to C^\infty(\partial D)$ extends to a continuous linear operator

$$T(\lambda) : B^{t,p}(\partial D) \longrightarrow B^{t-1,p}(\partial D)$$

for all $t \in \mathbf{R}$, we can introduce a densely defined, closed linear operator

$$\mathcal{T}_p(\lambda) : B^{2-1/p,p}(\partial D) \longrightarrow B^{2-1/p,p}(\partial D)$$

as follows.

(α) The domain $\mathcal{D}\left(\mathcal{T}_p(\lambda)\right)$ of $\mathcal{T}_p(\lambda)$ is the space

$$\mathcal{D}\left(\mathcal{T}_p(\lambda)\right) = \left\{\varphi \in B^{2-1/p,p}(\partial D) : T(\lambda)\varphi \in B^{2-1/p,p}(\partial D)\right\}.$$

(β) $\mathcal{T}_p(\lambda)\varphi = T(\lambda)\varphi$ for every $\varphi \in \mathcal{D}\left(\mathcal{T}_p(\lambda)\right)$.

Then we can obtain the following three results (I), (II) and (III) (see [114, Section 8.3]):

(I) The null space $\mathcal{N}(A_p - \lambda I)$ of $A_p - \lambda I$ has finite dimension if and only if the null space $\mathcal{N}\left(\mathcal{T}_p(\lambda)\right)$ of $\mathcal{T}_p(\lambda)$ has finite dimension, and we have the formula

$$\dim \mathcal{N}\left(A_p - \lambda I\right) = \dim \mathcal{N}\left(\mathcal{T}_p(\lambda)\right).$$

(II) The range $\mathcal{R}(A_p - \lambda I)$ of $A_p - \lambda I$ is closed if and only if the range $\mathcal{R}\left(\mathcal{T}_p(\lambda)\right)$ of $\mathcal{T}_p(\lambda)$ is closed; and $\mathcal{R}(A_p - \lambda I)$ has finite codimension if and only if $\mathcal{R}\left(\mathcal{T}_p(\lambda)\right)$ has finite codimension, and we have the formula

$$\operatorname{codim} \mathcal{R}\left(A_p - \lambda I\right) = \operatorname{codim} \mathcal{R}\left(\mathcal{T}_p(\lambda)\right).$$

(III) The operator $A_p - \lambda I$ is a Fredholm operator if and only if the operator $\mathcal{T}_p(\lambda)$ is a Fredholm operator, and we have the formula

$$\text{ind}\,(A_p - \lambda I) = \text{ind}\,\mathcal{T}_p(\lambda).$$

Therefore, the desired assertion (9.3) is reduced to the following assertion:

$$\text{ind}\,\mathcal{T}_p(\lambda) = 0 \quad \text{for all } \lambda \in \Sigma_p(\varepsilon). \tag{9.5}$$

Step II-2: To prove assertion (9.5), we shall make use of Agmon's method just as in Chapter 8.

Let $\widetilde{T}(\theta)$ be the classical pseudo-differential operator of first order on the boundary $\partial D \times S$ introduced in Chapter 8 (see formula (8.5)):

$$\widetilde{T}(\theta) = L\widetilde{P}(\theta) = \mu(x')\,\widetilde{\Pi}(\theta) + \gamma(x') \quad \text{for } -\pi < \theta < \theta,$$

where $\widetilde{\Pi}(\theta)$ is the Dirichlet-to-Neumann operator defined by the formula

$$\widetilde{\Pi}(\theta)\colon C^\infty(\partial D \times S) \longrightarrow C^\infty(\partial D \times S)$$

$$\tilde{\varphi} \longmapsto \frac{\partial}{\partial \mathbf{n}}\left(\widetilde{P}(\theta)\tilde{\varphi}\right)\Big|_{\partial D \times S}.$$

We define a densely defined, closed linear operator

$$\widetilde{\mathcal{T}}_p(\theta)\colon B^{2-1/p,p}(\partial D \times S) \longrightarrow B^{2-1/p,p}(\partial D \times S)$$

as follows.

($\tilde{\alpha}$) The domain $\mathcal{D}\left(\widetilde{\mathcal{T}}_p(\theta)\right)$ of $\widetilde{\mathcal{T}}_p(\theta)$ is the space

$$\mathcal{D}\left(\widetilde{\mathcal{T}}_p(\theta)\right) = \left\{\tilde{\varphi} \in B^{2-1/p,p}(\partial D \times S) : \widetilde{T}(\theta)\tilde{\varphi} \in B^{2-1/p,p}(\partial D \times S)\right\}.$$

($\tilde{\beta}$) $\widetilde{\mathcal{T}}_p(\theta)\tilde{\varphi} = \widetilde{T}(\theta)\tilde{\varphi}$ for every $\tilde{\varphi} \in \mathcal{D}\left(\widetilde{\mathcal{T}}_p(\theta)\right)$.

Then the most fundamental relationship between the operators $\widetilde{\mathcal{T}}_p(\theta)$ and $\mathcal{T}_p(\lambda)$ is stated as follows:

Proposition 9.2. *If* $\text{ind}\,\widetilde{\mathcal{T}}_p(\theta)$ *is finite, then there exists a* finite subset K *of* $\mathbf{Z}$ *such that the operator* $\mathcal{T}_p(\lambda')$ *is bijective for all* $\lambda' = \ell^2 e^{i\theta}$ *satisfying* $\ell \in \mathbf{Z} \setminus K$.

Granting Proposition 9.2 for the moment, we shall prove Theorem 9.1.

Step III: **End of Proof of Theorem 9.1**

Step III-1: We show that if conditions (A) and (B) are satisfied, then we have the assertion

$$\text{ind}\,\widetilde{\mathcal{T}}_p(\theta) = \dim \mathcal{N}(\widetilde{\mathcal{T}}_p(\theta)) - \text{codim}\,\mathcal{R}(\widetilde{\mathcal{T}}_p(\theta)) < \infty. \tag{9.6}$$

To this end, we need a useful criterion for Fredholm operators due to Peetre [89] (cf. [114, Theorem 3.7.6]):

Lemma 9.3 (Peetre). *Let X, Y, Z be Banach spaces such that $X \subset Z$ is a compact injection, and let T be a closed linear operator with $\mathcal{D}(T)$ from X into Y with domain $\mathcal{D}(T)$. Then the following two conditions (i) and (ii) are equivalent:*

(i) The null space $\mathcal{N}(T)$ of T has finite dimension and the range $\mathcal{R}(T)$ of $T is closed in Y.
(ii) There is a positive constant C such that

$$\|x\|_X \leq C\left(\|Tx\|_Y + \|x\|_Z\right) \quad \text{for all } x \in \mathcal{D}(T). \tag{9.7}$$

Proof. (i) $\implies$ (ii): Since the null space $\mathcal{N}(T)$ has finite dimension, we can find a closed topological complement X_0 in X:

$$X = \mathcal{N}(T) \oplus X_0. \tag{9.8}$$

This gives that

$$\mathcal{D}(T) = \mathcal{N}(T) \oplus (\mathcal{D}(T) \cap X_0).$$

Namely, every element x of $\mathcal{D}(T)$ can be written in the form (see Figure 9.1 below)

$$x = x_0 + x_1, \quad x_0 \in \mathcal{D}(T) \cap X_0, \; x_1 \in \mathcal{N}(T).$$

Moreover, since the range $\mathcal{R}(T)$ is closed in Y, it follows from an application of the closed graph theorem (see [147, Chapter II, Section 6, Theorem 1]) that there exists a positive constant C such that

$$\|x_0\|_X \leq C \|Tx_0\|_Y = \|Tx\|_Y. \tag{9.9}$$

Here and in the following the letter C denotes a generic positive constant independent of x.

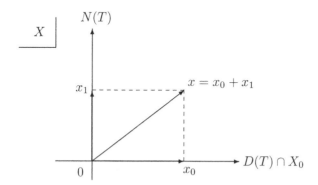

Fig. 9.1. The topological decomposition of $\mathcal{N}(T)$ in the domain $\mathcal{D}(T)$

On the other hand, it should be noticed that all norms on a finite dimensional linear space are equivalent. This gives that

$$\|x_1\|_X \leq C\|x_1\|_Z. \tag{9.10}$$

Moreover, since the injection $X \to Z$ is compact and hence is continuous, we obtain that

$$\|x_1\|_Z \leq \|x\|_Z + \|x_0\|_Z \leq \|x\|_Z + C\|x_0\|_X. \tag{9.11}$$

Thus we have, by inequalities (9.10) and (9.11),

$$\|x_1\|_X \leq C(\|x\|_Z + \|x_0\|_X). \tag{9.12}$$

Therefore, by combining inequalities (9.9) and (9.12) we obtain the desired inequality

$$\|x\|_X \leq \|x_0\|_X + \|x_1\|_X \leq C(\|Tx\|_Y + \|x\|_Z) \quad \text{for all } x \in \mathcal{D}(T), \tag{9.7}$$

since $Tx_0 = Tx$.

(ii) $\implies$ (i): Conversely, by inequality (9.7) it follows that

$$\|x\|_X \leq C\|x\|_Z \quad \text{for all } x \in \mathcal{N}(T). \tag{9.13}$$

However, the null space $\mathcal{N}(T)$ is closed in X, and so it is a Banach space. Since the injection $X \to Z$ is compact, we obtain from inequality (9.13) that the closed unit ball

$$\{x \in \mathcal{N}(T) : \|x\|_X \leq 1\}$$

of the Banach space $\mathcal{N}(T)$ is compact. Hence it follows from an application of [147, Chapter III, Section 2, Corollary 2] that

$$\dim \mathcal{N}(T) < \infty.$$

Let X_0 be a closed topological complement of $\mathcal{N}(T)$ as in the topological decomposition (9.8).

To prove the closedness of $\mathcal{R}(T)$, it suffices to show that

$$\|x\|_X \leq C\|Tx\|_Y \quad \text{for all } x \in \mathcal{D}(T) \cap X_0.$$

The proof is based on a reduction to absurdity. Assume, to the contrary, that:

For every $n \in \mathbf{N}$, there is an element x_n of $\mathcal{D}(T) \cap X_0$ such that

$$\|x_n\|_X > n\|Tx_n\|_Y.$$

If we let

$$x_n' := \frac{x_n}{\|x_n\|_X},$$

then we have the assertions

$$x_n' \in \mathcal{D}(T) \cap X_0, \quad \|x_n'\|_X = 1, \tag{9.14a}$$

$$\|Tx'_n\|_Y < \frac{1}{n}. \tag{9.14b}$$

Since the injection $X \to Z$ is compact, by passing to a subsequence we may assume that the sequence $\{x'_n\}$ is a Cauchy sequence in Z. Then, by combining inequalities (9.14b) and (9.7) we find that the sequence $\{x'_n\}$ is a Cauchy sequence in X, and hence it converges to some element x' of $X_0 \subset X$. Thus, by assertions (9.14a) and (9.14b) it follows that

$$\|x'\|_X = \lim_n \|x'_n\|_X = 1,$$

and further that

$$\begin{cases} x' \in \mathcal{D}(T), \\ Tx' = 0, \end{cases}$$

since the operator T is closed.

Summing up, we have proved that

$$x' \in \mathcal{N}(T),$$
$$\|x'\|_X = 1.$$

However, this is a contradiction. Indeed, we then have the assertion

$$x' \in \mathcal{N}(T) \cap X_0 = \{0\}.$$

The proof of Lemma 9.3 is complete. □

By using estimate (8.8) with $s := 2 - 1/p$, we obtain that

$$|\tilde{\varphi}|_{2-1/p,p} \leq \tilde{C}_t \left(|\tilde{T}(\theta)\tilde{\varphi}|_{2-1/p,p} + |\tilde{\varphi}|_{t,p} \right) \quad \text{for all } \tilde{\varphi} \in \mathcal{D}(\tilde{T}_p(\theta)), \tag{9.15}$$

where $t < 2 - 1/p$. However, it follows from an application of the Rellich–Kondrachov theorem (Theorem 4.10) that the injection

$$B^{2-1/p,p}(\partial D \times S) \longrightarrow B^{t,p}(\partial D \times S)$$

is *compact* for $t < 2 - 1/p$. Thus, by applying Peetre's lemma (Lemma 9.3) with

$$X = Y := B^{2-1/p,p}(\partial D \times S),$$
$$Z := B^{t,p}(\partial D \times S),$$
$$T := \tilde{T}_p(\theta),$$

we obtain that the range $\mathcal{R}\left(\tilde{T}_p(\theta)\right)$ is closed in the space $B^{2-1/p,p}(\partial D \times S)$ and further that

$$\dim \mathcal{N}\left(\tilde{T}_p(\theta)\right) < \infty. \tag{9.16}$$

On the other hand, by formula (8.7) we find that the complete symbol of the adjoint $\widetilde{T}(\theta)^*$ is given by the following formula (see Theorem 4.41):

$$\mu(x')\left(\tilde{p}_1(x',\xi',y,\eta;\theta) - \sqrt{-1}\,\tilde{q}_1(x',\xi',y,\eta;\theta)\right)$$

$$+\left(\left[\gamma(x') + \mu(x')\tilde{p}_0(x',\xi',y,\eta;\theta) - \sum_{j=1}^{n-1}\partial_{x_j}\left(\mu(x')\cdot\partial_{\xi_j}\tilde{q}_1(x',\xi',y,\eta;\theta)\right)\right]\right.$$

$$\left.-\sqrt{-1}\left[\mu(x')\tilde{q}_0(x',\xi',y,\eta;\theta) + \sum_{j=1}^{n-1}\partial_{x_j}\left(\mu(x')\cdot\partial_{\xi_j}\tilde{p}_1(x',\xi',y,\eta;\theta)\right)\right]\right)$$

$$+ \text{ terms of order } \leq -1.$$

However, it follows from an application of Lemma 7.3 that

$$\partial_{x_j}\mu(x') = 0 \quad \text{on } M = \{x' \in \partial D : \mu(x') = 0\}.$$

Thus we can easily verify that the pseudo-differential operator $\widetilde{T}(\theta)^*$ satisfies conditions (4.22a) and (4.22b)) of Theorem 4.49 with $\mu := 0$, $\rho := 1$ and $\delta := 1/2$. This implies that estimate (9.15) holds true for the adjoint operator $\widetilde{\mathcal{T}}_p(\theta)^*$ of $\widetilde{\mathcal{T}}_p(\theta)$:

$$\left|\tilde{\psi}\right|_{-2+1/p,p'} \leq \widetilde{C}_\tau\left(\left|\widetilde{T}(\theta)^*\tilde{\psi}\right|_{-2+1/p,p'} + \left|\tilde{\psi}\right|_{\tau,p'}\right) \quad \text{for all } \tilde{\psi} \in \mathcal{D}\left(\widetilde{\mathcal{T}}_p(\theta)^*\right),$$

where $\tau < -2+1/p$ and $p' = p/(p-1)$ is the exponent conjugate to p. Hence we have, by the closed range theorem ([147, Chapter VII, Section 5, Theorem]) and Peetre's lemma (Lemma 9.3),

$$\text{codim}\,\mathcal{R}\left(\widetilde{\mathcal{T}}_p(\theta)\right) = \dim\mathcal{N}\left(\widetilde{\mathcal{T}}_p(\theta)^*\right) < \infty, \tag{9.17}$$

since the injection $B^{-2+1/p,p'}(\partial D \times S) \to B^{\tau,p'}(\partial D \times S)$ is compact for $\tau < -2+1/p$.

Therefore, the desired assertion (9.6) follows by combining assertions (9.16) and (9.17).

Step III-2: By assertion (9.6), we can apply Proposition 9.2 to obtain that the operator

$$\mathcal{T}_p(\ell^2 e^{i\theta})\colon B^{2-1/p,p}(\partial D) \longrightarrow B^{2-1/p,p}(\partial D)$$

is bijective if $\ell \in \mathbf{Z} \setminus K$ for some *finite subset* K of $\mathbf{Z}$. In particular, we have the assertion

$$\text{ind}\,\mathcal{T}_p(\lambda_0) = 0 \quad \text{for all } \lambda_0 = \ell^2 e^{i\theta} \text{ with } \ell \in \mathbf{Z} \setminus K. \tag{9.18}$$

However, in view of formula (7.3) it follows that, for any given λ, $\lambda_0 \in \Sigma_p(\varepsilon)$, we can find a classical pseudo-differential operator $K(\lambda,\lambda_0)$ of order -1 on the boundary ∂D such that

$$T(\lambda) = T(\lambda_0) + K(\lambda, \lambda_0).$$

Furthermore, it follows from an application of the Rellich–Kondrachov theorem (Theorem 4.10) that the operator

$$K(\lambda, \lambda_0): B^{2-1/p,p}(\partial D) \longrightarrow B^{2-1/p,p}(\partial D)$$

is *compact*. Hence we have the assertion

$$\operatorname{ind} \mathcal{T}_p(\lambda) = \operatorname{ind} \mathcal{T}_p(\lambda_0) \quad \text{for all } \lambda, \lambda_0 \in \Sigma_p(\varepsilon), \tag{9.19}$$

since the index is stable under *compact perturbations* (see [51, Theorem 2.6], [100, Theorem 5.10]).

Therefore, the desired assertion (9.5) (and hence assertion (9.3)) follows by combining assertions (9.18) and (9.19).

Step III-3: Summing up, we have proved that the operator $A_p - \lambda I$ is bijective for all $\lambda \in \Sigma_p(\varepsilon)$ and further that its inverse $(A_p - \lambda I)^{-1}$ satisfies estimate (1.9).

Theorem 9.1 is proved, apart from the proof of Proposition 9.2. The proof of Proposition 9.2 will be given in the next subsection, due to its length. □

9.1.1 Proof of Proposition 9.2

The proof of Proposition 9.2 is divided into three steps.

Step 1: First, we study the null spaces $\mathcal{N}\left(\widetilde{\mathcal{T}}_p(\theta)\right)$ and $\mathcal{N}(\mathcal{T}_p(\lambda'))$ when $\lambda' = \ell^2 e^{i\theta}$ with $\ell \in \mathbf{Z}$:

$$\mathcal{N}\left(\widetilde{\mathcal{T}}_p(\theta)\right) = \left\{ \widetilde{\varphi} \in B^{2-1/p,p}(\partial D \times S) : \widetilde{T}(\theta)\widetilde{\varphi} = 0 \right\},$$
$$\mathcal{N}(\mathcal{T}_p(\lambda')) = \left\{ \varphi \in B^{2-1/p,p}(\partial D) : T(\lambda')\varphi = 0 \right\}.$$

Since the pseudo-differential operators $\widetilde{T}(\theta)$ and $T(\lambda')$ are both *hypoelliptic*, it follows that

$$\mathcal{N}\left(\widetilde{\mathcal{T}}_p(\theta)\right) = \left\{ \widetilde{\varphi} \in C^\infty(\partial D \times S) : \widetilde{T}(\theta)\widetilde{\varphi} = 0 \right\}, \tag{9.20a}$$
$$\mathcal{N}(\mathcal{T}_p(\lambda')) = \left\{ \varphi \in C^\infty(\partial D) : T(\lambda')\varphi = 0 \right\}. \tag{9.20b}$$

Therefore, we can apply [114, Proposition 8.4.6] to obtain the following most important relationship between the null spaces $\mathcal{N}\left(\widetilde{\mathcal{T}}_p(\theta)\right)$ and $\mathcal{N}(\mathcal{T}_p(\lambda'))$ when $\lambda' = \ell^2 e^{i\theta}$ with $\ell \in \mathbf{Z}$:

Lemma 9.4. *The following two conditions (1) and (2) are equivalent:*

(1) $\dim \mathcal{N}\left(\widetilde{\mathcal{T}}_p(\theta)\right) < \infty.$

(2) There exists a finite *subset I of $\mathbf{Z}$ such that*

$$\begin{cases} \dim \mathcal{N}\left(\mathcal{T}_p(\ell^2 e^{i\theta})\right) < \infty & \text{if } \ell \in I, \\ \dim \mathcal{N}\left(\mathcal{T}_p(\ell^2 e^{i\theta})\right) = 0 & \text{if } \ell \notin I. \end{cases}$$

Moreover, in this case we have the formulas

$$\mathcal{N}\left(\widetilde{\mathcal{T}}_p(\theta)\right) = \bigoplus_{\ell \in I} \mathcal{N}\left(\mathcal{T}_p(\ell^2 e^{i\theta})\right) \otimes e^{i\ell y},$$

$$\dim \mathcal{N}\left(\widetilde{\mathcal{T}}_p(\theta)\right) = \sum_{\ell \in I} \dim \mathcal{N}\left(\mathcal{T}_p(\ell^2 e^{i\theta})\right).$$

Remark 9.5. By virtue of assertions (9.20), we find that $\dim \mathcal{N}(\widetilde{\mathcal{T}}_p(\theta))$ and $\dim \mathcal{N}(\mathcal{T}_p(\lambda'))$ are *independent* of p, for $1 < p < \infty$.

Step 2: Secondly, we study the ranges $\mathcal{R}\left(\widetilde{\mathcal{T}}_p(\theta)\right)$ and $\mathcal{R}\left(\mathcal{T}_p(\lambda')\right)$ when $\lambda' = \ell^2 e^{i\theta}$ with $\ell \in \mathbf{Z}$. To do this, we consider the adjoint operators $\widetilde{\mathcal{T}}_p(\theta)^*$ and $\mathcal{T}_p(\lambda')^*$ of $\widetilde{\mathcal{T}}_p(\theta)$ and $\mathcal{T}_p(\lambda')$, respectively.

The next lemma allows us to give a characterization of the adjoint operators $\widetilde{\mathcal{T}}_p(\theta)^*$ and $\mathcal{T}_p(\lambda')^*$ in terms of pseudo-differential operators (cf. [114, Lemma 8.4.8]):

Lemma 9.6. *Let M be a compact smooth manifold without boundary. If T is a classical pseudo-differential operator of order m on M, we define a densely defined, closed linear operator*

$$\mathcal{T}: B^{s,p}(M) \longrightarrow B^{s-m+1,p}(M) \quad \text{for } s \in \mathbf{R}$$

as follows.

(a) The domain $\mathcal{D}(\mathcal{T})$ of $\mathcal{T}$ is the space

$$\mathcal{D}(\mathcal{T}) = \left\{ \varphi \in B^{s,p}(M) : T\varphi \in B^{s-m+1,p}(M) \right\}.$$

(b) $\mathcal{T}\varphi = T\varphi$ for every $\varphi \in \mathcal{D}(\mathcal{T})$.

Then the adjoint operator $\mathcal{T}^*$ of $\mathcal{T}$ is characterized as follows:

(c) The domain $\mathcal{D}(\mathcal{T}^*)$ of $\mathcal{T}^*$ is contained in the space

$$\left\{ \psi \in B^{-s+m-1,p'}(M) : T^*\psi \in B^{-s,p'}(M) \right\},$$

where $p' = p/(p-1)$ and $T^* \in L^m_{\mathrm{cl}}(M)$ is the adjoint of T.
(d) $\mathcal{T}^*\psi = T^*\psi$ for every $\psi \in \mathcal{D}(\mathcal{T}^*)$.

Proof. Let ψ be an arbitrary element of $\mathcal{D}(\mathcal{T}^*) \subset B^{-s+m-1,p'}(M)$, and let $\{\psi_j\}$ be a sequence in $C^\infty(M)$ such that

$$\psi_j \longrightarrow \psi \quad \text{in } B^{-s+m-1,p'}(M).$$

Then we have, for all $\varphi \in C^\infty(M) \subset \mathcal{D}(\mathcal{T})$,

$$\begin{aligned}
(\mathcal{T}^*\psi, \varphi) = (\psi, \mathcal{T}\varphi) &= (\psi, T\varphi) \\
&= \lim_{j\to\infty} (\psi_j, T\varphi) = \lim_{j\to\infty} (T^*\psi_j, \varphi) \\
&= (T^*\psi, \varphi).
\end{aligned}$$

This proves that

$$T^*\psi = \mathcal{T}^*\psi \in B^{-s,p'}(M).$$

The proof of Lemma 9.6 is complete. $\quad\square$

We remark that the pseudo-differential operators $T(\lambda)^*$ and $\widetilde{T}(\theta)^*$ also satisfy conditions (4.22a)) and (4.22b) of Theorem 4.49 with $\mu := 0$, $\rho := 1$ and $\delta := 1/2$; hence they are *hypoelliptic*.

Therefore, by applying Lemma 9.6 to the operators $\widetilde{T}(\theta)$ and $T(\lambda')$ we obtain the following lemma:

Lemma 9.7. *The null spaces* $\mathcal{N}\left(\widetilde{\mathcal{T}}_p(\theta)^*\right)$ *and* $\mathcal{N}(\mathcal{T}_p(\lambda')^*)$ *are characterized respectively as follows:*

$$\mathcal{N}\left(\widetilde{\mathcal{T}}_p(\theta)^*\right) = \left\{ \tilde{\psi} \in C^\infty(\partial D \times S) : \widetilde{T}(\theta)^*\tilde{\psi} = 0 \right\}. \tag{9.21a}$$

$$\mathcal{N}(\mathcal{T}_p(\lambda')^*) = \left\{ \psi \in C^\infty(\partial D) : T(\lambda')^*\psi = 0 \right\}. \tag{9.21b}$$

By using Lemma 9.7, we find that Lemma 9.4 remains valid for the adjoint operators $\widetilde{\mathcal{T}}_p(\theta)^*$ and $\mathcal{T}_p(\lambda')^*$ (cf. [114, Lemma 8.4.10]):

Lemma 9.8. *The following two conditions (1) and (2) are equivalent:*

(1) $\dim \mathcal{N}\left(\widetilde{\mathcal{T}}_p(\theta)^*\right) < \infty$.
(2) There exists a 4finite subset J of $\mathbf{Z}$ such that

$$\begin{cases}
\dim \mathcal{N}\left(\mathcal{T}_p(\ell^2 e^{i\theta})^*\right) < \infty & \text{if } \ell \in J, \\
\dim \mathcal{N}\left(\mathcal{T}_p(\ell^2 e^{i\theta})^*\right) = 0 & \text{if } \ell \notin J.
\end{cases}$$

Moreover, in this case we have the formula

$$\dim \mathcal{N}\left(\widetilde{\mathcal{T}}_p(\theta)^*\right) = \sum_{\ell \in J} \dim \mathcal{N}\left(\mathcal{T}_p(\ell^2 e^{i\theta})^*\right).$$

Hence, by combining Lemma 9.8 and the closed range theorem ([147, Chapter VII, Section 5, Theorem]) we obtain the most important relationship between $\operatorname{codim} \mathcal{R}\left(\widetilde{\mathcal{T}}_p(\theta)\right)$ and $\operatorname{codim} \mathcal{R}\left(\mathcal{T}_p(\lambda')\right)$ when $\lambda' = \ell^2 e^{i\theta}$ with $\ell \in \mathbf{Z}$ (cf. [114, Proposition 8.4.11]):

Lemma 9.9. *The following two conditions (1) and (2) are equivalent:*

(1) $\operatorname{codim} \mathcal{R}\left(\widetilde{\mathcal{T}}_p(\theta)\right) < \infty$.

(2) There exists a finite subset J of $\mathbf{Z}$ such that

$$\begin{cases} \operatorname{codim} \mathcal{R}\left(\mathcal{T}_p(\ell^2 e^{i\theta})\right) < \infty & \text{if } \ell \in J, \\ \operatorname{codim} \mathcal{N}\left(\mathcal{T}_p(\ell^2 e^{i\theta})\right) = 0 & \text{if } \ell \notin J. \end{cases}$$

Moreover, in this case we have the formula

$$\operatorname{codim} \mathcal{R}\left(\widetilde{\mathcal{T}}_p(\theta)\right) = \sum_{\ell \in J} \operatorname{codim} \mathcal{R}\left(\mathcal{T}_p(\ell^2 e^{i\theta})^*\right).$$

Remark 9.10. By virtue of assertions (9.21), we find that $\operatorname{codim} \mathcal{R}(\widetilde{\mathcal{T}}_p(\theta))$ and $\operatorname{codim} \mathcal{R}(\mathcal{T}_p(\lambda'))$ are *independent* of p, for $1 < p < \infty$.

Step 3: Proposition 9.2 is an immediate consequence of Lemmas 9.4 and 9.9, with $K := I \cup J$.

The proof of Proposition 9.2 is now complete. □

Summing up, we have proved Theorem 9.1 and hence part (i) of Theorem 1.4. □

9.2 Proof of Theorem 1.4, Part (ii)

Part (ii) of Theorem 1.4 may be proved as follows. Theorem 9.1 asserts that, for sufficiently large $\mu_\varepsilon > 0$, the operator $A_p - \mu_\varepsilon I$ satisfies conditions (2.1) and (2.2) (see Figure 9.2 below).

Thus, by applying Theorem 2.2 (and Remark 2.3) to the operator $A_p - \mu_\varepsilon I$ we obtain part (ii) of Theorem 1.4:

Theorem 9.11. *Assume that conditions (A) and (B) are satisfied. Then the operator A_p generates a semigroup U_z on $L^p(D)$ which is analytic in the sector*

$$\Delta_\varepsilon = \{z = t + is : z \neq 0, \ |\arg z| < \pi/2 - \varepsilon\}$$

for any $0 < \varepsilon < \pi/2$, and enjoys the following three properties (a), (b) and (c):

(a) The operators $A_p U_z$ and $\frac{dU_z}{dz}$ are bounded operators on $L^p(D)$ for each $z \in \Delta_\varepsilon$, and satisfy the relation

$$\frac{dU_z}{dz} = A_p U_z \quad \text{for all } z \in \Delta_\varepsilon.$$

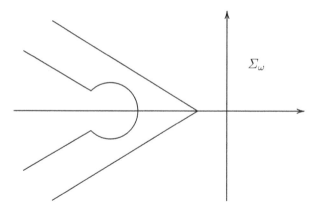

Fig. 9.2. The operator $A_p - \mu_\varepsilon$ satisfies conditions (2.1) and (2.2)

(b) For each $0 < \varepsilon < \pi/2$, there exist positive constants $\widetilde{M}_0(\varepsilon)$, $\widetilde{M}_1(\varepsilon)$ and μ_ε such that

$$\|U_z\| \leq \widetilde{M}_0(\varepsilon)e^{\mu_\varepsilon \cdot \mathrm{Re}\, z} \qquad \textit{for all } z \in \Delta_\varepsilon,$$

$$\|A_p U_z\| \leq \frac{\widetilde{M}_1(\varepsilon)}{|z|}e^{\mu_\varepsilon \cdot \mathrm{Re}\, z} \quad \textit{for all } z \in \Delta_\varepsilon.$$

(c) For each $u_0 \in L^p(D)$, we have, as $z \to 0, z \in \Delta_\varepsilon$,

$$U_z u_0 \longrightarrow u_0 \quad \textit{in } L^p(D).$$

The proof of Theorem 1.4 is now complete. $\square$

9.3 Notes and Comments

This chapter is a revised and expanded version of Chapter 7 of the second edition [121]. The material is based on the lecture entitled *Spectral Analysis of the Subelliptic Oblique Derivative Problem* delivered at Partial Differential Equations and Applications, Università di Bologna, Bologna, Italy, on May 24, 2017.

Section 9.1: For the detailed proof of Proposition 9.2, the reader might be referred to [113] and [114, Section 8.3].

Waldenfels Operators, Boundary Operators
and Maximum Principles

10

Elliptic Waldenfels Operators and Maximum Principles

Part IV (Chapters 10 and 11) is devoted to the general study of the maximum principles for second-order, elliptic Waldenfels operators in terms of pseudo-differential operators.

In this chapter, following Bony–Courrège–Priouret [15] we prove various maximum principles for second-order, elliptic Waldenfels operators which play an essential role throughout the book.

In Section 10.1 we give complete characterizations of linear operators $W = A + S$ which satisfy the positive maximum principle (PM) closely related to condition (β') given in the Hille–Yosida–Ray theorem (Theorem 3.36):

$$x_0 \in \overset{\circ}{D}, \; v \in C_0^2(D) \text{ and } v(x_0) = \sup_{x \in D} v(x) \geq 0 \Longrightarrow Wv(x_0) \leq 0 \qquad \text{(PM)}$$

(Theorems 10.1, 10.2 and 10.4).

In Section 10.2 we prove the weak and strong maximum principles and Hopf's boundary point lemma for second-order, elliptic Waldenfels operators $W = A + S$ (Theorems 10.5, 10.7 and Lemma 10.11) that play an important role in Part V.

10.1 Borel Kernels and Maximum Principles

Let D be an open subset of Euclidean space $\mathbf{R}^N$ or of the half-space $\overline{\mathbf{R}_+^N}$. More precisely, if D is an open set in $\overline{\mathbf{R}_+^N}$ in the topology induced on $\overline{\mathbf{R}_+^N}$ from $\mathbf{R}^N$, then we define the *boundary* ∂D of D to be the intersection of D with $\mathbf{R}^{N-1} \times \{0\}$ and the *interior* $\overset{\circ}{D}$ of D to be the complement of ∂D in D, that is,

$$\partial D = D \cap \{x \in \mathbf{R}^N : x_N = 0\},$$

$$\overset{\circ}{D} = D \cap \{x \in \mathbf{R}^N : x_N > 0\}.$$

© Springer Nature Switzerland AG 2020
K. Taira, *Boundary Value Problems and Markov Processes*, Lecture Notes in Mathematics 1499,
https://doi.org/10.1007/978-3-030-48788-1_10

We remark (see Figures 10.1 and 10.2 below) that

$$D = \begin{cases} \overset{\circ}{D} & \text{if } D \text{ is open in } \mathbf{R}^N, \\ \overset{\circ}{D} \cup \partial D & \text{if } D \text{ is open in } \overline{\mathbf{R}^N_+}. \end{cases}$$

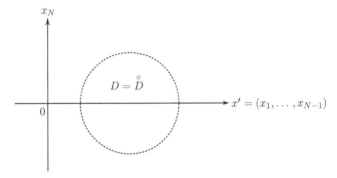

Fig. 10.1. D is open in $\mathbf{R}^N$

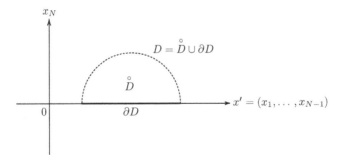

Fig. 10.2. D is open in $\overline{\mathbf{R}^N_+}$

Then we let

$$B_{\text{loc}}(\overset{\circ}{D}) = \text{the space of Borel measurable functions in } \overset{\circ}{D}$$
$$\text{which are bounded on compact subsets of } \overset{\circ}{D},$$

and

$$C_0(D) = \text{the space of continuous functions in } D$$
$$\text{with compact support in } D.$$

Let $\mathcal{B}_D$ and $\mathcal{B}_{\overset{\circ}{D}}$ be the σ-algebra of all Borel sets in D and the σ-algebra of all Borel sets in $\overset{\circ}{D}$, respectively.

A *positive Borel kernel* of $\overset{\circ}{D}$ into D is a mapping

$$\overset{\circ}{D} \ni x \longmapsto s(x, dy)$$

of $\overset{\circ}{D}$ into the space of non-negative measures on $\mathcal{B}_D$ such that, for each $X \in \mathcal{B}_D$ the function

$$\overset{\circ}{D} \ni x \longmapsto s(x, X) = \int_X s(x, dy) \tag{10.1}$$

is Borel measurable in $\overset{\circ}{D}$.

A *local unity function* on D is a smooth function $\sigma(x, y)$ in $D \times D$ which satisfies the following three conditions (LU1), (LU2) and (LU3) (see Figure 10.3 below):

(LU1) $0 \le \sigma(x, y) \le 1$ in $D \times D$.
(LU2) $\sigma(x, y) = 1$ in a neighborhood of the *diagonal*

$$\Delta_D = \{(x, x) : x \in D\}$$

in the product space $D \times D$.
(LU3) For any compact subset K of D, there exists a compact subset K' of D such that the functions $\sigma_x(\cdot) = \sigma(x, \cdot)$ for all $x \in K$ have their support in K'.

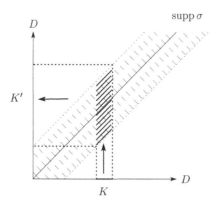

Fig. 10.3. The condition (LU3)

We can construct a local unity function $\sigma(x, y)$ in the following way (see [41, Proposition (8.15)]): Let $\{U_i\}_{i \in I}$ be a locally finite open covering of D and let $\{\varphi_i\}_{i \in I}$ be a partition of unity subordinate to the covering $\{U_i\}$. Namely, the family $\{\varphi_i\}_{i \in I}$ satisfies the following three conditions (PU1), (PU2) and (PU3):

(PU1) $0 \leq \varphi_i(x) \leq 1$ for all $x \in D$ and $i \in I$.

(PU2) $\operatorname{supp} \varphi_i \subset U_i$ for each $i \in I$.

(PU3) The collection $\{\operatorname{supp} \varphi_i\}_{i \in I}$ is locally finite and

$$\sum_{i \in I} \varphi_i(x) = 1 \quad \text{for every } x \in D.$$

Here $\operatorname{supp} \varphi_i$ is the support of φ_i, that is, the closure in D of the set $\{x \in D : \varphi_i(x) \neq 0\}$.

If we take a smooth function $\psi_i(x)$ in D such that

$$\begin{cases} 0 \leq \psi_i(x) \leq 1 & \text{for all } x \in D, \\ \psi_i(x) = 1 & \text{on } \operatorname{supp} \varphi_i, \end{cases}$$

then it is easy to verify that the function

$$\sigma(x, y) = \sum_{i \in I} \varphi_i(x) \, \psi_i(y) \quad \text{for } (x, y) \in D \times D,$$

satisfies the desired conditions (LU1), (LU2) and (LU3).

10.1.1 Linear Operators having Positive Borel Kernel

Now we assume that a positive Borel kernel $s(x, dy)$ satisfies the following two conditions (NS.1) and (NS.2):

(NS.1) $s(x, \{x\}) = 0$ for every $x \in \overset{\circ}{D}$.

(NS.2) For every non-negative function $f(x)$ in $C_0(D)$, the function

$$\overset{\circ}{D} \ni x \longmapsto \int_D s(x, dy)|y - x|^2 f(y)$$

belongs to the space $B_{\mathrm{loc}}(\overset{\circ}{D})$ of locally bounded, Borel measurable functions on $\overset{\circ}{D}$.

By using Taylor's formula and condition (NS.2), we can define a linear operator

$$\boxed{S \colon C_0^2(D) \longrightarrow B_{\mathrm{loc}}(\overset{\circ}{D})}$$

by the formula (see Example 10.1)

$$Su(x) \tag{10.2}$$

$$= \int_D s(x, dy) \left[u(y) - \sigma(x, y) \left(u(x) + \sum_{i=1}^{N} \frac{\partial u}{\partial x_i}(x)(y_i - x_i) \right) \right]$$

for every $x \in \overset{\circ}{D}$ $(u \in C_0^2(D))$.

Here $C_0^2(D)$ is the space of functions in $C^2(D)$ with compact support in D.

(I) First, we give a complete characterization of linear continuous operators

$$W \colon C_0^2(D) \longrightarrow B_{\mathrm{loc}}(\overset{\circ}{D})$$

that have positive Borel kernels in the case where D is an open subset of $\mathbf{R}^N$ or of $\overline{\mathbf{R}_+^N}$ (see [15, Théorème I], [117, Theorem 8.2]):

Theorem 10.1. *Let D be an open subset of $\mathbf{R}^N$ or of $\overline{\mathbf{R}_+^N}$. If W is a linear operator from $C_0^2(D)$ into $B_{\mathrm{loc}}(\overset{\circ}{D})$, then the following two assertions (p_0) and (w) are equivalent:*

(p_0) *The operator*

$$\boxed{W \colon C_0^2(D) \longrightarrow B_{\mathrm{loc}}(\overset{\circ}{D})}$$

is continuous and satisfies the condition

$$x \in \overset{\circ}{D},\ u \in C_0^2(D),\ u \geq 0 \ \text{in}\ D \ \text{and}\ x \notin \mathrm{supp}\, u \qquad (10.3)$$
$$\Longrightarrow W u(x) \geq 0.$$

(w) *There exist a second-order,* differential operator

$$\boxed{A \colon C^2(D) \longrightarrow B_{\mathrm{loc}}(\overset{\circ}{D})}$$

and positive Borel kernels $s(x, dy)$, *having properties (NS.1) and (NS.2), such that the operator W is written in the form*

$$W u(x) = A u(x) + S u(x) \qquad (10.4)$$

$$= \left(\sum_{i,j=1}^N a^{ij}(x) \frac{\partial^2 u}{\partial x_i \partial x_j}(x) + \sum_{i=1}^N b^i(x) \frac{\partial u}{\partial x_i}(x) + c(x) u(x) \right)$$

$$+ \int_D s(x, dy) \left[u(y) - \sigma(x,y) \left(u(x) + \sum_{i=1}^N \frac{\partial u}{\partial x_i}(x)(y_i - x_i) \right) \right],$$

for every $x \in \overset{\circ}{D}$ $(u \in C_0^2(D))$.

Here the coefficients $a^{ij}(x)$, $b^i(x)$ and $c(x)$ belong to the space $B_{\mathrm{loc}}(\overset{\circ}{D})$.

(II) Secondly, we give a useful characterization of linear continuous operators

$$W \colon C_0^2(D) \longrightarrow B_{\mathrm{loc}}(\overset{\circ}{D})$$

that have positive Borel kernels in the case where D is an open subset of $\mathbf{R}^N$ or of $\overline{\mathbf{R}_+^N}$ (see [15, Théorème II], [117, Theorem 8.4]):

Theorem 10.2. *Let D be an open subset of $\mathbf{R}^N$ or of $\overline{\mathbf{R}^N_+}$. Let $\mathcal{V}$ be a linear subspace of $C_0^2(D)$ that contains $C_0^\infty(D)$:*

$$C_0^\infty(D) \subset \mathcal{V} \subset C_0^2(D).$$

Assume that W is a linear operator from $\mathcal{V}$ into $B_{\mathrm{loc}}(\overset{\circ}{D})$ and satisfies the condition

$$x \in \overset{\circ}{D}, \ u \in \mathcal{V}, \ u \geq 0 \ \text{in } D \ \text{and} \ u(x) = 0 \Longrightarrow Wu(x) \geq 0. \tag{10.5}$$

Then the operator W can be extended uniquely to a linear operator from $C_0^2(D)$ into $B_{\mathrm{loc}}(\overset{\circ}{D})$ in such a way that:

(p_1) The extended operator

$$W : C_0^2(D) \longrightarrow B_{\mathrm{loc}}(\overset{\circ}{D})$$

is continuous *and satisfies condition* (10.5) *for all $u \in C_0^2(D)$:*

$$x \in \overset{\circ}{D}, \ u \in C_0^2(D), \ u \geq 0 \ \text{in } D \ \text{and} \ u(x) = 0 \Longrightarrow Wu(x) \geq 0. \tag{10.5$'$}$$

By combining Theorems 10.1 and 10.2, we have the following simple characterization of W (see [117, Corollary 8.7]):

Corollary 10.3. *Let D be an open subset of $\mathbf{R}^N$ or of $\overline{\mathbf{R}^N_+}$. If a linear operator*

$$W : C_0^2(D) \longrightarrow B_{\mathrm{loc}}(\overset{\circ}{D})$$

satisfies condition (p_1), then it is continuous and can be written in the form (10.4)

$$\boxed{W = A + S,}$$

where the coefficients $a^{ij}(x)$, $b^i(x)$ and $c(x)$ of A belong to the space $B_{\mathrm{loc}}(\overset{\circ}{D})$ and the positive Borel kernels $s(x, dy)$ of S enjoy properties (NS.1) and (NS.2).

Proof. If $W : C_0^2(D) \to B_{\mathrm{loc}}(\overset{\circ}{D})$ is a linear operator, then it follows from an application of Theorem 4.2 that condition (p_1) implies the continuity of W and condition (p_0). Therefore, we obtain from Theorem 4.1 that condition (w) is satisfied.

The proof of Corollary 10.3 is complete.   □

We remark that if the integro-differential operator

$$W = A + S$$

enjoys property (w), then the Borel kernel

$$s(x, dy)$$

and the principal part

$$\left(a^{ij}(x)\right)_{1 \leq i,j \leq n}$$

of A are uniquely determined by W. Indeed, it suffices to note the following two formulas (10.6) and (10.7):

$$\int_D s(x, dy)\, u(y) = Wu(x) \text{ if } u \in C_0^2(D) \text{ and } x \in \overset{\circ}{D} \setminus \operatorname{supp} u. \qquad (10.6)$$

$$2\, u(x) \sum_{i,j=1}^N a^{ij}(x)\xi_i\xi_j = W\left(\Phi_x^\xi u\right)(x) - \int_D s(x, dy)\, \Phi_x^\xi(y)u(y) \qquad (10.7)$$

$$\text{if } x \in \overset{\circ}{D},\ u \in C_0^2(D) \text{ and } \xi = (\xi_i) \in \mathbf{R}^N,$$

where

$$\Phi_x^\xi(y) = \left(\sum_{i=1}^N \xi_i\,(y_i - x_i)\right)^2 \qquad \text{for } x,\ y \in D \text{ and } \xi = (\xi_i) \in \mathbf{R}^N.$$

(III) Finally, we characterize a class of linear operators

$$\boxed{W = A + S \colon C_0^2(D) \longrightarrow B_{\mathrm{loc}}(\overset{\circ}{D})}$$

that satisfy the positive maximum principle (PM) (see [15, Théorème III], [117, Theorem 8.8]):

Theorem 10.4. *Let D be an open subset of $\mathbf{R}^N$ or of $\overline{\mathbf{R}_+^N}$. If W is a linear operator from $C_0^2(D)$ into $B_{\mathrm{loc}}(\overset{\circ}{D})$ of the form (10.4), then we have the following two assertions (i) and (ii):*

(i) The operator W satisfies condition (10.5) if and only if the principal symbol $-\sum_{i,j=1}^N a^{ij}(x)\xi_i\xi_j$ of A is non-positive on $\overset{\circ}{D} \times \mathbf{R}^N$.
(ii) The operator W satisfies the positive maximum principle *(PM)*

$$x_0 \in \overset{\circ}{D},\ v \in C_0^2(D) \text{ and } v(x_0) = \sup_{x \in D} v(x) \geq 0 \Longrightarrow Wv(x_0) \leq 0 \quad (PM)$$

if and only if the principal symbol of A is non-positive on $\overset{\circ}{D} \times \mathbf{R}^N$ and the following two conditions (10.8a) and (10.8b) hold true for all $x \in \overset{\circ}{D}$:

$$(A1)(x) = c(x) \leq 0, \qquad\qquad\qquad (10.8a)$$

$$(W1)(x) = c(x) + \int_D s(x, dy)[1 - \sigma(x, y)] \leq 0. \qquad (10.8b)$$

In particular, the positive Borel kernels $s(x, dy)$ enjoy the following property (NS.3):

(NS.3) For any open subset D' of $\overset{\circ}{D}$, the function

$$D' \ni x \longmapsto s(x, D \setminus D') = \int_{D \setminus D'} s(x, dy)$$

belongs to the space $B_{\mathrm{loc}}(D')$.

10.1.2 Borel Kernels and Pseudo-Differential Operators

Let D be a bounded domain of Euclidean space $\mathbf{R}^N$ with smooth boundary ∂D.

In this subsection we give two important examples of positive Borel kernels in terms of pseudo-differential operators:

Example 10.1. Let $s(x, y)$ be the distribution kernel of a properly supported, pseudo-differential operator $S \in L_{1,0}^{2-\kappa}(\mathbf{R}^N)$ with $\kappa > 0$, and $s(x, y) \geq 0$ off the diagonal $\Delta_{\mathbf{R}^N} = \{(x, x) : x \in \mathbf{R}^N\}$ in the product space $\mathbf{R}^N \times \mathbf{R}^N$. Then the integro-differential operator S_r, defined by the formula (see formula (10.2))

$$S_r u(x)$$
$$= \int_D s(x, y) \left[u(y) - \sigma(x, y) \left(u(x) + \sum_{j=1}^{N} (y_j - x_j) \frac{\partial u}{\partial x_j}(x) \right) \right] dy,$$

is absolutely convergent for every $x \in D$.

Proof. Indeed, we can write the integral $S_r u(x)$ in the form

$$S_r u(x)$$
$$= \int_D s(x, y) \left[1 - \sigma(x, y) \right] u(y) \, dy$$
$$+ \int_D s(x, y) \sigma(x, y) \left(u(y) - u(x) - \sum_{j=1}^{N} (y_j - x_j) \frac{\partial u}{\partial x_j}(x) \right) dy.$$

Then, by using Taylor's formula

$$u(y) - u(x) - \sum_{j=1}^{N} (y_j - x_j) \frac{\partial u}{\partial x_j}(x)$$
$$= \sum_{i,j=1}^{N} (y_i - x_i)(y_j - x_j) \left(\int_0^1 (1 - t) \frac{\partial^2 u}{\partial x_i \partial x_j}(x + t(y - x)) dt \right),$$

we can find a constant $C_1 > 0$ such that

$$\left| u(y) - u(x) - \sum_{j=1}^{N} (y_j - x_j) \frac{\partial u}{\partial x_j}(x) \right| \leq C_1 |x - y|^2 \quad \text{for all } x, y \in \overline{D}.$$

On the other hand, by using Theorem 4.51 with $n := N$ we can prove that the distribution kernel $s(x, y)$ of a pseudo-differential operator $S \in L_{1,0}^m(\mathbf{R}^N)$ satisfies the estimate

$$|s(x, y)| \leq \frac{C}{|x - y|^{m+N}} \quad \text{for all } x, y \in \mathbf{R}^N \text{ and } x \neq y.$$

Therefore, by taking the compact set $\overline{D} \subset \mathbf{R}^N$ and $m := 2 - \kappa$ we can find a constant $C_2 > 0$ such that the distribution kernel $s(x, y)$ of $S \in L_{1,0}^{2-\kappa}(\mathbf{R}^N)$ satisfies the estimate

$$0 \leq s(x, y) \leq \frac{C_2}{|x - y|^{N+2-\kappa}} \quad \text{for all } x, y \in \overline{D} \text{ and } x \neq y.$$

Hence we have the estimate

$$\left| \int_D s(x, y)\sigma(x, y) \left(u(y) - u(x) - \sum_{j=1}^{N} (y_j - x_j) \frac{\partial u}{\partial x_j}(x) \right) dy \right|$$

$$\leq C_1 C_2 \|u\|_{C^2(\overline{D})} \int_D \frac{1}{|x - y|^{N+2-\kappa}} \cdot |x - y|^2 \, dy$$

$$= C_1 C_2 \|u\|_{C^2(\overline{D})} \int_D \frac{1}{|x - y|^{N-\kappa}} \, dy.$$

Similarly, we have, with some constant $C_3 > 0$,

$$\left| \int_D s(x, y) [1 - \sigma(x, y)] u(y) \, dy \right| \leq C_3 \|u\|_{C(\overline{D})} \int_D \frac{1}{|x - y|^{N-\kappa}} \, dy,$$

since we have the formula

$$\sigma(x, y) - 1$$

$$= \sigma(x, y) - \sigma(x, x) - \sum_{j=1}^{N} (y_j - x_j) \frac{\partial \sigma}{\partial x_j}(x, x)$$

$$= \sum_{i,j=1}^{N} (y_i - x_i)(y_j - x_j) \left(\int_0^1 (1 - t) \frac{\partial^2 \sigma}{\partial x_i \partial x_j}(x, x + t(y - x)) dt \right).$$

Therefore, we obtain that the integral $S_r u(x)$ is absolutely convergent.

The proof of Example 10.1 is complete. $\square$

Example 10.2. Let $r(x', y')$ be the distribution kernel of a pseudo-differential operator $R \in L_{1,0}^{2-\kappa_1}(\partial D)$ with $\kappa_1 > 0$, and $r(x', y') \geq 0$ off the diagonal $\Delta_{\partial D} = \{(x', x') : x' \in \partial D\}$ in the product space $\partial D \times \partial D$. Let $t(x, y)$ be the distribution kernel of a properly supported, pseudo-differential operator $T \in L_{1,0}^{1-\kappa_2}(\mathbf{R}^N)$ with $\kappa_2 > 0$, and $t(x, y) \geq 0$ off the diagonal $\Delta_{\mathbf{R}^N} = \{(x, x) : x \in \mathbf{R}^N\}$ in the product space $\mathbf{R}^N \times \mathbf{R}^N$. Then the integro-differential operator Γ_r, defined by the formula

$$
\Gamma_r u(x')
$$
$$
= \int_{\partial D} r(x', y') \left[u(y') - \tau(x', y') \left(u(x') + \sum_{j=1}^{N-1} (y_j - x_j) \frac{\partial u}{\partial x_j}(x') \right) \right] dy'
$$
$$
+ \int_D t(x', y) \left[u(y) - u(x') \right] dy,
$$

is absolutely convergent for every $x' \in \partial D$.

Proof. Since $R \in L_{1,0}^{2-\kappa_1}(\partial D)$ and $T \in L_{1,0}^{1-\kappa_2}(\mathbf{R}^N)$, it follows from an application of [122, Section 7.8, Theorem 7.36] with $n := N$ that the kernels $r(x', y')$ and $t(x', y)$ satisfy respectively the estimates

$$
0 \leq r(x', y') \leq \frac{C'}{|x' - y'|^{(N-1)+2-\kappa_1}} \quad \text{for all } x', \, y' \in \partial D \text{ and } x' \neq y',
$$
$$
0 \leq t(x', y) \leq \frac{C''}{|x' - y|^{(N-1)+2-\kappa_2}} \quad \text{for all } x' \in \partial D \text{ and } y \in D,
$$

where $|x' - y'|$ denotes the geodesic distance between x' and y' with respect to the Riemannian metric of ∂D.

Therefore, by arguing just as in Example 10.1 we find that the integrals

$$
R_r u(x')
$$
$$
= \int_{\partial D} r(x', y') \left[u(y') - \tau(x', y') \left(u(x') + \sum_{j=1}^{N-1} (y_j - x_j) \frac{\partial u}{\partial x_j}(x') \right) \right] dy'
$$

and

$$
T_r u(x') = \int_D t(x', y) \left[u(y) - u(x') \right] dy
$$

are both absolutely convergent for every $x' \in \partial D$.

The proof of Example 10.2 is complete. $\square$

10.2 Maximum Principles for Elliptic Waldenfels Operators

In this section we prove the weak and strong maximum principles and Hopf's boundary point lemma for second-order, elliptic Waldenfels integro-differential operators which play an essential role in Chapters 12 and 13.

Let D be a bounded domain of Euclidean space $\mathbf{R}^N$, with smooth boundary ∂D. We consider a second-order, elliptic *Waldenfels operator* W with real coefficients such that

$$Wu(x) = Au(x) + Su(x) \tag{1.2}$$

$$:= \sum_{i,j=1}^{N} a^{ij}(x) \frac{\partial^2 u}{\partial x_i \partial x_j}(x) + \sum_{i=1}^{N} b^i(x) \frac{\partial u}{\partial x_i}(x) + c(x)u(x)$$

$$+ \int_{\mathbf{R}^N \setminus \{0\}} \left(u(x+z) - u(x) - \sum_{j=1}^{N} z_j \frac{\partial u}{\partial x_j}(x) \right) s(x,z) \, m(dz).$$

Here:

(1) $a^{ij}(x) \in C^\infty(\overline{D})$, $a^{ij}(x) = a^{ji}(x)$ for all $x \in \overline{D}$ and $1 \leq i, j \leq N$, and there exists a constant $a_0 > 0$ such that

$$\sum_{i,j=1}^{N} a^{ij}(x)\xi_i\xi_j \geq a_0|\xi|^2 \quad \text{for all } (x,\xi) \in \overline{D} \times \mathbf{R}^N.$$

(2) $b^i(x) \in C^\infty(\overline{D})$ for all $1 \leq i \leq N$.
(3) $c(x) \in C^\infty(\overline{D})$, and $c(x) \leq 0$ in D, but $c(x) \not\equiv 0$ in D.
(4) $s(x,z) \in L^\infty(\mathbf{R}^N \times \mathbf{R}^N)$ and $0 \leq s(x,z) \leq 1$ almost everywhere in $\mathbf{R}^N \times \mathbf{R}^N$, and there exist constants $C_0 > 0$ and $0 < \theta < 1$ such that

$$|s(x,z) - s(y,z)| \leq C_0|x-y|^\theta \tag{1.3a}$$

$$\text{for all } x, y \in \overline{D} \text{ and almost all } z \in \mathbf{R}^N,$$

and

$$s(x,z) = 0 \quad \text{if } x \in D \text{ and } x + z \notin \overline{D}. \tag{1.3b}$$

Probabilistically, condition (1.3b) implies that all jumps from D are within $\overline{D}$. Analytically, condition (1.3b) guarantees that the integral operator S may be considered as an operator acting on functions u defined on the closure $\overline{D}$ (see Garroni–Menaldi [48, Chapter II, Remark 1.19]).
(5) The measure $m(dz)$ is a Radon measure on $\mathbf{R}^N \setminus \{0\}$ which has a density with respect to the Lebesgue measure dz on $\mathbf{R}^N$, and satisfies the *moment condition* (see Example 1.1)

$$\int_{\{0<|z|\leq 1\}} |z|^2 \, m(dz) + \int_{\{|z|>1\}} |z| \, m(dz) < \infty. \tag{1.4}$$

The moment condition implies that the measure $m(\cdot)$ admits a singularity of order 2 at the origin, and this singularity at the origin is produced by the accumulation of *small jumps* of Markovian particles, while the measure $m(\cdot)$ admits a singularity of order 1 at infinity, and this singularity at infinity is produced by the accumulation of *large jumps* of Markovian particles.

Finally, it should be noticed that

$$(W1)(x) = (A1)(x) + (S1)(x) = c(x) \leq 0 \quad \text{in } D. \tag{1.5}$$

10.2.1 Weak Maximum Principle

First, we prove the *weak maximum principle* for elliptic, Waldenfels integro-differential operators (see [136, p. 191, Lemma 3.25], [122, Theorem 8.11], [127, Theorem 8.1]):

Theorem 10.5 (the weak maximum principle). *Let W be a second-order, elliptic Waldenfels operator. Then we have the following two assertions (i) and (ii):*

(i) If a function $u(x) \in C(\overline{D}) \cap C^2(D)$ satisfies the conditions

$$\begin{cases} Wu(x) \geq 0 & \text{in } D, \\ W1(x) < 0 & \text{in } D, \end{cases}$$

then the function $u(x)$ may take its positive *maximum only on the boundary ∂D.*

(ii) If a function $u(x) \in C(\overline{D}) \cap C^2(D)$ satisfies the conditions

$$\begin{cases} Wu(x) > 0 & \text{in } D, \\ W1(x) \leq 0 & \text{in } D, \end{cases}$$

then the function $u(x)$ may take its non-negative *maximum only on the boundary ∂D.*

Proof. Our proof is based on a reduction to absurdity. Assume, to the contrary, that there exists a point x_0 of D such that

$$u(x_0) = \max_{x \in \overline{D}} u(x).$$

Then we have the assertions

- $\dfrac{\partial u}{\partial x_i}(x_0) = 0 \quad \text{for } 1 \leq i \leq N,$

- $\displaystyle\sum_{i,j=1}^{N} a^{ij}(x_0) \dfrac{\partial^2 u}{\partial x_i \partial x_j}(x_0) \leq 0,$

and hence the inequalities

- $$Au(x_0) = \sum_{i,j=1}^{N} a^{ij}(x_0)\frac{\partial^2 u}{\partial x_i \partial x_j}(x_0) + c(x_0)u(x_0) \leq c(x_0)u(x_0), \quad (10.9)$$

- $$Su(x_0) = \int_{\mathbf{R}^N\setminus\{0\}} (u(x_0 + z) - u(x_0))\, s(x_0, z)\, m(dz) \leq 0. \quad (10.10)$$

Assertion (i): If $Wu(x) \geq 0$ in D, $c(x) = W1(x) < 0$ in D and if $u(x_0) = \max_{\overline{D}} u > 0$, then it follows from inequalities (10.9) and (10.10) that

$$0 \leq Wu(x_0) = Au(x_0) + Su(x_0) \leq \max_{x\in\overline{D}} u(x) \cdot W1(x_0) < 0.$$

This is a contradiction.

Assertion (ii): Similarly, if $Wu(x) > 0$, $c(x) = W1(x) \leq 0$ in D and if $u(x_0) = \max_{\overline{D}} u \geq 0$, then it follows from inequalities (10.9) and (10.10) that

$$0 < Wu(x_0) = Au(x_0) + Su(x_0) \leq \max_{x\in\overline{D}} u(x) \cdot W1(x_0) \leq 0.$$

This is also a contradiction.

The proof of Theorem 10.5 is complete. □

As an application of the weak maximum principle, we can obtain a point-wise estimate for solutions of the non-homogeneous equation $Wu = f$:

Theorem 10.6. *Let W be a second-order, elliptic Waldenfels operator. Assume that*

$$W1(x) < 0 \quad \text{on } \overline{D} = D \cup \partial D.$$

Then we have, for all $u \in C(\overline{D}) \cap C^2(D)$,

$$\max_{\overline{D}} |u| \leq \max\left\{\left(\frac{1}{\min_{\overline{D}}(-W1)}\right) \sup_{D} |Wu|,\ \max_{\partial D} |u|\right\}. \quad (10.11)$$

Proof. We let

$$M = \max\left\{\left(\frac{1}{\min_{\overline{D}}(-W1)}\right) \sup_{D} |Wu|,\ \max_{\partial D} |u|\right\},$$

and consider two functions

$$v_{\pm}(x) = M \pm u(x).$$

Then it follows that

$$Wv_{\pm}(x) = M \cdot W1(x) \pm Wu(x) \leq 0 \quad \text{in } D.$$

Hence, by applying part (i) of Theorem 10.5 to the functions $-v_{\pm}(x)$ we obtain that the functions $v_{\pm}(x)$ may take their negative minimums only on the boundary ∂D. However, it follows that

$$v_\pm(x) = M \pm u(x) \geq 0 \quad \text{on the boundary } \partial D.$$

Hence we have the inequality

$$v_\pm(x) \geq 0 \quad \text{on } \overline{D}.$$

This proves the desired estimate (10.11).

The proof of Theorem 10.6 is complete.   □

10.2.2 Strong Maximum Principle

The next theorem is a generalization of the *strong maximum principle* for the Laplacian to the integro-differential operator case (see [15, Théorème VII], [136, p. 193, Theorem 3.27], [122, Theorem 8.13], [128, Theorem 6.2]):

Theorem 10.7 (the strong maximum principle). *Let W be a second order, elliptic Waldenfels operator. Assume that a function $u(x) \in C^2(\overline{D})$ satisfies the conditions*

$$\begin{cases} Wu(x) \geq 0 & \text{in } D, \\ \max_{x \in \overline{D}} u(x) \geq 0. \end{cases}$$

If the function $u(x)$ takes its non-negative maximum at an interior point of D, then it is a constant.

Proof. The proof is divided into four steps.

Step (I): Our proof is based on a reduction to absurdity. First, we let

$$M = \max_{x \in \overline{D}} u(x) \geq 0,$$

$$F = \{x \in D : u(x) = M\},$$

and assume, to the contrary, that

$$F \subsetneq D.$$

Since F is closed in D, we can find a point x_0 of F and an open ball V contained in the set $D \setminus F$, centered at x_1, such that (see Figure 10.4 below)

(a) $V \subset D \setminus F$;
(b) x_0 is on the boundary ∂V of V.

Step (II): Secondly, we study the integral operator S in the framework of Hölder spaces. To do this, we need the following elementary estimates (10.12a), (10.12b) and (10.12c) for the measure $m(dz)$:

Claim 10.8. *For each $\varepsilon > 0$, we let*

$$\sigma(\varepsilon) = \int_{\{0 < |z| \leq \varepsilon\}} |z|^2 \, m(dz),$$

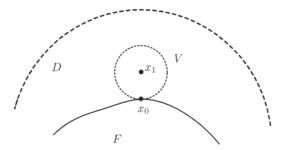

Fig. 10.4. The open ball V contained in $D \setminus F$

$$\delta(\varepsilon) = \int_{\{|z|>\varepsilon\}} |z| \, m(dz),$$

$$\tau(\varepsilon) = \int_{\{|z|>\varepsilon\}} m(dz).$$

Then we have, as $\varepsilon \downarrow 0$,

$$\sigma(\varepsilon) \longrightarrow 0, \qquad\qquad (10.12a)$$

$$\delta(\varepsilon) \leq \frac{C_1}{\varepsilon} + C_2, \qquad\qquad (10.12b)$$

$$\tau(\varepsilon) \leq \frac{C_1}{\varepsilon^2} + C_2, \qquad\qquad (10.12c)$$

where

$$C_1 = \int_{\{0<|z|\leq 1\}} |z|^2 \, m(dz),$$

$$C_2 = \int_{\{|z|>1\}} |z| \, m(dz).$$

Proof. Assertion (10.12a) follows immediately from condition (1.4). The term $\delta(\varepsilon)$ can be estimated as follows:

$$\begin{aligned}
\delta(\varepsilon) &= \int_{\{|z|>1\}} |z| \, m(dz) + \int_{\{\varepsilon<|z|\leq 1\}} |z| \, m(dz) \\
&\leq \int_{\{|z|>1\}} |z| \, m(dz) + \frac{1}{\varepsilon} \int_{\{\varepsilon<|z|\leq 1\}} |z|^2 \, m(dz) \\
&\leq \int_{\{|z|>1\}} |z| \, m(dz) + \frac{1}{\varepsilon} \int_{\{0<|z|\leq 1\}} |z|^2 \, m(dz).
\end{aligned}$$

The term $\tau(\varepsilon)$ is estimated in a similar way.
The proof of Claim 10.8 is complete. $\square$

Step (III): The next claim on the existence of "barriers" is an essential step in the proof of Theorem 10.7:

Claim 10.9. *There exists a function $v(x) \in C^\infty(\overline{D})$ that satisfies the following four properties (i) through (iv):*

(i) $v(x) > 0$ in V.
(ii) $v(x) < 0$ on $\overline{D} \setminus \overline{V}$.
(iii) $v(x) = 0$ on ∂V.
(iv) $Wv(x_0) > 0$.

Remark 10.10. In Lemma 13.3, we shall prove the following assertion:

$$v \in C^{2+\theta_0}(\overline{D}) \implies Sv \in C^{\theta_0}(\overline{D})$$
$$\implies Wv = Av + Sv \in C^{\theta_0}(\overline{D}).$$

Proof. In order to prove Claim 10.9, we define a function $v(x)$ by the formula

$$v(x) = \exp\left[-q|x - x_1|^2\right] - \exp\left[-q\rho^2\right], \quad \rho = |x_0 - x_1|,$$

where q is a positive constant to be chosen later on. Then it is easy to see that the function $v(x)$ satisfies conditions (i) through (iii). Hence it suffices to show that $v(x)$ satisfies condition (iv) for q sufficiently large.

(1) First, we estimate the function $Av(x_0)$ from below. To do this, it should be noticed that

$$v(x_0) = 0, \tag{10.13a}$$
$$\nabla v(x_0) = 2q(x_1 - x_0)\exp\left[-q\rho^2\right] \neq 0. \tag{10.13b}$$

Hence we have the formula

$$Av(x_0) = \left\{ 4q^2 \sum_{i,j=1}^{N} a^{ij}(x_0)(x_1^i - x_0^i)(x_1^j - x_0^j) \right.$$
$$\left. - 2q \sum_{i=1}^{N} \left(a^{ii}(x_0) + b^i(x_0)(x_0^i - x_1^i)\right) \right\} \exp\left[-q\rho^2\right].$$

Since the matrix (a^{ij}) is positive definite, we can estimate the function $Av(x_0)$ from below as follows:

$$Av(x_0) \geq \left(4a_0\rho^2 q^2 - Cq\right)\exp\left[-q\rho^2\right], \tag{10.14}$$

where $C > 0$ is a constant independent of q.

(2) Secondly, in order to estimate the function $Sv(x_0)$ we study the kernel $s(x_0, z)$. We recall that

$$u(x_0) = M = \max_{x \in \overline{D}} u(x) \geq 0.$$

Hence we have the inequalities

$$Au(x_0) = \sum_{i,j=1}^{N} a^{ij}(x_0)\frac{\partial^2 u}{\partial x_i \partial x_j}(x_0) + c(x_0)u(x_0) \le c(x_0)u(x_0) \le 0,$$

and

$$Su(x_0) = \int_{\mathbf{R}^N \setminus \{0\}} s(x_0, z)\,(u(x_0 + z) - u(x_0))\,m(dz) \le 0.$$

This implies that

$$Au(x_0) = 0,$$
$$Su(x_0) = 0.$$

Indeed, it suffices to note that

$$0 \le Wu(x_0) = Au(x_0) + Su(x_0) \le 0.$$

Thus we obtain that

$$0 = Su(x_0) = \int_{\mathbf{R}^N \setminus \{0\}} s(x_0, z)\,(u(x_0 + z) - u(x_0))\,m(dz)$$
$$= \int_{x_0 + z \notin F} s(x_0, z)\,(u(x_0 + z) - u(x_0))\,m(dz),$$

so that

$$s(x_0, z) = 0 \quad \text{if } x_0 + z \notin F,$$

since we have, for all $x_0 + z \notin F$,

$$u(x_0 + z) - u(x_0) < 0.$$

Therefore, we can write the function $Sv(x_0)$ in the form (see Figure 10.5 below)

$$Sv(x_0)$$
$$= \int_{\substack{x_0 + z \in F \\ z \ne 0}} s(x_0, z)\left(v(x_0 + z) - v(x_0) - \sum_{j=1}^{N} z_j \frac{\partial v}{\partial x_j}(x_0)\right) m(dz).$$

For each $\varepsilon > 0$, we decompose the function $Sv(x_0)$ into the two terms

$$Sv(x_0)$$
$$= \int_{\substack{x_0 + z \in F \\ 0 < |z| \le \varepsilon}} s(x_0, z)\left(v(x_0 + z) - v(x_0) - \sum_{j=1}^{N} z_j \frac{\partial v}{\partial x_j}(x_0)\right) m(dz)$$
$$+ \int_{\substack{x_0 + z \in F \\ |z| > \varepsilon}} s(x_0, z)\left(v(x_0 + z) - v(x_0) - \sum_{j=1}^{N} z_j \frac{\partial v}{\partial x_j}(x_0)\right) m(dz)$$

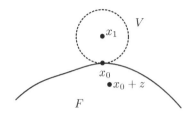

Fig. 10.5. $s(x_0, z)$ for $x_0 + z \in F$

$$:= S_1^{(\varepsilon)} v(x_0) + S_2^{(\varepsilon)} v(x_0).$$

Then, by using formulas (10.13) and Claim 10.8 we can estimate the second term $S_2^{(\varepsilon)} v(x_0)$ as follows:

$$\left| S_2^{(\varepsilon)} v(x_0) \right| \tag{10.15}$$

$$\leq \int_{\substack{x_0+z \in F \\ |z|>\varepsilon}} s(x_0, z) \left| v(x_0 + z) - v(x_0) - \sum_{j=1}^{N} z_j \frac{\partial v}{\partial x_j}(x_0) \right| m(dz)$$

$$\leq \left\{ \int_{|z|>\varepsilon} m(dz) + 2q\rho \int_{|z|>\varepsilon} |z|\, m(dz) \right\} \exp\left[-q\rho^2\right]$$

$$= \left\{ \tau(\varepsilon) + 2q\rho\delta(\varepsilon) \right\} \exp\left[-q\rho^2\right]$$

$$\leq \left\{ \left(\frac{C_1}{\varepsilon^2} + C_2 \right) + 2q\rho \left(\frac{C_1}{\varepsilon} + C_2 \right) \right\} \exp\left[-q\rho^2\right].$$

Indeed, it suffices to note (see Figure 10.5) that

$$|x_0 + z - x_1| \geq |x_0 - x_1| = \rho \quad \text{for all } x_0 + z \in F,$$

so that

$$\exp\left[-q|x + z_0 - x_1|^2\right]$$
$$\leq \exp\left[-q|x_0 - x_1|^2\right] = \exp\left[-q\rho^2\right] \quad \text{for all } x_0 + z \in F.$$

Similarly, we can estimate the first term $S_1^{(\varepsilon)} v(x_0)$ as follows:

$$\left| S_1^{(\varepsilon)} v(x_0) \right| \tag{10.16}$$

$$\leq \int_{\substack{x_0+z \in F \\ 0<|z|\leq\varepsilon}} s(x_0, z) \left| v(x_0 + z) - v(x_0) - \sum_{j=1}^{N} z_j \frac{\partial v}{\partial x_j}(x_0) \right| m(dz)$$

$$\leq \left\{ \int_{0<|z|\leq\varepsilon} m(dz) + 2q\rho \int_{0<|z|\leq\varepsilon} |z|\, m(dz) \right\} \exp\left[-q\rho^2\right]$$

$$\leq \left\{ \left(\frac{C_1}{\varepsilon^2} + C_2 \right) + 2q\rho \left(\frac{C_1}{\varepsilon} + C_2 \right) \right\} \exp\left[-q\rho^2 \right].$$

Hence, by combining estimates (10.15) and (10.16) with $\varepsilon := 1$ we obtain that, for every $\epsilon > 0$,

$$|Sv(x_0)| \leq \{2 (C_1 + C_2) + 4q\rho (C_1 + C_2)\} \exp\left[-q\rho^2 \right]. \tag{10.17}$$

Therefore, if we take a positive constant q so large that

- $a_0 \rho^2 q^2 > C_1 + C_2$,
- $2a_0 \rho^2 q > C + 4\rho (C_1 + C_2)$,

then we obtain from estimates (10.14) and (10.17) that

$$
\begin{aligned}
Wv(x_0) &= Av(x_0) + Sv(x_0) \\
&\geq Av(x_0) - |Sv(x_0)| \\
&\geq \left(4a_0 \rho^2 q^2 - Cq - 2 (C_1 + C_2) - 4q\rho (C_1 + C_2) \right) \exp\left[-q\rho^2 \right] \\
&= \left[4a_0 \rho^2 q^2 - (C + 4\rho (C_1 + C_2)) q - 2 (C_1 + C_2) \right] \exp\left[-q\rho^2 \right] \\
&> 0,
\end{aligned}
$$

if q is sufficiently large.

The proof of Claim 10.9 is complete. $\square$

Step (IV): Now we introduce a function

$$u_\lambda(x) = u(x) + \lambda v(x),$$

where λ is a positive constant to be chosen later on. Then, by Claim 10.9 we have the following four assertions (a) through (d) (see Figure 10.6 below):

(a) There exists a neighborhood V' of x_0 such that

$$Wv > 0 \quad \text{in } V',$$

since $Wv(x)$ is a continuous function.
(b) $Wu_\lambda = Wu + \lambda Wv \geq \lambda Wv > 0$ in V'.
(c) $u_\lambda = u + \lambda v \leq u \leq M$ on $\overline{D} \setminus V$, since $v \leq 0$ on $\overline{D} \setminus V$.
(d) $u_\lambda = u + \lambda v \leq M$ on $\overline{V} \setminus V'$ for λ sufficiently small, since $u < M$ on $\overline{V} \setminus V'$.

Therefore, we obtain that (see Figure 10.7 below)

$$
\begin{cases}
Wu_\lambda > 0 & \text{in } V', \\
u_\lambda \leq M & \text{on } \partial V', \\
u_\lambda(x_0) = M.
\end{cases}
$$

However, this contradicts part (ii) of Theorem 10.5 (the weak maximum principle).

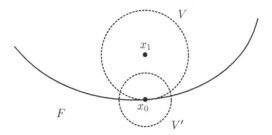

Fig. 10.6. The neighborhood V' of x_0 and the open ball V

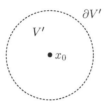

Fig. 10.7. $Wu_\lambda > 0$ in V', $u_\lambda \leq M$ on $\partial V'$ and $u_\lambda(x_0) = M$

Summing up, we have proved that

$$F = D,$$

that is,

$$u(x) \equiv M \quad \text{in } D.$$

Now the proof of Theorem 10.7 is complete.   □

10.2.3 Hopf's Boundary Point Lemma

Finally, we consider the inward normal derivative $(\partial u)/(\partial \mathbf{n})$ at a boundary point where the function $u(x)$ takes its non-negative maximum (cf. Hopf [55], Oleĭnik [85]).

The Hopf boundary point lemma for elliptic, Waldenfels integro-differential operators reads as follows (see [15, Théorème VIII], [136, p. 192, Lemma 3.26], [122, Theorem 8.15], [128, Lemma 6.1]):

Lemma 10.11 (the Hopf boundary point lemma). *Let W be a second-order elliptic Waldenfels operator. Assume that a function $u \in C^2(\overline{D})$ satisfies the conditions*

$$\begin{cases} Wu(x) \geq 0 & \text{in } D, \\ \max_{x \in \overline{D}} u(x) \geq 0, \end{cases}$$

and further that there exists a point $x_0' \in \partial D$ such that

$$u(x_0') = \max_{x \in \overline{D}} u(x).$$

Then the inward normal derivative $(\partial u)/(\partial \mathbf{n})(x_0')$ at x_0' satisfies the condition (see Figure 10.8 below)

$$\frac{\partial u}{\partial \mathbf{n}}(x_0') < 0,$$

unless the function $u(x)$ is a constant in D.

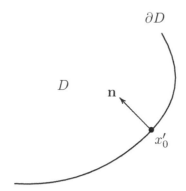

Fig. 10.8. The inward normal **n** at x_0'

Proof. By Theorem 10.7, it suffices to consider the following case:

$$\begin{cases} u(x_0') = \max_{x \in \overline{D}} u(x) \geq 0, \\ u(y) < u(x_0'), \quad y \in D. \end{cases} \tag{10.18}$$

The proof is divided into three steps.

Step (I): Our proof is based on a reduction to absurdity. We assume, to the contrary, that

$$\frac{\partial u}{\partial \mathbf{n}}(x_0') = 0. \tag{10.19}$$

We can find an open ball V contained in the domain D, centered at x_1, such that (see Figure 10.9 below)

(a) The point x_0' is on the boundary ∂V of V;
(b) $\mathbf{n} = s(x_1 - x_0')$ for some $s > 0$.

Step (II): The next claim on the existence of "barriers" is an essential step in the proof of Lemma 10.11, just as in the proof of Theorem 10.7:

Claim 10.12. *There exists a function $v(x) \in C^\infty(\overline{D})$ which satisfies the following five properties (i) through (v):*

(i) $v(x) > 0$ in V.
(ii) $v(x) < 0$ on $\overline{D} \setminus \overline{V}$.

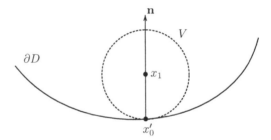

Fig. 10.9. The open ball V and the inward normal $\mathbf{n}$ at x_0'

(iii) $v(x) = 0$ *on* ∂V.
(iv) $Wv(x_0') > 0$.
(v) $(\partial v/\partial \mathbf{n})(x_0') > 0$.

Proof. Near the boundary point x_0', we introduce local coordinate systems (x', x_N) such that $x' = (x_1, x_2, \ldots, x_{N-1})$ give local coordinates for the boundary ∂D and that (see Figure 10.10 below)

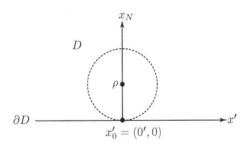

Fig. 10.10. The local coordinate system (x', x_N)

$$D = \{(x', x_N) : x_N > 0\},$$
$$\partial D = \{(x', x_N) : x_N = 0\},$$
$$x_0' = (0', 0) = (0, \ldots, 0, 0),$$
$$x_1 = (0', \rho) = (0, \ldots, 0, \rho).$$

We take a function $\psi(x) \in C^\infty(\overline{D})$ such that

$$\begin{cases} 0 \leq \psi(x) \leq 1 & \text{on } \overline{D}, \\ \psi(x) = 1 & \text{in a tubular neighborhood of } \partial D. \end{cases}$$

Then we define a function $v(x)$ by the formula

$$v(x) = v(x', x_N) = \left(\exp\left[-q(|x'|^2 + (x_N - \rho)^2)\right] - \exp\left[-q\rho^2\right] \right) \psi(x),$$

$$\rho = |x_0' - x_1|,$$

where q is a positive constant to be chosen later on. Then it is easy to see that the function $v(x)$ satisfies conditions (i) through (iii) and (v). Hence it suffices to show that $v(x)$ satisfies condition (iv) for q sufficiently large.

(1) First, we estimate the function $Av(x_0')$ from below. To do this, it should be noticed that

$$v(x_0') = 0, \tag{10.20a}$$

$$\frac{\partial v}{\partial x_i}(x_0') = 0 \quad \text{for } 1 \le i \le N - 1, \tag{10.20b}$$

$$\frac{\partial v}{\partial x_N}(x_0') = 2\rho q \exp\left[-q\rho^2\right], \tag{10.20c}$$

and also

- $\dfrac{\partial^2 v}{\partial x_i \partial x_j}(x_0') = -2q\delta_{ij}\exp\left[-q\rho^2\right]$ for $1 \le i, \ j \le N - 1,$

- $\dfrac{\partial^2 v}{\partial x_N^2}(x_0') = (4q^2\rho^2 - 2q)\exp\left[-q\rho^2\right].$

Hence we have the formula

$$Av(x_0')$$
$$= \left\{ 4a^{NN}(x_0')\rho^2 q^2 + 2\left(b^{NN}(x_0')\rho - \sum_{i=1}^{N} a^{ii}(x_0') \right) q \right\} \exp\left[-q\rho^2\right].$$

Since the matrix (a^{ij}) is positive definite, we can estimate the function $Av(x_0')$ from below as follows:

$$Av(x_0') \ge \left(4a_0\rho^2 q^2 - Cq \right)\exp\left[-q\rho^2\right], \tag{10.21}$$

where $C > 0$ is a constant independent of q.

(2) Secondly, in order to estimate the function $Sv(x_0')$ we study the kernel $s(x_0', z)$: By conditions (10.18) and (10.19), it follows that

- $\dfrac{\partial u}{\partial x_i}(x_0') = 0$ for $1 \le i \le N,$

- $\dfrac{\partial^2 u}{\partial x_N^2}(x_0') \le 0.$

Hence we have the inequality

$$Au(x_0') = \sum_{i,j=1}^{N} a^{ij}(x_0')\frac{\partial^2 u}{\partial x_i \partial x_j}(x_0') + c(x_0')u(x_0')$$

$$= a^{NN}(x_0') \frac{\partial^2 u}{\partial x_N^2}(x_0') + \sum_{i,j=1}^{N-1} a^{ij}(x_0') \frac{\partial^2 u}{\partial x_i \partial x_j}(x_0') + c(x_0')u(x_0')$$

$$\leq 0,$$

and also

$$Su(x_0') = \int_{\mathbf{R}^N \setminus \{0\}} s(x_0', z) \left(u(x_0' + z) - u(x_0') \right) m(dz) \leq 0.$$

This implies that

$$Au(x_0') = 0,$$
$$Su(x_0') = 0.$$

Indeed, it suffices to note that

$$0 \leq Wu(x_0') = Au(x_0') + Su(x_0') \leq 0.$$

Thus we obtain that

$$0 = Su(x_0')$$
$$= \int_{\mathbf{R}^N \setminus \{0\}} s(x_0', z) \left(u(x_0' + z) - u(x_0') \right) m(dz)$$
$$= \int_{x_0' + z \notin G} s(x_0', z) \left(u(x_0' + z) - u(x_0') \right) m(dz),$$

so that

$$s(x_0', z) = 0 \quad \text{if } x_0' + z \in \partial D \setminus G,$$

since we have, for all $x_0 + z \in \partial D \setminus G$,

$$u(x_0' + z) - u(x_0') < 0.$$

Here

$$G = \left\{ x' \in \partial D : u(x') = \max_{x \in \overline{D}} u(x) \right\}.$$

Moreover, it should be noticed that

$$x_0' + z \in G \Longleftrightarrow z = (z', 0) \in G.$$

Therefore, we can write the function $Sv(x_0')$ in the form

$$Sv(x_0')$$
$$= \int_{\substack{x_0' + z \in G \\ z \neq 0}} s(x_0', z) \left(v(x_0' + z) - v(x_0') - \sum_{j=1}^{N} z_j \frac{\partial v}{\partial x_j}(x_0') \right) m(dz).$$

For each $\varepsilon > 0$, we decompose the function $Sv(x_0')$ into the two terms

$$Sv(x_0')$$

$$= \int_{\substack{x_0'+z\in G \\ 0<|z|\le\varepsilon}} s(x_0', z) \left(v(x_0' + z) - v(x_0') - \sum_{j=1}^{N} z_j \frac{\partial v}{\partial x_j}(x_0') \right) m(dz)$$

$$+ \int_{\substack{x_0'+z\in G \\ |z|>\varepsilon}} s(x_0', z) \left(v(x_0' + z) - v(x_0') - \sum_{j=1}^{N} z_j \frac{\partial v}{\partial x_j}(x_0') \right) m(dz)$$

$$:= S_1^{(\varepsilon)}v(x_0') + S_2^{(\varepsilon)}v(x_0').$$

By using formulas (10.20), we can estimate the second term $S_2^{(\varepsilon)}v(x_0')$ as follows:

$$\left| S_2^{(\varepsilon)}v(x_0') \right| \tag{10.22}$$

$$\le \int_{\substack{x_0'+z\in G \\ |z|>\varepsilon,\, z_N=0}} s(x_0', z) \Big| \left[\exp\left[-q(|z'|^2 + (z_N - \rho)^2) \right] - \exp\left[-q\rho^2 \right] \right]$$

$$- 2\rho z_N q \exp\left[-q\rho^2 \right] \Big| m(dz)$$

$$\le \int_{\substack{|z|>\varepsilon \\ z_N=0}} m(dz) \exp\left[-q\rho^2 \right]$$

$$\le \tau(\varepsilon) \exp\left[-q\rho^2 \right]$$

$$\le \left(\frac{C_1}{\varepsilon^2} + C_2 \right) \exp\left[-q\rho^2 \right].$$

Similarly, we can estimate the first term $S_1^{(\varepsilon)}v(x_0')$ as follows:

$$\left| S_1^{(\varepsilon)}v(x_0') \right| \tag{10.23}$$

$$\le \int_{\substack{x_0'+z\in G \\ 0<|z|\le\varepsilon,\, z_N=0}} s(x_0', z)v(x_0' + z)\, m(dz)$$

$$\le \int_{\substack{0<|z'|\le\varepsilon \\ z_N=0}} s(0, z', 0) \left[\exp\left[-q(|z'|^2 + \rho^2) \right] - \exp\left[-q\rho^2 \right] \right] m(dz)$$

$$\le \int_{\substack{0<|z'|\le\varepsilon \\ z_N=0}} m(dz) \exp\left[-q\rho^2 \right]$$

$$\le \tau(\varepsilon) \exp\left[-q\rho^2 \right]$$

$$\le \left(\frac{C_1}{\varepsilon^2} + C_2 \right) \exp\left[-q\rho^2 \right].$$

By combining estimates (10.22) and (10.23) with $\varepsilon := 1$, we have proved that

$$|Sv(x_0')| \le 2\,(C_1 + C_2)\exp\left[-q\rho^2\right]. \tag{10.24}$$

Therefore, it follows from estimates (10.21) and (10.24) that

$$
\begin{aligned}
Wv(x_0') = Av(x_0') + Sv(x_0') &\ge Av(x_0') - |Sv(x_0')| \\
&\ge \left(4\,a_0\,\rho^2\,q^2 - C\,q\right)\exp\left[-q\rho^2\right] - 2\,(C_1 + C_2)\exp\left[-q\rho^2\right] \\
&= \left[4\,a_0\,\rho^2\,q^2 - C\,q - 2\,(C_1 + C_2)\right]\exp\left[-q\rho^2\right] \\
&> 0,
\end{aligned}
$$

if q is sufficiently large.

The proof of Claim 10.12 is complete. □

Step (III): If we introduce a function

$$u_\lambda(x) = u(x) + \lambda v(x)$$

for a positive constant λ, then we have, by Claim 10.12, the following four assertions (a) through (d) (see Figure 10.11 below):

(a) There exists a neighborhood V' of x_0' such that

$$Wv > 0 \quad \text{in } V' \cap \overline{D},$$

since $Wv(x)$ is a continuous function on $\overline{D}$.

(b) $Wu_\lambda = Wu + \lambda Wv \ge \lambda Wv > 0$ in $V' \cap \overline{D}$.

(c) $u_\lambda = u + \lambda v \le u \le M$ on $\overline{D} \setminus V$, since $v \le 0$ on $\overline{D} \setminus V$.

(d) $u_\lambda = u + \lambda v \le M$ on $\overline{V} \setminus V'$ for λ sufficiently small, since $u < M$ on $\overline{V} \setminus V'$.

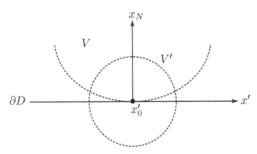

Fig. 10.11. The open ball V and the neighborhood V' of x_0'

Hence it follows from an application of part (ii) of Theorem 10.5 that

$$u_\lambda \le M \quad \text{in } V' \cap D,$$

so that

$$u_\lambda(y) = u(y) + \lambda v(y) \le M = u(x_0') + \lambda v(x_0'), \quad y \in V' \cap D,$$

or equivalently

$$u(y) - u(x_0') \leq -\lambda(v(y) - v(x_0')), \quad y \in V' \cap D.$$

Therefore, we obtain that

$$\frac{\partial u}{\partial \mathbf{n}}(x_0') \leq -\lambda \frac{\partial v}{\partial \mathbf{n}}(x_0') < 0.$$

This contradicts hypothesis (10.19).

Now the proof of Lemma 10.11 is complete.   □

10.3 Notes and Comments

This chapter is a modern version of the Bony–Courrège–Priouret theory ([15]) from the viewpoint of pseudo-differential operators with the special emphasis on the *transmission property* introduced by Boutet de Monvel [19].

Protter–Weinberger [91] is the classic for maximum principles. For a general study of maximum principles for Waldenfels integro-differential operators, the reader might refer to Bony–Courrège–Priouret [15], Troianiello [136, Section 3.7.2], Taira [122, Chapter 8], [127] and [128]. See also Taira [125] for a strong maximum principle for globally hypoelliptic operators.

This chapter is based on the lecture entitled *A class of hypoelliptic Vishik–Wentzell boundary vale problems* delivered at Functional Analysis Methods for Partial Differential Equations, Centro Polifunzionale, Cesena, Italy, on June 27, 2019.

11

Boundary Operators and Boundary Maximum Principles

Let D be a bounded domain of Euclidean space $\mathbf{R}^N$, with smooth boundary ∂D; its closure $\overline{D} = D \cup \partial D$ is an N-dimensional, compact smooth manifold with boundary (see Figure 11.1 below). In this chapter, following Bony–Courrège–Priouret ([15, Chapter II]) we characterize Ventcel'–Lévy boundary operators T (Theorem 11.3) and Ventcel' boundary operators $\Gamma = \Lambda + T$ (Theorem 11.4) defined on the compact smooth manifold $\overline{D}$ with boundary ∂D in terms of the positive boundary maximum principle:

$$x_0' \in \partial D, \ u \in C^2(\overline{D}) \text{ and } u(x_0') = \max_{x \in \overline{D}} u(x) \geq 0 \Longrightarrow (\Gamma u)(x_0') \leq 0.$$

$$\text{(PMB)}$$

This chapter will be very useful in the study of Markov processes with general Ventcel' boundary conditions in the last Chapter 16.

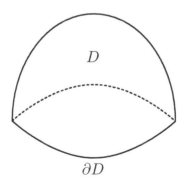

D

∂D

Fig. 11.1. The compact smooth manifold $\overline{D} = D \cup \partial D$ with boundary ∂D

© Springer Nature Switzerland AG 2020
K. Taira, *Boundary Value Problems and Markov Processes*, Lecture Notes in Mathematics 1499,
https://doi.org/10.1007/978-3-030-48788-1_11

11.1 Ventcel'–Lévy Boundary Operators

Let $\{(U_\beta, \chi_\beta)\}_{\beta=1}^m$ be a finite open covering of $M = \overline{D}$ by local charts. We choose a family $\{\tau_\beta\}_{\beta=1}^m$ of functions in $C^\infty(\overline{D} \times \overline{D})$ such that

$$\operatorname{supp} \tau_\beta \subset U_\beta \times U_\beta$$

and that

$$\tau(x, y) = \sum_{\beta=1}^m \tau_\beta(x, y) = 1$$

in a neighborhood of the diagonal $\Delta_{\overline{D}} = \{(x, x) : x \in \overline{D}\}$.

We study a positive Borel kernel $t(x', dy)$ which satisfies the following three conditions (NV.1), (NV.2) and (NV.3):

(NV.1) $t(x', \{x'\}) = 0$ for every $x' \in \partial D$.
(NV.2) For every local chart (U_β, χ_β) of $\overline{D}$ such that $U_\beta \cap \partial D \neq \emptyset$ and for every non-negative function $f \in C(\overline{D})$ with $\operatorname{supp} f \subset U_\beta$, the function

$$U_\beta \cap \partial D \ni x' \mapsto \int_{U_\beta} t(x', dy)\, f(y) \Big[\chi_N(y) + \sum_{j=1}^{N-1} \big(\chi_j^\beta(y) - \chi_j^\beta(x')\big)^2\Big]$$

belongs to the space $B_{\mathrm{loc}}(U_\beta \cap \partial D)$ of locally bounded, Borel measurable functions on $U_\beta \cap \partial D$.
(NV.3) For every non-negative function $f \in C(\overline{D})$, the function

$$\partial D \ni x' \longmapsto \int_{\overline{D}} t(x', dy)\, f(y) \Big(1 - \sum_{\beta=1}^m \tau_\beta(x', y)\Big) f(y) \tag{11.1}$$

belongs to the space $B(\partial D)$ of bounded, Borel measurable functions on ∂D.

The Borel kernel $t(x', dy)$ is called a *Ventcel' kernel* on the manifold $\overline{D}$.

We can give a global equivalent version of the conditions (NV.2) and (NV.3) in the following way: For points $x' \in \partial D$ and $y \in \overline{D}$, we introduce a function

$$\widetilde{x'\, y} = \sum_{\beta=1}^m \tau_\beta(x', y) \Big[\chi_N^\beta(y) + \sum_{j=1}^{N-1} \big(\chi_j^\beta(y) - \chi_j^\beta(x')\big)^2\Big]. \tag{11.2}$$

Then it is easy to see that the conditions (NV.2) and (NV.3) are equivalent to the following global condition (NV.4):

(NV.4) For every non-negative function $f \in C(\overline{D})$, the function

$$\partial D \ni x' \longmapsto \int_{\overline{D}} t(x', dy)\, \widetilde{x'\, y}\, f(y)$$

belongs to the space $B(\partial D)$.

Now we are in a position to define Ventcel'–Lévy boundary operators;

Definition 11.1. *We are given the following:*

(1) $t(x', dy)$ is a Ventcel' kernel on the manifold $\overline{D}$.
(2) $\zeta(x')$ is a bounded, Borel measurable vector field on the boundary ∂D:
More precisely, we have, in a local chart (U_β, χ_β),

$$\zeta(x') \cdot u = \sum_{j=1}^{N-1} \zeta_\beta^j(x') \frac{\partial u}{\partial \chi_j^\beta} \quad \text{with } \zeta_\beta^j \in B_{\mathrm{loc}}(U_\beta \cap \partial D).$$

(3) $\eta(x')$ is a bounded, Borel measurable function on the boundary ∂D such that

$$\eta(x') + \int_{\overline{D}} t(x', dy)\left(1 - \sum_{\beta=1}^m \tau_\beta(x', y)\right) \leq 0 \quad \text{on } \partial D. \tag{11.3}$$

Then we define a continuous boundary operator

$$\boxed{T\colon C^2(\overline{D}) \longrightarrow B(\partial D)}$$

by the formula

$$(Tu)(x') = \eta(x')u(x') + \zeta(x') \cdot u(x') + \int_{\overline{D}} t(x', dy)\Bigg[u(y) \tag{11.4}$$

$$- \sum_{\beta=1}^m \tau_\beta(x', y)\left(u(x') + \sum_{j=1}^{N-1} (\chi_j^\beta(y) - \chi_j^\beta(x')) \frac{\partial u}{\partial \chi_j^\beta}(x')\right)\Bigg].$$

We remark that condition (11.3) is equivalent to the following global one:

$$(T1)(x') \leq 0 \quad \text{on } \partial D. \tag{11.5}$$

The boundary operator T is called a Ventcel'–Lévy *boundary operator.*

Remark 11.2. In terms of a local chart (U, χ) of $\overline{D}$ such that $U \cap \partial D \neq \emptyset$, every Ventcel'–Lévy boundary operator T can be characterized as follows (see [15, p. 440, Lemme]):

(V1) For every point $x' \in \partial D$, we have the formula

$$(Tu)(x') = \int_{\overline{D}} t(x', dy)\, u(y) \quad \text{whenever } x' \notin \operatorname{supp} u.$$

(V2) For every function $u \in C^2(\overline{D})$ with $\operatorname{supp} u \subset U$, we have the local expression

$$(Tu)(x')$$

$$= \eta(x')u(x') + \sum_{j=1}^{N-1} \zeta^j(x') \frac{\partial u}{\partial \chi_j}(x')$$

$$+ \int_U t(x', dy) \left[u(y) - \left(u(x') + \sum_{j=1}^{N-1} (\chi_j(y) - \chi_j(x')) \frac{\partial u}{\partial \chi_j}(x') \right) \right]$$

for every $x' \in U \cap \partial D$.

(V3) $\eta(x') + t(x', \overline{D} \setminus U) \leq 0$ for every $x' \in U \cap \partial D$.

11.2 Positive Boundary Maximum Principles

In this section we characterize a general boundary operator of the form

$$\boxed{\Gamma = \Lambda + T \colon C^2(\overline{D}) \longrightarrow B(\partial D)}$$

in terms of the positive boundary maximum principle (see [15, p. 441, Section II.2.6]).

11.2.1 Boundary Maximum Principles for Ventcel'–Lévy operators

First, we prove the following theorem for Ventcel'–Lévy boundary operators T:

Theorem 11.3. *If $T \colon C^2(\overline{D}) \to B(\partial D)$ is a Ventcel'–Lévy boundary operator, then T satisfies the* positive boundary maximum principle*:*

$$x_0' \in \partial D, \ u \in C^2(\overline{D}) \ \text{and} \ u(x_0') = \max_{x \in \overline{D}} u(x) \geq 0 \implies (Tu)(x_0') \leq 0. \quad \text{(PMB)}$$

Proof. Indeed, we have, by formula (11.4) and condition (11.5),

$$(Tu)(x_0') = \eta(x_0')u(x_0') + \zeta(x_0') \cdot u(x_0') + \int_{\overline{D}} t(x_0', dy) \Bigg[u(y)$$

$$- \sum_{\beta=1}^{m} \tau_\beta(x_0', y) \left(u(x_0') + \sum_{j=1}^{N-1} \left(\chi_j^\beta(y) - \chi_j^\beta(x_0') \right) \frac{\partial u}{\partial \chi_j^\beta}(x_0') \right) \Bigg]$$

$$= \eta(x_0')u(x_0') + \int_{\overline{D}} t(x_0', dy) \Bigg[u(y) - \sum_{\beta=1}^{m} \tau_\beta(x_0', y) \cdot u(x_0') \Bigg]$$

$$= \left(\eta(x_0') + \int_{\overline{D}} t(x_0', dy) \left(1 - \sum_{\beta=1}^{m} \tau_\beta(x_0', y) \right) \right) \cdot u(x_0')$$

$$+ \int_{\overline{D}} t(x_0', dy) \left(u(y) - u(x_0') \right)$$

$$\leq 0.$$

This proves the positive boundary maximum principle (PMB) for the boundary operator T.

The proof of Theorem 11.3 is complete. $\quad\square$

11.2.2 Boundary Maximum Principles for Ventcel' operators

Secondly, we can prove the following theorem for Ventcel' boundary operators Γ:

Theorem 11.4. *For a linear boundary operator*

$$\Gamma \colon C^2(\overline{D}) \longrightarrow B(\partial D),$$

the following two assertions (I) and (II) are equivalent:

(I) The boundary operator Γ satisfies the positive boundary maximum principle*:*

$$x_0' \in \partial D, \ u \in C^2(\overline{D}) \ \text{and} \ u(x_0') = \max_{x \in \overline{D}} u(x) \geq 0 \Longrightarrow (\Gamma u)(x_0') \leq 0. \quad \text{(PMB)}$$

(II) The boundary operator Γ is of the form

$$(\Gamma u)(x') := (\Lambda u)(x') + (Tu)(x') \tag{11.6}$$

$$= \left(\mu(x') \frac{\partial u}{\partial \mathbf{n}}(x') + Qu(x') \right) + (Tu)(x')$$

$$= \mu(x') \frac{\partial u}{\partial \mathbf{n}}(x') + \sum_{i,j=1}^{N-1} \alpha^{ij}(x') \frac{\partial^2 u}{\partial \chi_i \partial \chi_j}(x') + \sum_{i=1}^{N-1} \beta^i(x') \frac{\partial u}{\partial \chi_i}(x')$$

$$+ \gamma(x')u(x') + \eta(x')u(x') + \sum_{j=1}^{N-1} \zeta^j(x') \frac{\partial u}{\partial \chi_j}(x') + \int_{\overline{D}} t(x', dy) \Bigg[u(y)$$

$$- \sum_{\beta=1}^{m} \tau_\beta(x', y) \left(u(x') + \sum_{j=1}^{N-1} \left(\chi_j^\beta(y) - \chi_j^\beta(x') \right) \frac{\partial u}{\partial \chi_j^\beta}(x') \right) \Bigg]$$

for every $x' \in U \cap \partial D$.

Here:

(a) The operator Q is a second-order, degenerate elliptic differential operator on the boundary ∂D with non-positive principal symbol. In other words, the α^{ij} are the components of a symmetric contravariant tensor of type $\binom{2}{0}$ on ∂D satisfying the condition

$$\sum_{i,j=1}^{N-1} \alpha^{ij}(x')\xi_i\xi_j \geq 0 \quad \text{for all } x' \in \partial D \text{ and } \xi' = \sum_{j=1}^{N-1} \xi_j dx_j \in T_{x'}^*(\partial D).$$

(b) $Q1(x') = \gamma(x') \in B(\partial D)$ and $\gamma(x') \leq 0$ on ∂D.
(c) $\mu(x') \in B(\partial D)$ and $\mu(x') \geq 0$ on ∂D.
(d) $\mathbf{n} = (n_1, n_2, \ldots, n_N)$ is the unit inward normal to ∂D.

(e) The operator T is a Ventcel'–Lévy boundary operator, that is,
(e.1) $t(x', dy)$ is a Ventcel' kernel on the manifold $\overline{D}$.
(e.2) $\zeta(x')$ is a bounded, Borel measurable vector field on ∂D; more precisely, we have, in a local chart (U, χ),

$$\zeta(x') \cdot u = \sum_{j=1}^{N-1} \zeta^j(x') \frac{\partial u}{\partial \chi_j} \quad \text{with } \zeta^j \in B_{\mathrm{loc}}(U \cap \partial D).$$

(e.3) $\eta(x')$ is a bounded, Borel measurable function on ∂D which satisfies condition (11.3).

Definition 11.5. *The boundary operator*

$$\boxed{\Lambda = \mu(x') \frac{\partial}{\partial \mathbf{n}} + Q \colon C^2(\overline{D}) \longrightarrow B(\partial D)}$$

is called a Ventcel'–Višik boundary operator *(see [15, p. 436, formule (II.2.1)], [138]).*

The boundary operator

$$\boxed{\Gamma = \Lambda + T = \mu(x') \frac{\partial}{\partial \mathbf{n}} + Q + T \colon C^2(\overline{D}) \longrightarrow B(\partial D)}$$

is called a Ventcel' boundary operator *(see [15, p. 442, formule (II.2.16)], [138]).*

The proof of Theorem 11.4 is given in the next Subsubsections 11.2.2 and 11.2.2, due to its length.

Proof of Theorem 11.4, Part (i)

- Assertion (II) $\Longrightarrow$ Assertion (I): Assume that

$$x_0' \in \partial D, \ u \in C^2(\overline{D}) \text{ and } u(x_0') = \max_{x \in \overline{D}} u(x) \geq 0.$$

Then we remark that

$$\frac{\partial u}{\partial \mathbf{n}}(x_0') \leq 0,$$

since $\mathbf{n}$ is the unit inward normal to ∂D.

Therefore, we have, by formula (11.6) and condition (11.5),

$$(\Gamma u)(x_0') = (\Lambda u)(x_0') + (Tu)(x_0')$$

$$= \left(\mu(x_0')\frac{\partial u}{\partial \mathbf{n}}(x_0') + Qu(x_0')\right) + (Tu)(x_0')$$

$$= \mu(x_0')\frac{\partial u}{\partial \mathbf{n}}(x_0') + \sum_{i,j=1}^{N-1} a^{ij}(x_0')\frac{\partial^2 u}{\partial \chi_i \partial \chi_j}(x_0') + \sum_{i=1}^{N-1} \beta^i(x_0')\frac{\partial u}{\partial \chi_i}(x_0')$$

$$+ \gamma(x_0')u(x_0') + \eta(x_0')u(x_0') + \sum_{j=1}^{N-1} \zeta^j(x_0')\frac{\partial u}{\partial \chi_j}(x_0') + \int_{\overline{D}} t(x_0', dy)\Big[u(y)$$

$$- \sum_{\beta=1}^{m} \tau_\beta(x_0', y)\left(u(x_0') + \sum_{j=1}^{N-1}\left(\chi_j^\beta(y) - \chi_j^\beta(x_0')\right)\frac{\partial u}{\partial \chi_j^\beta}(x_0')\right)\Big]$$

$$= \mu(x_0')\frac{\partial u}{\partial \mathbf{n}}(x') + \sum_{i,j=1}^{N-1} a^{ij}(x_0')\frac{\partial^2 u}{\partial \chi_i \partial \chi_j}(x_0') + \gamma(x_0')u(x_0')$$

$$+ \eta(x_0')u(x_0') + \int_{\overline{D}} t(x_0', dy)\Big[u(y) - \sum_{\beta=1}^{m} \tau_\beta(x_0', y)\cdot u(x_0')\Big]$$

$$\leq \left(\eta(x_0') + \int_{\overline{D}} t(x_0', dy)\left(1 - \sum_{\beta=1}^{m} \tau_\beta(x_0', y)\right)\right)\cdot u(x_0')$$

$$+ \int_{\overline{D}} t(x_0', dy)\,(u(y) - u(x_0'))$$

$$\leq 0.$$

This proves the positive boundary maximum principle (PMB) for the boundary operator Γ. □

Proof of Theorem 11.4, Part (ii)

• Assertion (I) $\Longrightarrow$ Assertion (II): The proof is divided into five steps.

   Step 1: First, by the positive boundary maximum principle (PMB) for $\Gamma = \Lambda + T$ we obtain that

$$u \in C_0^\infty(\overline{D} \setminus \{x'\}), \ u \geq 0 \text{ on } \overline{D} \implies (\Gamma u)(x') \geq 0.$$

Hence there exists a positive Borel kernel $t(x', dy)$ of ∂D into $\overline{D}$ which satisfies the condition (NV.1) and the formula

$$(\Gamma u)(x') = \int_{\overline{D}} t(x', dy)\, u(y) \quad \text{whenever } u \in C^2(\overline{D}) \text{ and } x' \notin \operatorname{supp} u. \quad (11.7)$$

Step 2: Moreover, the kernel $t(x', dy)$ satisfies the condition (NV.3). Indeed, let f be an arbitrary non-negative function on $\overline{D}$. Then, for each point $x' \in \partial D$ there exist a neighborhood V of x' and a non-negative function $u \in C^2(\overline{D})$ vanishing on V such that

$$(1 - \sigma(x', y))\, f(y) \le u(y) \quad \text{for all } y \in M \text{ and } x' \in V.$$

Hence we have, by formula (11.7),

$$\int_{\overline{D}} t(x', dy)\, (1 - \sigma(x', y))\, f(y) \le \int_{\overline{D}} t(x', dy)\, u(y) = (\Gamma u)(x') \quad \text{for all } x' \in V.$$

This proves the desired condition (NV.3), since $\Gamma u \in B(\partial D)$.

Step 3: Now we verify the condition (NV.2) for the kernel $t(x', dy)$. Let (U, χ) be an arbitrary local chart on $\overline{D}$ (see Figure 11.2 below) such that

$$U \cap \partial D \ne \emptyset,$$
$$\chi(U) \text{ is bounded in } \mathbf{R}^N.$$

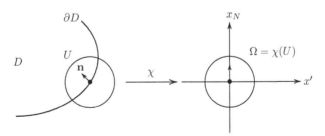

Fig. 11.2. The local chart (U, χ) on $\overline{D}$

Substep 3.1: First, we show the following two assertions (i) and (ii):

(i) For every non-negative function $f \in C(\overline{D})$ with $\operatorname{supp} f \subset U$, we have the assertion

$$\int_U t(x', dy)\, f(y) \left[\chi_N(y) + \sum_{j=1}^{N-1} \left(\chi_j(y) - \chi_j(x') \right)^2 \right] \in B_{\mathrm{loc}}(U \cap \partial D). \quad (11.8)$$

(ii) There exist functions $b^{ij}(x')$, $b^i(x')$ and $b(x') \in B_{\mathrm{loc}}(U \cap \partial D)$ such that we have, for every function $u \in C^2(\overline{D})$ with $\operatorname{supp} u \subset U$,

$$(\Gamma u)(x') \qquad\qquad\qquad\qquad\qquad\qquad\qquad\qquad (11.9)$$

$$= \mu(x') \frac{\partial u}{\partial \chi_N}(x') + \sum_{i,j=1}^{N-1} b^{ij}(x') \frac{\partial^2 u}{\partial \chi_i \partial \chi_j}(x') + \sum_{i=1}^{N-1} b^i(x') \frac{\partial u}{\partial \chi_i}(x')$$

$$+ b(x')u(x')$$

$$+ \int_U t(x', dy) \left[u(y) - u(x') - \sum_{j=1}^{N-1} \frac{\partial u}{\partial \chi_j}(x') \big(\chi_j(y) - \chi_j(x') \big) \right].$$

Here:

$$\sum_{i,j=1}^{N-1} b^{ij}(x') \xi_i \xi_j \geq 0 \quad \text{for all } \xi' = (\xi_1, \ldots, \xi_{N-1}) \in \mathbf{R}^{N-1}, \qquad (11.10)$$

and

$$\mu(x') \geq 0. \qquad (11.11)$$

Now we let

$$\Omega = \chi(U),$$

and choose an open subset $\widetilde{\Omega}$ of $\mathbf{R}^N$ (see Figure 11.3 below) such that

$$\chi(U) = \widetilde{\Omega} \cap \overline{\mathbf{R}_+^N}.$$

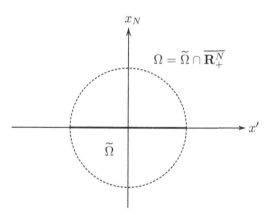

Fig. 11.3. The set $\Omega = \chi(U)$ and the open set $\widetilde{\Omega}$

Then we define a linear operator (see Figure 11.4 below)

$$\boxed{\widetilde{\Gamma} \colon C_0^2(\widetilde{\Omega}) \longrightarrow B(\widetilde{\Omega})}$$

by the formula

$$
\begin{array}{ccc}
C_0^2(\widetilde{\Omega}) & \xrightarrow{\ \widetilde{\Gamma}\ } & B(\widetilde{\Omega}) \\
\chi_* \downarrow & & \uparrow (\chi^{-1})_* \\
C_{(0)}^2(U) & \xrightarrow{\ \ \Gamma\ \ } & B_{\mathrm{loc}}(U \cap \partial D)
\end{array}
$$

Fig. 11.4. The operators Γ and $\widetilde{\Gamma}$

$$
\widetilde{\Gamma}\varphi(z) = \begin{cases} \Gamma\left(\widetilde{\varphi}\circ\chi\right)\left(\chi^{-1}(z)\right) & \text{for } z \in \partial\Omega, \\ 0 & \text{for } z \in \widetilde{\Omega}\setminus\partial\Omega. \end{cases}
$$

Here $\widetilde{\varphi}$ is the restriction of φ to the set $\Omega = \chi(U)$.

Since $\Gamma \colon C_{(0)}^2(U) \to B_{\mathrm{loc}}(U\cap\partial D)$ satisfies the condition (PMB), it is easy to see that $\widetilde{\Gamma}$ satisfies the positive maximum principle

$$
z \in \overset{\circ}{\widetilde{\Omega}},\ \varphi \in C_0^2(\widetilde{\Omega})\ \text{and}\ \varphi(z) = \max_{w\in\widetilde{\Omega}}\varphi(w) \ge 0 \Longrightarrow \widetilde{\Gamma}\varphi(z) \le 0. \qquad \text{(PM)}
$$

Therefore, by applying Theorem 4.6 to our situation we obtain that the function $\widetilde{\Gamma}\varphi(z)$ for $\varphi \in C_0^2(\widetilde{\Omega})$ and $z \in \widetilde{\Omega}$ is of the form

$$
\widetilde{\Gamma}\varphi(z) = \sum_{i,j=1}^{N} \widetilde{b}^{ij}(z)\frac{\partial^2\varphi}{\partial z_i \partial z_j}(z) + \sum_{i=1}^{N} \widetilde{b}^i(z)\frac{\partial\varphi}{\partial z_i}(z) + \widetilde{b}(z)\varphi(z) \qquad (11.12)
$$

$$
+ \int_{\widetilde{\Omega}} \widetilde{t}(z,dw)\left[\varphi(w) - \sigma(z,w)\left(\varphi(z) + \sum_{j=1}^{N}\frac{\partial\varphi}{\partial z_j}(z)(w_j - z_j)\right)\right].
$$

Here:

(a) $\widetilde{b}^{ij}(z)$, $\widetilde{b}^i(z)$ and $\widetilde{b}(z) \in B_{\mathrm{loc}}(\widetilde{\Omega})$ such that

$$
\sum_{i,j=1}^{N} \widetilde{b}^{ij}(z) \ge 0 \quad \text{for all } (z,\xi) \in \widetilde{\Omega}\times\mathbf{R}^N. \qquad (11.13)
$$

(b) $\sigma(z,w)$ is a local unity function on $\widetilde{\Omega}$.

(c) $\widetilde{t}(z,dw)$ is a Lévy kernel on $\widetilde{\Omega}$ such that

$$
\widetilde{b}(z) + \int_{\widetilde{\Omega}} \widetilde{t}(z,dw)\,(1 - \sigma(z,w)) \le 0 \quad \text{for all } z \in \overset{\circ}{\widetilde{\Omega}}. \qquad (11.14)
$$

We remark that the kernels $t(x',dy)$ and $\widetilde{t}(z,dw)$ are related as follows:

$$
\widetilde{t}(z,Z) = t\left(\chi^{-1}(z),\chi^{-1}(Z)\right) \quad \text{for all } z \in \partial\Omega \text{ and } Z \in \mathcal{B}_{\widetilde{\Omega}}.
$$

Indeed, it suffices to note the following formulas with $w = \chi(y)$ and $x' = \chi^{-1}(z) \in \partial D$:

$$\int_{\widetilde{\Omega}} \widetilde{t}(z, dw)\, \varphi(w) = \widetilde{\Gamma}\varphi(z) = \Gamma(\widetilde{\varphi} \circ \chi)(x') = \int_{\overline{D}} t(x', dy)\, (\widetilde{\varphi} \circ \chi)(y)$$

$$= \int_{U} t(x', dy)\, (\widetilde{\varphi} \circ \chi)(y)$$

for all $\varphi \in C_0^2(\widetilde{\Omega})$ with $z \notin \operatorname{supp}\varphi$.

In particular, it should be emphasized that the measure $\widetilde{t}(z, \cdot)$ for $z \in \partial\Omega$ is supported on the $\Omega = \chi(U)$.

Moreover, since the Lévy kernel $\widetilde{t}(z, \cdot)$ on $\widetilde{\Omega}$ satisfies condition (NS2), we have the assertion

$$\int_{U} t(x', dy)\, (\chi_j(y) - \chi_j(x'))^2 \tag{11.15}$$

$$= \int_{\Omega} \widetilde{t}(z, dw)\, (w_j - z_j)^2 \in B_{\mathrm{loc}}(U \cap \partial D) \quad \text{for all } 1 \leq j \leq N-1.$$

Substep 3.2: We show that

$$\int_{U} t(x', dy)\, f(y)\, \chi_N(y) \in B(U \cap \partial D) \tag{11.16}$$

for every non-negative function $f \in C(\overline{D})$ with $\operatorname{supp} f \subset U$.

First, we prove that

$$\widetilde{b}^N(z) \geq 0 \quad \text{on } \partial\Omega. \tag{11.17}$$

To do so, for $z \in \partial\Omega$ we let

$$\varphi_1(w) = \sigma(z, w)\, w_N \quad \text{for } w \in \widetilde{\Omega}.$$

Then it is easy to see the following formulas:

- $\varphi_1(z) = \sigma(z, z)\, 0 = 0,$

- $\dfrac{\partial \varphi_1}{\partial w_i}(z) = 0 \quad \text{for all } 1 \leq i \leq N-1,$

- $\dfrac{\partial \varphi_1}{\partial w_N}(z) = 1,$

and

$$w - z = (w_1 - z_1, \ldots, w_{N-1} - z_{N-1}, w_N).$$

Hence, by applying the positive boundary maximum principle (PMB) to the function

$$-\varphi_1(w) = -\sigma(z, w)\, w_N \quad \text{for } w \in \widetilde{\Omega},$$

we obtain that

$$0 \leq \widetilde{\Gamma}\varphi_1(z)$$

$$= \widetilde{b}^N(z)$$

$$+ \int_\Omega \widetilde{t}(z, dw) \left[\varphi_1(w) - \sigma(z, w)(\varphi_1(z) + \sum_{j=1}^N \frac{\partial \varphi_1}{\partial z_j}(z)(w_j - z_j)) \right]$$

$$= \widetilde{b}^N(z) + \int_\Omega \widetilde{t}(z, dw) \left[\sigma(z, w) w_N - \sigma(z, w) w_N \right]$$

$$= \widetilde{b}^N(z) \quad \text{on } \partial\Omega,$$

so that,

$$\widetilde{b}^N(z) \geq 0 \quad \text{on } \partial\Omega.$$

Secondly, for $z \in \partial\Omega$ we let

$$\varphi_2(w) = \sigma_1(z, w) w_N, \quad w \in \widetilde{\Omega},$$

where $\sigma_1(z, w)$ is a local unity function on $\widetilde{\Omega}$ that will be chosen later on. Then it is easy to see the following formulas:

- $\varphi_2(z) = \sigma(z, z) 0 = 0,$
- $\dfrac{\partial \varphi_2}{\partial w_i}(z) = 0 \quad \text{for all } 1 \leq i \leq N - 1,$
- $\dfrac{\partial \varphi_2}{\partial w_N}(z) = 1.$

Hence we have the formula

$$\widetilde{\Gamma}\varphi_2(z) \tag{11.18}$$

$$= \widetilde{b}^N(z) + \int_\Omega \widetilde{t}(z, dw) \left[\varphi_2(w) - \sigma(z, w) \left(\varphi_1(z) + \sum_{j=1}^N \frac{\partial \varphi_2}{\partial z_j}(z)(w_j - z_j) \right) \right]$$

$$= \widetilde{b}^N(z) + \int_\Omega \widetilde{t}(z, dw) \left[\sigma_1(z, w) w_N - \sigma(z, w) w_N \right] \quad \text{on } \partial\Omega.$$

However, by virtue of inequality (11.17) it follows that

$$\widetilde{\Gamma}\varphi_2(z) \geq \int_\Omega \widetilde{t}(z, dw) \left(\sigma_1(z, w) - \sigma(z, w) \right) w_N \quad \text{on } \partial\Omega.$$

Therefore, by shrinking the support $\operatorname{supp} \sigma$ of σ to the diagonal

$$\Delta_{\widetilde{\Omega}} = \left\{ (z, z) : z \in \widetilde{\Omega} \right\},$$

we obtain from an application of the Lebesgue dominated convergence theorem that

$$\widetilde{\Gamma}\varphi_2(z) \geq \int_\Omega \widetilde{t}(z, dw)\,\sigma_1(z, w)\,w_N \quad \text{on } \partial\Omega. \tag{11.19}$$

Indeed, it suffices to note that

$$\sigma(z, w)\,w_N \longrightarrow 0 \quad \text{as } w \to z.$$

Hence, by assertion (11.19) we find that

$$\int_U t(x', dy)\,f(y)\,\chi_N(y)$$

$$= \int_{\chi(U)} \widetilde{t}(z, dw)\,f(\chi^{-1}(w))\,w_N$$

$$= \int_\Omega \widetilde{t}(z, dw)\,\sigma_1(z, w)\,w_N\,f(\chi^{-1}(w)) \in B(U \cap \partial D)$$

for every non-negative function $f \in C(\overline{D})$ with supp $f \subset U$, if we choose the local unity function $\sigma_1(z, w)$ such that

$$\sigma_1(z, w) = 1 \quad \text{on supp}\,(f \circ \chi^{-1}).$$

The desired condition (NV.2) for the kernel $t(x', dy)$ follows by combining assertions (11.15) and (11.16). That is, for every local chart (U, χ) of $\overline{D}$ such that $U \cap \partial D \neq \emptyset$ and for every non-negative function $f \in C(\overline{D})$ with supp $f \subset U$, the function

$$U \cap \partial D \ni x' \longmapsto \int_U t(x', dy)\,f(y)\left[\chi_N(y) + \sum_{j=1}^{N-1}\left(\chi_j(y) - \chi_j(x')\right)^2\right]$$

belongs to the space $B_{\mathrm{loc}}\,(U \cap \partial D)$.

Summing up, we have proved that $t(x', dy)$ is a Lévy kernel on the manifold $\overline{D}$.

Step 4: Now we prove assertion (ii). By condition (NV.2), we can rewrite formula (11.12) in the form

$$\widetilde{\Gamma}\varphi(z) = \left(\widetilde{b}^N(z) - \int_\Omega \widetilde{t}(z, dw)\,w_N\right)\frac{\partial\varphi}{\partial z_N}(z) \tag{11.20}$$

$$+ \sum_{i,j=1}^N \widetilde{b}^{ij}(z)\frac{\partial^2\varphi}{\partial z_i \partial z_j}(z) + \sum_{i=1}^{N-1} \widetilde{b}^i(z)\frac{\partial\varphi}{\partial z_i}(z) + \widetilde{b}(z)\varphi(z)$$

$$+ \int_\Omega \widetilde{t}(z, dw)\left[\varphi(w) - \sigma(z, w)\left(\varphi(z) + \sum_{j=1}^{N-1}\frac{\partial\varphi}{\partial z_j}(z)(w_j - z_j)\right)\right].$$

Substep 4-1: We show assertion (11.11), that is,

$$\widetilde{\mu}(z) = \widetilde{b}^N(z) - \int_\Omega \widetilde{t}(z, dw)\,\sigma(z, w)\,w_N \geq 0 \quad \text{on } \partial\Omega. \tag{11.21}$$

Indeed, by applying the positive boundary maximum principle (PMB) to the function $-\varphi_2(w)$ we obtain that

$$0 \leq \widetilde{\Gamma}\varphi_2(z) = \widetilde{b}^N(z)$$
$$+ \int_\Omega \widetilde{t}(z, dw)\left[\varphi_2(w) - \sigma(z, w)\left(\varphi_2(z) + \sum_{j=1}^N \frac{\partial \varphi_2}{\partial z_j}(z)(w_j - z_j)\right)\right]$$
$$= \widetilde{b}^N(z) + \int_\Omega \widetilde{t}(z, dw)\left[\sigma_1(z, w)\, w_N - \sigma(z, w)\, w_N\right]$$
$$= \widetilde{\mu}(z) + \int_\Omega \widetilde{t}(z, dw)\,\sigma_1(z, w)\, w_N \quad \text{on } \partial\Omega,$$

so that

$$\widetilde{\mu}(z) \geq -\int_\Omega \widetilde{t}(z, dw)\,\sigma_1(z, w)\, w_N \quad \text{on } \partial\Omega.$$

Therefore, by shrinking the support $\operatorname{supp}\sigma_1$ of σ_1 to the diagonal $\Delta_{\widetilde{\Omega}}$ we find from the condition (NV.2) that

$$\widetilde{\mu}(z) \geq 0 \quad \text{on } \partial\Omega.$$

Substep 4-2: We show assertion (11.10). First, we prove that

$$\widetilde{b}^{NN}(z) \equiv 0 \quad \text{on } \partial\Omega. \tag{11.22}$$

In view of condition (11.13), we assume, to the contrary, that

$$\widetilde{b}^{NN}(z_0) > 0 \quad \text{for some point } z_0 \in \partial\Omega.$$

For $z \in \partial\Omega$, we let

$$\varphi_3(w) = \sigma_1(z, w)\,\theta(w_N) \quad \text{for every } w \in \widetilde{\Omega},$$

where $\sigma_1(z, w)$ is a local unity function on $\widetilde{\Omega}$ and $\theta(w_N)$ is a non-negative function in $C^2[0, \infty)$ that will be chosen later on. By applying the positive boundary maximum principle (PMB) to the function $-\varphi_3(w)$, we obtain from formula (11.20) that

$$\widetilde{\Gamma}\varphi_3(z_0) = \widetilde{\mu}(z_0)\,\theta'(0) + \widetilde{b}(z_0)\,\theta(0) + \widetilde{b}^{NN}(z_0)\,\theta''(0) \tag{11.23}$$
$$+ \int_\Omega \widetilde{t}(z_0, dw)\,\sigma_1(z_0, w)\,(\theta(w_N) - \theta(0))$$
$$\geq 0.$$

However, we can choose a function $\theta(w_N)$ such that

$$\theta(0) = \sup_{[0,\infty)]} \theta > 0,$$

$$\theta'(0) = 0,$$
$$\theta''(0) < 0,$$

and further that

$$\widetilde{b}(z_0)\,\theta(0) + \widetilde{b}^{NN}(z_0)\,\theta''(0) < 0, \tag{11.24}$$

if $\theta(0)$ is sufficiently small and $|\theta''(0)|$ is sufficiently large. Therefore, by shrinking the support $\operatorname{supp}\sigma_1$ of σ_1 to the diagonal $\Delta_{\widetilde{\Omega}}$ we obtain from inequality (11.24) that

$$\widetilde{\Gamma}\varphi_3(z_0)$$
$$= \widetilde{b}(z_0)\,\theta(0) + \widetilde{b}^{NN}(z_0)\,\theta''(0) + \int_\Omega \widetilde{t}(z_0, dw)\,\sigma_1(z_0, w)\,(\theta(w_N) - \theta(0))$$
$$< 0.$$

This contradicts inequality (11.23).

Secondly, we show that

$$\widetilde{b}^{N\,1}(z) = \ldots = \widetilde{b}^{N\,N-1}(z) \equiv 0 \quad \text{on } \partial\Omega. \tag{11.25}$$

We assume, to the contrary, that

$$\widetilde{b}^{Nj}(z_0) \neq 0 \quad \text{for some } z_0 = (z_1, \ldots, z_{N-1}, 0) \in \partial\Omega \text{ and } 1 \leq j \leq N - 1.$$

For $z \in \partial\Omega$, we let

$$\varphi_4(w) = \sigma_1(z, w)\,\theta(w_j)\,w_N \quad \text{for every } w \in \widetilde{\Omega},$$

where $\sigma_1(z, w)$ is a local unity function on $\widetilde{\Omega}$ and $\theta(w_j)$ is a non-negative function in $C^2[0, \infty)$ that will be chosen later on. By applying the positive boundary maximum principle (PMB) to the function $-\varphi_4(w)$, we obtain from formula (11.20) that

$$\widetilde{\Gamma}\varphi_4(z_0) \tag{11.26}$$
$$= \widetilde{\mu}(z_0)\,\theta(z_j) + \widetilde{b}^{Nj}(z_0)\,\theta'(z_j) + \int_\Omega \widetilde{t}(z_0, dw)\,\sigma_1(z, w)\,\theta(w_j)\,w_N$$
$$\geq 0.$$

However, we can choose a function $\theta(w_j)$ such that

$$\theta'(z_j) < 0 \quad \text{if } \widetilde{b}^{Nj}(z) > 0,$$
$$\theta'(z_j) > 0 \quad \text{if } \widetilde{b}^{Nj}(z) < 0,$$

and further that

$$\widetilde{\mu}(z_0)\,\theta(z_j) + \widetilde{b}^{Nj}(z_0)\,\theta'(z_j) < 0, \tag{11.27}$$

if $\theta(z_j)$ is sufficiently small and $|\theta'(z_j)|$ is sufficiently large. Therefore, by shrinking the support $\operatorname{supp}\sigma_1$ of σ_1 to the diagonal $\Delta_{\widehat{\Omega}}$ we obtain from inequality (11.27) that

$$\widetilde{\Gamma}\varphi_4(z_0)$$
$$= \widetilde{\mu}(z_0)\,\theta(z_j) + \widetilde{b}^{Nj}(z_0)\,\theta'(z_j) + \int_\Omega \widetilde{t}(z_0, dw)\,\sigma_1(z_0, w)\,\theta(w_j)\,w_N$$
$$< 0.$$

This contradicts inequality (11.26).

Therefore, the desired assertion (11.10) follows by combining assertions (11.13), (11.22) and (11.25).

Step 5: Finally, we prove assertion (11.9). Now we consider the following:

(1) Let $t(x', dy)$ be the Ventcel' kernel on the manifold $\overline{D}$ constructed in Step 3.

(2) Let $\zeta(x')$ be a bounded, Borel measurable vector field on the boundary ∂D. More precisely, we have, in a local chart (U_β, χ_β),

$$\zeta(x') \cdot u = \sum_{j=1}^{N-1} \zeta^j(x')\frac{\partial u}{\partial \chi_j^\beta} \quad \text{with } \zeta^j \in B_{\mathrm{loc}}(U_\beta \cap \partial D).$$

(3) Let $\eta(x')$ be a bounded, Borel measurable function on the boundary ∂D such that

$$\eta(x') + \int_{\overline{D}} t(x', dy)\left(1 - \sum_{\beta=1}^m \tau_\beta(x', y)\right) \le 0 \quad \text{on } \partial D.$$

Then we can define a Ventcel'–Lévy boundary operator

$$\boxed{\widetilde{T}: C^2(\overline{D}) \longrightarrow B(\partial D)}$$

by formula (11.4)

$$\widetilde{T}u(x') = \eta(x')u(x') + \zeta(x') \cdot u(x') + \int_{\overline{D}} t(x', dy)\Big[u(y)$$
$$- \sum_{\beta=1}^m \tau_\beta(x', y)\left(u(x') + \sum_{j=1}^{N-1}(\chi_j^\beta(y) - \chi_j^\beta(x'))\frac{\partial u}{\partial \chi_j^\beta}(x')\right)\Big].$$

We remark that the above condition (3) is equivalent to the following:

$$(\widetilde{T}1)(x') \le 0 \quad \text{on } \partial D.$$

If we let

$$\begin{cases} Tu := \widetilde{T}u + \left(\varGamma 1 - \widetilde{T}1 \right) u, \\ \varLambda u := \varGamma u - \widetilde{T}u - \left(\varGamma 1 - \widetilde{T}1 \right) u, \end{cases}$$

then we can obtain the desired decomposition (11.9) of the boundary operator

$$\varGamma = T + \varLambda.$$

Now the proof of Theorem 11.4 is complete. □

11.3 Notes and Comments

Section 11.1: Theorem 11.3 is taken from Bony–Courrège–Priouret [15, p. 441, assertion (PMB)].

Section 11.2: Theorem 11.4 is taken from Bony–Courrège–Priouret [15, p. 441, Théorème X].

This chapter is based on the lecture entitled *A class of hypoelliptic Vishik–Wentzell boundary vale problems* delivered at Functional Analysis Methods for Partial Differential Equations, Centro Polifunzionale, Cesena, Italy, on June 27, 2019.

Part V

Feller Semigroups for Elliptic Waldenfels Operators

Proof of Theorem 1.5 - Part (i) -

Part V (Chapters 12 through 14) is devoted to the proof of generation theorems of Feller semigroup for second-order, elliptic Waldenfels integro-differential operators.

This Chapter 12 and the next Chapter 13 are devoted to the proof of Theorem 1.5 and Theorem 1.6 (cf. [121], [123]). In this chapter we prove part (i) of Theorem 1.5. In the proof we make use of Sobolev's imbedding theorems (Theorems 4.8 and 4.9) and a λ-*dependent localization* argument due to Masuda [78] (Lemma 12.2) in order to adjust the resolvent estimate

$$\left\| (A_p - \lambda I)^{-1} \right\| \leq \frac{c_p(\varepsilon)}{|\lambda|} \quad \text{for all } \lambda \in \Sigma_p(\varepsilon) \tag{1.9}$$

to obtain the desired estimate

$$\left\| (\mathfrak{A} - \lambda I)^{-1} \right\| \leq \frac{c(\varepsilon)}{|\lambda|} \quad \text{for all } \lambda \in \Sigma(\varepsilon). \tag{1.11}$$

Here recall that the operator

$$\boxed{A_p \colon L^p(D) \longrightarrow L^p(D)}$$

is a densely defined, closed linear operator defined by the following:

(a) The domain of definition $\mathcal{D}(A_p)$ is the set

$$\mathcal{D}(A_p) = \left\{ u \in H^{2,p}(D) = W^{2,p}(D) : Lu = 0 \text{ on } \partial D \right\}. \tag{1.8}$$

(b) $A_p u = Au$ for every $u \in \mathcal{D}(A_p)$.

Furthermore, the operator

$$\boxed{\mathfrak{A} \colon C_0\left(\overline{D} \setminus M\right) \longrightarrow C_0\left(\overline{D} \setminus M\right)}$$

© Springer Nature Switzerland AG 2020
K. Taira, *Boundary Value Problems and Markov Processes*, Lecture Notes in Mathematics 1499,
https://doi.org/10.1007/978-3-030-48788-1_12

is a densely defined, closed linear operator defined by the following:

(c) The domain of definition $\mathcal{D}(\mathfrak{A})$ is the set

$$\mathcal{D}(\mathfrak{A}) = \left\{ u \in C_0\left(\overline{D} \setminus M\right) : Au \in C_0\left(\overline{D} \setminus M\right), \ Lu = 0 \text{ on } \partial D \right\}. \quad (1.10)$$

(d) $\mathfrak{A}u = Au$ for every $u \in \mathcal{D}(\mathfrak{A})$.

The proof of part (i) of Theorem 1.5 can be flowcharted as in Table 12.1 below.

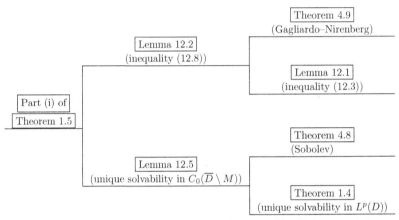

Table 12.1. A flowchart for the proof of part (i) of Theorem 1.5

12.1 Space $C_0\left(\overline{D} \setminus M\right)$

First, we consider a one-point compactification $K_\partial = K \cup \{\partial\}$ of the space $K = \overline{D} \setminus M$.

We say that two points x and y of $\overline{D}$ are equivalent modulo M if $x = y$ or $x, y \in M$; we then write $x \sim y$. It is easy to verify that this relation $\sim$ enjoys the so-called equivalence laws. We denote by $\overline{D}/M$ the totality of equivalence classes modulo M. On the set $\overline{D}/M$ we define the quotient topology induced by the projection

$$q \colon \overline{D} \longrightarrow \overline{D}/M.$$

Namely, a subset O of $\overline{D}/M$ is defined to be open if and only if the inverse image $q^{-1}(O)$ of O is open in $\overline{D}$. It is easy to see that the topological space $\overline{D}/M$ is a *one-point compactification* of the space $\overline{D} \setminus M$ and that the *point at infinity* ∂ corresponds to the set M (see Figure 12.1 below):

$$\begin{cases} K_\partial := \overline{D}/M, \\ \partial := M. \end{cases}$$

Furthermore, we obtain the following two assertions (i) and (ii):

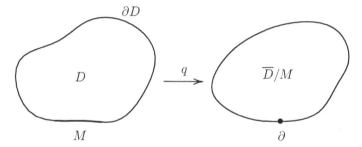

Fig. 12.1. The compactification $\overline{D}/M$ of $\overline{D} \setminus M$

(i) If $\tilde{u}$ is a continuous function defined on K_∂, then the function $\tilde{u} \circ q$ is continuous on $\overline{D}$ and constant on M.

(ii) Conversely, if u is a continuous function defined on $\overline{D}$ and constant on M, then it defines a continuous function $\tilde{u}$ on K_∂.

In other words, we have the following isomorphism:

$$C(K_\partial) \cong \left\{ u \in C(\overline{D}) : u(x) \text{ is constant on } M \right\}. \tag{12.1}$$

Now we introduce a closed subspace of $C(K_\partial)$ as in Subsection 4.1.1:

$$C_0(K) = \left\{ u \in C(K_\partial) : u(\partial) = 0 \right\}.$$

Then we have, by assertion (12.1),

$$C_0(K) \cong C_0\left(\overline{D} \setminus M \right) = \left\{ u \in C(\overline{D}) : u(x) = 0 \text{ on } M \right\}. \tag{12.2}$$

12.2 Proof of Part (i) of Theorem 1.5

The proof is carried out in a chain of auxiliary lemmas.

Step (I): We begin with the following fundamental inequality (see [121, p. 115, Lemma 8.3]):

Lemma 12.1. *Let $N < p < \infty$. Assume that the following two conditions (A) and (B) are satisfied:*

(A) $\mu(x') \geq 0$ and $\gamma(x') \leq 0$ on ∂D.
(B) $\mu(x') - \gamma(x') = \mu(x') + |\gamma(x')| > 0$ on ∂D.

Then, for every $\varepsilon > 0$, there exists a constant $r'_p(\varepsilon) > 0$ such that if $\lambda = r^2 e^{i\theta}$ with $r \geq r'_p(\varepsilon)$ and $-\pi + \varepsilon \leq \theta \leq \pi - \varepsilon$, we have, for all $u \in \mathcal{D}(A_p)$,

$$|\lambda|^{1/2} |u|_{C^1(\overline{D})} + |\lambda| \cdot |u|_{C(\overline{D})} \leq c'_p(\varepsilon) |\lambda|^{N/2p} \|(A - \lambda)u\|_p, \tag{12.3}$$

with a constant $c'_p(\varepsilon) > 0$. Here

$$\mathcal{D}(A_p) = \left\{ u \in H^{2,p}(D) : Lu = \mu(x') \frac{\partial u}{\partial \mathbf{n}} + \gamma(x')u = 0 \right\}.$$

Proof. First, we know from [121, p. 102, estimate (7.1)] that the estimate

$$|u|_{2,p} + |\lambda|^{1/2} |u|_{1,p} + |\lambda| \, \|u\|_p \le c_p'(\varepsilon) \, \|(A_p - \lambda I)u\|_p \qquad (12.4)$$

holds true for all $u \in \mathcal{D}(A_p)$, where

$$\|u\|_p = \|u\|_{L^p(D)}, \quad |u|_{1,p} = \|\nabla u\|_{L^p(D)}, \quad |u|_{2,p} = \|\nabla^2 u\|_{L^p(D)}.$$

On the other hand, by applying Theorem 4.9 with $p := r > N$, $\theta := N/p$ and $\nu := 0$ we obtain from the Gagliardo–Nirenberg inequality (4.6) that

$$|u|_{C(\overline{D})} \le C \, |u|_{1,p}^{N/p} \, \|u\|_p^{1-N/p} \quad \text{for all } u \in H^{2,p}(D). \qquad (12.5)$$

Here and in the following the letter C denotes a generic positive constant depending on p and ε, but independent of u and λ.

By using estimate (12.4), we obtain from inequality (12.5) that

$$\begin{aligned}
|u|_{C(\overline{D})} &\le C \, |u|_{1,p}^{N/p} \, \|u\|_p^{1-N/p} \\
&\le C \left(|\lambda|^{-1/2} \, \|(A - \lambda)u\|_p \right)^{N/p} \left(|\lambda|^{-1} \, \|(A - \lambda)u\|_p \right)^{1-N/p} \\
&= C \, |\lambda|^{-1+N/2p} \, \|(A - \lambda)u\|_p.
\end{aligned}$$

This proves that

$$|\lambda| \cdot |u|_{C(\overline{D})} \le C \, |\lambda|^{N/2p} \, \|(A - \lambda)u\|_p \quad \text{for all } u \in \mathcal{D}(A_p). \qquad (12.6)$$

Similarly, by applying inequality (12.5) to the functions

$$D_i u \in H^{1,p}(D) \quad \text{for } 1 \le i \le n,$$

we obtain that

$$\begin{aligned}
|D_i u|_{C(\overline{D})} &\le C \, |D_i u|_{1,p}^{N/p} \, \|D_i u\|_p^{1-N/p} \le C \, |u|_{2,p}^{N/p} \, |u|_{1,p}^{1-N/p} \\
&\le C \left(\|(A - \lambda)u\|_p \right)^{N/p} \left(|\lambda|^{-1/2} \, \|(A - \lambda)u\|_p \right)^{1-N/p} \\
&= C \, |\lambda|^{-1/2+N/2p} \, \|(A - \lambda)u\|_p.
\end{aligned}$$

This proves that

$$|\lambda|^{1/2} \, |u|_{C^1(\overline{D})} \le C \, |\lambda|^{N/2p} \, \|(A - \lambda)u\|_p \quad \text{for all } u \in \mathcal{D}(A_p). \qquad (12.7)$$

Therefore, the desired inequality (12.3) follows by combining inequalities (12.6) and (12.7).

The proof of Lemma 12.1 is complete. $\square$

The next lemma proves the resolvent estimate (1.7):

Lemma 12.2. *Assume that conditions (A) and (B) are satisfied. Then, for every $\varepsilon > 0$, there exists a constant $r(\varepsilon) > 0$ such that if $\lambda = r^2 e^{i\theta}$ with $r \geq r(\varepsilon)$ and $-\pi + \varepsilon \leq \theta \leq \pi - \varepsilon$, we have, for all $u \in \mathcal{D}(\mathfrak{A})$,*

$$|\lambda|^{1/2}|u|_{C^1(\overline{D})} + |\lambda| \cdot |u|_{C(\overline{D})} \leq c(\varepsilon)|(A - \lambda)u|_{C(\overline{D})}, \tag{12.8}$$

with a constant $c(\varepsilon) > 0$. Here

$$\mathcal{D}(\mathfrak{A}) = \left\{ u \in C_0\left(\overline{D} \setminus M\right) : Au \in C_0\left(\overline{D} \setminus M\right), \ Lu = 0 \ on \ \partial D \right\}.$$

Proof. We shall make use of a λ-*dependent localization* argument due to Masuda [78] in order to adjust the term $\|(A - \lambda)u\|_p$ in inequality (12.3) to obtain inequality (12.8).

First, we remark that

$$\mathfrak{A} \subset A_p \quad \text{for all} \quad 1 < p < \infty.$$

Indeed, since we have, for any $u \in \mathcal{D}(\mathfrak{A})$,

$$u \in C(\overline{D}) \subset L^p(D), \ Au \in C(\overline{D}) \subset L^p(D) \quad \text{and} \quad Lu = 0 \text{ on } \partial D,$$

it follows from an application of Theorem 6.21 and [123, Theorem 7.1] (see also [68, Theorem 3.1]) that

$$u \in H^{2,p}(D).$$

(1) Let x_0 be an arbitrary point of the closure $\overline{D} = D \cup \partial D$.

If x_0' is a *boundary* point and if χ is a smooth coordinate transformation such that χ maps $B(x_0, \eta_0) \cap D$ into $B(0, \delta) \cap \mathbf{R}_+^N$ and flattens a part of the boundary ∂D into the plane $x_N = 0$ (see Figure 12.2 below), then we let

- $G_0 = B(x_0', \eta_0) \cap D,$
- $G' = B(x_0', \eta) \cap D$ for $0 < \eta < \eta_0,$
- $G'' = B(x_0', \eta/2) \cap D$ for $0 < \eta < \eta_0,$

where $B(x, \eta)$ denotes the open ball of radius η about x (see Figure 12.3 below).

Similarly, if x_0 is an *interior* point and if χ is a smooth coordinate transformation such that χ maps $B(x_0, \eta_0)$ into $B(0, \delta)$, then we let (see Figure 12.4 below)

- $G_0 = B(x_0, \eta_0),$
- $G' = B(x_0, \eta)$ for $0 < \eta < \eta_0,$
- $G'' = B(x_0, \eta/2)$ for $0 < \eta < \eta_0.$

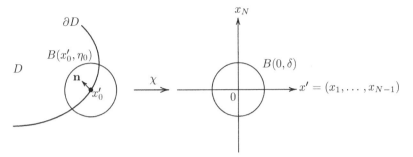

Fig. 12.2. The coordinate transformation χ maps $B(x_0, \eta_0)$ into $B(0, \delta)$

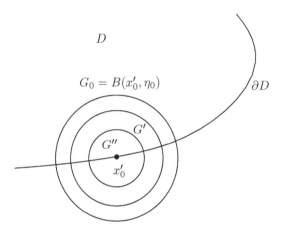

Fig. 12.3. The half-balls G_0, G' and G''

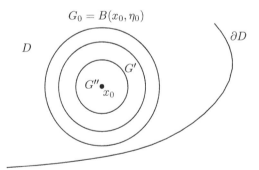

Fig. 12.4. The open balls G_0, G' and G''

(2) Now we take a function $\theta(t)$ in $C_0^\infty(\mathbf{R})$ such that $\theta(t)$ equals one near the origin, and define a function

$$\varphi(x) = \theta(|x'|^2)\,\theta(x_N) \quad \text{for } x = (x', x_N).$$

Here we may assume that the function $\varphi(x)$ is chosen so that

$$\begin{cases} \operatorname{supp} \varphi \subset B(0,1), \\ \varphi(x) = 1 \text{ on } B(0,1/2). \end{cases}$$

We introduce a localizing function

$$\varphi_0(x, \eta) \equiv \varphi\left(\frac{x - x_0}{\eta}\right) = \theta\left(\frac{|x' - x_0'|^2}{\eta^2}\right) \theta\left(\frac{x_N - t}{\eta}\right)$$

$$\text{for } x_0 = (x_0', t) \in \overline{D}.$$

We remark that

$$\begin{cases} \operatorname{supp} \varphi_0 \subset B(x_0, \eta), \\ \varphi_0(x, \eta) = 1 \text{ on } B(x_0, \eta/2). \end{cases}$$

Then, for the localizing function $\varphi_0(x, \eta)$ we have the following claim:

Claim 12.3. *If $u \in \mathcal{D}(\mathfrak{A})$, then it follows that $\varphi_0(x, \eta)u \in \mathcal{D}(A_p)$ for all $1 < p < \infty$.*

Proof. (i) First, we recall that

$$u \in H^{2,p}(D) \quad \text{for all } 1 < p < \infty.$$

Hence we have the assertion

$$\varphi_0(x, \eta)u \in H^{2,p}(D).$$

(ii) Secondly, it is easy to verify (see Figure 12.5 below) that the function $\varphi_0(x, \eta)u$, $x \in \overline{D}$, satisfies the boundary condition

$$L(\varphi_0(x', \eta)u) = 0 \quad \text{on } \partial D.$$

Indeed, this is obvious if we have the condition

$$\operatorname{supp}(\varphi_0(x, \eta)u) \subset B(x_0, \eta) \quad \text{for } x_0 \in D.$$

Moreover, if we have the condition

$$\operatorname{supp}(\varphi_0(x, \eta)u) \subset B(x_0, \eta) \cap \overline{D} \quad \text{for } x_0 \in \partial D,$$

then it follows that

$$\left.\frac{\partial}{\partial x_N}(\varphi_0(x, \eta))\right|_{x_N=0} = \frac{1}{\eta} \theta'(0) \cdot \theta\left(\frac{|x' - x_0'|^2}{\eta^2}\right) = 0,$$

since $\theta'(0) = 0$. This proves that

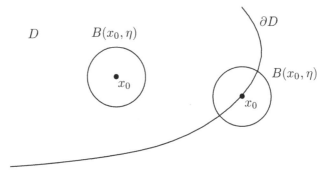

Fig. 12.5. The open balls $B(x_0, \eta)$ in D and on ∂D

$$\frac{\partial}{\partial \mathbf{n}}(\varphi_0(x', \eta)) = 0 \quad \text{on } \partial D.$$

Therefore, we have the assertion

$$L(\varphi_0(x, \eta)u) = \mu(x')\frac{\partial}{\partial \mathbf{n}}(\varphi_0(x', \eta)u) + \gamma(x')\varphi_0(x', \eta)u$$

$$= \varphi_0(x', \eta)(Lu) + \mu(x')\left(\frac{\partial}{\partial \mathbf{n}}(\varphi_0(x', \eta))\right)u$$

$$= 0 \quad \text{on } \partial D,$$

since $Lu = 0$ on ∂D.

Summing up, we have proved that

$$\varphi_0(x, \eta)u \in \mathcal{D}(A_p) \quad \text{for all } 1 < p < \infty.$$

The proof of Claim 12.3 is complete.   □

(3) Now we take a positive number p such that

$$N < p < \infty.$$

By virtue of Claim 12.3, we can apply the fundamental inequality (12.3) to $\varphi_0(x, \eta)\,u$ with $u \in \mathcal{D}(\mathfrak{A})$ to obtain that

$$|\lambda|^{1/2}|u|_{C^1(\overline{G''})} + |\lambda| \cdot |u|_{C(\overline{G''})} \tag{12.9}$$

$$\leq |\lambda|^{1/2}|\varphi_0(x, \eta)u|_{C^1(\overline{G'})} + |\lambda| \cdot |\varphi_0(x, \eta)u|_{C(\overline{G'})}$$

$$= |\lambda|^{1/2}|\varphi_0(x, \eta)u|_{C^1(\overline{D})} + |\lambda| \cdot |\varphi_0(x, \eta)u|_{C(\overline{D})}$$

$$\leq C|\lambda|^{N/2p}\|(A - \lambda)(\varphi_0(x, \eta)u)\|_{L^p(D)}$$

$$= C|\lambda|^{N/2p}\|(A - \lambda)(\varphi_0(x, \eta)u)\|_{L^p(G')} \quad \text{for } 0 < \eta < \eta_0,$$

since we have the assertions

$$\begin{cases} \varphi_0(x,\eta) = 1 \quad \text{on } G'', \\ \text{supp}\,(\varphi_0(x,\eta)u) \subset \overline{G'}. \end{cases}$$

However, we have the formula

$$(A - \lambda)\,(\varphi_0(x,\eta)u) = \varphi_0(x,\eta)\,((A - \lambda)u) + [A,\varphi_0(x,\eta)]\,u, \qquad (12.10)$$

where $[A,\varphi_0(x,\eta)]$ is the commutator of A and $\varphi_0(x,\eta)$ defined by the formula

$$[A,\varphi_0(x,\eta)]\,u \qquad\qquad\qquad (12.11)$$
$$= A\,(\varphi_0(x,\eta)u) - \varphi_0(x,\eta)Au$$
$$= 2\sum_{i,j=1}^{N} a^{ij}(x)\frac{\partial\varphi_0}{\partial x_i}\frac{\partial u}{\partial x_j} + \left(\sum_{i,j=1}^{N} a^{ij}(x)\frac{\partial^2\varphi_0}{\partial x_i\partial x_j} + \sum_{i=1}^{N} b^i(x)\frac{\partial\varphi_0}{\partial x_i}\right)u.$$

Here we need the following elementary inequality:

Claim 12.4. *We have, for all $v \in C^j(\overline{G'})$, $j = 0, 1, 2$,*

$$\|v\|_{H^{j,p}(G')} \le |G'|^{1/p}\,\|v\|_{C^j(\overline{G'})},$$

where $|G'|$ denotes the measure of G'.

Proof. It suffices to note that we have, for all $w \in C(\overline{G'})$,

$$\int_{G'} |w(x)|^p dx \le |G'|\,|w|_{C(\overline{G'})}^p.$$

This proves Claim 12.4.   □

Since we have (see Figure 12.3), for some positive constant c,

$$|G'| \le |B\,(x_0,\eta)| \le c\eta^N,$$

it follows from an application of Claim 12.4 that

$$\|\varphi_0(x,\eta)\,((A - \lambda)\,u)\|_{L^p(G')} \le c^{1/p}\eta^{N/p}\,|(A - \lambda)\,u|_{C(\overline{G'})} \qquad (12.12)$$
$$\text{for } 0 < \eta < \eta_0.$$

Furthermore, we remark that

$$|D^\alpha\varphi_0(x,\eta)| = O\left(\eta^{-|\alpha|}\right) \quad \text{as } \eta \downarrow 0.$$

Hence it follows from an application of Claim 12.4 that we have, for $0 < \eta < \eta_0$,

$$\bullet\ \left\|\frac{\partial\varphi_0}{\partial x_i}\frac{\partial u}{\partial x_j}\right\|_{L^p(G')} \le \frac{C}{\eta}\,|u|_{1,p,G'} \le C\eta^{-1+N/p}\,|u|_{C^1(\overline{G'})}, \qquad (12.13)$$

- $$\left\|\frac{\partial^2 \varphi_0}{\partial x_i \partial x_j} u\right\|_{L^p(G')} \leq \frac{C}{\eta^2} |u|_{L^p(G')} \leq C\eta^{-2+N/p} |u|_{C(\overline{G'})}, \qquad (12.14)$$

- $$\left\|\frac{\partial \varphi_0}{\partial x_i} u\right\|_{L^p(G')} \leq \frac{C}{\eta} |u|_{L^p(G')} \leq C\eta^{-1+N/p} |u|_{C(\overline{G'})}. \qquad (12.15)$$

By using inequalities (12.13), (12.14) and (12.15), we obtain from formula (12.11) that

$$\|[A, \varphi_0(x,\eta)] u\|_{L^p(G')} \qquad (12.16)$$
$$\leq C\left(\eta^{-1+N/p} |u|_{C^1(\overline{G'})} + \eta^{-2+N/p} |u|_{C(\overline{G'})} + \eta^{-1+N/p} |u|_{C(\overline{G'})}\right)$$
$$\leq C\left(\eta^{-1+N/p} |u|_{C^1(\overline{D})} + \eta^{-2+N/p} |u|_{C(\overline{D})}\right) \quad \text{for } 0 < \eta < \eta_0.$$

In view of formula (12.10), it follows from inequalities (12.12) and (12.16) that

$$\|(A - \lambda)(\varphi_0(x,\eta)u)\|_{L^p(G')} \qquad (12.17)$$
$$\leq \|\varphi_0(x,\eta)((A - \lambda)u)\|_{L^p(G')} + \|[A, \varphi_0(x,\eta)]u\|_{L^p(G')}$$
$$\leq C\eta^{N/p}\left(|(A - \lambda)u|_{C(\overline{G'})} + \eta^{-1} |u|_{C^1(\overline{D})} + \eta^{-2} |u|_{C(\overline{D})}\right)$$
$$\text{for } 0 < \eta < \eta_0.$$

Therefore, by combining inequalities (12.9) and (12.17) we obtain that

$$|\lambda|^{1/2} |u|_{C^1(\overline{G''})} + |\lambda| \cdot |u|_{C(\overline{G''})} \qquad (12.18)$$
$$\leq C |\lambda|^{N/2p} \|(A - \lambda)(\varphi_0(x,\eta)u)\|_{L^p(G')}$$
$$\leq C |\lambda|^{N/2p} \eta^{N/p}\left(|(A - \lambda)u|_{C(\overline{G'})} + \eta^{-1} |u|_{C^1(\overline{G'})} + \eta^{-2} |u|_{C(\overline{G'})}\right)$$
$$\leq C |\lambda|^{N/2p} \eta^{N/p}\left(|(A - \lambda)u|_{C(\overline{D})} + \eta^{-1} |u|_{C^1(\overline{D})} + \eta^{-2} |u|_{C(\overline{D})}\right)$$
$$\text{for } 0 < \eta < \eta_0.$$

We remark (see Figure 12.6 below) that the closure $\overline{D} = D \cup \partial D$ can be covered by a finite number of sets of the forms

$$B(x_0', \eta/2) \cap \overline{D} \quad x_0' \in \partial D,$$

and

$$B(x_0, \eta/2) \quad x_0 \in D.$$

Hence, by taking the supremum of inequality (12.18) over $x \in \overline{D}$ we find that

$$|\lambda|^{1/2} |u|_{C^1(\overline{D})} + |\lambda| \cdot |u|_{C(\overline{D})} \qquad (12.19)$$

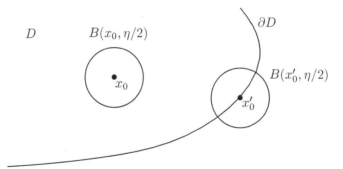

Fig. 12.6. The open ball $B(x_0, \eta/2)$ in D and the open ball $B(x_0', \eta/2)$ on ∂D

$$\leq C\,|\lambda|^{N/2p}\,\eta^{N/p}\left(|(A-\lambda)u|_{C(\overline{D})} + \eta^{-1}\,|u|_{C^1(\overline{D})} + \eta^{-2}\,|u|_{C(\overline{D})}\right)$$

for all $u \in \mathcal{D}(\mathfrak{A})$.

Here we remark that

$$0 < \eta < \eta_0. \tag{12.20}$$

(4) We now choose the localization parameter η. To do so, we take a constant K so that

$$0 < K < r_p'(\varepsilon). \tag{12.21}$$

For a complex number $\lambda = r^2\,e^{i\theta}$ with $r \geq r_p'(\varepsilon)$, we let

$$\eta := \frac{\eta_0}{|\lambda|^{1/2}}K, \tag{12.22}$$

where K is chosen later on.

Then the parameter η satisfies condition (12.20), since we have, by condition (12.21),

$$\eta = \frac{\eta_0}{|\lambda|^{1/2}}K = \frac{\eta_0}{r}K \leq \frac{\eta_0}{r_p'(\varepsilon)}K < \eta_0.$$

Hence it follows from inequality (12.19) that

$$|\lambda|^{1/2}\,|u|_{C^1(\overline{D})} + |\lambda| \cdot |u|_{C(\overline{D})} \tag{12.23}$$

$$\leq C\,\eta_0^{N/p}K^{N/p}\,|(A-\lambda)u|_{C(\overline{D})} + \left(C\,\eta_0^{N/p-1}K^{-1+N/p}\right)|\lambda|^{1/2} \cdot |u|_{C^1(\overline{D})}$$

$$+ \left(C\,\eta_0^{N/p-2}K^{-2+N/p}\right)|\lambda| \cdot |u|_{C(\overline{D})} \quad \text{for all } u \in \mathcal{D}(\mathfrak{A}).$$

However, since the exponents $-1 + N/p$ and $-2 + N/p$ are both negative for $N < p < \infty$, we can choose the constant K so large that

$$C\,\eta_0^{N/p-1}K^{-1+N/p} < 1,$$

and

$$C\,\eta_0^{N/p-2}K^{-2+N/p} < 1.$$

For example, we may take

$$K > \widetilde{C} := \frac{1}{\eta_0}\,\max\left\{C^{1/\sigma}, C^{1/(\sigma+1)}\right\}, \quad \sigma = 1 - \frac{N}{p} > 0. \tag{12.24}$$

Then the desired inequality (12.8) follows from inequality (12.23). Indeed, if we let

$$r(\varepsilon) := \max\left\{r_p'(\varepsilon), \widetilde{C} + 1\right\},$$

and choose the constant K such that

$$\widetilde{C} < K < r(\varepsilon),$$

then, for all complex numbers $\lambda = r^2\,e^{i\theta}$ with $r \geq r(\varepsilon)$ we have, by conditions (12.22) and (12.24),

- $0 < \eta < \eta_0$,
- $0 < K < |\lambda|^{1/2}$,
- $C\,\eta_0^{N/p-1}K^{-1+N/p} < 1$,
- $C\,\eta_0^{N/p-1}K^{-1+N/p} < 1$.

Now the proof of Lemma 12.2 is complete. □

Step (II): The next lemma, together with Lemma 12.2, proves that the resolvent set of $\mathfrak{A}$ contains the set

$$\Sigma(\varepsilon) = \left\{\lambda = r^2 e^{i\theta} : r \geq r(\varepsilon), \; -\pi + \varepsilon \leq \theta \leq \pi - \varepsilon\right\},$$

that is, the resolvent $(\mathfrak{A} - \lambda I)^{-1}$ exists for all $\lambda \in \Sigma(\varepsilon)$.

Lemma 12.5. *If $\lambda \in \Sigma(\varepsilon)$, then, for any $f \in C_0\left(\overline{D} \setminus M\right)$, there exists a unique function $u \in \mathcal{D}(\mathfrak{A})$ such that $(\mathfrak{A} - \lambda I)u = f$.*

Proof. Since we have the assertion

$$f \in C_0\left(\overline{D} \setminus M\right) \subset L^p(D) \quad \text{for all} \quad 1 < p < \infty,$$

it follows from an application of Theorem 1.4 that if $\lambda \in \Sigma(\varepsilon)$ there exists a unique function $u(x) \in H^{2,p}(D)$ such that

$$(A - \lambda)\,u = f \quad \text{in } D, \tag{12.25}$$

and

$$Lu = \mu(x')\frac{\partial u}{\partial \mathbf{n}} + \gamma(x')u\Big|_{\partial D} = 0 \quad \text{on } \partial D. \tag{12.26}$$

However, part (ii) of Theorem 4.8 asserts that

$$u \in H^{2,p}(D) \subset C^{2-N/p}(\overline{D}) \subset C^1(\overline{D}) \quad \text{if} \quad N < p < \infty.$$

Hence we have, by formula (12.26) and condition (B),

$$u = 0 \text{ on } M = \{x' \in \partial D : \mu(x') = 0\},$$

so that

$$u \in C_0\left(\overline{D} \setminus M\right).$$

Furthermore, in view of formula (12.25) it follows that

$$Au = f + \lambda u \in C_0\left(\overline{D} \setminus M\right).$$

Summing up, we have proved that

$$\begin{cases} u \in \mathcal{D}\left(\mathfrak{A}\right), \\ (\mathfrak{A} - \lambda I)u = f. \end{cases}$$

The proof of Lemma 12.5 is complete. □

Now the proof of part (i) of Theorem 1.5 is complete. □

12.3 Notes and Comments

This chapter is adapted from Taira [122, Chapter 11], which is an expanded and revised version of Chapter 8 of the second edition [121].

13

Proofs of Theorem 1.5, Part (ii) and Theorem 1.6

In this chapter we prove Theorem 1.6 in Section 13.3 and part (ii) of Theorem 1.5 in Section 13.4. This chapter is the heart of the subject. In Section 13.1 general existence theorems for Feller semigroups are formulated in terms of elliptic boundary value problems with spectral parameter (Theorem 13.14). In Section 13.2 we study Feller semigroups with reflecting barrier (Theorem 13.17) and then, by using these Feller semigroups we construct Feller semigroups corresponding to such a diffusion phenomenon that either absorption or reflection phenomenon occurs at each point of the boundary (Theorem 13.22). Our proof is based on the generation theorems of Feller semigroups discussed in Chapter 3.

We prove Theorem 1.6 in Section 13.3. To do so, we apply part (ii) of Theorem 3.34 (the Hille–Yosida theorem) to the operator $\mathfrak{A}$ defined by formula (1.11). The proof of Theorem 1.6 can be flowcharted as in Table 13.1 below.

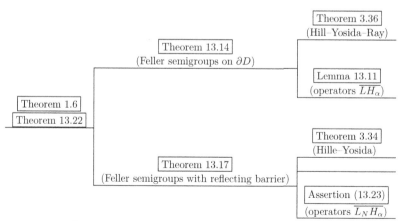

Table 13.1. A flowchart for the proof of Theorem 1.6

© Springer Nature Switzerland AG 2020
K. Taira, *Boundary Value Problems and Markov Processes*, Lecture Notes in Mathematics 1499,
https://doi.org/10.1007/978-3-030-48788-1_13

The proof of part (ii) of Theorem 1.5 in Section 13.4 can be flowcharted as in Table 13.2 below.

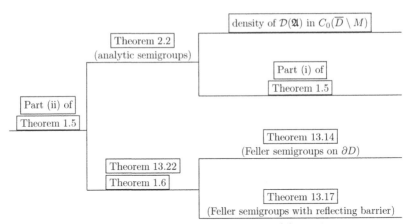

Table 13.2. A flowchart for the proof of part (ii) of Theorem 1.5

13.1 General Existence Theorem for Feller Semigroups

The purpose of this section is to give a general existence theorem for Feller semigroups in terms of boundary value problems (Theorem 13.14), following Taira [122, Chapter 10].

Let D be a bounded domain of Euclidean space $\mathbf{R}^N$, with smooth boundary ∂D; its closure $\overline{D} = D \cup \partial D$ is an N-dimensional, compact smooth manifold with boundary. We let

$$A = \sum_{i,j=1}^{N} a^{ij}(x) \frac{\partial^2}{\partial x_i \partial x_j} + \sum_{i=1}^{N} b^i(x) \frac{\partial}{\partial x_i} + c(x)$$

be a second-order, *uniformly elliptic* differential operator with real coefficients such that:

(1) $a^{ij} \in C^\infty(\overline{D})$ and $a^{ij}(x) = a^{ji}(x)$ for all $x \in \overline{D}$ and $1 \le i, j \le N$, and there exists a positive constant a_0 such that

$$\sum_{i,j=1}^{N} a^{ij}(x)\xi_i\xi_j \ge a_0|\xi|^2 \quad \text{for all } (x, \xi) \in \overline{D} \times \mathbf{R}^N.$$

(2) $b^i \in C^\infty(\overline{D})$.
(3) $c \in C^\infty(\overline{D})$ and $c(x) \le 0$ on $\overline{D}$.

The diffusion operator A describes analytically a strong Markov process with continuous paths in the interior D such as Brownian motion (see Figure 1.4). The functions $a^{ij}(x)$, $b^i(x)$ and $c(x)$ are called the diffusion coefficients, the drift coefficients and the termination coefficient, respectively.

Let L be a first-order boundary condition such that

$$Lu = \mu(x')\frac{\partial u}{\partial \mathbf{n}} + \gamma(x')u,$$

where:

(4) $\mu \in C^\infty(\partial D)$ and $\mu(x') \geq 0$ on ∂D.
(5) $L1 = \gamma \in C^\infty(\partial D)$ and $\gamma(x') \leq 0$ on ∂D.
(6) $\mathbf{n} = (n_1, n_2, \ldots, n_N)$ is the unit inward normal to the boundary ∂D (see Figure 1.1).

The boundary condition L is called a first-order *Ventcel' boundary condition* (cf. [143]). Its terms $\mu(x')(\partial u)/(\partial \mathbf{n})$ and $\gamma(x')u$ are supposed to correspond to the reflection and absorption phenomena, respectively (see Figure 1.3).

We are interested in the following problem:

Problem. Given analytic data (A, L), can we construct a Feller semigroup $\{T_t\}_{t\geq 0}$ on the state space $\overline{D}$ whose infinitesimal generator $\mathfrak{A}$ is characterized by (A, L) ?

First, we consider the following Dirichlet problem: Given functions $f(x)$ and $\varphi(x')$ defined in D and on ∂D, respectively, find a function $u(x)$ in D such that

$$\begin{cases} Au = f & \text{in } D, \\ \gamma_0 u = u|_{\partial D} = \varphi & \text{on } \partial D. \end{cases} \tag{13.1}$$

The next theorem summarizes the basic facts about the Dirichlet problem in the framework of *Hölder spaces* (cf. [50]):

Theorem 13.1. *(i) (Existence and Uniqueness) If $f \in C^\theta(D)$ with $0 < \theta < 1$ and if $\varphi \in C(\partial D)$, then the Dirichlet problem (13.1) has a unique solution $u(x)$ in $C(\overline{D}) \cap C^{2+\theta}(D)$.*

(ii) (Interior Regularity) If $u \in C^2(D)$ and $Au = f \in C^{k+\theta}(D)$ for some non-negative integer k, then it follows that $u \in C^{k+2+\theta}(D)$.

(iii) (Global Regularity) If $f \in C^{k+\theta}(\overline{D})$ and $\varphi \in C^{k+2+\theta}(\partial D)$ for some non-negative integer k, then a solution $u \in C(\overline{D}) \cap C^2(D)$ of the Dirichlet problem (13.1) belongs to the space $C^{k+2+\theta}(\overline{D})$.

In the following we shall use the notation:

$$\|f\|_\infty = \max_{x \in \overline{D}} |f(x)| \quad \text{for } f \in C(\overline{D}),$$

$$|\varphi|_\infty = \max_{x' \in \partial D} |\varphi(x')| \quad \text{for } \varphi \in C(\partial D).$$

Secondly, we consider the following Dirichlet problem with spectral parameter: For given functions $f(x)$ and $\varphi(x')$ defined in D and on ∂D, respectively, find a function $u(x)$ in D such that

$$\begin{cases} (\alpha - A)\,u = f & \text{in } D, \\ \gamma_0 u = u|_{\partial D} = \varphi & \text{on } \partial D, \end{cases} \tag{13.2}$$

where α is a positive parameter.

By applying Theorem 13.1 with $A := A - \alpha$, we obtain that problem (13.2) has a unique solution $u(x)$ in $C^{2+\theta}(\overline{D})$ for any $f \in C^{\theta}(\overline{D})$ and any $\varphi \in C^{2+\theta}(\partial D)$ with $0 < \theta < 1$. Therefore, we can introduce linear operators

$$\boxed{G^0_\alpha : C^\theta(\overline{D}) \longrightarrow C^{2+\theta}(\overline{D})}$$

and

$$\boxed{H_\alpha : C^{2+\theta}(\partial D) \longrightarrow C^{2+\theta}(\overline{D})}$$

as follows.

(a) For any $f \in C^\theta(\overline{D})$, the function $G^0_\alpha f \in C^{2+\theta}(\overline{D})$ is the unique solution of the problem

$$\begin{cases} (\alpha - A)\,G^0_\alpha f = f & \text{in } D, \\ G^0_\alpha f|_{\partial D} = 0 & \text{on } \partial D. \end{cases} \tag{13.3}$$

(b) For any $\varphi \in C^{2+\theta}(\partial D)$, the function $H_\alpha \varphi \in C^{2+\theta}(\overline{D})$ is the unique solution of the problem

$$\begin{cases} (\alpha - A)\,H_\alpha \varphi = 0 & \text{in } D, \\ H_\alpha \varphi|_{\partial D} = \varphi & \text{on } \partial D. \end{cases} \tag{13.4}$$

The operator G^0_α is called the *Green operator* and the operator H_α is called the *harmonic operator*, respectively.

Then we have the following lemma:

Lemma 13.2. *The operator G^0_α for $\alpha > 0$, considered from $C(\overline{D})$ into itself, is non-negative and continuous (bounded) with norm*

$$\left\| G^0_\alpha \right\| = \left\| G^0_\alpha 1 \right\|_\infty = \max_{x \in \overline{D}}(G^0_\alpha 1)(x).$$

Proof. Let $f(x)$ be an arbitrary function in $C^\theta(\overline{D})$ such that $f(x) \geq 0$ on $\overline{D}$. Then, by applying the weak maximum principle (see Theorem 10.1) with $A := A - \alpha$ to the function $-G^0_\alpha f$ we obtain from formula (13.3) that

$$G^0_\alpha f \geq 0 \quad \text{on } \overline{D}.$$

This proves the non-negativity of G^0_α.

Since G_α^0 is non-negative, we have, for all $f \in C^\theta(\overline{D})$,

$$-G_\alpha^0 \|f\|_\infty \leq G_\alpha^0 f \leq G_\alpha^0 \|f\|_\infty \quad \text{on } \overline{D}.$$

This implies the continuity of G_α^0 with norm

$$\|G_\alpha^0\| = \|G_\alpha^0 1\|_\infty.$$

The proof of Lemma 13.2 is complete. □

Similarly, we have the following lemma:

Lemma 13.3. *The operator H_α for $\alpha > 0$, considered from $C(\partial D)$ into $C(\overline{D})$, is non-negative and continuous (bounded) with norm*

$$\|H_\alpha\| = \|H_\alpha 1\|_\infty = \max_{x \in \overline{D}}(H_\alpha 1)(x).$$

More precisely, we have the following fundamental theorem:

Theorem 13.4. *(i) (a) The operator G_α^0 for $\alpha > 0$ can be uniquely extended to a non-negative, bounded linear operator on $C(\overline{D})$ into itself, denoted again by G_α^0, with norm*

$$\|G_\alpha^0\| = \|G_\alpha^0 1\|_\infty \leq \frac{1}{\alpha}. \tag{13.5}$$

(b) For any $f \in C(\overline{D})$, we have the assertion

$$G_\alpha^0 f(x') = 0 \quad \text{on } \partial D.$$

(c) For all $\alpha, \beta > 0$, the resolvent equation holds true:

$$G_\alpha^0 f - G_\beta^0 f + (\alpha - \beta) G_\alpha^0 G_\beta^0 f = 0 \quad \text{for every } f \in C(\overline{D}). \tag{13.6}$$

(d) For any $f \in C(\overline{D})$, we have the assertion

$$\lim_{\alpha \to +\infty} \alpha G_\alpha^0 f(x) = f(x) \quad \text{for all } x \in D. \tag{13.7}$$

Furthermore, if $f(x') = 0$ on ∂D, then this convergence is uniform in $x \in \overline{D}$, that is, we have the assertion

$$\lim_{\alpha \to +\infty} \alpha G_\alpha^0 f = f \quad \text{in } C(\overline{D}). \tag{13.8}$$

(e) The operator G_α^0 maps $C^{k+\theta}(\overline{D})$ into $C^{k+2+\theta}(\overline{D})$ for any non-negative integer k.

(ii) (a') The operator H_α for $\alpha > 0$ can be uniquely extended to a non-negative, bounded linear operator on $C(\partial D)$ into $C(\overline{D})$, denoted again by H_α, with norm $\|H_\alpha\| = 1$.

(b') For any $\varphi \in C(\partial D)$, we have the assertion

$$H_\alpha \varphi|_{\partial D} = \varphi \quad \text{on } \partial D.$$

(c′) For all α, $\beta > 0$, we have the equation

$$H_\alpha \varphi - H_\beta \varphi + (\alpha - \beta) G_\alpha^0 H_\beta \varphi = 0 \quad \text{for every } \varphi \in C(\partial D). \tag{13.9}$$

(d′) The operator H_α maps $C^{k+2+\theta}(\partial D)$ into $C^{k+2+\theta}(\overline{D})$ for any non-negative integer k.

Proof. (i) (a) By making use of Friedrichs' mollifiers ([122, Subsection 5.2.6], [136, Subsection 1.3.2]), we find that the Hölder space $C^\theta(\overline{D})$ is dense in $C(\overline{D})$ and further that non-negative functions can be approximated by non-negative smooth functions. Hence, by Lemma 13.2 it follows that the operator $G_\alpha^0 : C^\theta(\overline{D}) \to C^{2+\theta}(\overline{D})$ can be uniquely extended to a non-negative, bounded linear operator

$$G_\alpha^0 : C(\overline{D}) \longrightarrow C(\overline{D})$$

with norm $\left\| G_\alpha^0 \right\| = \left\| G_\alpha^0 1 \right\|_\infty$. Furthermore, since the function $G_\alpha^0 1$ satisfies the conditions

$$\begin{cases} (A - \alpha) G_\alpha^0 1 = -1 & \text{in } D, \\ G_\alpha^0 1|_{\partial D} = 0 & \text{on } \partial D, \end{cases}$$

by applying Theorem 10.2 with $A := A - \alpha$ we obtain that

$$\left\| G_\alpha^0 \right\| = \left\| G_\alpha^0 1 \right\|_\infty \le \frac{1}{\alpha}.$$

(b) This assertion follows from formula (13.3), since the space $C^\theta(\overline{D})$ is dense in $C(\overline{D})$ and since the operator $G_\alpha^0 : C(\overline{D}) \to C(\overline{D})$ is bounded.

(c) We find from the uniqueness theorem for problem (13.2) (Theorem 13.1) that equation (13.6) holds true for all $f \in C^\theta(\overline{D})$. Indeed, it suffices to note that the function

$$v := G_\alpha^0 f - G_\beta^0 f + (\alpha - \beta) G_\alpha^0 G_\beta^0 f \in C^{2+\theta}(\overline{D})$$

satisfies the conditions

$$\begin{cases} (\alpha - A) v = 0 & \text{in } D, \\ v|_{\partial D} = 0 & \text{on } \partial D, \end{cases}$$

so that

$$v = 0 \quad \text{in } D.$$

Therefore, we obtain that the resolvent equation (13.6) holds true for all $f \in C(\overline{D})$, since the space $C^\theta(\overline{D})$ is dense in $C(\overline{D})$ and since the operators G_α^0 and G_β^0 are bounded.

(d) First, let $f(x)$ be an arbitrary function in $C^\theta(\overline{D})$ satisfying the boundary condition $f|_{\partial D} = 0$. Then it follows from an application of the uniqueness theorem for problem (13.2) (Theorem 13.1) that we have, for all α, β,

$$f - \alpha G_\alpha^0 f = G_\alpha^0 \left((\beta - A)f\right) - \beta G_\alpha^0 f.$$

Indeed, the both sides satisfy the same equation $(\alpha - A)u = -Af$ in D and have the same boundary value 0 on ∂D. Thus we have, by estimate (13.5),

$$\left\| f - \alpha G_\alpha^0 f \right\|_\infty \leq \frac{1}{\alpha} \left\| (\beta - A)f \right\|_\infty + \frac{\beta}{\alpha} \left\| f \right\|_\infty,$$

so that

$$\lim_{\alpha \to +\infty} \left\| f - \alpha G_\alpha^0 f \right\|_\infty = 0. \tag{13.10}$$

Now let $f(x)$ be an arbitrary function in $C(\overline{D})$ satisfying the boundary condition $f|_{\partial D} = 0$. By means of mollifiers ([122, Subsection 5.2.6], [136, Subsection 1.3.2]), we can find a sequence $\{f_j\}_{j=1}^\infty$ in $C^\theta(\overline{D})$ such that

$$\begin{cases} f_j \longrightarrow f & \text{in } C(\overline{D}) \text{ as } j \to \infty, \\ f_j|_{\partial D} = 0 & \text{on } \partial D. \end{cases}$$

Then we have, by estimate (13.5) and assertion (13.10) with $f := f_j$,

$$\left\| f - \alpha G_\alpha^0 f \right\|_\infty \leq \left\| f - f_j \right\|_\infty + \left\| f_j - \alpha G_\alpha^0 f_j \right\|_\infty + \left\| \alpha G_\alpha^0 (f_j - f) \right\|_\infty$$
$$\leq 2 \left\| f - f_j \right\|_\infty + \left\| f_j - \alpha G_\alpha^0 f_j \right\|_\infty,$$

and hence

$$\limsup_{\alpha \to +\infty} \left\| f - \alpha G_\alpha^0 f \right\|_\infty \leq 2 \left\| f - f_j \right\|_\infty.$$

This proves the desired assertion (13.8), since $\left\| f - f_j \right\|_\infty \to 0$ as $j \to \infty$.

To prove assertion (13.7), let $f(x)$ be an arbitrary function in $C(\overline{D})$ and let x be an arbitrary point of D. If we take a function $\psi(y)$ in $C(\overline{D})$ such that

$$\begin{cases} 0 \leq \psi(y) \leq 1 & \text{on } \overline{D}, \\ \psi(y) = 0 & \text{in a neighborhood of } x, \\ \psi(y) = 1 & \text{near the set } \partial D, \end{cases}$$

then it follows from the non-negativity of G_α^0 and estimate (13.5) that

$$0 \leq \alpha G_\alpha^0 \psi(x) + \alpha G_\alpha^0 (1 - \psi)(x) = \alpha G_\alpha^0 1(x) \leq 1. \tag{13.11}$$

However, by applying assertion (13.8) to the function $1 - \psi(y)$ we have the assertion

$$\lim_{\alpha \to +\infty} \alpha G_\alpha^0 (1 - \psi)(x) = (1 - \psi)(x) = 1 \quad \text{for all } x \in D.$$

In view of inequalities (13.11), this implies that

$$\lim_{\alpha \to +\infty} \alpha G_\alpha^0 \psi(x) = 0 \quad \text{for all } x \in D.$$

Thus, since we have the inequalities

$$- \|f\|_\infty \, \psi(x) \leq f(x)\psi(x) \leq \|f\|_\infty \, \psi(x) \quad \text{on } \overline{D},$$

it follows that, for $x \in D$,

$$\left| \alpha G_\alpha^0 (f\psi)(x) \right| \leq \|f\|_\infty \cdot \alpha G_\alpha^0 \psi(x) \longrightarrow 0 \quad \text{as } \alpha \to +\infty.$$

Therefore, by applying assertion (13.8) to the function $(1 - \psi(y))f(y)$ we obtain that

$$f(x) = ((1 - \psi)f)(x) = \lim_{\alpha \to +\infty} \alpha G_\alpha^0 ((1 - \psi)f)(x)$$

$$= \lim_{\alpha \to +\infty} \alpha G_\alpha^0 f(x) \quad \text{for all } x \in D.$$

This proves the desired assertion (13.7).

(ii) (a') Since the space $C^{2+\theta}(\partial D)$ is dense in $C(\partial D)$, by Lemma 13.3 it follows that the operator $H_\alpha \colon C^{2+\theta}(\partial D) \to C^{2+\theta}(\overline{D})$ can be uniquely extended to a non-negative, bounded linear operator

$$H_\alpha \colon C(\partial D) \longrightarrow C(\overline{D}).$$

Furthermore, by applying Theorem 10.2 with $A := A - \alpha$ we have the assertion

$$\|H_\alpha\| = \|H_\alpha 1\|_\infty = 1.$$

(b') This assertion follows from formula (13.4), since $C^{2+\theta}(\partial D)$ is dense in $C(\partial D)$ and since the operator $H_\alpha \colon C(\partial D) \to C(\overline{D})$ is bounded.

(c') We find from the uniqueness theorem for problem (13.2) that equation (13.9) holds true for all $\varphi \in C^{2+\theta}(\partial D)$. Indeed, it suffices to note that the function

$$w := H_\alpha \varphi - H_\beta \varphi + (\alpha - \beta) G_\alpha^0 \, H_\beta \varphi \in C^{2+\theta}(\overline{D})$$

satisfies the conditions

$$\begin{cases} (\alpha - A)\, w = 0 & \text{in } D, \\ w|_{\partial D} = 0 & \text{on } \partial D, \end{cases}$$

so that

$$w = 0 \quad \text{in } D.$$

Therefore, we obtain that the desired equation (13.9) holds true for all $\varphi \in C(\partial D)$, since the space $C^{2+\theta}(\partial D)$ is dense in $C(\partial D)$ and since the operators G_α^0 and H_α are bounded.

The proof of Theorem 13.4 is now complete. $\square$

Summing up, we have the mapping properties of the operators G_α^0 and H_α as in Figures 13.1 and 13.2 below.

$$
\begin{array}{ccc}
C(\overline{D}) & \xrightarrow{\;G^0_\alpha\;} & C(\overline{D}) \\
\uparrow & & \uparrow \\
\mathcal{D}(G^0_\alpha) = C^\theta(\overline{D}) & \xrightarrow[\;G^0_\alpha\;]{} & C^{2+\theta}(\overline{D})
\end{array}
$$

Fig. 13.1. The mapping properties of the operator G^0_α

$$
\begin{array}{ccc}
C(\partial D) & \xrightarrow{\;H_\alpha\;} & C(\overline{D}) \\
\uparrow & & \uparrow \\
\mathcal{D}(H_\alpha) = C^{2+\theta}(\partial D) & \xrightarrow[\;H_\alpha\;]{} & C^{2+\theta}(\overline{D})
\end{array}
$$

Fig. 13.2. The mapping properties of the operator H_α

Now we consider the following Ventcel' boundary value problem in the framework of the spaces of *continuous functions*.

$$
\begin{cases}
(\alpha - A)\,u = f & \text{in } D, \\
Lu = 0 & \text{on } \partial D.
\end{cases}
\tag{13.12}
$$

To do this, we introduce three operators associated with the Ventcel' boundary value problem (13.12).

Step (I): First, we introduce a linear operator

$$
A\colon C(\overline{D}) \longrightarrow C(\overline{D})
$$

as follows.

(a) The domain $\mathcal{D}(A)$ of A is the space $C^2(\overline{D})$.

(b) $Au = \sum_{i,j=1}^{N} a^{ij}(x)\dfrac{\partial^2 u}{\partial x_i \partial x_j} + \sum_{i=1}^{N} b^i(x)\dfrac{\partial u}{\partial x_i} + c(x)u$, $u \in \mathcal{D}(A)$.

Then we have the following lemma:

Lemma 13.5. *The operator A has its minimal closed extension $\overline{A}$ in the space* $C(\overline{D})$.

Proof. We apply part (i) of Theorem 3.36 (the Hille–Yosida–Ray theorem) to the operator A.

Assume that a function $u \in C^2(\overline{D})$ takes a positive maximum at an interior point x_0 of D:

$$u(x_0) = \max_{x \in \overline{D}} u(x) > 0.$$

Then it follows that

$$\bullet \quad \frac{\partial u}{\partial x_i}(x_0) = 0 \quad \text{for } 1 \leq i \leq N,$$

$$\bullet \quad \sum_{i,j=1}^{N} a^{ij}(x_0) \frac{\partial^2 u}{\partial x_i \partial x_j}(x_0) \leq 0,$$

since the matrix $(a^{ij}(x))$ is positive definite. Hence we have the assertion

$$Au(x_0) = \sum_{i,j=1}^{N} a^{ij}(x_0) \frac{\partial^2 u}{\partial x_i \partial x_j}(x_0) + c(x_0)u(x_0) \leq 0,$$

This implies that the operator A satisfies condition (β) of Theorem 3.36 with $K_0 := D$ and $K := \overline{D}$. Therefore, Lemma 13.5 follows from an application of the same theorem.

The proof of Lemma 13.5 is complete.   □

Remark 13.6. Since the injection: $C(\overline{D}) \to \mathcal{D}'(D)$ is continuous, we have the formula

$$\overline{A}u = \sum_{i,j=1}^{N} a^{ij}(x) \frac{\partial^2 u}{\partial x_i \partial x_j} + \sum_{i=1}^{N} b^i(x) \frac{\partial u}{\partial x_i} + c(x)u \quad \text{for every } u \in C(\overline{D}),$$

where the right-hand side is taken in the sense of *distributions*. The operators A and $\overline{A}$ can be visualized as in Figure 13.3 below.

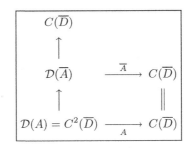

Fig. 13.3. The operators A and $\overline{A}$

The extended operators

$$G^0_\alpha : C(\overline{D}) \longrightarrow C(\overline{D}),$$
$$H_\alpha : C(\partial D) \longrightarrow C(\overline{D})$$

still satisfy formulas (13.3) and (13.4) respectively in the following sense:

Lemma 13.7. *Let $\alpha > 0$. Then we have the following two assertions (i) and (ii):*

(i) For any $f \in C(\overline{D})$, we have the formulas

$$\begin{cases} G_\alpha^0 f \in \mathcal{D}\left(\overline{A}\right), \\ \left(\alpha I - \overline{A}\right) G_\alpha^0 f = f \ \text{in } D. \end{cases}$$

(ii) For any $\varphi \in C(\partial D)$, we have the formulas

$$\begin{cases} H_\alpha \varphi \in \mathcal{D}\left(\overline{A}\right), \\ \left(\alpha I - \overline{A}\right) H_\alpha \varphi = 0 \ \text{in } D. \end{cases}$$

Here $\mathcal{D}\left(\overline{A}\right)$ is the domain of the closed extension $\overline{A}$.

Proof. (i) By making use of Friedrichs' mollifiers ([122, Subsection 5.2.6], [136, Subsection 1.3.2]), we can choose a sequence $\{f_j\}_{j=1}^\infty$ in $C^\theta(\overline{D})$ such that $f_j \to f$ in $C(\overline{D})$ as $j \to \infty$. Then it follows from the boundedness of G_α^0 that

$$G_\alpha^0 f_j \longrightarrow G_\alpha^0 f \quad \text{in } C(\overline{D}),$$

and further that

$$(\alpha - A) G_\alpha^0 f_j = f_j \longrightarrow f \quad \text{in } C(\overline{D}).$$

Hence we have the assertions

$$\begin{cases} G_\alpha^0 f \in \mathcal{D}\left(\overline{A}\right), \\ \left(\alpha I - \overline{A}\right) G_\alpha^0 f = f \quad \text{in } D. \end{cases}$$

since the operator $\overline{A} \colon C(\overline{D}) \to C(\overline{D})$ is closed.

(ii) Similarly, part (ii) is proved, since the space $C^{2+\theta}(\partial D)$ is dense in $C(\partial D)$ and since the operator $H_\alpha \colon C(\partial D) \to C(\overline{D})$ is bounded.

The proof of Lemma 13.7 is complete. □

Then we have the following corollary:

Corollary 13.8. *Every function u in $\mathcal{D}\left(\overline{A}\right)$ can be written in the following form:*

$$u = G_\alpha^0 \left((\alpha I - \overline{A})u\right) + H_\alpha(u|_{\partial D}) \quad \text{for all } \alpha > 0. \tag{13.13}$$

Proof. We let

$$w := u - G_\alpha^0 \left((\alpha I - \overline{A})u\right) - H_\alpha(u|_{\partial D}).$$

Then it follows from Lemma 13.7 that the function w is in $\mathcal{D}\left(\overline{A}\right)$ and satisfies the conditions

$$\begin{cases} \left(\alpha I - \overline{A}\right) w = 0 & \text{in } D, \\ w|_{\partial D} = 0 & \text{on } \partial D. \end{cases}$$

However, in light of Remark.13.6, by applying Lemma 7.1 (cf. [68, Theorem 3.1], [123, Theorem 7.1]) and Theorem 6.22 with $A := A - \alpha$ to the Dirichlet case ($\mu(x') \equiv 0$ and $\gamma(x') \equiv -1$ on ∂D) we obtain that

$$w \in C^\infty(\overline{D}).$$

Therefore, it follows from an application of part (i) of Theorem 13.1 with $A := A - \alpha$ that

$$w = 0.$$

This proves the desired formula (13.13).

The proof of Corollary 13.8 is complete. $\square$

Therefore, we can express the relationships between the operators $\overline{A}$, G_α^0 and H_α in matrix form:

$$\begin{pmatrix} \alpha I - \overline{A} \\ \gamma_0 \end{pmatrix} \left(G_\alpha^0 \ H_\alpha\right) = \begin{pmatrix} I & 0 \\ 0 & I \end{pmatrix}.$$

Step (II): Secondly, we introduce a linear operator

$$\boxed{LG_\alpha^0 \colon C(\overline{D}) \longrightarrow C(\partial D)}$$

as follows.

(a) The domain $\mathcal{D}\left(LG_\alpha^0\right)$ of LG_α^0 is the Hölder space $C^\theta(\overline{D})$ with $0 < \theta < 1$.
(b) $LG_\alpha^0 f = L\left(G_\alpha^0 f\right)$ for every $f \in \mathcal{D}\left(LG_\alpha^0\right)$.

Then we have the following lemma:

Lemma 13.9. *The operator LG_α^0 for $\alpha > 0$ can be uniquely extended to a non-negative, bounded linear operator $\overline{LG_\alpha^0} \colon C(\overline{D}) \to C(\partial D)$.*

Proof. Let $f(x)$ be an arbitrary function in $\mathcal{D}\left(LG_\alpha^0\right) = C^\theta(\overline{D})$ such that $f(x) \geq 0$ on $\overline{D}$. Then we have the assertions

$$\begin{cases} G_\alpha^0 f \in C^{2+\theta}(\overline{D}), \\ G_\alpha^0 f \geq 0 & \text{on } \overline{D}, \\ G_\alpha^0 f|_{\partial D} = 0 & \text{on } \partial D, \end{cases}$$

and hence

$$LG_\alpha^0 f = \mu(x') \frac{\partial}{\partial \mathbf{n}}(G_\alpha^0 f) + \gamma(x')(G_\alpha^0 f)\bigg|_{\partial D}$$

$$= \mu(x') \frac{\partial}{\partial \mathbf{n}}(G_\alpha^0 f) \geq 0 \quad \text{on } \partial D.$$

This proves that the operator LG_α^0 is non-negative.

By the non-negativity of LG_α^0, we have, for all $f \in \mathcal{D}(LG_\alpha^0)$,

$$-LG_\alpha^0 \|f\|_\infty \le LG_\alpha^0 f \le LG_\alpha^0 \|f\|_\infty \quad \text{on } \partial D.$$

This implies the boundedness of LG_α^0 with norm

$$\left\|LG_\alpha^0\right\| = \left|LG_\alpha^0 1\right|_\infty = \max_{x' \in \partial D} (LG_\alpha^0 1)(x').$$

Recall that the space $C^\theta(\overline{D})$ is dense in $C(\overline{D})$ and that non-negative functions can be approximated by non-negative smooth functions. Hence we find that the operator LG_α^0 can be uniquely extended to a non-negative, bounded linear operator $\overline{LG_\alpha^0} \colon C(\overline{D}) \to C(\partial D)$.

The proof of Lemma 13.9 is complete. □

The operators LG_α^0 and $\overline{LG_\alpha^0}$ can be visualized in Figure 13.4 below.

$$
\begin{array}{ccc}
D(\overline{LG_\alpha^0}) = C(\overline{D}) & \xrightarrow{\overline{LG_\alpha^0}} & C(\partial D) \\
\uparrow & & \uparrow \\
\mathcal{D}(LG_\alpha^0) = C^\theta(\overline{D}) & \xrightarrow[LG_\alpha^0]{} & C^{1+\theta}(\partial D)
\end{array}
$$

Fig. 13.4. The operators LG_α^0 and $\overline{LG_\alpha^0}$

The next lemma states a fundamental relationship between the operators $\overline{LG_\alpha^0}$ and $\overline{LG_\beta^0}$ for α, $\beta > 0$:

Lemma 13.10. *For any $f \in C(\overline{D})$, we have the formula*

$$\overline{LG_\alpha^0}f - \overline{LG_\beta^0}f + (\alpha - \beta)\overline{LG_\alpha^0} G_\beta^0 f = 0 \quad \text{for all } \alpha, \beta > 0. \tag{13.14}$$

Proof. Choose a sequence $\{f_j\}$ in $C^\theta(\overline{D})$ such that $f_j \to f$ in $C(\overline{D})$ as $j \to \infty$, just as in Lemma 13.7. Then, by using the resolvent equation (13.6) with $f := f_j$ we have the formula

$$LG_\alpha^0 f_j - LG_\beta^0 f_j + (\alpha - \beta)LG_\alpha^0 G_\beta^0 f_j = 0.$$

Hence, the desired formula (13.14) follows by letting $j \to \infty$, since the operators $\overline{LG_\alpha^0}$, $\overline{LG_\beta^0}$ and G_β^0 are all bounded.

The proof of Lemma 13.10 is complete. □

Step (III): Finally, we introduce a linear operator

$$\boxed{LH_\alpha : C(\partial D) \longrightarrow C(\partial D)}$$

as follows.

(a) The domain $\mathcal{D}(LH_\alpha)$ of LH_α is the space $C^{2+\theta}(\partial D)$.
(b) $LH_\alpha\psi = L(H_\alpha\psi)$ for every $\psi \in \mathcal{D}(LH_\alpha)$.

Then we have the following lemma:

Lemma 13.11. *The operator LH_α for $\alpha > 0$ has its minimal closed extension $\overline{LH_\alpha}$ in the space $C(\partial D)$.*

Proof. We apply part (i) of Theorem 3.36 (the Hille–Yosida–Ray theorem) to the operator LH_α. To do this, it suffices to show that the operator LH_α satisfies condition (β') with $K := \partial D$ (or condition (β) with $K := K_0 = \partial D$) of the same theorem.

Assume that a function ψ in $\mathcal{D}(LH_\alpha) = C^{2+\theta}(\partial D)$ takes its positive maximum at some point x' of ∂D. Since the function $H_\alpha\psi$ is in $C^{2+\theta}(\overline{D})$ and satisfies

$$\begin{cases} (A - \alpha) H_\alpha\psi = 0 & \text{in } D, \\ H_\alpha\psi|_{\partial D} = \psi & \text{on } \partial D, \end{cases}$$

by applying the weak maximum principle (Theorem 10.1) with $A := A - \alpha$ to the function $H_\alpha\psi$ we find that the function $H_\alpha\psi$ takes its positive maximum at the boundary point $x' \in \partial D$. Thus we can apply the Hopf boundary point lemma (Lemma 10.11) with $A := A - \alpha$ to obtain that

$$\frac{\partial}{\partial \mathbf{n}}(H_\alpha\psi)(x') < 0.$$

Hence we have the inequality

$$LH_\alpha\psi(x') = \sum_{i,j=1}^{N-1} a^{ij}(x')\frac{\partial^2\psi}{\partial x_i \partial x_j}(x') + \mu(x')\frac{\partial}{\partial \mathbf{n}}(H_\alpha\psi)(x') + \gamma(x')\psi(x')$$
$$\leq 0.$$

This verifies condition (β') of Theorem 3.36. Therefore, Lemma 13.11 follows from an application of the same theorem.

The proof of Lemma 13.11 is complete. □

The operators LH_α and $\overline{LH_\alpha}$ can be visualized in Figure 13.5 below.

Remark 13.12. The operator $\overline{LH_\alpha}$ enjoys the following property:

$$\text{If a function } \psi \in \mathcal{D}\left(\overline{LH_\alpha}\right) \text{ takes its } positive\ maximum \qquad (13.15)$$

at some point x' of ∂D, then we have the inequality

$$\overline{LH_\alpha}\psi(x') \leq 0.$$

The next lemma states a fundamental relationship between the operators $\overline{LH_\alpha}$ and $\overline{LH_\beta}$ for α, $\beta > 0$:

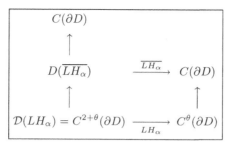

Fig. 13.5. The operators LH_α and $\overline{LH_\alpha}$

Lemma 13.13. *The domain $\mathcal{D}\left(\overline{LH_\alpha}\right)$ of $\overline{LH_\alpha}$ does not depend on $\alpha > 0$; so we denote by $\mathcal{D}$ the common domain. Then we have the formula*

$$\overline{LH_\alpha}\psi - \overline{LH_\beta}\psi + (\alpha - \beta)\overline{LG_\alpha^0}\,H_\beta\psi = 0 \tag{13.16}$$

for all α, $\beta > 0$ and $\psi \in \mathcal{D}$.

Proof. Let $\psi(x')$ be an arbitrary function in $\mathcal{D}\left(\overline{LH_\beta}\right)$, and choose a sequence $\{\psi_j\}$ in $\mathcal{D}\left(LH_\beta\right) = C^{2+\theta}(\partial D)$ such that, as $j \to \infty$,

$$\begin{cases} \psi_j \longrightarrow \psi & \text{in } C(\partial D), \\ LH_\beta\psi_j \longrightarrow \overline{LH_\beta}\psi & \text{in } C(\partial D). \end{cases}$$

Then it follows from the boundedness of H_β and $\overline{LG_\alpha^0}$ that

$$LG_\alpha^0(H_\beta\psi_j) = \overline{LG_\alpha^0}(H_\beta\psi_j) \longrightarrow \overline{LG_\alpha^0}(H_\beta\psi) \quad \text{in } C(\partial D).$$

Therefore, by using formula (13.9) with $\varphi := \psi_j$ we obtain that, as $j \to \infty$,

$$\begin{aligned} LH_\alpha\psi_j &= LH_\beta\psi_j - (\alpha - \beta)LG_\alpha^0(H_\beta\psi_j) \\ &\longrightarrow \overline{LH_\beta}\psi - (\alpha - \beta)\overline{LG_\alpha^0}(H_\beta\psi) \quad \text{in } C(\partial D). \end{aligned}$$

This implies that

$$\begin{cases} \psi \in \mathcal{D}\left(\overline{LH_\alpha}\right), \\ \overline{LH_\alpha}\psi = \overline{LH_\beta}\psi - (\alpha - \beta)\overline{LG_\alpha^0}(H_\beta\psi), \end{cases}$$

since the operator $\overline{LH_\alpha} \colon C(\partial D) \to C(\partial D)$ is closed.

Conversely, by interchanging α and β we have the assertion

$$\mathcal{D}\left(\overline{LH_\alpha}\right) \subset \mathcal{D}\left(\overline{LH_\beta}\right),$$

and so

$$\mathcal{D}\left(\overline{LH_\alpha}\right) = \mathcal{D}\left(\overline{LH_\beta}\right).$$

The proof of Lemma 13.13 is complete. $\square$

Now we can prove a general existence theorem for Feller semigroups on the state space ∂D in terms of boundary value problem (13.12). The next theorem asserts that the operator $\overline{LH_\alpha}$ is the infinitesimal generator of some Feller semigroup on the state space ∂D if and only if problem (13.12) is solvable for *sufficiently many* functions φ in the Banach space $C(\partial D)$:

Theorem 13.14. *(i) If the operator $\overline{LH_\alpha}$ for $\alpha > 0$ is the infinitesimal generator of a Feller semigroup on the state space ∂D, then, for each positive constant λ, the boundary value problem*

$$\begin{cases} (\alpha - A)\,u = 0 & in\ D, \\ (\lambda - L)\,u = \varphi & on\ \partial D \end{cases} \tag{13.17}$$

has a solution $u \in C^{2+\theta}(\overline{D})$ for any φ in some dense *subset of $C(\partial D)$.*

(ii) Conversely, if, for some non-negative constant λ, problem (13.17) has a solution $u \in C^{2+\theta}(\overline{D})$ for any φ in some dense *subset of $C(\partial D)$, then the operator $\overline{LH_\alpha}$ is the infinitesimal generator of some Feller semigroup on the state space ∂D.*

Proof. (i) If the operator $\overline{LH_\alpha}$ generates a Feller semigroup on the state space ∂D, by applying part (i) of Theorem 3.36 (the Hille–Yosida–Ray theorem) with $K := \partial D$ to the operator $\overline{LH_\alpha}$ we obtain that

$$\mathcal{R}\left(\lambda I - \overline{LH_\alpha}\right) = C(\partial D) \quad \text{for each } \lambda > 0.$$

This implies that the range $\mathcal{R}\left(\lambda I - LH_\alpha\right)$ is a dense subset of $C(\partial D)$ for each $\lambda > 0$. However, if $\varphi \in C(\partial D)$ is in the range $\mathcal{R}\left(\lambda I - LH_\alpha\right)$ and if $\varphi = (\lambda I - LH_\alpha)\,\psi$ with $\psi \in C^{2+\theta}(\partial D)$, then the function $u = H_\alpha \psi \in C^{2+\theta}(\overline{D})$ is a solution of problem (13.17). This proves part (i) of Theorem 13.14.

(ii) We apply part (ii) of Theorem 3.36 with $K := \partial D$ to the operator LH_α. To do this, it suffices to show that the operator LH_α satisfies condition (γ) of the same theorem, since it satisfies condition (β'), as is shown in the proof of Lemma 13.11.

By the uniqueness theorem for problem (13.2), it follows that any function $u \in C^{2+\theta}(\overline{D})$ which satisfies the homogeneous equation

$$(\alpha - A)\,u = 0 \quad \text{in } D$$

can be written in the form:

$$u = H_\alpha\left(u|_{\partial D}\right), \quad u|_{\partial D} \in C^{2+\theta}(\partial D) = \mathcal{D}\left(LH_\alpha\right).$$

Thus we find that if there exists a solution $u \in C^{2+\theta}(\overline{D})$ of problem (13.17) for a function $\varphi \in C(\partial D)$, then we have the assertion

$$(\lambda I - LH_\alpha)\left(u|_{\partial D}\right) = \varphi,$$

and so

$$\varphi \in \mathcal{R}\left(\lambda I - LH_\alpha\right).$$

Hence, if, for some non-negative constant λ, problem (13.12) has a solution $u \in C^{2+\theta}(\overline{D})$ for any φ in some dense subset of $C(\partial D)$, then the range $\mathcal{R}\left(\lambda I - LH_\alpha\right)$ is dense in $C(\partial D)$. This verifies condition (γ) (with $\alpha_0 := \lambda$) of Theorem 3.36. Therefore, part (ii) of Theorem 13.14 follows from an application of the same theorem.

Now the proof of Theorem 13.14 is complete. $\square$

Remark 13.15. Intuitively, Theorem 13.14 asserts that we can "piece together" a Markov process (Feller semigroup) on the boundary ∂D with A-diffusion in the interior D to construct a Markov process (Feller semigroup) on the closure $\overline{D} = D \cup \partial D$. The situation may be represented schematically as in Figure 13.6 below.

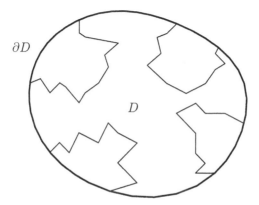

Fig. 13.6. A Markov process on ∂D pieced together with an A-diffusion in D

We conclude this section by giving a precise meaning to the boundary conditions Lu for functions u in $\mathcal{D}\left(\overline{A}\right)$.

We let

$$\mathcal{D}(L) = \left\{u \in \mathcal{D}\left(\overline{A}\right) : u|_{\partial D} \in \mathcal{D}\right\},$$

where $\mathcal{D}$ is the common domain of the operators $\overline{LH_\alpha}$ for all $\alpha > 0$. We remark that the space $\mathcal{D}(L)$ contains $C^{2+\theta}(\overline{D})$, since $C^{2+\theta}(\partial D) = \mathcal{D}\left(LH_\alpha\right) \subset \mathcal{D}$. Corollary 13.8 asserts that every function u in $\mathcal{D}(L) \subset \mathcal{D}\left(\overline{A}\right)$ can be written in the form

$$u = G_\alpha^0\left((\alpha I - \overline{A})u\right) + H_\alpha\left(u|_{\partial D}\right) \quad \text{for all } \alpha > 0. \tag{13.13}$$

Then we define the boundary condition Lu by the formula

$$Lu = \overline{LG_\alpha^0}\left((\alpha I - \overline{A})u\right) + \overline{LH_\alpha}\left(u|_{\partial D}\right) \quad \text{for every } u \in \mathcal{D}(L). \tag{13.18}$$

The next lemma justifies definition (13.18) of Lu for each $u \in \mathcal{D}(L)$:

Lemma 13.16. *The right-hand side of formula* (13.18) *depends only on* u, *not on the choice of expression* (13.13).

Proof. Assume that

$$u = G_\alpha^0 \left(\left(\alpha I - \overline{A} \right) u \right) + H_\alpha \left(u|_{\partial D} \right)$$
$$= G_\beta^0 \left(\left(\beta I - \overline{A} \right) u \right) + H_\beta \left(u|_{\partial D} \right),$$

where α, $\beta > 0$. Then it follows from formula (13.14) with

$$f := \left(\alpha I - \overline{A} \right) u$$

and formula (13.18) with

$$\psi := u|_{\partial D}$$

that

$$\overline{LG_\alpha^0} \left(\left(\alpha I - \overline{A} \right) u \right) + \overline{LH_\alpha} \left(u|_{\partial D} \right) \tag{13.19}$$
$$= \overline{LG_\beta^0} \left(\left(\alpha I - \overline{A} \right) u \right) - (\alpha - \beta) \overline{LG_\alpha^0} G_\beta^0 \left(\left(\alpha I - \overline{A} \right) u \right)$$
$$+ \overline{LH_\beta} \left(u|_{\partial D} \right) - (\alpha - \beta) \overline{LG_\alpha^0} H_\beta \left(u|_{\partial D} \right)$$
$$= \overline{LG_\beta^0} \left((\beta I - A) u \right) + \overline{LH_\beta} \left(u|_{\partial D} \right)$$
$$+ (\alpha - \beta) \left\{ \overline{LG_\beta^0} u - \overline{LG_\alpha^0} G_\beta^0 \left(\alpha I - \overline{A} \right) u - \overline{LG_\alpha^0} H_\beta \left(u|_{\partial D} \right) \right\}.$$

However, the last term of formula (13.19) vanishes. Indeed, it follows from formula (13.13) with $\alpha := \beta$ and formula (13.14) with $f := u$ that

$$\overline{LG_\beta^0} u - \overline{LG_\alpha^0} \left(G_\beta^0 \left(\alpha I - \overline{A} \right) u \right) - \overline{LG_\alpha^0} H_\beta \left(u|_{\partial D} \right)$$
$$= \overline{LG_\beta^0} u - \overline{LG_\alpha^0} \left(G_\beta^0 \left(\beta I - \overline{A} \right) u + H_\beta \left(u|_{\partial D} \right) + (\alpha - \beta) G_\beta^0 u \right)$$
$$= \overline{LG_\beta^0} u - \overline{LG_\alpha^0} u - (\alpha - \beta) \overline{LG_\alpha^0} G_\beta^0 u$$
$$= 0.$$

Therefore, we obtain from formula (13.19) that

$$\overline{LG_\alpha^0} \left(\left(\alpha I - \overline{A} \right) u \right) + \overline{LH_\alpha} \left(u|_{\partial D} \right) = \overline{LG_\beta^0} \left(\left(\beta I - \overline{A} \right) u \right) + \overline{LH_\beta} \left(u|_{\partial D} \right).$$

The proof of Lemma 13.16 is complete.   □

13.2 Feller Semigroups with Reflecting Barrier

Now we consider the Neumann boundary condition

$$L_N u \equiv \left. \frac{\partial u}{\partial \mathbf{n}} \right|_{\partial D}.$$

We recall that the boundary condition L_N is supposed to correspond to the *reflection phenomenon*.

The next theorem (formula (13.21)) asserts that we can "piece together" a Markov process on ∂D with A-diffusion in D to construct a Markov process on $\overline{D} = D \cup \partial D$ with reflecting barrier (cf. [15, Théorème XIX]):

Theorem 13.17 (the reflecting diffusion). *We define a linear operator*

$$\boxed{\mathfrak{A}_N \colon C(\overline{D}) \longrightarrow C(\overline{D})}$$

as follows.

(a) The domain $\mathcal{D}(\mathfrak{A}_N)$ of $\mathfrak{A}_N$ is the space

$$\mathcal{D}(\mathfrak{A}_N) = \left\{ u \in \mathcal{D}(\overline{A}) : u|_{\partial D} \in \mathcal{D}_N,\ L_N u = 0 \right\}, \tag{13.20}$$

where $\mathcal{D}_N$ is the common domain of the operators $\overline{L_N H_\alpha}$ for all $\alpha > 0$.
(b) $\mathfrak{A}_N u = \overline{A} u$ for every $u \in \mathcal{D}(\mathfrak{A}_N)$.

Then the operator $\mathfrak{A}_N$ is the infinitesimal generator of some Feller semi-group on the state space $\overline{D}$, and the Green operator $G_\alpha^N = (\alpha I - \mathfrak{A}_N)^{-1}$ for $\alpha > 0$ is given by the following formula:

$$G_\alpha^N f = G_\alpha^0 f - H_\alpha \left(\overline{L_N H_\alpha}^{-1} \left(\overline{L_N G_\alpha^0} f \right) \right) \quad \text{for every } f \in C(\overline{D}). \tag{13.21}$$

Remark 13.18. (1) We can express the relationships between the operators $\overline{A}$, G_α^N, H_α and $\overline{L_N H_\alpha}$ in matrix form:

$$\begin{pmatrix} \alpha I - \overline{A} \\ L_N \end{pmatrix} \begin{pmatrix} G_\alpha^N & H_\alpha \left(\overline{L_N H_\alpha}^{-1} \right) \end{pmatrix} = \begin{pmatrix} I & 0 \\ 0 & I \end{pmatrix}.$$

(2) In terms of the Boutet de Monvel calculus, we can express the Green operator G_α^N in the matrix form via the Boutet de Monvel calculus (see Figure 13.7 below)

$$\begin{pmatrix} G_\alpha^0 & -H_\alpha \\ \overline{L_N G_\alpha^0} & \overline{L_N H_\alpha}^{-1} \end{pmatrix} : \begin{matrix} C(\overline{D}) \\ \oplus \\ C(\partial D) \end{matrix} \longrightarrow \begin{matrix} C(\overline{D}) \\ \oplus \\ C(\partial D). \end{matrix}$$

Proof. In order to prove Theorem 13.17, we apply part (ii) of Theorem 3.34 (the Hille–Yosida theorem) to the operator $\mathfrak{A}_N$ defined by formula (14.20). The proof is divided into eight steps.

Step 1: First, we prove that:

> The operator $\overline{L_N H_\alpha}$ is the generator of some Feller semigroup, on the state space ∂D for any *sufficiently large* $\alpha > 0$.

$$
\begin{array}{ccc}
C(\overline{D}) & \xrightarrow{\;\overline{L_N G_\alpha^0}\;} & C(\partial D) \\[4pt]
{\scriptstyle G_\alpha^0}\Big\downarrow & {\scriptstyle \overline{L_N H_\alpha}^{\,-1}}\Big\downarrow & \\[4pt]
C(\overline{D}) & \xleftarrow{\;\;-H_\alpha\;\;} & \mathcal{D}_N = D(\overline{L_N H_\alpha})
\end{array}
$$

Fig. 13.7. The mapping properties of the terms of the Green operator G_α^N defined by formula (13.21)

We introduce a linear operator

$$
\boxed{\,\mathcal{T}_N(\alpha) = L_N H_\alpha : B^{1-1/p,p}(\partial D) \longrightarrow B^{1-1/p,p}(\partial D)\,}
$$

for $N < p < \infty$ as follows.

(a) The domain $\mathcal{D}\left(\mathcal{T}_N(\alpha)\right)$ of $\mathcal{T}_N(\alpha)$ is the space

$$
\mathcal{D}\left(\mathcal{T}_N(\alpha)\right) = \left\{\varphi \in B^{1-1/p,p}(\partial D) : L_N H_\alpha \varphi \in B^{1-1/p,p}(\partial D)\right\}.
$$

(b) $\mathcal{T}_N(\alpha)\varphi = L_N H_\alpha \varphi$ for every $\varphi \in \mathcal{D}\left(\mathcal{T}_N(\alpha)\right)$.

Here it should be emphasized that the harmonic operator H_α is essentially the same as the Poisson operator $P(\alpha)$ introduced in [121, p. 89, Section 5.2].

Then, by arguing just as in the proof of [121, p. 101, Theorem 7.1] with $\mu(x') := 1$ and $\gamma(x') := 0$ on ∂D we obtain that

> The operator $\mathcal{T}_N(\alpha)$ is *bijective* for all sufficiently large $\alpha > 0$.   (13.22)
>
> Furthermore, $\mathcal{T}_N(\alpha)$ maps the space $C^\infty(\partial D)$ *onto* itself.

Since we have the assertion

$$
L_N H_\alpha = \mathcal{T}_N(\alpha) \quad \text{on } C^\infty(\partial D),
$$

it follows from assertion (13.22) that the operator $L_N H_\alpha$ also maps $C^\infty(\partial D)$ onto itself, for any sufficiently large $\alpha > 0$. This implies that the range $\mathcal{R}(L_N H_\alpha)$ is a *dense* subset of $C(\partial D)$. Hence, by applying part (ii) of Theorem 13.14 we obtain that the operator $\overline{L_N H_\alpha}$ generates a Feller semigroup on the state space ∂D, for any sufficiently large $\alpha > 0$. The situation can be visualized in Figure 13.8 below.

Step 2: Next we prove that:

> The operator $\overline{L_N H_\beta}$ generates a Feller semigroup,
>
> on the state space ∂D for *any* $\beta > 0$.

$$\begin{array}{ccc}
\mathcal{D}_N = D(\overline{L_N H_\alpha}) & \xrightarrow{\overline{L_N H_\alpha}} & C(\partial D) \\
\uparrow & & \uparrow \\
\mathcal{D}(\mathcal{T}_N(\alpha)) & \xrightarrow{\mathcal{T}_N(\alpha)} & B^{1-1/p,p}(\partial D) \\
\uparrow & & \uparrow \\
D(L_N H_\alpha) = C^\infty(\partial D) & \xrightarrow[\overline{L_N H_\alpha}]{} & C^\infty(\partial D)
\end{array}$$

Fig. 13.8. The mapping properties of the operators $\mathcal{T}_N(\alpha)$ and $\overline{L_N H_\alpha}$ for $N < p < \infty$

We take a positive constant α so large that the operator $\overline{L_N H_\alpha}$ generates a Feller semigroup on the state space ∂D. We apply Corollary 3.37 with $K := \partial D$ to the operator $\overline{L_N H_\beta}$ for $\beta > 0$. By formula (13.16), it follows that the operator $\overline{L_N H_\beta}$ can be written as

$$\overline{L_N H_\beta} = \overline{L_N H_\alpha} + N_{\alpha\beta},$$

where

$$N_{\alpha\beta} = (\alpha - \beta)\overline{L_N G_\alpha^0}\, H_\beta$$

is a bounded linear operator on $C(\partial D)$ into itself. Furthermore, assertion (13.16) implies that the operator $\overline{L_N H_\beta}$ satisfies condition (β') of Theorem 3.36 (the Hille–Yosida–Ray theorem). Therefore, it follows from an application of Corollary 3.37 that the operator $\overline{L_N H_\beta}$ also generates a Feller semigroup on the state space ∂D.

Step 3: Now we prove that:

The equation   (13.23)
$$\overline{L_N H_\alpha}\,\psi = \varphi$$
has a unique solution ψ in $\mathcal{D}\left(\overline{L_N H_\alpha}\right)$ for any $\varphi \in C(\partial D)$; hence the inverse $\overline{L_N H_\alpha}^{-1}$ of $\overline{L_N H_\alpha}$ can be defined on the whole space $C(\partial D)$. Furthermore, the operator $-\overline{L_N H_\alpha}^{-1}$ is *non-negative* and *bounded* on the space $C(\partial D)$.

Since the function $H_\alpha 1(x)$ takes its positive maximum 1 only on the boundary ∂D, we can apply the Hopf boundary point lemma (Lemma 10.11) to obtain that

$$\frac{\partial}{\partial \mathbf{n}}(H_\alpha 1) < 0 \quad \text{on } \partial D. \tag{13.24}$$

Hence the Neumann boundary condition implies that

$$L_N H_\alpha 1 = L_N (H_\alpha 1) = \frac{\partial}{\partial \mathbf{n}} (H_\alpha 1) < 0 \quad \text{on } \partial D,$$

and so

$$\ell_\alpha = \min_{x' \in \partial D} (-L_N H_\alpha 1)(x') = -\max_{x' \in \partial D} (L_N H_\alpha 1)(x') > 0.$$

Furthermore, by using Corollary 3.35 with

$$\begin{cases} K := \partial D, \\ A := \overline{L_N H_\alpha}, \\ c := \ell_\alpha, \end{cases}$$

we obtain that the operator $\overline{L_N H_\alpha} + \ell_\alpha I$ is the infinitesimal generator of some Feller semigroup on the state space ∂D. Therefore, since $\ell_\alpha > 0$, it follows from an application of part (i) of Theorem 3.34 (the Hille–Yosida theorem) with

$$\mathfrak{A} := \overline{L_N H_\alpha} + \ell_\alpha I$$

that the equation

$$-\overline{L_N H_\alpha}\, \psi = \left(\ell_\alpha I - (\overline{L_N H_\alpha} + \ell_\alpha I)\right) \psi = \varphi$$

has a unique solution $\psi \in \mathcal{D}\left(\overline{L_N H_\alpha}\right)$ for any $\varphi \in C(\partial D)$, and further that the operator $-\overline{L_N H_\alpha}^{-1} = \left(\ell_\alpha I - (\overline{L_N H_\alpha} + \ell_\alpha I)\right)^{-1}$ is non-negative and bounded on the space $C(\partial D)$ with norm

$$\left\|-\overline{L_N H_\alpha}^{-1}\right\| = \left\|\left(\ell_\alpha I - (\overline{L_N H_\alpha} + \ell_\alpha I)\right)^{-1}\right\| \le \frac{1}{\ell_\alpha}.$$

Step 4: By assertion (13.23), we can define the right-hand side of formula (13.21) for all $\alpha > 0$. We prove that:

$$G_\alpha^N = (\alpha I - \mathfrak{A}_N)^{-1} \quad \text{for all } \alpha > 0. \tag{13.25}$$

In view of Lemmas 13.7 and 13.16 with $L := L_N$, it follows that we have, for all $f \in C(\overline{D})$,

$$\begin{cases} G_\alpha^N f = G_\alpha^0 f - H_\alpha \left(\overline{L_N H_\alpha}^{-1} \left(\overline{L_N G_\alpha^0 f}\right)\right) \in \mathcal{D}\left(\overline{A}\right), \\ G_\alpha^N f|_{\partial D} = -\overline{L_N H_\alpha}^{-1} \left(\overline{L_N G_\alpha^0 f}\right) \in \mathcal{D}\left(\overline{L_N H_\alpha}\right) = \mathcal{D}_N, \\ L_N(G_\alpha^N f) = \overline{L_N G_\alpha^0} f - \overline{L_N H_\alpha} \left(\overline{L_N H_\alpha}^{-1} \left(\overline{L_N G_\alpha^0 f}\right)\right) = 0, \end{cases}$$

and

$$(\alpha I - \overline{A})(G_\alpha^N f) = f.$$

Hence we have proved that, for all $f \in C(\overline{D})$,

$$\begin{cases} G_\alpha^N f \in \mathcal{D}\left(\mathfrak{A}_N\right), \\ (\alpha I - \mathfrak{A}_N)\, G_\alpha^N f = f. \end{cases}$$

This proves that

$$(\alpha I - \mathfrak{A}_N) G_\alpha^N = I \quad \text{on } C(\overline{D}).$$

Therefore, in order to prove formula (13.25) it suffices to show the injectivity of the operator $\alpha I - \mathfrak{A}_N$ for $\alpha > 0$.

Assume that:

$$u \in \mathcal{D}(\mathfrak{A}_N) \quad \text{and} \quad (\alpha I - \mathfrak{A}_N) u = 0.$$

Then, by Corollary 13.8 it follows that the function u can be written as

$$u = H_\alpha(u|_{\partial D}), \quad u|_{\partial D} \in \mathcal{D}_N = \mathcal{D}\left(\overline{L_N H_\alpha}\right).$$

Thus we have the assertion

$$\overline{L_N H_\alpha}(u|_{\partial D}) = L_N u = 0.$$

In view of assertion (13.23), this implies that

$$u|_{\partial D} = 0,$$

so that

$$u = H_\alpha(u|_{\partial D}) = 0 \quad \text{in } D.$$

This proves the injectivity of $\alpha I - \mathfrak{A}_N$ for $\alpha > 0$.

Step 5: The non-negativity of G_α^N for $\alpha > 0$ follows immediately from formula (13.21), since the operators G_α^0, H_α, $-\overline{L_N H_\alpha}^{-1}$ and $\overline{L_N G_\alpha^0}$ are all non-negative.

Step 6: We prove that the operator G_α^N is bounded on the space $C(\overline{D})$ with norm

$$\|G_\alpha^N\| \le \frac{1}{\alpha} \quad \text{for all } \alpha > 0. \tag{13.26}$$

To do this, it suffices to show that, for all $\alpha > 0$,

$$G_\alpha^N 1 \le \frac{1}{\alpha} \quad \text{on } \overline{D}. \tag{13.27}$$

since G_α^N is non-negative on $C(\overline{D})$.

First, it follows from the uniqueness property of solutions of problem (13.2) that

$$\alpha G_\alpha^0 1 + H_\alpha 1 = 1 + G_\alpha^0(c(x)) \quad \text{on } \overline{D}. \tag{13.28}$$

Indeed, the both sides satisfy the same equation $(\alpha - A)u = \alpha$ in D and have the same boundary value 1 on ∂D.

By applying the boundary operator L_N to the both hand sides of equality (13.28), we obtain that

$$-L_N H_\alpha 1 = -L_N(H_\alpha 1) = -L_N 1 - L_N G_\alpha^0(c(x)) + \alpha L_N G_\alpha^0 1$$

$$= -\frac{\partial}{\partial \mathbf{n}} \left(G_\alpha^0(c(x)) \right) \Big|_{\partial D} + \alpha L_N G_\alpha^0 1$$

$$\geq \alpha L_N G_\alpha^0 1 \big|_{\partial D} \quad \text{on } \partial D,$$

since $G_\alpha^0(c(x))\big|_{\partial D} = 0$ on ∂D and $G_\alpha^0(c(x)) \leq 0$ on $\overline{D}$. Hence we have, by the non-negativity of $-\overline{L_N H_\alpha}^{-1}$,

$$-\overline{L_N H_\alpha}^{-1}\left(L_N G_\alpha^0 1 \right) \leq \frac{1}{\alpha} \quad \text{on } \partial D. \tag{13.29}$$

By using formula (13.21) with $f := 1$, inequality (13.29) and equality (13.28), we obtain that

$$G_\alpha^N 1 = G_\alpha^0 1 + H_\alpha \left(-\overline{L_N H_\alpha}^{-1}\left(L_N G_\alpha^0 1 \right) \right)$$

$$\leq G_\alpha^0 1 + \frac{1}{\alpha} H_\alpha 1 = \frac{1}{\alpha} + \frac{1}{\alpha} G_\alpha^0(c(x))$$

$$\leq \frac{1}{\alpha} \quad \text{on } \overline{D},$$

since the operators H_α and G_α^0 are non-negative and since $G_\alpha^0(c(x)) \leq 0$ on $\overline{D}$.

Therefore, we have proved the desired assertion (13.27) for all $\alpha > 0$.

Step 7: Finally, we prove that:

$$\text{The domain } \mathcal{D}(\mathfrak{A}_N) \text{ is } \textit{dense} \text{ in the space } C(\overline{D}). \tag{13.30}$$

Step 7-1: Before the proof, we need some lemmas on the behavior of G_α^0, H_α and $-\overline{L_N H_\alpha}^{-1}$ as $\alpha \to +\infty$:

Lemma 13.19. *For all $f \in C(\overline{D})$, we have the assertion*

$$\lim_{\alpha \to +\infty} \left[\alpha G_\alpha^0 f + H_\alpha(f|_{\partial D}) \right] = f \quad \text{in } C(\overline{D}). \tag{13.31}$$

Proof. Choose a positive constant β and let

$$g := f - H_\beta(f|_{\partial D}).$$

Then, by using formula (13.9) with $\varphi := f|_{\partial D}$ we obtain that

$$\alpha G_\alpha^0 g - g = \left[\alpha G_\alpha^0 f + H_\alpha(f|_{\partial D}) - f \right] - \beta G_\alpha^0 H_\beta(f|_{\partial D}). \tag{13.32}$$

However, we have, by estimate (13.5),

$$\lim_{\alpha \to +\infty} G_\alpha^0 H_\beta(f|_{\partial D}) = 0 \quad \text{in } C(\overline{D}),$$

and, by assertion (13.8),

$$\lim_{\alpha \to +\infty} \alpha G_\alpha^0 g = g \quad \text{in } C(\overline{D}),$$

since $g|_{\partial D} = 0$.

Therefore, the desired formula (13.31) follows by letting $\alpha \to +\infty$ in formula (13.32).

The proof of Lemma 13.19 is complete. □

Lemma 13.20. *The function*

$$\frac{\partial}{\partial \mathbf{n}} (H_\alpha 1) \Big|_{\partial D}$$

diverges to $-\infty$ *uniformly and monotonically as* $\alpha \to +\infty$.

Proof. First, formula (13.9) with $\varphi := 1$ gives that

$$H_\alpha 1 = H_\beta 1 - (\alpha - \beta) G_\alpha^0 H_\beta 1.$$

Thus, in view of the non-negativity of G_α^0 and H_α it follows that

$$\alpha \geq \beta \implies H_\alpha 1 \leq H_\beta 1 \quad \text{on } \overline{D}.$$

Since $H_\alpha 1|_{\partial D} = H_\beta 1|_{\partial D} = 1$, this implies that the functions

$$\frac{\partial}{\partial \mathbf{n}} (H_\alpha 1) \Big|_{\partial D}$$

are monotonically non-increasing in α. Furthermore, by using formula (13.7) with $f := H_\beta 1$ we find that the function

$$H_\alpha 1(x) = H_\beta 1(x) - \left(1 - \frac{\beta}{\alpha}\right) \alpha G_\alpha^0 H_\beta 1(x)$$

converges to zero monotonically as $\alpha \to +\infty$, for each interior point x of D.

Now, for any given positive constant K we can construct a function $u \in C^2(\overline{D})$ such that

$$u|_{\partial D} = 1 \quad \text{on } \partial D, \tag{13.33a}$$

$$\frac{\partial u}{\partial \mathbf{n}} \Big|_{\partial D} \leq -K \quad \text{on } \partial D. \tag{13.33b}$$

Indeed, it follows from an application of Theorem 13.1 that, for any integer $m > 0$, the function

$$u = (H_{\alpha_0} 1)^m$$

belongs to $C^{2+\theta}(\overline{D})$ and satisfies condition (13.33a). Furthermore, we have the assertion

$$\frac{\partial u}{\partial \mathbf{n}}\Big|_{\partial D} = m \cdot \frac{\partial}{\partial \mathbf{n}} (H_{\alpha_0} 1)\Big|_{\partial D}$$

$$\leq m \cdot \max_{x' \in \partial D} \left(\frac{\partial}{\partial \mathbf{n}} (H_{\alpha_0} 1)(x') \right).$$

In view of inequality (13.24) with $\alpha := \alpha_0$, this implies that the function $u = (H_{\alpha_0} u)^m$ satisfies condition (13.33b) for m sufficiently large.

We take a function $u \in C^2(\overline{D})$ which satisfies conditions (13.33a) and (13.33b), and choose a neighborhood U of ∂D, relative to $\overline{D}$, with smooth boundary ∂U such that (see Figure 13.9 above)

$$u \geq \frac{1}{2} \quad \text{on } U. \tag{13.34}$$

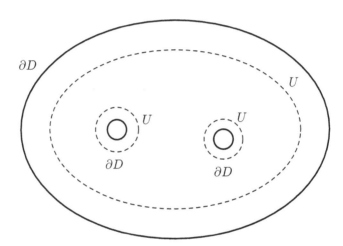

Fig. 13.9. The neighborhood U of ∂D

Recall that the function $H_\alpha 1$ converges to zero in the interior D monotonically as $\alpha \to +\infty$. Since $u = H_\alpha 1 = 1$ on the boundary ∂D, by using Dini's theorem we can find a positive constant α (depending on u and hence on K) such that

$$H_\alpha 1 \leq u \quad \text{on } \partial U \setminus \partial D, \tag{13.35a}$$
$$\alpha > 2 \|Au\|_\infty. \tag{13.35b}$$

It follows from inequalities (13.34) and (13.35b) that

$$(A - \alpha)(H_\alpha 1 - u) = \alpha u - Au \geq \frac{\alpha}{2} - \|Au\|_\infty$$
$$> 0 \quad \text{in } U.$$

Thus, by applying the weak maximum principle (Theorem 10.1) with $A :=$ $A - \alpha$ to the function $H_\alpha 1 - u$ we obtain that the function $H_\alpha 1 - u$ may take its positive maximum only on the boundary ∂U. However, conditions (13.33a) and (13.35a) imply that

$$H_\alpha 1 - u \leq 0 \quad \text{on } \partial U = (\partial U \setminus \partial D) \cup \partial D.$$

Therefore, we have the assertion

$$H_\alpha 1 \leq u \quad \text{on } \overline{U} = U \cup \partial U,$$

and hence

$$\frac{\partial}{\partial \mathbf{n}}(H_\alpha 1)\Big|_{\partial D} \leq \frac{\partial u}{\partial \mathbf{n}}\Big|_{\partial D} \leq -K \quad \text{on } \partial D,$$

since $u|_{\partial D} = H_\alpha 1|_{\partial D} = 1$.

The proof of Lemma 13.20 is complete. □

Now we can study the behavior of the operator norm $\| - \overline{L_N H_\alpha}^{-1}\|$ as $\alpha \to +\infty$:

Corollary 13.21. $\lim_{\alpha \to +\infty} \| - \overline{L_N H_\alpha}^{-1}\| = 0$.

Proof. By Lemma 13.20, it follows that the function

$$L_N H_\alpha 1(x') = L_N (H_\alpha 1)(x') = \frac{\partial}{\partial \mathbf{n}}(H_\alpha 1)(x') \quad \text{for } x' \in \partial D,$$

diverges to $-\infty$ monotonically as $\alpha \to +\infty$. By Dini's theorem, this convergence is uniform in $x' \in \partial D$. Hence the function

$$\frac{1}{L_N H_\alpha 1(x')}$$

converges to zero uniformly in $x' \in \partial D$ as $\alpha \to +\infty$. This implies that

$$\left\| -\overline{L_N H_\alpha}^{-1} \right\| = \left| -\overline{L_N H_\alpha}^{-1} 1 \right|_\infty$$
$$\leq \left| \frac{1}{L_N H_\alpha 1} \right|_\infty \longrightarrow 0 \quad \text{as } \alpha \to +\infty.$$

Indeed, it suffices to note that

$$1 = \frac{-L_N H_\alpha 1(x')}{|L_N H_\alpha 1(x')|} \leq \left| \frac{1}{L_N H_\alpha 1} \right|_\infty (-L_N H_\alpha 1(x')) \quad \text{for all } x' \in \partial D.$$

The proof of Corollary 13.21 is complete. □

Step 7-2: **Proof of Assertion** (13.30)

In view of formula (13.25) and inequality (13.26), it suffices to prove that

$$\lim_{\alpha \to +\infty} \left\| \alpha G_\alpha^N f - f \right\|_\infty = 0 \quad \text{for every } f \in C^\infty(\overline{D}), \tag{13.36}$$

since the space $C^\infty(\overline{D})$ is dense in $C(\overline{D})$.

First, we remark that

$$\begin{aligned}
\left\| \alpha G_\alpha^N f - f \right\|_\infty &= \left\| \alpha G_\alpha^0 f - \alpha H_\alpha \left(\overline{L_N H_\alpha}^{-1} \left(L_N G_\alpha^0 f \right) \right) - f \right\|_\infty \\
&\le \left\| \alpha G_\alpha^0 f + H_\alpha \left(f|_{\partial D} \right) - f \right\|_\infty \\
&\quad + \left\| -\alpha H_\alpha \left(\overline{L_N H_\alpha}^{-1} \left(L_N G_\alpha^0 f \right) \right) - H_\alpha \left(f|_{\partial D} \right) \right\|_\infty \\
&\le \left\| \alpha G_\alpha^0 f + H_\alpha \left(f|_{\partial D} \right) - f \right\|_\infty \\
&\quad + \left| -\alpha \overline{L_N H_\alpha}^{-1} \left(L_N G_\alpha^0 f \right) - \left(f|_{\partial D} \right) \right|_\infty.
\end{aligned}$$

Thus, in view of assertion (13.31) it suffices to show that

$$\lim_{\alpha \to +\infty} \left[-\alpha \overline{L_N H_\alpha}^{-1} \left(L_N G_\alpha^0 f \right) - \left(f|_{\partial D} \right) \right] = 0 \quad \text{in } C(\partial D). \tag{13.37}$$

We take a constant β such that $0 < \beta < \alpha$, and write

$$f = G_\beta^0 g + H_\beta \varphi,$$

where (cf. formula (13.13)):

$$\begin{cases} g = (\beta - A)f \in C^\theta(\overline{D}), \\ \varphi = f|_{\partial D} \in C^{2+\theta}(\partial D). \end{cases}$$

Then, by using equations (13.6) (with $f := g$) and (13.9) we obtain that

$$\begin{aligned}
G_\alpha^0 f &= G_\alpha^0 G_\beta^0 g + G_\alpha^0 H_\beta \varphi \\
&= \frac{1}{\alpha - \beta} \left(G_\beta^0 g - G_\alpha^0 g + H_\beta \varphi - H_\alpha \varphi \right).
\end{aligned}$$

Hence we have the assertion

$$\begin{aligned}
&\left| -\alpha \overline{L_N H_\alpha}^{-1} \left(L_N G_\alpha^0 f \right) - \left(f|_{\partial D} \right) \right|_\infty \\
&= \left| \frac{\alpha}{\alpha - \beta} \left(-\overline{L_N H_\alpha}^{-1} \right) \left(L_N G_\beta^0 g - L_N G_\alpha^0 g + L_N H_\beta \varphi \right) + \frac{\alpha}{\alpha - \beta} \varphi - \varphi \right|_\infty \\
&\le \frac{\alpha}{\alpha - \beta} \left\| -\overline{L_N H_\alpha}^{-1} \right\| \cdot \left| L_N G_\beta^0 g + L_N H_\beta \varphi \right|_\infty \\
&\quad + \frac{\alpha}{\alpha - \beta} \left\| -\overline{L_N H_\alpha}^{-1} \right\| \cdot \left\| L_N G_\alpha^0 \right\| \cdot \left\| g \right\|_\infty + \frac{\beta}{\alpha - \beta} \left| \varphi \right|_\infty.
\end{aligned}$$

By Corollary 13.21, it follows that the first term on the last inequality converges to zero as $\alpha \to +\infty$:

$$\frac{\alpha}{\alpha - \beta} \left\| -\overline{L_N H_\alpha}^{-1} \right\| \cdot \left| L_N G_\beta^0 g + L_N H_\beta \varphi \right|_\infty \longrightarrow 0 \quad \text{as } \alpha \to +\infty.$$

For the second term, by using formula (13.6) with $f := 1$ and the non-negativity of G_β^0 and $L_N G_\alpha^0$ we find that

$$\left\| L_N G_\alpha^0 \right\| = \left| L_N G_\alpha^0 1 \right|_\infty = \left| L_N G_\beta^0 1 - (\alpha - \beta) L_N G_\alpha^0 G_\beta^0 1 \right|_\infty$$
$$\leq \left| L_N G_\beta^0 1 \right|_\infty = \left\| L_N G_\beta^0 \right\| \quad \text{for all } \alpha > 0.$$

Hence, by Corollary 13.21 it follows that the second term also converges to zero as $\alpha \to +\infty$:

$$\frac{\alpha}{\alpha - \beta} \left\| -\overline{L_N H_\alpha}^{-1} \right\| \cdot \left\| L_N G_\alpha^0 \right\| \cdot \|g\|_\infty \longrightarrow 0 \quad \text{as } \alpha \to +\infty.$$

It is clear that the third term converges to zero as $\alpha \to +\infty$:

$$\frac{\beta}{\alpha - \beta} |\varphi|_\infty \longrightarrow 0 \quad \text{as } \alpha \to +\infty.$$

Therefore, we have proved assertion (13.37) and hence the desired assertion (13.36).

The proof of assertion (13.30) is complete.

Step 8: Summing up, we have proved that the operator $\mathfrak{A}_N$, defined by formula (13.20), satisfies conditions (a) through (d) in Theorem 3.34. Hence it follows from an application of the same theorem that the operator $\mathfrak{A}_N$ is the infinitesimal generator of some Feller semigroup on the state space $\overline{D}$.

Now the proof of Theorem 13.17 is complete. $\square$

13.3 Proof of Theorem 1.6

We apply part (ii) of Theorem 3.34 (the Hille–Yosida theorem) to the operator $\mathfrak{A}$ defined by formula (1.6).

First, we simplify the boundary condition

$$Lu = \mu(x') \frac{\partial u}{\partial \mathbf{n}} + \gamma(x') u \bigg|_{\partial D} = 0 \quad \text{on } \partial D.$$

We remark that the original two conditions (A) and (B) are equivalent to the following two conditions (A) and (B'):

(A) $\mu(x') \geq 0$ and $\gamma(x') \leq 0$ on ∂D.
(B') $\gamma(x') < 0$ on $M = \{x' \in \partial D : \mu(x') = 0\}$.

Since we have the inequality

$$\mu(x') - \gamma(x') > 0 \quad \text{on } \partial D,$$

we find that the boundary condition

$$\mu(x')\frac{\partial u}{\partial \mathbf{n}} + \gamma(x')u\bigg|_{\partial D} = 0 \quad \text{on } \partial D$$

is equivalent to the boundary condition

$$\left(\frac{\mu(x')}{\mu(x') - \gamma(x')}\right)\frac{\partial u}{\partial \mathbf{n}} + \left(\frac{\gamma}{\mu(x') - \gamma(x')}\right)u\bigg|_{\partial D} = 0 \quad \text{on } \partial D.$$

Hence, if we let

$$\widetilde{\mu}(x') := \frac{\mu(x')}{\mu(x') - \gamma(x')},$$

$$\widetilde{\gamma}(x') := \frac{\gamma(x')}{\mu(x') - \gamma(x')},$$

then we have the assertions

$$\widetilde{\mu}(x')\frac{\partial u}{\partial \mathbf{n}} + \widetilde{\gamma}(x')u\bigg|_{\partial D} = 0 \quad \text{on } \partial D$$

and

- $0 \leq \widetilde{\mu}(x') \leq 1$ on ∂D,
- $\widetilde{\gamma}(x') = \widetilde{\mu}(x') - 1$ on ∂D.

Namely, we may assume that the boundary condition L is of the form

$$Lu = \mu(x')\frac{\partial u}{\partial \mathbf{n}} + (\mu(x') - 1)u\bigg|_{\partial D} = 0 \quad \text{on } \partial D,$$

with

$$0 \leq \mu(x') \leq 1 \quad \text{on } \partial D.$$

Next we express the boundary condition L in terms of the Dirichlet and Neumann conditions.

It follows from an application of Lemmas 13.9 and 13.11 that

$$\overline{LG_\alpha^0} = \mu(x')\,\overline{L_N G_\alpha^0},$$

and that

$$\overline{LH_\alpha} = \mu(x')\,\overline{L_N H_\alpha} + \mu(x') - 1.$$

Hence, in view of definition (13.18) we obtain that

$$Lu$$
$$= \overline{LG_\alpha^0}\left((\alpha I - \overline{A})\,u\right) + \overline{LH_\alpha}(u|_{\partial D})$$
$$= \mu(x')\overline{L_N G_\alpha^0}\left((\alpha I - \overline{A})\,u\right) + \mu(x')\overline{L_N H_\alpha}(u|_{\partial D}) + (\mu(x') - 1)(u|_{\partial D})$$
$$= \mu(x')\left(\overline{L_N G_\alpha^0}\left((\alpha I - \overline{A})\,u\right) + \overline{L_N H_\alpha}(u|_{\partial D})\right) + (\mu(x') - 1)(u|_{\partial D})$$
$$= \mu(x')L_N u + (\mu(x') - 1)\,(u|_{\partial D}) \quad \text{for every } u \in \mathcal{D}(L).$$

This proves the desired formula

$$L = \mu(x')L_N + \mu(x') - 1. \tag{13.38}$$

Therefore, the next theorem proves Theorem 1.6:

Theorem 13.22. *We define a linear operator*

$$\boxed{\mathfrak{A}\colon C_0\left(\overline{D} \setminus M\right) \longrightarrow C_0\left(\overline{D} \setminus M\right)}$$

as follows (cf. formula (13.15)).

(a) The domain $\mathcal{D}\left(\mathfrak{A}\right)$ of $\mathfrak{A}$ is the space

$$\mathcal{D}\left(\mathfrak{A}\right) = \left\{ u \in C_0\left(\overline{D} \setminus M\right) : \overline{A}u \in C_0\left(\overline{D} \setminus M\right), \right. \tag{13.39}$$

$$\left. Lu = \mu(x')\,L_N u + (\mu(x') - 1)\,(u|_{\partial D}) = 0 \right\}.$$

(b) $\mathfrak{A}u = \overline{A}u$ for every $u \in \mathcal{D}\left(\mathfrak{A}\right)$.

Assume that the following condition (A') is satisfied:

(A') $0 \le \mu(x') \le 1$ on ∂D.

Then the operator $\mathfrak{A}$ is the infinitesimal generator of some Feller semigroup $\{T_t\}_{t \ge 0}$ on the state space $\overline{D} \setminus M$, and the Green operator $G_\alpha = (\alpha I - \mathfrak{A})^{-1}$ for $\alpha > 0$ is given by the following formula:

$$G_\alpha f = G_\alpha^N f - H_\alpha\left(\overline{LH_\alpha}^{-1}\left(LG_\alpha^N f\right)\right) \tag{13.40}$$

$$\text{for every } f \in C_0\left(\overline{D} \setminus M\right).$$

Here G_α^N is the Green operator for the Neumann condition L_N given by formula

$$G_\alpha^N f = G_\alpha^0 f - H_\alpha\left(\overline{L_N H_\alpha}^{-1}\left(\overline{L_N G_\alpha^0} f\right)\right) \tag{13.21}$$

$$\text{for every } f \in C(\overline{D}),$$

and LG_α^N is a boundary operator given by the formula

$$LG_\alpha^N f = (\mu(x') - 1)\left(G_\alpha^N f|_{\partial D}\right) \quad \text{for every } f \in C(\overline{D}).$$

Remark 13.23. (1) We can express the relationships between the operators $\overline{A}$, G_α, H_α and $\overline{LH_\alpha}$ in matrix form:

$$\begin{pmatrix} \alpha I - \overline{A} \\ L \end{pmatrix} \left(G_\alpha \quad H_\alpha \left(\overline{LH_\alpha}^{-1} \right) \right) = \begin{pmatrix} I & 0 \\ 0 & I \end{pmatrix}.$$

(2) In terms of the Boutet de Monvel calculus, we can express the Green operator G_α in the matrix form (see Figure 13.10 below)

$$\begin{pmatrix} G_\alpha^N & -H_\alpha \\ LG_\alpha^N & \overline{LH_\alpha}^{-1} \end{pmatrix} : \begin{matrix} C(\overline{D}) \\ \oplus \\ C(\partial D) \end{matrix} \longrightarrow \begin{matrix} C(\overline{D}) \\ \oplus \\ C(\partial D). \end{matrix}$$

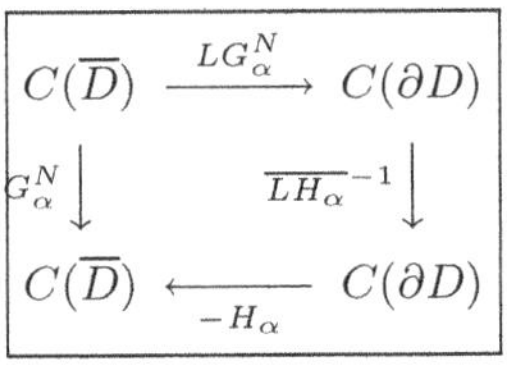

$$\begin{array}{ccc} C(\overline{D}) & \xrightarrow{\ LG_\alpha^N\ } & C(\partial D) \\ {\scriptstyle G_\alpha^N} \downarrow & & \downarrow {\scriptstyle \overline{LH_\alpha}^{-1}} \\ C(\overline{D}) & \xleftarrow[\ -H_\alpha\ ]{} & C(\partial D) \end{array}$$

Fig. 13.10. The mapping properties of the terms of the Green operator G_α defined by formula (13.40)

Proof. In order to prove Theorem 13.22, we apply part (ii) of Theorem 3.34 (the Hille–Yosida theorem) to the operator $\mathfrak{A}$. The proof is divided into six steps.

Step 1: First, we prove that:

If condition (A′) is satisfied, then the closed operator $\overline{LH_\alpha}$ is the generator of some Feller semigroup on the state space ∂D, for any *sufficiently large* $\alpha > 0$.

We introduce a linear operator

$$\boxed{\mathcal{T}(\alpha) = LH_\alpha \colon B^{1-1/p,p}(\partial D) \longrightarrow B^{1-1/p,p}(\partial D)}$$

for $N < p < \infty$ as follows.

(a) The domain $\mathcal{D}(\mathcal{T}(\alpha))$ of $\mathcal{T}(\alpha)$ is the space

$$\mathcal{D}(\mathcal{T}(\alpha)) = \left\{ \varphi \in B^{1-1/p,p}(\partial D) : LH_\alpha \varphi \in B^{1-1/p,p}(\partial D) \right\}.$$

(b) $\mathcal{T}(\alpha)\varphi = LH_\alpha \varphi$ for every $\varphi \in \mathcal{D}(\mathcal{T}(\alpha))$.

Then, by arguing just as in the proof of [121, p. 101, Theorem 7.1] with

$$\gamma(x') := \mu(x') - 1,$$

we obtain that

The operator $\mathcal{T}(\alpha)$ is *bijective* for any sufficiently large $\alpha > 0$.   (13.41)
Furthermore, it maps the space $C^\infty(\partial D)$ onto itself.

Since we have the assertion

$$LH_\alpha = \mathcal{T}(\alpha) \quad \text{on } C^\infty(\partial D),$$

it follows from assertion (13.41) that the operator

$$LH_\alpha : C^\infty(\partial D) \longrightarrow C^\infty(\partial D)$$

is *surjective* for any sufficiently large $\alpha > 0$. This implies that the range $\mathcal{R}(LH_\alpha)$ is a *dense* subset of $C(\partial D)$. Hence, by applying part (ii) of Theorem 13.14 we obtain that the operator $\overline{LH_\alpha}$ generates a Feller semigroup on the state space ∂D, for any sufficiently large $\alpha > 0$. The situation can be visualized in Figure 13.11 below.

$$
\begin{array}{ccc}
C(\partial D) & \xrightarrow{\;\overline{LH_\alpha}\;} & C(\partial D) \\[2mm]
\uparrow & & \uparrow \\[2mm]
\mathcal{D}(\mathcal{T}(\alpha)) & \xrightarrow{\;\mathcal{T}(\alpha)\;} & B^{1-1/p,p}(\partial D) \\[2mm]
\uparrow & & \uparrow \\[2mm]
C^\infty(\partial D) & \xrightarrow[\;LH_\alpha\;]{} & C^\infty(\partial D)
\end{array}
$$

Fig. 13.11. The mapping properties of the operator $\mathcal{T}(\alpha)$ and $\overline{LH_\alpha}$ for $N < p < \infty$

Step 2: Next we prove that:

The operator $\overline{LH_\beta}$ generates a Feller semigroup,
on the state space ∂D for *any* $\beta > 0$.

We take a positive constant α so large that the operator $\overline{LH_\alpha}$ generates a Feller semigroup on the state space ∂D. We apply Corollary 3.37 with $K := \partial D$ to the operator $\overline{LH_\beta}$ for $\beta > 0$. By formula (13.16), it follows that the operator $\overline{LH_\beta}$ can be written as

$$\overline{LH_\beta} = \overline{LH_\alpha} + M_{\alpha\beta},$$

where

$$M_{\alpha\beta} = (\alpha - \beta)\overline{LG_\alpha^0}\, H_\beta$$

is a bounded linear operator on $C(\partial D)$ into itself. Furthermore, assertion (13.15) implies that the operator $\overline{LH_\beta}$ satisfies condition (β') of Theorem 3.36 (the Hille–Yosida–Ray theorem). Therefore, it follows from an application of Corollary 3.37 that the operator $\overline{LH_\beta}$ also generates a Feller semigroup on the state space ∂D.

Step 3: Now we prove that:

If condition (A') is satisfied, then the equation $\qquad$ (13.42)

$$\overline{LH_\alpha}\,\psi = \varphi$$

has a unique solution ψ in $\mathcal{D}\left(\overline{LH_\alpha}\right)$ for any $\varphi \in C(\partial D)$; hence the inverse $\overline{LH_\alpha}^{\,-1}$ of $\overline{LH_\alpha}$ can be defined on the whole space $C(\partial D)$. Furthermore, the operator $-\overline{LH_\alpha}^{\,-1}$ is *non-negative* and *bounded* on the space $C(\partial D)$.

Since we have, by inequality (13.24),

$$LH_\alpha 1 = \left. \mu(x')\frac{\partial}{\partial \mathbf{n}}(H_\alpha 1) + (\mu(x') - 1)\right|_{\partial D} < 0 \quad \text{on } \partial D,$$

it follows that

$$k_\alpha = \min_{x' \in \partial D}\left(-LH_\alpha 1(x')\right) = - \max_{x' \in \partial D} LH_\alpha 1(x') > 0.$$

In view of Lemma 13.20, we find that the constants k_α are increasing in $\alpha > 0$:

$$\alpha \geq \beta > 0 \implies k_\alpha \geq k_\beta.$$

Furthermore, by using Corollary 3.35 with

$$\begin{cases} K := \partial D, \\ A := \overline{LH_\alpha}, \\ c := k_\alpha, \end{cases}$$

we obtain that the operator $\overline{LH_\alpha} + k_\alpha I$ is the infinitesimal generator of some Feller semigroup on the state space ∂D. Therefore, since $k_\alpha > 0$, it follows from an application of part (i) of Theorem 3.34 with

$$\mathfrak{A} := \overline{LH_\alpha} + k_\alpha I$$

that the equation

$$-\overline{LH_\alpha}\,\psi = \left(k_\alpha I - (\overline{LH_\alpha} + k_\alpha I)\right)\psi = \varphi$$

has a unique solution $\psi \in \mathcal{D}\left(\overline{LH_\alpha}\right)$ for any $\varphi \in C(\partial D)$, and further that the operator $-\overline{LH_\alpha}^{-1} = \left(k_\alpha I - (\overline{LH_\alpha} + k_\alpha I)\right)^{-1}$ is non-negative and bounded on the space $C(\partial D)$ with norm

$$\left\|-\overline{LH_\alpha}^{-1}\right\| = \left\|\left(k_\alpha I - (\overline{LH_\alpha} + k_\alpha I)\right)^{-1}\right\| \le \frac{1}{k_\alpha}.$$

Step 4: By assertion (13.42), we can define the operator G_α by formula (13.40) for all $\alpha > 0$. We prove that:

$$G_\alpha = (\alpha I - \mathfrak{A})^{-1} \quad \text{for all } \alpha > 0. \tag{13.43}$$

By Lemma 13.7 and Theorem 13.17, it follows that we have, for all $f \in C_0\left(\overline{D}\setminus M\right)$,

$$G_\alpha f \in \mathcal{D}\left(\overline{A}\right),$$

and

$$\overline{A}(G_\alpha f) = \alpha G_\alpha f - f.$$

Furthermore, we obtain that the function $G_\alpha f$ satisfies the boundary condition

$$L(G_\alpha f) = LG_\alpha^N f - \overline{LH_\alpha}\left(\overline{LH_\alpha}^{-1}\left(LG_\alpha^N f\right)\right) = 0 \quad \text{on } \partial D. \tag{13.44}$$

However, we recall that

$$Lu = \mu(x')L_N u + (\mu(x') - 1)\,(u|_{\partial D}) \quad \text{for every } u \in \mathcal{D}(L). \tag{13.45}$$

Hence it follows that the boundary condition (13.44) is equivalent to the following:

$$L(G_\alpha f) = \mu(x')L_N(G_\alpha f) + (\mu(x') - 1)\,(G_\alpha f|_{\partial D}) = 0 \quad \text{on } \partial D. \tag{13.46}$$

In particular, we have, for all $f \in C_0\left(\overline{D}\setminus M\right)$,

$$G_\alpha f = 0 \text{ on } M = \{x' \in \partial D : \mu(x') = 0\},$$

and so

$$\overline{A}(G_\alpha f) = \alpha G_\alpha f - f = 0 \quad \text{on } M.$$

Summing up, we have proved that

$$f \in C_0\left(\overline{D}\setminus M\right)$$
$$\Longrightarrow$$
$$G_\alpha f \in \mathcal{D}\left(\mathfrak{A}\right) = \left\{u \in C_0\left(\overline{D}\setminus M\right) : \overline{A}u \in C_0\left(\overline{D}\setminus M\right),\ Lu = 0 \text{ on } \partial D\right\},$$

and further that

$$(\alpha I - \mathfrak{A})G_\alpha f = f, \quad f \in C_0\left(\overline{D} \setminus M\right).$$

This proves that

$$(\alpha I - \mathfrak{A})G_\alpha = I \quad \text{on } C_0\left(\overline{D} \setminus M\right).$$

Therefore, in order to prove formula (13.43), it suffices to show the injectivity of the operator $\alpha I - \mathfrak{A}$ for $\alpha > 0$.

Assume that, for $\alpha > 0$,

$$u \in \mathcal{D}(\mathfrak{A}) \quad \text{and} \quad (\alpha I - \mathfrak{A})u = 0.$$

Then, by Corollary 13.8 it follows that the function u can be written as follows:

$$u = H_\alpha\left(u|_{\partial D}\right), \quad u|_{\partial D} \in \mathcal{D} = \mathcal{D}\left(\overline{LH_\alpha}\right).$$

Thus we have the formula

$$\overline{LH_\alpha}\left(u|_{\partial D}\right) = Lu = 0 \text{ on } \partial D.$$

In view of assertion (13.42), this implies that

$$u|_{\partial D} = 0,$$

so that

$$u = H_\alpha\left(u|_{\partial D}\right) = 0 \quad \text{in } D.$$

This proves the injectivity of $\alpha I - \mathfrak{A}$ for $\alpha > 0$.

Step 5: Now we prove the following three assertions (i), (ii) and (iii):

(i) The operator G_α is non-negative on the space $C_0\left(\overline{D} \setminus M\right)$:

$$f \in C_0\left(\overline{D} \setminus M\right), f \geq 0 \quad \text{on } \overline{D} \setminus M \implies G_\alpha f \geq 0 \quad \text{on } \overline{D} \setminus M. \quad (13.47)$$

(ii) The operator G_α is bounded on the space $C_0\left(\overline{D} \setminus M\right)$ with norm

$$\|G_\alpha\| \leq \frac{1}{\alpha} \quad \text{for all } \alpha > 0. \tag{13.48}$$

(iii) The domain $\mathcal{D}(\mathfrak{A})$ is dense in the space $C_0\left(\overline{D} \setminus M\right)$.

Step 5-1: In order to prove assertion (i), we have only to show the non-negativity of the operator G_α on the space $C(\overline{D})$:

$$f \in C(\overline{D}), f \geq 0 \quad \text{on } \overline{D} \implies G_\alpha f \geq 0 \quad \text{on } \overline{D}. \tag{13.49}$$

Recall that the Dirichlet problem

$$\begin{cases} (\alpha - A)u = f & \text{in } D, \\ u|_{\partial D} = \varphi & \text{on } \partial D \end{cases} \tag{13.2}$$

is uniquely solvable. Hence it follows that

$$G_\alpha^N f = H_\alpha \left(G_\alpha^N f|_{\partial D} \right) + G_\alpha^0 f \quad \text{on } \overline{D}. \tag{13.50}$$

Indeed, the both sides satisfy the same equation $(\alpha - A)u = f$ in D and have the same boundary values $G_\alpha^N f|_{\partial D}$ on ∂D.

Thus, by applying the boundary operator L to the both sides of formula (13.50) we obtain that

$$LG_\alpha^N f = \overline{LH_\alpha} \left(G_\alpha^N f|_{\partial D} \right) + \overline{LG_\alpha^0} f.$$

Since the operators $-\overline{LH_\alpha}^{-1}$ and $\overline{LG_\alpha^0}$ are non-negative, it follows that

$$f \geq 0 \quad \text{on } \overline{D}$$

$$\Longrightarrow$$

$$\left(-\overline{LH_\alpha}^{-1} \right) (LG_\alpha^N f) = -G_\alpha^N f|_{\partial D} + \left(-\overline{LH_\alpha}^{-1} \right) \left(\overline{LG_\alpha^0} f \right)$$

$$\geq -G_\alpha^N f|_{\partial D} \quad \text{on } \partial D.$$

Therefore, by the non-negativity of H_α and G_α^0 we find from formulas (13.40) and (13.50) that

$$G_\alpha f = G_\alpha^N f + H_\alpha \left(-\overline{LH_\alpha}^{-1} (LG_\alpha^N f) \right)$$

$$\geq G_\alpha^N f + H_\alpha \left(-G_\alpha^N f|_{\partial D} \right) = G_\alpha^0 f$$

$$\geq 0 \quad \text{on } \overline{D}.$$

This proves the desired assertion (13.49) and hence assertion (i).

Step 5-2: Next we prove assertion (ii). To do this, it suffices to show the boundedness of the operator G_α on the space $C(\overline{D})$:

$$G_\alpha 1 \leq \frac{1}{\alpha} \quad \text{on } \overline{D}, \tag{13.51}$$

since G_α is non-negative on the space $C(\overline{D})$.

We remark (cf. formula (13.45)) that

$$LG_\alpha^N f = \mu(x') L_N G_\alpha^N f + (\mu(x') - 1) \left(G_\alpha^N f|_{\partial D} \right)$$

$$= (\mu(x') - 1) \left(G_\alpha^N f|_{\partial D} \right),$$

so that

$$G_\alpha f = G_\alpha^N f - H_\alpha \left(\overline{LH_\alpha}^{-1} (LG_\alpha^N f) \right)$$

$$= G_\alpha^N f + H_\alpha \left(-\overline{LH_\alpha}^{-1} ((\mu(x') - 1) G_\alpha^N f|_{\partial D}) \right).$$

Hence, by using this formula with $f := 1$ we obtain that

$$G_\alpha 1 = G_\alpha^N 1 - H_\alpha \left(-\overline{LH_\alpha}^{-1} \left((1 - \mu(x')) G_\alpha^N 1|_{\partial D} \right) \right).$$

However, we have, by inequality (13.27),

$$0 \leq G_\alpha^N 1 \leq \frac{1}{\alpha} \quad \text{on } \overline{D},$$

and also

$$H_\alpha \left(-\overline{LH_\alpha}^{-1} \left((1 - \mu(x')) G_\alpha^N 1|_{\partial D} \right) \right) \geq 0 \quad \text{on } \overline{D},$$

since the operators H_α and $-\overline{LH_\alpha}^{-1}$ are non-negative and since $1 - \mu(x') \geq 0$ on ∂D.

Therefore, we obtain that

$$0 \leq G_\alpha 1 \leq G_\alpha^N 1 \leq \frac{1}{\alpha} \quad \text{on } \overline{D}.$$

This proves the desired assertion (13.51) and hence assertion (ii).

Step 5-3: Finally, we prove assertion (iii). In view of formula (13.43), it suffices to show that

$$\lim_{\alpha \to +\infty} \| \alpha G_\alpha f - f \|_\infty = 0 \tag{13.52}$$

$$\text{for every } f \in C_0 \left(\overline{D} \setminus M \right) \cap C^\infty(\overline{D}),$$

since the space $C_0 \left(\overline{D} \setminus M \right) \cap C^\infty(\overline{D})$ is dense in $C_0 \left(\overline{D} \setminus M \right)$.

We remark that

$$\alpha G_\alpha f - f = \alpha G_\alpha^N f - f - \alpha H_\alpha \left(\overline{LH_\alpha}^{-1} \left(L G_\alpha^N f \right) \right) \tag{13.53}$$

$$= \left(\alpha G_\alpha^N f - f \right) + H_\alpha \left(\overline{LH_\alpha}^{-1} \left(\alpha(1 - \mu(x')) G_\alpha^N f|_{\partial D} \right) \right).$$

We estimate the last two terms of formula (13.53) as follows:

(1) By assertion (13.36), it follows that the first term of formula (13.53) tends to zero as $\alpha \to +\infty$:

$$\lim_{\alpha \to +\infty} \| \alpha G_\alpha^N f - f \|_\infty = 0. \tag{13.54}$$

(2) To estimate the second term of formula (13.53), we remark that

$$H_\alpha \left(\overline{LH_\alpha}^{-1} \left(\alpha \left(1 - \mu(x') \right) G_\alpha^N f|_{\partial D} \right) \right) \tag{13.55}$$

$$= H_\alpha \left(\overline{LH_\alpha}^{-1} \left((1 - \mu(x')) f|_{\partial D} \right) \right)$$

$$+ H_\alpha \left(\overline{LH_\alpha}^{-1} \left((1 - \mu(x')) \left(\alpha G_\alpha^N f - f \right) |_{\partial D} \right) \right).$$

However, it follows that the second term in the right-hand side of formula (13.55) tends to zero as $\alpha \to +\infty$. Indeed, we have, by assertion (13.54),

$$\left\|H_\alpha\left(\overline{LH_\alpha}^{-1}\left((1-\mu(x'))\left(\alpha G_\alpha^N f - f\right)|_{\partial D}\right)\right)\right\|_\infty \tag{13.56}$$

$$\leq \left\|-\overline{LH_\alpha}^{-1}\right\| \cdot \left|(1-\mu(x'))\left(\alpha G_\alpha^N f - f\right)|_{\partial D}\right|_\infty$$

$$\leq \frac{1}{k_\alpha}\left|(1-\mu(x'))\left(\alpha G_\alpha^N f - f\right)|_{\partial D}\right|_\infty$$

$$\leq \frac{1}{k_1}\left\|\alpha G_\alpha^N f - f\right\|_\infty \longrightarrow 0 \quad \text{as } \alpha \to +\infty.$$

Here we have used the following two inequalities (cf. the proof of assertion (13.42)):

- $\left\|-\overline{LH_\alpha}^{-1}\right\| \leq \dfrac{1}{k_\alpha}$ for all $\alpha > 0$.

- $k_1 = \min\limits_{x'\in\partial D}(-LH_1 1)(x') \leq k_\alpha = \min\limits_{x'\in\partial D}(-LH_\alpha 1)(x')$ for all $\alpha \geq 1$.

Thus we are reduced to the study of the first term of the right-hand side of formula (13.55)

$$H_\alpha\left(\overline{LH_\alpha}^{-1}\left((1-\mu(x'))f|_{\partial D}\right)\right).$$

Now, for any given $\varepsilon > 0$, we can find a function $h(x')$ in $C^\infty(\partial D)$ such that

$$\begin{cases} h = 0 \quad \text{near } M = \{x' \in \partial D : \mu(x') = 0\}, \\ \|(1-\mu(x'))f|_{\partial D} - h\|_\infty < \varepsilon. \end{cases}$$

Then we have, for all $\alpha \geq 1$,

$$\left\|H_\alpha\left(\overline{LH_\alpha}^{-1}\left((1-\mu(x'))f|_{\partial D}\right)\right) - H_\alpha\left(\overline{LH_\alpha}^{-1}h\right)\right\|_\infty \tag{13.57}$$

$$\leq \left\|-\overline{LH_\alpha}^{-1}\right\| \cdot \left|(1-\mu(x'))f|_{\partial D} - h\right|_\infty \leq \frac{\varepsilon}{k_\alpha}$$

$$\leq \frac{\varepsilon}{k_1}.$$

Furthermore, we can find a function $\theta(x')$ in $C_0^\infty(\partial D)$ such that

$$\begin{cases} \theta(x') = 1 & \text{near } M, \\ (1-\theta(x'))h(x') = h(x') & \text{on } \partial D. \end{cases}$$

Then we have the assertion

$$h(x') = (1-\theta(x'))\,h(x') = (-LH_\alpha 1(x'))\left(\frac{1-\theta(x')}{-LH_\alpha 1(x')}\right)h(x')$$

$$\leq \left\|\frac{1-\theta}{-LH_\alpha 1}\right\|_\infty \cdot |h|_\infty\,(-LH_\alpha 1(x')).$$

Since the operator $-\overline{LH_\alpha}^{-1}$ is non-negative on the space $C(\partial D)$, it follows that

$$-\overline{LH_\alpha}^{-1} h \le \left\|\frac{1-\theta}{-LH_\alpha 1}\right\|_\infty \cdot |h|_\infty \quad \text{on } \partial D,$$

so that

$$\left\|H_\alpha\left(\overline{LH_\alpha}^{-1} h\right)\right\| \le \left|-\overline{LH_\alpha}^{-1} h\right|_\infty \le \left|\frac{1-\theta}{-LH_\alpha 1}\right|_\infty \cdot |h|_\infty . \tag{13.58}$$

However, there exists a constant $\delta_0 > 0$ such that

$$0 \le \frac{1-\theta(x')}{\mu(x')} \le \delta_0 \quad \text{for all } x' \in \partial D.$$

Thus it follows that

$$\frac{1-\theta(x')}{-LH_\alpha 1(x')} = \frac{1-\theta(x')}{\mu(x')\left(-\dfrac{\partial}{\partial \mathbf{n}}\left(H_\alpha 1(x')\right)\right) + (1-\mu(x'))}$$

$$\le \left(\frac{1-\theta(x')}{\mu(x')}\right)\frac{1}{\left(-\dfrac{\partial}{\partial \mathbf{n}}\left(H_\alpha 1(x')\right)\right)}$$

$$\le \delta_0 \frac{1}{\min_{x'\in\partial D}\left(-\dfrac{\partial}{\partial \mathbf{n}}\left(H_\alpha 1(x')\right)\right)} \quad \text{for all } x' \in \partial D,$$

and hence from Lemma 13.20 that

$$\lim_{\alpha\to+\infty}\left|\frac{1-\theta}{-LH_\alpha 1}\right|_\infty = 0. \tag{13.59}$$

Summing up, we obtain from inequalities (13.57) and (13.58) and assertion (13.59) that

$$\limsup_{\alpha\to+\infty}\left\|H_\alpha\left(\overline{LH_\alpha}^{-1}\left((1-\mu(x'))f|_{\partial D}\right)\right)\right\|_\infty$$

$$\le \limsup_{\alpha\to+\infty}\left[\left\|H_\alpha\left(\overline{LH_\alpha}^{-1} h\right)\right\|_\infty\right.$$

$$\left.+ \left\|H_\alpha\left(\overline{LH_\alpha}^{-1}\left((1-\mu(x'))f|_{\partial D}\right)\right) - H_\alpha\left(\overline{LH_\alpha}^{-1} h\right)\right\|_\infty\right]$$

$$\le \lim_{\alpha\to+\infty}\left|\frac{1-\theta}{-LH_\alpha 1}\right|_\infty \cdot |h|_\infty + \frac{\varepsilon}{k_1}$$

$$= \frac{\varepsilon}{k_1}.$$

Since ε is arbitrary, this proves that the first term of the right-hand side of formula (13.55) tends to zero as $\alpha \to +\infty$:

$$\lim_{\alpha \to +\infty} \left\| H_\alpha \left(\overline{LH_\alpha}^{-1} \left((1 - \mu(x'))f|_{\partial D} \right) \right) \right\|_\infty = 0. \tag{13.60}$$

By assertions (13.56) and (13.60), we obtain that the last term of formula (13.53) also tends to zero:

$$\lim_{\alpha \to +\infty} \left\| H_\alpha \left(\overline{LH_\alpha}^{-1} \left(\alpha(1 - \mu(x'))G_\alpha^N f|_{\partial D} \right) \right) \right\|_\infty = 0. \tag{13.61}$$

Therefore, the desired assertion (13.52) follows by combining assertions (13.54) and (13.61).

The proof of assertion (iii) is complete.

Step 6: Summing up, we have proved that the operator $\mathfrak{A}$, defined by formula (13.39), satisfies conditions (a) through (d) in Theorem 3.34. Hence, in view of assertion (13.36), it follows from an application of part (ii) of the same theorem that the operator $\mathfrak{A}$ is the infinitesimal generator of some Feller semigroup $\{T_t\}_{t \geq 0}$ on the state space $\overline{D} \setminus M$.

Now the proof of Theorem 13.22 and hence that of Theorem 1.6 is complete. □

13.4 Proof of Part (ii) of Theorem 1.5

We apply Theorem 2.2 to the operator $\mathfrak{A}$.

In the proof of Theorem 13.22, we have proved that the domain $\mathcal{D}(\mathfrak{A})$ is *dense* in the space $C_0\left(\overline{D} \setminus M\right)$. Furthermore, part (i) of Theorem 1.5 verifies conditions (2.1) and (2.2). Therefore, it follows from an application of Theorem 2.2 that:

The semigroup T_t can be extended to a semigroup T_z that is *analytic* in the sector $\Delta_\varepsilon = \{z = t + is : z \neq 0, |\arg z| < \pi/2 - \varepsilon\}$ for any $0 < \varepsilon < \pi/2$.

This (together with Theorem 1.6) proves part (ii) of Theorem 1.5. □

13.5 Notes and Comments

This chapter is an expanded and revised version of Chapter 9 of the second edition [121], which is adapted from Bony–Courrège–Priouret [15], Sato–Ueno [98] and Taira [114], [116] and [122].

Proofs of Theorems 1.8, 1.9, 1.10 and 1.11

In this chapter we prove Theorems 1.8, 1.9, 1.10 and 1.11, generalizing Theorems 1.4, 1.5 and 1.6 for second-order, elliptic Waldenfels operators. More precisely, we consider a second-order, *elliptic Waldenfels operator* W with real coefficients such that

$$W u(x) = A u(x) + S u(x) \tag{1.2}$$

$$:= \left(\sum_{i,j=1}^{N} a^{ij}(x) \frac{\partial^2 u}{\partial x_i \partial x_j}(x) + \sum_{i=1}^{N} b^i(x) \frac{\partial u}{\partial x_i}(x) + c(x) u(x) \right)$$

$$+ \int_{\mathbf{R}^N \setminus \{0\}} \left(u(x+z) - u(x) - \sum_{j=1}^{N} z_j \frac{\partial u}{\partial x_j}(x) \right) s(x,z) \, m(dz).$$

Here:

(1) $a^{ij} \in C^\infty(\overline{D})$, $a^{ij}(x) = a^{ji}(x)$ for all $1 \le i,j \le N$, and there exists a constant $a_0 > 0$ such that

$$\sum_{i,j=1}^{N} a^{ij}(x) \xi_i \xi_j \ge a_0 |\xi|^2 \quad \text{for all } x \in \overline{D} \text{ and } \xi \in \mathbf{R}^N.$$

(2) $b^i \in C^\infty(\overline{D})$ for all $1 \le i \le N$.

(3) $c \in C^\infty(\overline{D})$, and $c(x) \le 0$ in D, but $c(x) \not\equiv 0$ in D.

(4) $s(x,z) \in L^\infty(\mathbf{R}^N \times \mathbf{R}^N)$ and $0 \le s(x,z) \le 1$ almost everywhere in the product space $\mathbf{R}^N \times \mathbf{R}^N$, and there exist constants $C_0 > 0$ and $0 < \theta_0 < 1$ such that

$$|s(x,z) - s(y,z)| \le C_0 |x-y|^{\theta_0} \tag{1.3a}$$

$$\text{for all } x, y \in \overline{D} \text{ and almost all } z \in \mathbf{R}^N,$$

and the *support condition*

© Springer Nature Switzerland AG 2020
K. Taira, *Boundary Value Problems and Markov Processes*, Lecture Notes in Mathematics 1499,
https://doi.org/10.1007/978-3-030-48788-1_14

$$s(x, z) = 0 \quad \text{if } x \in D \text{ and } x + z \notin \overline{D}. \tag{1.3b}$$

Probabilistically, the support condition (1.3b) implies that all jumps from D are within $\overline{D}$. Analytically, the support condition (1.3b) guarantees that the integral operator S may be considered as an operator acting on functions u defined on the closure $\overline{D}$ (see [48, Chapter II, Remark 1.19]).

(5) The measure $m(dz)$ is a Radon measure on $\mathbf{R}^N \setminus \{0\}$ which has a density with respect to the Lebesgue measure dz on $\mathbf{R}^N$, and satisfies the *moment condition* (see Example 1.1)

$$\int_{\{0<|z|\leq 1\}} |z|^2 \, m(dz) + \int_{\{|z|>1\}} |z| \, m(dz) < \infty. \tag{1.4}$$

(6) Finally, we assume that

$$W1(x) = A1(x) + S1(x) = c(x) \leq 0 \text{ and } c(x) \not\equiv 0 \text{ in } D. \tag{1.5}$$

In Section 14.1, by using the Hölder space theory of pseudo-differential operators we study the non-homogeneous boundary value problem

$$\begin{cases} Au = f & \text{in } D, \\ Lu = \mu(x')\dfrac{\partial u}{\partial \mathbf{n}} + \gamma(x')u = \varphi & \text{on } \partial D, \end{cases} \tag{$*$}$$

in the framework of *Hölder spaces*, and prove an existence and uniqueness theorem for the problem $(*)$ (Theorem 14.1). More precisely, we prove that if the two conditions

(A) $\mu(x') \geq 0$ and $\gamma(x') \leq 0$ on ∂D,
(B) $\mu(x') - \gamma(x') = \mu(x') + |\gamma(x')| > 0$ on ∂D,

are satisfied, then the mapping

$$\boxed{(A, L) \colon C^{2+\theta}(\overline{D}) \longrightarrow C^{\theta}(\overline{D}) \oplus C_*^{1+\theta}(\partial D)}$$

is an algebraic and topological *isomorphism* for all $0 < \theta < 1$. The function space $C_*^{1+\theta}(\partial D)$ is introduced in Subsection 1.2.2 under conditions (A) and (B).

In Section 14.2, we prove an existence and uniqueness theorem for the non-homogeneous boundary value problem (Theorem 1.8)

$$\begin{cases} Wu = f & \text{in } D, \\ Lu = \varphi & \text{on } \partial D. \end{cases} \tag{$**$}$$

in the framework of *Hölder spaces*. Namely, we prove that if conditions (A) and (B) and the condition

(H) The integral operator S satisfies conditions (1.3a), (1.3b), (1.4) and (1.5), are satisfied, then the mapping

$$\boxed{(W, L) = (A + S, L) : C^{2+\theta}(\overline{D}) \longrightarrow C^{\theta}(\overline{D}) \oplus C_{\star}^{1+\theta}(\partial D)}$$

is an algebraic and topological *isomorphism* for all $0 < \theta \leq \theta_0$. The proof of Theorem 1.8 is flowcharted (Table 14.1).

Due to the non-local character of the Waldenfels integro-differential operator $W = A + S$, we find more difficulties in the bounded domain D than in the whole space $\mathbf{R}^N$. In fact, when considering the Dirichlet problem in D, it is natural to use the *zero extension* of functions in the interior D outside of the closure $\overline{D} = D \cup \partial D$. This extension has a probabilistic interpretation. Namely, this corresponds to stopping the diffusion process with jumps in the whole space $\mathbf{R}^N$ at the *first exit time* of the closure $\overline{D}$.

However, the zero extension produces a singularity of solutions at the boundary ∂D. In order to remove this singularity, we introduce various conditions on the structure of jumps for the Waldenfels integro-differential operator $W = A + S$ such as conditions (1.3a), (1.3b) and (1.4). More precisely, we can estimate the Lévy integro-differential operator S in terms of Hölder norms, just as in [48, Chapter II, Lemmas 1.2 and 1.5] (Lemmas 14.4 and 14.5), and prove that the Lévy operator

$$S : C^{2+\theta}(\overline{D}) \longrightarrow C^{\theta}(\overline{D})$$

is *compact* for all $0 < \theta \leq \theta_0$ under conditions (1.3a), (1.3b) and (1.4) (Lemma 14.6). This implies that the mapping

$$(W, L) = (A, L) + (S, 0) : C^{2+\theta}(\overline{D}) \longrightarrow C^{\theta}(\overline{D}) \oplus C_{\star}^{1+\theta}(\partial D)$$

is a perturbation of a compact operator to the mapping

$$(A, L) : C^{2+\theta}(\overline{D}) \longrightarrow C^{\theta}(\overline{D}) \oplus C_{\star}^{1+\theta}(\partial D).$$

In this way, the proof of Theorem 1.8 can be reduced to the differential operator case (Theorem 14.1). Hence we have, by Theorem 14.1 and Lemma 14.6,

$$\text{ind}\,(W, L) = \text{ind}\,(A, L) = 0,$$

since the index is *stable* under compact perturbations ([51]).

On the other hand, by using the maximum principle (Theorem 10.7 and Lemma 10.11) we find that if conditions (A), (B) and (1.5) are satisfied, then the mapping (W, L) is *injective* (Proposition 14.7).

Therefore, we can prove that if conditions (A), (B) and (H) are satisfied, then the mapping (W, L) is *bijective* (Theorem 1.8).

In Section 14.3 we prove Theorem 1.9 (Theorem 14.8). We estimate the integral operator S in terms of L^p norms, and show that S is an A_p-completely continuous operator in the sense of Gohberg–Kreĭn [51] (Lemmas 14.9 and 14.10). The proof of Theorem 1.9 is flowcharted (Table 14.2).

Section 14.4 is devoted to the proof of Theorem 1.10. Theorem 1.10 follows from Theorem 1.9 by using Sobolev's imbedding theorems and a λ-dependent localization argument. The proof is carried out in a chain of auxiliary lemmas (Lemmas 14.11, 14.12 and 14.15). The proof of Theorem 1.10 is flowcharted (Table 14.3).

In Section 14.5, as an application, we construct a Feller semigroup corresponding to such a diffusion phenomenon that a Markovian particle moves both by jumps and continuously in the state space until it "dies" at the time when it reaches the set where the particle is definitely absorbed, generalizing Theorem 1.6 (Theorem 1.11). The proof of Theorem 1.11 is flowcharted (Table 14.4).

14.1 Existence and Uniqueness Theorem in Hölder Spaces

In this section, by using the Hölder space theory of pseudo-differential operators we study the boundary value problem $(*)$ in the framework of *Hölder spaces*. First, we prove an existence and uniqueness theorem for the non-homogeneous boundary value problem $(*)$ ([117, Theorem 1.1]):

Theorem 14.1 (the existence and uniqueness theorem). *Let* $0 < \theta < 1$. *Assume that the following two conditions (A) and (B) are satisfied:*

(A) $\mu(x') \geq 0$ *and* $\gamma(x') \leq 0$ *on* ∂D.
(B) $\mu(x') - \gamma(x') = \mu(x') + |\gamma(x')| > 0$ *on* ∂D.

Then the mapping

$$(A, L) : C^{2+\theta}(\overline{D}) \longrightarrow C^{\theta}(\overline{D}) \oplus C_\star^{1+\theta}(\partial D)$$

is an algebraic and topological isomorphism. In particular, for any $f \in C^{\theta}(\overline{D})$ *and any* $\varphi \in C_\star^{1+\theta}(\partial D)$, *there exists a unique solution* $u \in C^{2+\theta}(\overline{D})$ *of problem* $(*)$.

Proof. In order to prove Theorem 14.1, it suffices to show that the mapping (A, L) is *bijective*. Indeed, the continuity of the inverse of (A, L) follows immediately from an application of Banach's open mapping theorem (see [20, Chapter 2, Corollary 2.7], [44, Chapter 4, Theorem 4.6.2], [147, Chapter II, Section 5, Corollary]), since (A, L) is a continuous operator.

The proof is divided into four steps.
Step 1: Let (f, φ) be an arbitrary element of the space

$$C^{\theta}(\overline{D}) \oplus C_\star^{1+\theta}(\partial D)$$

with

$$\varphi = \mu(x')\varphi_1 - \gamma(x')\varphi_2 \quad \text{for } \varphi_1 \in C^{1+\theta}(\partial D) \text{ and } \varphi_2 \in C^{2+\theta}(\partial D).$$

First, we show that the boundary value problem $(*)$ can be reduced to the study of an operator on the boundary, just as in Section 6.3.

To do this, we consider the following Neumann problem:

$$\begin{cases} Av = f & \text{in } D, \\ \dfrac{\partial v}{\partial \mathbf{n}} = \varphi_1 & \text{on } \partial D. \end{cases} \tag{14.1}$$

Recall that the existence and uniqueness theorem for problem (14.1) is well established in the framework of Hölder spaces (see Gilbarg–Trudinger [50, Theorem 6.31]). Thus we find that a function $u \in C^{2+\theta}(\overline{D})$ is a solution of problem $(*)$ if and only if the function

$$w = u - v \in C^{2+\theta}(\overline{D})$$

is a solution of the problem

$$\begin{cases} Aw = 0 & \text{in } D, \\ Lw = \varphi - Lv = \mu(x')\varphi_1 - \gamma(x')\varphi_2 - Lv & \text{on } \partial D. \end{cases}$$

Here it should be noticed that

$$Lv = \mu(x')\frac{\partial v}{\partial \mathbf{n}} + \gamma(x')\,(v|_{\partial D}) = \mu(x')\varphi_1 + \gamma(x')\,(v|_{\partial D}),$$

so that

$$Lw = -\gamma(x')\,(\varphi_2 + (v|_{\partial D})) \in C^{2+\theta}(\partial D).$$

However, we know that every solution $w \in C^{2+\theta}(\overline{D})$ of the homogeneous equation: $Aw = 0$ in D can be expressed as follows (see [50, Theorem 6.14]):

$$w = P\psi \quad \text{for } \psi \in C^{2+\theta}(\partial D).$$

Thus we can reduce the study of problem $(*)$ to that of the pseudo-differential equation

$$T\psi := L\,(P\psi) = Lw = -\gamma(x')\,(\varphi_2 + (v|_{\partial D})) \quad \text{on } \partial D. \tag{14.2}$$

We remark that the pseudo-differential equation (14.2) is a generalization of the classical *Fredholm integral equation*.

Summing up, we have proved the following proposition:

Proposition 14.2. *For given functions $f \in C^\theta(\overline{D})$ and $\varphi \in C_\star^{1+\theta}(\partial D)$, there exists a solution $u \in C^{2+\theta}(\overline{D})$ of problem $(*)$ if and only if there exists a solution $\psi \in C^{2+\theta}(\partial D)$ of the pseudo-differential equation (14.2).*

Step 2: We study the operator $T = LP$ in question. The next proposition is an essential step in the proof of Theorem 14.1 (see Lemma 7.1):

Proposition 14.3. *If conditions (A) and (B) are satisfied, then there exists a parametrix E in the* Hörmander *class $L^0_{1,1/2}(\partial D)$ for T which maps the* Hölder *space $C^{k+\theta}(\partial D)$ continuously into itself, for any non-negative integer k.*

Proof. Indeed, by virtue of [121, p. 91, Lemma 5.2] (cf. [68, Theorem 3.1]) we can construct a parametrix E in the Hörmander class $L^0_{1,1/2}(\partial D)$ for T. Furthermore, it follows from an application of the Besov space boundedness theorem (Theorem 4.47) that the parametrix E maps the Hölder space $C^{k+\theta}(\partial D)$ continuously into itself, for any nonnegative integer k, since we have the formula (see [135, Theorem 2.5.7])

$$C^{k+\theta}(\partial D) = B^{k+\theta}_{\infty,\infty}(\partial D).$$

The proof of proposition 14.3 is complete. □

Step 3: Now we remark that

$$\begin{cases} C^\theta(\overline{D}) \subset L^p(D), \\ C^{1+\theta}_\star(\partial D) \subset B^{1-1/p,p}_\star(\partial D) \end{cases}$$

for $1 < p < \infty$. Thus, by applying [123, Theorem 1.1] to our situation we find that problem $(*)$ has a unique solution $u \in H^{2,p}(D)$ for any $f \in C^\theta(\overline{D})$ and any $\varphi \in C^{1+\theta}_\star(\partial D)$. Furthermore, by virtue of Proposition 14.3 it follows that the solution u can be written in the form

$$u = v + P\psi \quad \text{for } v \in C^{2+\theta}(\overline{D}) \text{ and } \psi \in B^{2-1/p,p}(\partial D). \tag{14.3}$$

However, we have, by equation (14.2) and Proposition 14.3,

$$\psi \in C^{2+\theta}(\partial D).$$

Indeed, it suffices to note that

$$\psi \equiv E(T\psi) = -E\left(\gamma(x')(\varphi_2 + (v|_{\partial D}))\right) \in C^{2+\theta}(\partial D) \bmod C^\infty(\partial D).$$

Therefore, we obtain from formula (14.3) that

$$u = v + P\psi \in C^{2+\theta}(\overline{D}).$$

Step 4: The injectivity of (A, L) follows from an application of the maximum principle (Theorem 10.7 and Lemma 10.11 with $S \equiv 0$).

The proof of Theorem 14.1 is now complete. □

14.2 Proof of Theorem 1.8

In this section we study problem $(**)$ in the framework of Hölder spaces, and prove Theorem 1.8 under conditions (A), (B) and (H). The essential point

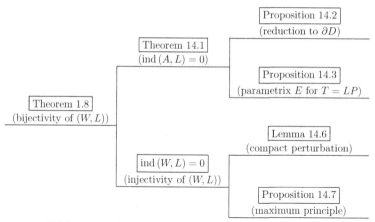

Table 14.1. A flowchart for the proof of Theorem 1.8

in the proof is to estimate the integral operator S in terms of Hölder norms. We show that the operator (W, L) may be considered as a perturbation of a *compact operator* to the operator (A, L) in the framework of Hölder spaces. Thus the proof of Theorem 1.8 is reduced to the differential operator case. The proof of Theorem 1.8 can be flowcharted as in Table 14.1 above.

The proof of Theorem 1.8 is divided into three steps.

Step (I): By virtue of Claim 10.8, we can estimate the term Su in terms of Hölder norms, just as in [48, Chapter II, Lemmas 1.2 and 1.5]:

Lemma 14.4. *Assume that condition (H) is satisfied. Then, for every $\eta > 0$ there exists a constant $C_\eta > 0$ such that we have, for all $u \in C^2(\overline{D})$,*

$$\|Su\|_\infty \leq \eta \left\|\nabla^2 u\right\|_\infty + C_\eta \left(\|u\|_\infty + \|\nabla u\|_\infty\right). \tag{14.4}$$

Here

$$\|u\|_\infty = \sup_{x \in D} |u(x)| = \max_{x \in \overline{D}} |u(x)|.$$

Proof. By using Taylor's formula, we decompose the integral term Su into the following two terms:

$$Su(x) \tag{14.5}$$

$$= \int_{\mathbf{R}^N \setminus \{0\}} \left(u(x + z) - u(x) - \sum_{j=1}^N z_j \frac{\partial u}{\partial x_j}(x) \right) s(x, z) \, m(dz)$$

$$= \sum_{i,j=1}^N \int_{\{0 < |z| \leq \varepsilon\}} z_i z_j \left(\int_0^1 \frac{\partial^2 u}{\partial x_i \partial x_j}(x + tz) \cdot (1 - t) \, dt \right) s(x, z) \, m(dz)$$

$$+ \int_{\{|z|>\varepsilon\}} \left(u(x+z) - u(x) - \sum_{j=1}^{N} z_j \frac{\partial u}{\partial x_j}(x) \right) s(x,z)\, m(dz)$$

$$= \int_0^1 (1-t)\, dt \int_{\{0<|z|\leq\varepsilon\}} z \cdot \nabla^2 u(x+tz) z\, s(x,z)\, m(dz)$$

$$+ \int_{\{|z|>\varepsilon\}} (u(x+z) - u(x) - z \cdot \nabla u(x))\, s(x,z)\, m(dz)$$

$$:= S_\varepsilon^{(1)} u(x) + S_\varepsilon^{(2)} u(x).$$

By Claim 10.8, we can estimate the terms $S_\varepsilon^{(1)} u$ and $S_\varepsilon^{(2)} u$ as follows:

$$\left\| S_\varepsilon^{(1)} u \right\|_\infty$$

$$\leq \frac{1}{2} \left(\int_{\{0<|z|\leq\varepsilon\}} |z|^2\, m(dz) \right) \left\| \nabla^2 u \right\|_\infty = \frac{1}{2} \sigma(\varepsilon) \left\| \nabla^2 u \right\|_\infty,$$

and

$$\left\| S_\varepsilon^{(2)} u \right\|_\infty$$

$$\leq 2 \left(\int_{\{|z|>\varepsilon\}} m(dz) \right) \| u \|_\infty + \left(\int_{\{|z|>\varepsilon\}} |z|\, m(dz) \right) \| \nabla u \|_\infty$$

$$= 2\tau(\varepsilon) \| u \|_\infty + \delta(\varepsilon) \| \nabla u \|_\infty$$

$$\leq 2 \left(\frac{C_1}{\varepsilon^2} + C_2 \right) \| u \|_\infty + \left(\frac{C_1}{\varepsilon} + C_2 \right) \| \nabla u \|_\infty.$$

Therefore, we have proved that

$$\| Su \|_\infty \leq \left\| S_\varepsilon^{(1)} u \right\|_\infty + \left\| S_\varepsilon^{(2)} u \right\|_\infty$$

$$\leq \frac{1}{2} \sigma(\varepsilon) \left\| \nabla^2 u \right\|_\infty + 2 \left(\frac{C_1}{\varepsilon^2} + C_2 \right) \| u \|_\infty + \left(\frac{C_1}{\varepsilon} + C_2 \right) \| \nabla u \|_\infty.$$

In view of assertion (10.12a), this proves the desired estimate (14.4) if we choose ε sufficiently small.

The proof of Lemma 14.4 is complete.   □

Lemma 14.5. *Assume that condition (H) is satisfied. Then, for every $\eta > 0$ there exists a constant $C_\eta > 0$ such that we have, for all $u \in C^{2+\theta_0}(\overline{D})$,*

$$\| Su \|_{C^{\theta_0}(\overline{D})} \leq \eta \left\| \nabla^2 u \right\|_{C^{\theta_0}(\overline{D})} + C_\eta \left(\| u \|_{C^{\theta_0}(\overline{D})} + \| \nabla u \|_{C^{\theta_0}(\overline{D})} \right). \quad (14.6)$$

Here

$$\| u \|_{C^{\theta_0}(\overline{D})} = \| u \|_\infty + [u]_{\theta_0}, \quad [u]_{\theta_0} = \sup_{\substack{x,y\in D \\ x\neq y}} \frac{|u(x) - u(y)|}{|x-y|^{\theta_0}}.$$

Proof. We estimate the terms $S_\varepsilon^{(1)}u$ and $S_\varepsilon^{(2)}u$ of formula (14.5) in terms of Hölder norms.

In order to estimate the term $S_\varepsilon^{(2)}u$, we write the difference $S_\varepsilon^{(2)}u(x) - S_\varepsilon^{(2)}u(y)$ in the following form:

$$
\begin{aligned}
&S_\varepsilon^{(2)}u(x) - S_\varepsilon^{(2)}u(y) \\
&= \int_{\{|z|>\varepsilon\}} (u(x+z) - u(y+z))\, s(x,z)\, m(dz) \\
&\quad + \int_{\{|z|>\varepsilon\}} u(y+z)\, (s(x,z) - s(y,z))\, m(dz) \\
&\quad + \int_{\{|z|>\varepsilon\}} (u(y) - u(x))\, s(x,z)\, m(dz) \\
&\quad - \int_{\{|z|>\varepsilon\}} u(y)\, (s(x,z) - s(y,z))\, m(dz) \\
&\quad - \int_{\{|z|>\varepsilon\}} (\nabla u(x) - \nabla u(y)) \cdot z\, s(x,z)\, m(dz) \\
&\quad - \int_{\{|z|>\varepsilon\}} \nabla u(y) \cdot z\, (s(x,z) - s(y,z))\, m(dz) \\
&:= A(x,y) + B(x,y) + C(x,y) - D(x,y) - E(x,y) - F(x,y).
\end{aligned}
$$

Then, by using estimates (10.12c), (10.12b) and condition (1.3a) we can estimate the terms $A(x,y)$ through $F(x,y)$ as follows:

- $|A(x,y)|,\ |C(x,y)| \le [u]_{\theta_0}\, \tau(\varepsilon)\, |x-y|^{\theta_0}$
$$\le \left(\frac{C_1}{\varepsilon^2} + C_2\right) [u]_{\theta_0}\, |x-y|^{\theta_0}.$$

- $|E(x,y)| \le [\nabla u]_{\theta_0}\, \delta(\varepsilon)\, |x-y|^{\theta_0}$
$$\le \left(\frac{C_1}{\varepsilon} + C_2\right) [\nabla u]_{\theta_0}\, |x-y|^{\theta_0}.$$

- $|B(x,y)|,\ |D(x,y)| \le \|u\|_\infty\, C_0\, \tau(\varepsilon)\, |x-y|^{\theta_0}$
$$\le C_0 \left(\frac{C_1}{\varepsilon^2} + C_2\right) \|u\|_\infty\, |x-y|^{\theta_0}.$$

- $|F(x,y)| \le \|\nabla u\|_\infty\, C_0\, \delta(\varepsilon)\, |x-y|^{\theta_0}$
$$\le C_0 \left(\frac{C_1}{\varepsilon} + C_2\right) \|\nabla u\|_\infty\, |x-y|^{\theta_0}.$$

Summing up, we have proved that

$$
\left[S_\varepsilon^{(2)}u\right]_{\theta_0} \le 2\left(\frac{C_1}{\varepsilon^2} + C_2\right)[u]_{\theta_0} + 2C_0\left(\frac{C_1}{\varepsilon^2} + C_2\right)\|u\|_\infty \tag{14.7}
$$

$$+ \left(\frac{C_1}{\varepsilon} + C_2 \right) [\nabla u]_{\theta_0} + C_0 \left(\frac{C_1}{\varepsilon} + C_2 \right) \| \nabla u \|_\infty$$

$$\leq 2(1 + C_0) \left(\frac{C_1}{\varepsilon^2} + C_2 \right) \| u \|_{C^{\theta_0}(\overline{D})}$$

$$+ (1 + C_0) \left(\frac{C_1}{\varepsilon} + C_2 \right) \| \nabla u \|_{C^{\theta_0}(\overline{D})}.$$

The term $S_\varepsilon^{(1)} u$ can be estimated in a similar way:

$$\left[S_\varepsilon^{(1)} u \right]_{\theta_0} \leq \frac{1}{2} \sigma(\varepsilon) \left[\nabla^2 u \right]_{\theta_0} + \frac{1}{2} \sigma(\varepsilon) C_0 \| \nabla^2 u \|_\infty \qquad (14.8)$$

$$\leq \left(\frac{1 + C_0}{2} \right) \sigma(\varepsilon) \| \nabla^2 u \|_{C^{\theta_0}(\overline{D})}.$$

Thus, by combining estimates (14.7) and (14.8) we obtain that

$$[Su]_{\theta_0} \leq \left[S_\varepsilon^{(1)} u \right]_{\theta_0} + \left[S_\varepsilon^{(2)} u \right]_{\theta_0} \qquad (14.9)$$

$$\leq \eta \| \nabla^2 u \|_{C^{\theta_0}(\overline{D})} + C_\eta \left(\| u \|_{C^{\theta_0}(\overline{D})} + \| \nabla u \|_{C^{\theta_0}(\overline{D})} \right),$$

if we choose ε sufficiently small.

Therefore, the desired estimate (14.6) follows by combining estimates (14.4) and (14.9).

The proof of Lemma 14.5 is complete. $\quad \Box$

Step (II): We find from Theorem 14.1 that

$$\text{ind}\,(A, L) = 0.$$

However, the next lemma asserts that the Lévy integro-differential operator

$$S \colon C^{2+\theta}(\overline{D}) \longrightarrow C^\theta(\overline{D})$$

is *compact* for all $0 < \theta \leq \theta_0$; hence that the mapping

$$(W, L) = (A, L) + (S, 0) \colon C^{2+\theta}(\overline{D}) \longrightarrow C^\theta(\overline{D}) \oplus C_\star^{1+\theta}(\partial D)$$

is a perturbation of a compact operator to the mapping (A, L) under conditions (1.3a), (1.3b) and (1.4).

Lemma 14.6. *If condition (H) is satisfied, then the Lévy integro-differential operator*

$$S \colon C^{2+\theta}(\overline{D}) \longrightarrow C^\theta(\overline{D})$$

is compact *for all* $0 < \theta \leq \theta_0$.

Proof. We consider the following two cases (1) and (2).

(1) The case where $\theta = \theta_0$: Let $\{u_j\}$ be an arbitrary bounded sequence in $C^{2+\theta_0}(\overline{D})$; hence there exists a constant $K > 0$ such that

$$\|u_j\|_{C^{2+\theta_0}(\overline{D})} \leq K.$$

Then it follows from an application of the Ascoli–Arzelà theorem that the injection

$$C^{2+\theta_0}(\overline{D}) \longrightarrow C^{1+\theta_0}(\overline{D})$$

is compact (see [50, Lemma 6.36]). Hence we may assume that the sequence $\{u_j\}$ itself is a Cauchy sequence in $C^{1+\theta_0}(\overline{D})$. Then, by applying estimate (14.6) to the sequence $\{u_j - u_k\}$ we obtain that

$$\|Su_j - Su_k\|_{C^{\theta_0}(\overline{D})} \leq \eta \|\nabla^2 u_j - \nabla^2 u_k\|_{C^{\theta_0}(\overline{D})}$$
$$+ C_\eta \left(\|u_j - u_k\|_{C^{\theta_0}(\overline{D})} + \|\nabla u_j - \nabla u_k\|_{C^{\theta_0}(\overline{D})} \right)$$
$$\leq 2\eta K + C_\eta \|u_j - u_k\|_{C^{1+\theta_0}(\overline{D})}.$$

Hence we have the inequality

$$\limsup_{j,k \to \infty} \|Su_j - Su_k\|_{C^{\theta_0}(\overline{D})} \leq 2\eta K.$$

This proves that the sequence $\{Su_j\}$ is a Cauchy sequence in the space $C^{\theta_0}(\overline{D})$, since η is arbitrary.

(2) The case where $0 < \theta < \theta_0$: We remark that

$$|x - y|^{\theta_0} = |x - y|^{\theta_0 - \theta} |x - y|^\theta \leq \operatorname{diam} D \cdot |x - y|^\theta \quad \text{for all } x, y \in \overline{D},$$

where $\operatorname{diam} D$ is the *diameter* of the domain D defined by the formula

$$\operatorname{diam} D = \max_{x,y \in \overline{D}} |x - y|.$$

Hence, just as in the proof of estimate (14.9) we can prove that

$$[Su]_\theta \leq \left[S_\varepsilon^{(1)} u \right]_\theta + \left[S_\varepsilon^{(2)} u \right]_\theta$$
$$\leq \eta \|\nabla^2 u\|_{C^\theta(\overline{D})} + C_\eta \left(\|u\|_{C^\theta(\overline{D})} + \|\nabla u\|_{C^\theta(\overline{D})} \right).$$

Indeed, it suffices to note that, in the proof of Lemma 14.5 we have the following four estimates:

- $|A(x,y|, |C(x,y)| \leq \tau(\varepsilon) \, [u]_\theta \, |x - y|^\theta.$
- $|E(x,y)| \leq \delta(\varepsilon) \, [\nabla u]_\theta \, |x - y|^\theta.$
- $|B(x,y)|, |D(x,y)| \leq \|u\|_\infty \, C_0 \, \tau(\varepsilon) \, |x - y|^{\theta_0}$

$$\leq \tau(\varepsilon) \, C_0 \, \|u\|_\infty \, \text{diam} \, D \, |x - y|^\theta \, .$$

- $|F(x, y)| \leq \|\nabla u\|_\infty \, C_0 \, \delta(\varepsilon) \, |x - y|^{\theta_0}$

$$\leq \delta(\varepsilon) \, C_0 \, \text{diam} \, D \, \|\nabla u\|_\infty \, |x - y|^\theta \, .$$

In this way, we can obtain the following estimate for $0 < \theta < \theta_0$, similar to estimate (14.6):

$$\|Su\|_{C^\theta(\overline{D})}$$

$$\leq \eta \, \|\nabla^2 u\|_{C^\theta(\overline{D})} + C_\eta \left(\|u\|_{C^\theta(\overline{D})} + \|\nabla u\|_{C^\theta(\overline{D})} \right) \quad \text{for all } u \in C^{2+\theta}(\overline{D}).$$

Therefore, just as in the case (1) we find that the operator

$$S \colon C^{2+\theta}(\overline{D}) \longrightarrow C^\theta(\overline{D})$$

is compact for all $0 < \theta < \theta_0$.

The proof of Lemma 14.6 is complete.   □

Therefore, by combining Theorem 14.1 and Lemma 14.6 we obtain from Gohberg–Kreĭn [51] that

$$\text{ind} \, (W, L) = \text{ind} \, (A, L) = 0.$$

Step (III): Therefore, in order to show the bijectivity of (W, L) it suffices to prove its *injectivity*:

$$\begin{cases} u \in C^{2+\theta}(\overline{D}), W u = 0 \text{ in } D, L u = 0 \text{ on } \partial D \\ \implies u = 0 \text{ in } D. \end{cases}$$

However, this is an immediate consequence of the following *maximum principle*:

Proposition 14.7 (the maximum principle). *If conditions (A), (B) and (1.5) are satisfied, then we have the assertion*

$$\begin{cases} u \in C^2(\overline{D}), W u \geq 0 \text{ in } D, L u \geq 0 \text{ on } \partial D \\ \implies u \leq 0 \text{ on } \overline{D}. \end{cases}$$

Proof. (1) If $u(x)$ is a constant m, then we have the assertion

$$0 \leq W u(x) = m \, c(x) \quad \text{in } D.$$

However, it follows from condition (1.5) that $u(x) \equiv m$ is non-positive, since $c(x) \leq 0$ and $c(x) \not\equiv 0$ in D.

(2) Now we consider the case where $u(x)$ is not a constant. Our proof is based on a reduction to absurdity. Assume, to the contrary, that

$$m = \max_{\overline{D}} u > 0.$$

Then, by applying the strong maximum principle (see Theorem 10.7) to the operator W we obtain that there exists a point x_0' of ∂D such that

$$\begin{cases} u(x_0') = m, \\ u(x) < u(x_0') \quad \text{for all } x \in D. \end{cases}$$

Furthermore, it follows from an application of the Hopf boundary point lemma (see Lemma 10.11) that

$$\frac{\partial u}{\partial \mathbf{n}}(x_0') < 0.$$

Therefore, since we have, by condition (A),

$$0 \le Lu(x_0') = \mu(x_0') \frac{\partial u}{\partial \mathbf{n}}(x_0') + \gamma(x_0')\, u(x_0'),$$

it follows that

$$\mu(x_0') = 0 \text{ and } \gamma(x_0') = 0.$$

This contradicts condition (B).

The proof of Proposition 14.7 is complete. □

Now the proof of Theorem 1.8 is complete. □

14.3 Proof of Theorem 1.9

In this section we prove Theorem 1.9 under conditions (A), (B) and (H). We estimate the integral operator S in terms of L^p norms, and show that S is an A_p-*completely continuous* operator in the sense of Gohberg–Kreĭn [51].

The proof of Theorem 1.9 can be flowcharted as in Table 14.2 below.

The next theorem proves Theorem 1.9:

Theorem 14.8. Let $1 < p < \infty$. Assume that the following three conditions (A), (B) and (H) are satisfied:

(A) $\mu(x') \ge 0$ and $\gamma(x') \le 0$ on ∂D.
(B) $\mu(x') - \gamma(x') = \mu(x') + |\gamma(x')| > 0$ on ∂D.
(H) The integral operator S satisfies conditions (1.3a), (1.3b), (1.4) and (1.5).

Then, for every $0 < \varepsilon < \pi/2$, there exists a constant $r_p(\varepsilon) > 0$ such that the resolvent set of W_p contains the set

$$\Sigma_p(\varepsilon) = \left\{ \lambda = r^2\, e^{i\vartheta} : r \ge r_p(\varepsilon), -\pi + \varepsilon \le \vartheta \le \pi - \varepsilon \right\},$$

and that the resolvent $(W_p - \lambda I)^{-1}$ satisfies estimate (1.14):

$$\left\| (W_p - \lambda I)^{-1} \right\| \le \frac{c_p(\varepsilon)}{|\lambda|} \quad \text{for all } \lambda \in \Sigma_p(\varepsilon).$$

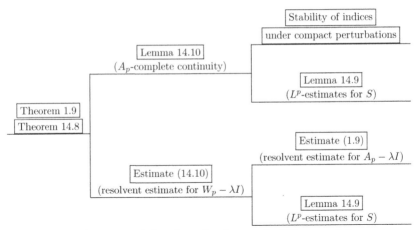

Table 14.2. A flowchart for the proof of Theorem 1.9

Proof. The proof is divided into three steps.

Step (i): We show that there exist constants $r_p(\varepsilon)$ and $c_p(\varepsilon)$ such that we have, for all $\lambda = r^2 \, e^{i \vartheta}$ satisfying $r \geq r_p(\varepsilon)$ and $-\pi + \varepsilon \leq \vartheta \leq \pi + \varepsilon$,

$$|u|_{2,p} + |\lambda|^{1/2}|u|_{1,p} + |\lambda| \, \|u\|_p \leq c_p(\varepsilon) \, \|(W_p - \lambda I) \, u\|_p \, . \tag{14.10}$$

Here

$$\|u\|_p = \|u\|_{L^p(D)} \, , \quad |u|_{1,p} = \|\nabla u\|_{L^p(D)} \, , \quad |u|_{2,p} = \|\nabla^2 u\|_{L^p(D)} \, .$$

However, we know from estimate (12.4) (see [121, p. 102, estimate (7.1)]) that the desired estimate (14.10) is proved for the differential operator A:

$$|u|_{2,p} + |\lambda|^{1/2} |u|_{1,p} + |\lambda| \, \|u\|_p \leq c_p'(\varepsilon) \, \|(A_p - \lambda I) \, u\|_p \tag{12.4}$$
$$\text{for all } u \in \mathcal{D}(A_p).$$

In order to replace the last term

$$\| (A_p - \lambda I) \, u\|_p$$

by the term

$$\| (W_p - \lambda I) \, u\|_p,$$

we need the following L^p-estimate (14.11) for the integral operator S:

Lemma 14.9. *Assume that condition (H) is satisfied. Then, for every $\eta > 0$ there exists a constant $C_\eta > 0$ such that we have, for all $u \in H^{2,p}(D)$,*

$$\|Su\|_p \leq \eta \, |u|_{2,p} + C_\eta \left(\|u\|_p + |u|_{1,p} \right) . \tag{14.11}$$

Proof. By using Taylor's formula,, we decompose the integral term Su into the following three terms:

$$Su(x) := S_1u(x) + S_2u(x) - S_3u(x)$$

$$= \int_0^1 (1-t)\,dt \int_{\{0<|z|\le\varepsilon\}} z\cdot\nabla^2 u(x+tz)z\, s(x,z)\,m(dz)$$

$$+ \int_{\{|z|>\varepsilon\}} (u(x+z) - u(x))s(x,z)\,m(dz)$$

$$- \int_{\{|z|>\varepsilon\}} z\cdot\nabla u(x)\, s(x,z)\,m(dz).$$

First, we estimate the L^p norm of the term S_3u. By using estimate (10.12b), we obtain that

$$\left| \int_{\{|z|>\varepsilon\}} z\cdot\nabla u(x)\,s(x,z)\,m(dz) \right| \le \delta(\varepsilon)\,|\nabla u(x)| \le \left(\frac{C_1}{\varepsilon} + C_2 \right) |\nabla u(x)|.$$

Hence we have the L^p estimate of the term S_3u:

$$\|S_3u\|_p \le \left(\frac{C_1}{\varepsilon} + C_2 \right) \|\nabla u\|_p.$$

Secondly, we have the inequality

$$\left\| \int_{\{|z|>\varepsilon\}} u(\cdot)\, s(\cdot, z)\,m(dz) \right\|_p \le \left(\frac{C_1}{\varepsilon^2} + C_2 \right) \|u\|_p.$$

Furthermore, by using Hölder's inequality and Fubini's theorem we obtain from the support condition (1.3b) that

$$\int_{\mathbf{R}^N} \left| \int_{\{|z|>\varepsilon\}} u(x+z)\, s(x,z)\,m(dz) \right|^p dx$$

$$\le \int_{\mathbf{R}^N} \left(\int_{\{|z|>\varepsilon\}} |u(x+z)|\, s(x,z)\,m(dz) \right)^p dx$$

$$\le \int_{\mathbf{R}^N} \left(\int_{\{|z|>\varepsilon\}} |u(x+z)|^p s(x,z)^p m(dz) \right) \left(\int_{\{|z|>\varepsilon\}} m(dz) \right)^{p/q} dx$$

$$= \tau(\varepsilon)^{p/q} \int_{\mathbf{R}^N} \int_{\{|z|>\varepsilon\}} |u(x+z)|^p s(x,z)^p\, m(dz)\, dx$$

$$= \tau(\varepsilon)^{p/q} \int_{\{|z|>\varepsilon\}} \left(\int_{\mathbf{R}^N} |u(x+z)|^p s(x,z)^p\, dx \right) m(dz)$$

$$\le \tau(\varepsilon)^{p/q} \left(\int_D |u(y)|^p\, dy \right) \left(\int_{\{|z|>\varepsilon\}} m(dz) \right)$$

$$= \tau(\varepsilon)^p \, \|u\|_p^p.$$

By estimate (10.12c), we have the L^p estimate of the term $S_2 u$:

$$\|S_2 u\|_p \leq \left(\frac{C_1}{\varepsilon^2} + C_2 \right) \|u\|_p.$$

Similarly, by using Hölder's inequality and Fubini's theorem we find that

$$\int_{\mathbf{R}^N} \left| \int_0^1 (1-t) \, dt \int_{\{0 < |z| \leq \varepsilon\}} z \cdot \nabla^2 u(x+tz) z \, s(x,z) \, m(dz) \right|^p dx$$

$$\leq \int_{\mathbf{R}^N} \left(\int_0^1 dt \int_{\{0 < |z| \leq \varepsilon\}} |z|^2 |\nabla^2 u(x+tz)| \, s(x,z) \, m(dz) \right)^p dx$$

$$\leq \int_{\mathbf{R}^N} \int_0^1 dt \left(\int_{\{0 < |z| \leq \varepsilon\}} |z|^2 \, |\nabla^2 u(x+tz)|^p \, s(x,z)^p \, m(dz) \right)$$

$$\times \left(\int_{\{0 < |z| \leq \varepsilon\}} |z|^2 \, m(dz) \right)^{p/q} dx$$

$$= \sigma(\varepsilon)^{p/q} \int_{\mathbf{R}^N} \int_0^1 dt \left(\int_{\{0 < |z| \leq \varepsilon\}} |z|^2 \, |\nabla^2 u(x+tz)|^p \, s(x,z)^p \, m(dz) \right) dx$$

$$= \sigma(\varepsilon)^{p/q} \int_0^1 dt \int_{\{0 < |z| \leq \varepsilon\}} |z|^2 \left(\int_{\mathbf{R}^N} |\nabla^2 u(x+tz)|^p \, s(x,z)^p \, dx \right) m(dz)$$

$$\leq \sigma(\varepsilon)^{p/q} \left(\int_D |\nabla^2 u(y)|^p \, dy \right) \left(\int_{\{0 < |z| \leq \varepsilon\}} |z|^2 \, m(dz) \right)$$

$$\leq \sigma(\varepsilon)^p \left(\int_D |\nabla^2 u(y)|^p \, dy \right).$$

Hence we have the L^p estimate of the term $S_1 u$:

$$\|S_1 u\|_p \leq \sigma(\varepsilon) \, \|\nabla^2 u\|_p.$$

Summing up, we have proved that

$$\|Su\|_p \leq \|S_1 u\|_p + \|S_2 u\|_p + \|S_3 u\|_p$$
$$\leq \sigma(\varepsilon) \, |u|_{2,p} + \left(\frac{C_1}{\varepsilon} + C_2 \right) |u|_{1,p} + \left(\frac{C_1}{\varepsilon^2} + C_2 \right) \|u\|_p.$$

In view of assertion (10.12a), this proves the desired estimate (14.11) if we choose ε so small that

$$\sigma(\varepsilon) \leq \eta.$$

The proof of Lemma 14.9 is complete. $\square$

Since we have the formula

$$(A - \lambda)\,u = (W - \lambda)\,u - Su,$$

it follows from estimate (14.11) that

$$\|(A_p - \lambda I)\,u\|_p \leq \|(W_p - \lambda I)\,u\|_p + \|Su\|_p$$
$$\leq \|(W_p - \lambda I)\,u\|_p + \eta\,|u|_{2,p} + C_\eta\left(|u|_{1,p} + \|u\|_p\right).$$

Thus, by carrying this estimate into estimate (12.4) we obtain that

$$|u|_{2,p} + |\lambda|^{1/2}\,|u|_{1,p} + |\lambda|\,\|u\|_p \tag{14.12}$$
$$\leq c_p'(\varepsilon)\,\|(A_p - \lambda I)\,u\|_p$$
$$\leq c_p'(\varepsilon)\,\|(W_p - \lambda I)\,u\|_p + \eta\,c_p'(\varepsilon)\,|u|_{2,p} + C_\eta\,c_p'(\varepsilon)\left(|u|_{1,p} + \|u\|_p\right).$$

Therefore, the desired estimate (14.10) follows from estimate (14.12) if we take the constant η so small that

$$\eta c_p'(\varepsilon) \leq \frac{1}{2}$$

and the parameter λ so large that

$$|\lambda| \geq |\lambda|^{1/2} \geq 2\,C_\eta\,c_p'(\varepsilon).$$

More precisely, we have the desired estimate

$$|u|_{2,p} + |\lambda|^{1/2}|u|_{1,p} + |\lambda|\,\|u\|_p \leq c_p(\varepsilon)\,\|(W_p - \lambda I)\,u\|_p,$$

with

$$c_p(\varepsilon) = 2\,c_p'(\varepsilon).$$

Step (ii): By estimate (14.10), we find that the operator $W_p - \lambda I$ is *injective* and its range $\mathcal{R}\,(W_p - \lambda I)$ is closed in $L^p(D)$, for all $\lambda \in \Sigma_p(\varepsilon)$.

We show that the operator $W_p - \lambda I$ is *surjective* for all $\lambda \in \Sigma_p(\varepsilon)$:

$$\mathcal{R}\,(W_p - \lambda I) = L^p(D), \quad \lambda \in \Sigma_p(\varepsilon).$$

To do this, it suffices to show that the operator $W_p - \lambda I$ is a Fredholm operator with

$$\operatorname{ind}\,(W_p - \lambda I) = 0 \quad \text{for all } \lambda \in \Sigma_p(\varepsilon), \tag{14.13}$$

since $W_p - \lambda I$ is injective for all $\lambda \in \Sigma_p(\varepsilon)$.

In order to prove assertion (14.13), we need the following lemma due to Gohberg–Kreĭn [51, Theorem 2.6]:

Lemma 14.10. *If condition (H) is satisfied, then the operator S is A_p-completely continuous, that is, the operator $S\colon \mathcal{D}(A_p) \to L^p(D)$ is completely continuous where the domain $\mathcal{D}(A_p)$ is endowed with the* graph norm *of A_p:*

$$\|u\|_{\mathcal{D}(A_p)} = \|u\|_p + \|A_p u\|_p \quad \text{for } u \in \mathcal{D}(A_p).$$

Proof. Let $\{u_j\}_{j=1}^{\infty}$ be an arbitrary bounded sequence in the domain $\mathcal{D}(A_p)$; hence there exists a constant $K > 0$ such that

$$\begin{cases} \|u_j\|_p \leq K, \\ \|A_p u_j\|_p \leq K. \end{cases}$$

Then, by using the *a priori* estimate (1.7) with

$$\lambda := 0, \quad \varphi := 0,$$

we obtain that

$$\|u_j\|_{2,p} \leq C \left(\|A_p u_j\|_p + \|u_j\|_p \right) \leq 2CK. \tag{14.14}$$

However, it follows from an application of the Rellich–Kondrachov theorem (Theorem 4.10) with $\Omega := D$ and $s := 2$, $t := 1$ that the injection

$$H^{2,p}(D) \longrightarrow H^{1,p}(D)$$

is *compact*. Hence we can find a subsequence $\{u_{j'}\}_{j'=1}^{\infty}$ that is a Cauchy sequence in the space $H^{1,p}(D)$. Then, by applying estimate (14.11) to the sequence $\{u_{j'} - u_{k'}\}$ and using estimate (14.14), we obtain that

$$\|Su_{j'} - Su_{k'}\|_p \leq \eta \, |u_{j'} - u_{k'}|_{2,p} + C_\eta \left(\|u_{j'} - u_{k'}\|_p + |u_{j'} - u_{k'}|_{1,p} \right)$$
$$\leq 4\eta \, CK + C_\eta \, \|u_{j'} - u_{k'}\|_{1,p}.$$

Hence we have the inequality

$$\limsup_{j',k'\to\infty} \|Su_{j'} - Su_{k'}\|_p \leq 4\eta CK.$$

This proves that the subsequence $\{Su_{j'}\}_{j'=1}^{\infty}$ is a Cauchy sequence in the space $L^p(D)$, since η is arbitrary.

The proof of Lemma 14.10 is complete. $\square$

In view of Lemma 14.10, the desired assertion (14.13) follows from an application of Gohberg–Kreĭn [51, Theorem 2.6] with

$$X = Y := L^p(D), \quad T := A_p.$$

Indeed, we have, by Theorem 1.4,

$$\text{ind}\,(W_p - \lambda I) = \text{ind}\,(A_p - \lambda I + S) = \text{ind}\,(A_p - \lambda I) = 0.$$

Step (iii): Summing up, we have proved that the operator $W_p - \lambda I$ is bijective for all $\lambda \in \Sigma_p(\varepsilon)$ and its inverse $(W_p - \lambda I)^{-1}$ satisfies estimate (1.14).

The proof of Theorem 14.8 is now complete. $\square$

14.4 Proof of Theorem 1.10

This section is devoted to the proof of Theorem 1.10 under conditions (A), (B) and (H). Theorem 1.10 follows from Theorem 1.9 by using Sobolev's imbedding theorems and a λ-dependent localization argument. The proof is carried out in a chain of auxiliary lemmas (Lemmas 14.11, 14.12 and 14.15).

The proof of Theorem 1.10 can be flowcharted as in Table 14.3 below.

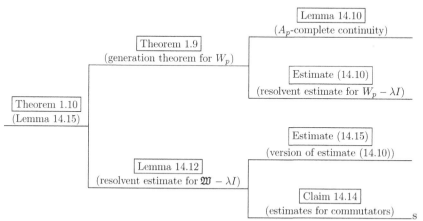

Table 14.3. A flowchart for the proof of Theorem 1.10

Step (I): We begin with a version of estimate (14.10):

Lemma 14.11. *Let $N < p < \infty$. Assume that the following three conditions (A), (B) and (H) are satisfied:*

(A) $\mu(x') \geq 0$ and $\gamma(x') \leq 0$ on ∂D.
(B) $\mu(x') - \gamma(x') = \mu(x') + |\gamma(x')| > 0$ on ∂D.
(H) The integral operator S satisfies conditions (1.3a), (1.3b), (1.4) and (1.5).

Then, for every $\varepsilon > 0$ there exists a constant $r_p(\varepsilon) > 0$ such that if $\lambda = r^2\, e^{i\vartheta}$ with $r \geq r_p(\varepsilon)$ and $-\pi + \varepsilon \leq \vartheta \leq \pi - \varepsilon$, we have, for all $u \in \mathcal{D}(W_p)$,

$$|\lambda|^{1/2} \|u\|_{C^1(\overline{D})} + |\lambda|\, \|u\|_{C(\overline{D})} \leq C_p(\varepsilon)\, |\lambda|^{N/2p} \|(W - \lambda)u\|_p, \qquad (14.15)$$

with a constant $C_p(\varepsilon) > 0$.

Proof. First, it follows from an application of the Gagliardo-Nirenberg inequality (12.3) with $p := r > N$, $q := \infty$ and $\theta := N/p$ that

$$\|u\|_{C(\overline{D})} \leq C\, |u|_{1,p}^{N/p} \|u\|_p^{1-N/p} \quad \text{for all } u \in H^{1,p}(D). \qquad (14.16)$$

Here and in the following the letter C denotes a generic positive constant depending on p and ε, but independent of u and λ.

By combining inequality (14.16) with inequality (14.10), we obtain that

$$
\begin{aligned}
\|u\|_{C(\overline{D})} &\leq C \, |u|_{1,p}^{N/p} \, \|u\|_p^{1-N/p} \\
&\leq C \left(|\lambda|^{-1/2} \, \|(W - \lambda)u\|_p \right)^{N/p} \left(|\lambda|^{-1} \, \|(W - \lambda)u\|_p \right)^{1-N/p} \\
&= C \, |\lambda|^{-1+N/2p} \, \|(W - \lambda)u\|_p \, ,
\end{aligned}
$$

so that

$$
|\lambda| \, \|u\|_{C(\overline{D})} \leq C \, |\lambda|^{N/2p} \, \|(W - \lambda)u\|_p \quad \text{for all } u \in \mathcal{D}\left(W_p\right). \tag{14.17}
$$

Similarly, by applying inequality (14.16) to the functions

$$
D_i u \in H^{1,p}(D) \quad \text{for } 1 \leq i \leq n,
$$

we obtain that

$$
\begin{aligned}
\|\nabla u\|_{C(\overline{D})} &\leq C \, |\nabla u|_{1,p}^{N/p} \, \|\nabla u\|_p^{1-N/p} \\
&\leq C \, |u|_{2,p}^{N/p} \, |u|_{1,p}^{1-N/p} \\
&\leq C \left(\|(W - \lambda)u\|_p \right)^{N/p} \left(|\lambda|^{-1/2} \, \|(W - \lambda)u\|_p \right)^{1-N/p} \\
&= C \, |\lambda|^{-1/2+N/2p} \, \|(W - \lambda)u\|_p \, .
\end{aligned}
$$

This proves that

$$
|\lambda|^{1/2} \, \|u\|_{C^1(\overline{D})} \leq C \, |\lambda|^{N/2p} \, \|(W - \lambda)u\|_p \quad \text{for all } u \in \mathcal{D}\left(W_p\right). \tag{14.18}
$$

Therefore, the desired inequality (14.15) follows by combining inequalities (14.17) and (14.18).

The proof of Lemma 14.11 is complete.   □

Step (II): The next lemma proves the resolvent estimate (1.16):

Lemma 14.12. *Let $N < p < \infty$. If conditions (A), (B) and (H) are satisfied, then, for every $\varepsilon > 0$ there exists a constant $r(\varepsilon) > 0$ such that if $\lambda = r^2 \, e^{i \vartheta}$ with $r \geq r(\varepsilon)$ and $-\pi + \varepsilon \leq \vartheta \leq \pi - \varepsilon$, we have the inequality*

$$
|\lambda|^{1/2} \, \|u\|_{C^1(\overline{D})} + |\lambda| \, \|u\|_{C(\overline{D})} \leq c(\varepsilon) \, \|(\mathfrak{W} - \lambda I) \, u\|_{C(\overline{D})} \tag{14.19}
$$
$$
\text{for all } u \in \mathcal{D}\left(\mathfrak{W}\right),
$$

with a constant $c(\varepsilon) > 0$.

Proof. (1) First, we show that the domain

$$
\mathcal{D}\left(\mathfrak{W}\right)
$$

$$= \left\{ u \in C_0 \left(\overline{D} \setminus M \right) \cap H^{2,p}(D) : Wu \in C_0 \left(\overline{D} \setminus M \right), \ Lu = 0 \text{ on } \partial D \right\}$$

is *independent* of $N < p < \infty$.

We let

$$D_p$$
$$= \left\{ u \in H^{2,p}(D) \cap C_0 \left(\overline{D} \setminus M \right) : Wu \in C_0 \left(\overline{D} \setminus M \right), \ Lu = 0 \text{ on } \partial D \right\}.$$

Since we have $L^{p_1}(D) \subset L^{p_2}(D)$ for $p_1 > p_2$, it follows that

$$D_{p_1} \subset D_{p_2} \quad \text{if } p_1 > p_2.$$

Conversely, let v be an arbitrary element of D_{p_2}:

$$v \in H^{2,p_2}(D) \cap C_0 \left(\overline{D} \setminus M \right), \quad Wv \in C_0 \left(\overline{D} \setminus M \right), \quad Lv = 0.$$

Then, since we have v, $Wv \in C_0 \left(\overline{D} \setminus M \right) \subset L^{p_1}(D)$, it follows from an application of Theorem 14.8 with $p := p_1$ that there exists a unique function $u \in H^{2,p_1}(D)$ such that

$$\begin{cases} (W - \lambda)u = (W - \lambda)v & \text{in } D, \\ Lu = 0 & \text{on } \partial D, \end{cases}$$

if we choose λ sufficiently large. Hence we have $u - v \in H^{2,p_2}(D)$ and

$$\begin{cases} (W - \lambda)(u - v) = 0 & \text{in } D, \\ L(u - v) = 0 & \text{on } \partial D. \end{cases}$$

Therefore, by applying again Theorem 14.8 with $p := p_2$ we obtain that $u - v = 0$ in D, so that

$$v = u \in H^{2,p_1}(D).$$

This proves that

$$v \in D_{p_1}.$$

(2) We shall make use of a λ-dependent localization argument in order to adjust the term $\|(W - \lambda)u\|_p$ in inequality (14.15) to obtain inequality (14.19), just as in Section 12.2.

(2-a) If x_0' is a point of ∂D and if χ is a smooth coordinate transformation such that χ maps $B(x_0', \eta_0) \cap D$ into $B(0, \delta) \cap \mathbf{R}_+^N$ and flattens a part of the boundary ∂D into the plane $x_N = 0$ (see Figure 12.2), then we let

- $G_0 = B(x_0', \eta_0) \cap D$,
- $G' = B(x_0', \eta) \cap D$ for $0 < \eta < \eta_0$,
- $G'' = B(x_0', \eta/2) \cap D$ for $0 < \eta < \eta_0$.

Similarly, if x_0 is a point of D, then we let

- $G_0 = B(x_0, \eta_0) \subset D$,
- $G' = B(x_0, \eta)$ for $0 < \eta < \eta_0$,
- $G'' = B(x_0, \eta/2)$ for $0 < \eta < \eta_0$.

(2-b) We take a function $\theta(t) \in C_0^\infty(\mathbf{R})$ such that θ equals 1 near the origin, and define a localizing function

$$\varphi_0(x, \eta) := \theta\left(\frac{|x' - x_0'|^2}{\eta^2}\right) \theta\left(\frac{x_N - t}{\eta}\right) \quad \text{for } x_0 = (x_0', t). \tag{14.20}$$

Then we have the following claim, analogous to Claim 12.3:

Claim 14.13. *If $u \in \mathcal{D}(\mathfrak{W})$, then it follows that $\varphi_0 u \in \mathcal{D}(W_p)$.*

Proof. First, we recall that

$$u \in H^{2,p}(D) \quad \text{for all } 1 < p < \infty.$$

Hence we have the assertion

$$\varphi_0 u \in H^{2,p}(D).$$

Furthermore, it is easy to verify that the function $\varphi_0 u$ satisfies the boundary condition

$$L(\varphi_0 u) = 0 \quad \text{on } \partial D.$$

Summing up, we have proved that

$$\varphi_0 u \in \mathcal{D}(W_p) \quad \text{for all } 1 < p < \infty.$$

The proof of Claim 14.13 is complete. □

(3) Now let u be an arbitrary element of $\mathcal{D}(\mathfrak{W})$. Then, by Claim 14.13 we can apply inequality (14.15) to the function $\varphi_0 u$ to obtain that

$$|\lambda|^{1/2} \|u\|_{C^1(\overline{G''})} + |\lambda| \|u\|_{C(\overline{G''})} \tag{14.21}$$

$$\leq |\lambda|^{1/2} \|\varphi_0 u\|_{C^1(\overline{G'})} + |\lambda| \|\varphi_0 u\|_{C(\overline{G'})}$$

$$= |\lambda|^{1/2} \|\varphi_0 u\|_{C^1(\overline{D})} + |\lambda| \|\varphi_0 u\|_{C(\overline{D})}$$

$$\leq C_p(\varepsilon) |\lambda|^{N/2p} \|(W - \lambda)(\varphi_0 u)\|_{L^p(D)} \quad \text{for all } u \in \mathcal{D}(\mathfrak{W}).$$

(3-a) We estimate the last term $\|(W - \lambda)(\varphi_0 u)\|_{L^p(D)}$ in terms of the supremum norm of $C(\overline{D})$.

First, we write the term $(W - \lambda)(\varphi_0 u)$ in the following form:

$$(W - \lambda)(\varphi_0 u) = \varphi_0 \left((W - \lambda)u\right) + [A, \varphi_0] u + [S, \varphi_0] u,$$

where $[A, \varphi_0]$ and $[S, \varphi_0]$ are the commutators of A and φ_0 and of S and φ_0, respectively:

- $[A, \varphi_0]\, u = A(\varphi_0 u) - \varphi_0 A u,$
- $[S, \varphi_0]\, u = S(\varphi_0 u) - \varphi_0 S u.$

Since we have, for some constant $c > 0$,

$$|G'| \le |B(x_0, \eta)| \le c\eta^N,$$

it follows from an application of Claim 12.4 that

$$\|\varphi_0 (W - \lambda)u\|_{L^p(D)} = \|\varphi_0 (W - \lambda)u\|_{L^p(G')} \tag{14.22}$$
$$\le c^{1/p}\eta^{N/p} \|(W - \lambda)u\|_{C(\overline{G'})}$$
$$\le c^{1/p}\eta^{N/p} \|(W - \lambda)u\|_{C(\overline{D})} \qquad \text{for all } u \in \mathcal{D}(\mathfrak{W}).$$

On the other hand, we can estimate the commutators $[A, \varphi_0]\, u$ and $[S, \varphi_0]\, u$ as follows:

Claim 14.14. *We have, as $\eta \downarrow 0$,*

- $\|[A, \varphi_0]\, u\|_{L^p(D)}$ \hfill (14.23a)
$$\le C\big(\eta^{-1+N/p} \|u\|_{C^1(\overline{D})} + \eta^{-2+N/p} \|u\|_{C(\overline{D})}\big),$$
- $\|[S, \varphi_0]\, u\|_{L^p(D)}$ \hfill (14.23b)
$$\le C\big(\eta^{-1+N/p} \|u\|_{C^1(\overline{D})} + \eta^{-2+N/p} \|u\|_{C(\overline{D})}\big).$$

Proof. By formula (14.20), we have, as $\eta \downarrow 0$,

$$|D^\alpha \varphi_0| = O\left(\eta^{-|\alpha|}\right).$$

Hence it follows from an application of Claim 12.4 that

$$\left\|\frac{\partial \varphi_0}{\partial x_i} \frac{\partial u}{\partial x_j}\right\|_{L^p(G')} \le C\frac{1}{\eta} |u|_{1,p,G'} \le C\eta^{-1+N/p} |u|_{C^1(\overline{G'})},$$

$$\left\|\frac{\partial^2 \varphi_0}{\partial x_i \partial x_j} u\right\|_{L^p(G')} \le C\frac{1}{\eta^2} |u|_{L^p(G')} \le C\eta^{-2+N/p} |u|_{C(\overline{G'})},$$

$$\left\|\frac{\partial \varphi_0}{\partial x_i} u\right\|_{L^p(G')} \le C\frac{1}{\eta} |u|_{L^p(G')} \le C\eta^{-1+N/p} |u|_{C(\overline{G'})}.$$

Therefore, we obtain that

$$\|[A, \varphi_0]\, u\|_{L^p(G')} \le C\eta^{-1+N/p} |u|_{C^1(\overline{G'})} + \eta^{-2+N/p} |u|_{C(\overline{G'})}$$
$$\le C\eta^{-1+N/p} |u|_{C^1(\overline{D})} + \eta^{-2+N/p} |u|_{C(\overline{D})}.$$

This proves estimate (14.23a).

In order to prove estimate (14.23b), we remark that

$$
S(\varphi_0 u)(x)
$$
$$
= \int_{\mathbf{R}^N \setminus \{0\}} \Big(\varphi_0(x+z)u(x+z) - \varphi_0(x)u(x)
$$
$$
\qquad - z \cdot \nabla(\varphi_0 u)(x) \Big) s(x,z)\, m(dz)
$$
$$
= \varphi_0(x) \int_{\mathbf{R}^N \setminus \{0\}} (u(x+z) - u(x) - z \cdot \nabla u(x))\, s(x,z)\, m(dz)
$$
$$
\quad + \left(\int_{\mathbf{R}^N \setminus \{0\}} (u(x+z) - u(x))z\, s(x,z)\, m(dz) \right) \cdot \nabla\varphi_0(x)
$$
$$
\quad + \int_{\mathbf{R}^N \setminus \{0\}} (\varphi_0(x+z) - \varphi_0(x) - z \cdot \nabla\varphi_0(x))\, u(x+z)\, s(x,z)\, m(dz)
$$
$$
= \varphi_0(x)Su(x) + \left(\int_{\mathbf{R}^N \setminus \{0\}} (u(x+z) - u(x))z\, s(x,z)\, m(dz) \right) \cdot \nabla\varphi_0(x)
$$
$$
\quad + \int_{\mathbf{R}^N \setminus \{0\}} (\varphi_0(x+z) - \varphi_0(x) - z \cdot \nabla\varphi_0(x))\, u(x+z)\, s(x,z)\, m(dz).
$$

Hence we can write the commutator $[S, \varphi_0]u$ in the following form:

$$
[S, \varphi_0]\, u(x)
$$
$$
:= S_0^{(1)} u(x) + S_0^{(2)} u(x)
$$
$$
= \left(\int_{\mathbf{R}^N \setminus \{0\}} (u(x+z) - u(x))z\, s(x,z)\, m(dz) \right) \cdot \nabla\varphi_0(x)
$$
$$
\quad + \int_{\mathbf{R}^N \setminus \{0\}} (\varphi_0(x+z) - \varphi_0(x) - z \cdot \nabla\varphi_0(x))\, u(x+z)\, s(x,z)\, m(dz).
$$

First, just as in Lemma 14.4 we can estimate the term $S_0^{(1)}u$ as follows:

$$
\left\| S_0^{(1)}u \right\|_{L^p(D)} = \left\| S_0^{(1)}u \right\|_{L^p(G')}
$$
$$
\leq 2 \left(\sigma(\eta)\, \|u\|_{C^1(\overline{D})} + \delta(\eta)\, \|u\|_{C(\overline{D})} \right) \|\nabla\varphi_0\|_{L^p(G')}
$$
$$
\leq 2 \left(\sigma(\eta)\, \|u\|_{C^1(\overline{D})} + \left(\frac{C_1}{\eta} + C_2 \right) \|u\|_{C(\overline{D})} \right) \|\nabla\varphi_0\|_{L^p(G')}.
$$

However, it follows from an application of Claim 12.4 that

- $\|\nabla\varphi_0\|_{L^p(G')} \leq C\eta^{N/p} \|\nabla\varphi_0\|_{C(\overline{G'})} \leq C'\eta^{-1+N/p}$,
- $\|\nabla^2\varphi_0\|_{L^p(G')} \leq C\eta^{N/p} \|\nabla^2\varphi_0\|_{C(\overline{G'})} \leq C'\eta^{-2+N/p}$,

since we have, as $\eta \downarrow 0$,

$$|\nabla \varphi_0| = O\left(\eta^{-1}\right), \quad |\nabla^2 \varphi_0| = O\left(\eta^{-2}\right).$$

Hence we obtain that

$$\left\|S_0^{(1)}u\right\|_{L^p(D)} \le C\left(\eta^{-1+N/p}\|u\|_{C^1(\overline{D})} + \eta^{-2+N/p}\|u\|_{C(\overline{D})}\right). \qquad (14.24)$$

Similarly, by arguing as in the proof of Lemma 14.15 we can estimate the term $S_0^{(2)}u$ as follows:

$$\left\|S_0^{(2)}u\right\|_{L^p(D)} \le C\|u\|_{C(\overline{D})}\left\|\nabla^2 \varphi_0\right\|_{L^p(G')} \qquad (14.25)$$

$$\le C\|u\|_{C(\overline{D})}\,\eta^{N/p}\left\|\nabla^2 \varphi_0\right\|_{C(\overline{G'})}$$

$$\le C\eta^{-2+N/p}\|u\|_{C(\overline{D})}.$$

Since we have the formula

$$[S, \varphi_0]\,u = S_0^{(1)}u + S_0^{(2)}u,$$

the desired estimate (14.23b) follows by combining estimates (14.24) and (14.25).

The proof of Claim 14.14 is complete. □

Therefore, by combining four estimates (14.21), (14.22), (14.23a) and (14.23b) we obtain that

$$|\lambda|^{1/2}\|u\|_{C^1(\overline{G''})} + |\lambda|\,\|u\|_{C(\overline{G''})} \qquad (14.26)$$

$$\le C_p(\varepsilon)\,|\lambda|^{N/2p}\|(W - \lambda)(\varphi_0 u)\|_{L^p(D)}$$

$$= C_p(\varepsilon)\,|\lambda|^{N/2p}\|\varphi_0\left((W - \lambda)u\right) + [A, \varphi_0]\,u + [S, \varphi_0]\,u\|_{L^p(D)}$$

$$\le C\,|\lambda|^{N/2p}\left(\eta^{N/p}\|(W - \lambda)u\|_{C(\overline{G'})} + \eta^{-1+N/p}\|u\|_{C^1(\overline{G'})}\right.$$

$$\left. + \eta^{-2+N/p}\|u\|_{C(\overline{G'})}\right)$$

$$\le C\,|\lambda|^{N/2p}\left(\eta^{N/p}\|(W - \lambda)u\|_{C(\overline{D})} + \eta^{-1+N/p}\|u\|_{C^1(\overline{D})}\right.$$

$$\left. + \eta^{-2+N/p}\|u\|_{C(\overline{D})}\right) \quad \text{for all } u \in \mathcal{D}(\mathfrak{W}).$$

Here and in the following the letter C denotes a generic positive constant depending on p and ε, but independent of u and λ.

(3-b) We recall (see Figure 12.3) that the closure $\overline{D} = D \cup \partial D$ can be covered by a finite number of sets of the forms

$$\begin{cases} B(x_0, \eta/2) & \text{for } x_0 \in D, \\ B(x_0', \eta/2) \cap \overline{D} & \text{for } x_0' \in \partial D. \end{cases}$$

Therefore, by taking the supremum of inequality (14.26) over $x \in \overline{D}$ we find that

$$|\lambda|^{1/2} \|u\|_{C^1(\overline{D})} + |\lambda| \|u\|_{C(\overline{D})} \tag{14.27}$$
$$\leq C |\lambda|^{N/2p} \eta^{N/p} \left(\|(W - \lambda)u\|_{C(\overline{D})} + \eta^{-1} \|u\|_{C^1(\overline{D})} + \eta^{-2} \|u\|_{C(\overline{D})} \right)$$

for all $u \in \mathcal{D}(\mathfrak{W})$.

(4) We now choose the localization parameter η. We let

$$\eta = \frac{\eta_0}{|\lambda|^{1/2}} K,$$

where K is a positive constant (to be chosen later on) satisfying the condition

$$0 < \eta = \frac{\eta_0}{|\lambda|^{1/2}} K < \eta_0,$$

that is,

$$0 < K < |\lambda|^{1/2}.$$

Then we have the following three formulas:

- $|\lambda|^{N/2p} \eta^{N/p} = |\lambda|^{N/2p} \dfrac{\eta_0^{N/p} K^{N/p}}{|\lambda|^{N/2p}} = \eta_0^{N/p} K^{N/p},$
- $|\lambda|^{N/2p} \eta^{N/p-1} = |\lambda|^{1/2} \eta_0^{N/p-1} K^{N/p-1},$
- $|\lambda|^{N/2p} \eta^{N/p-2} = |\lambda| \eta_0^{N/p-2} K^{N/p-2}.$

Therefore, we obtain from inequality (14.27) that

$$|\lambda|^{1/2} \|u\|_{C^1(\overline{D})} + |\lambda| \|u\|_{C(\overline{D})} \tag{14.28}$$
$$\leq C \eta_0^{N/p} K^{N/p} \|(W - \lambda)u\|_{C(\overline{D})}$$
$$+ \left(C \eta_0^{N/p-1} K^{-1+N/p} \right) |\lambda|^{1/2} \|u\|_{C^1(\overline{D})}$$
$$+ \left(C \eta_0^{N/p-2} K^{-2+N/p} \right) |\lambda| \|u\|_{C(\overline{D})} \quad \text{for all } u \in \mathcal{D}(\mathfrak{W}).$$

However, since the exponents $-1 + N/p$ and $-2 + N/p$ are negative, we can choose the constant K so large that

$$C \eta_0^{N/p-1} K^{-1+N/p} \leq \frac{1}{2},$$

and

$$C \, \eta_0^{N/p-2} \, K^{-2+N/p} \leq \frac{1}{2}.$$

Then the desired inequality (14.19) follows from inequality (14.28), with

$$c(\varepsilon) := 2 \, C \, \eta_0^{N/p} \, K^{N/p}.$$

Recall that the letter C denotes a generic positive constant depending on p and ε, but independent of u and λ.

The proof of Lemma 14.12 is now complete. □

Step (III): The next lemma, together with Lemma 14.12, proves that the resolvent set of $\mathfrak{W}$ contains the set

$$\Sigma(\varepsilon) = \left\{ \lambda = r^2 \, e^{i\vartheta} : r \geq r(\varepsilon), \ -\pi + \varepsilon \leq \vartheta \leq \pi - \varepsilon \right\}.$$

Lemma 14.15. *If* $\lambda \in \Sigma(\varepsilon)$, *then, for any* $f \in C_0 \left(\overline{D} \setminus M \right)$ *there exists a unique function* $u \in \mathcal{D}(\mathfrak{W})$ *such that* $(\mathfrak{W} - \lambda I) \, u = f$.

Proof. Since we have, for all $1 < p < \infty$,

$$f \in C_0 \left(\overline{D} \setminus M \right) \subset L^p(D),$$

it follows from an application of Theorem 1.9 that if $\lambda \in \Sigma_p(\varepsilon)$, there exists a unique function $u \in H^{2,p}(D)$ such that

$$(W - \lambda) \, u = f \quad \text{in } D, \tag{14.29}$$

and

$$Lu = \mu(x') \frac{\partial u}{\partial \mathbf{n}} + \gamma(x') \, u = 0 \quad \text{on } \partial D. \tag{14.30}$$

However, by Sobolev's imbedding theorem it follows that

$$u \in H^{2,p}(D) \subset C^{2-N/p}(\overline{D}) \subset C^1(\overline{D}) \quad \text{if } N < p < \infty.$$

Hence we have, by formula (14.30) and condition (B),

$$u = 0 \quad \text{on } M = \{x' \in \partial D : \mu(x') = 0\},$$

so that

$$u \in C_0 \left(\overline{D} \setminus M \right).$$

Furthermore, in view of equation (14.29) we find that

$$Wu = f + \lambda u \in C_0 \left(\overline{D} \setminus M \right).$$

Summing up, we have proved that

$$\begin{cases} u \in \mathcal{D}(\mathfrak{W}), \\ (\mathfrak{W} - \lambda I) \, u = f. \end{cases}$$

The proof of Lemma 14.15 is complete. □

Now the proof of Theorem 1.10 is complete. □

14.5 Proof of Theorem 1.11

This section is devoted to the proof of Theorem 1.11 under conditions (A), (B) and (H). The proof of Theorem 1.11 can be flowcharted as in Table 14.4 below.

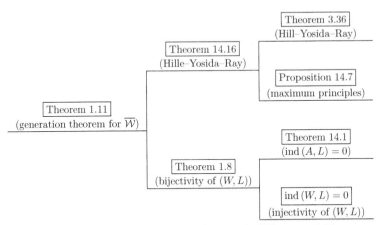

Table 14.4. A flowchart for the proof of Theorem 1.11

The proof is based on the following version of the Hille–Yosida theorem in terms of the maximum principle (cf. [15, Théorème de Hille–Yosida–Ray]):

Theorem 14.16 (Hille–Yosida–Ray). *Let $\mathcal{A}$ be a linear operator from the space $C_0\left(\overline{D} \setminus M\right)$ into itself, and assume that the following three conditions (a), (b) and (c) are satisfied:*

(a) The domain $\mathcal{D}(\mathcal{A})$ is dense in the space $C_0\left(\overline{D} \setminus M\right)$.

(b) For any $u \in \mathcal{D}(\mathcal{A})$ such that $\max_{\overline{D}} u > 0$, there exists a point $x \in \overline{D} \setminus M$ such that

$$\begin{cases} u(x) = \max_{\overline{D}} u, \\ \mathcal{A}u(x) \leq 0. \end{cases}$$

(c) For all $\alpha > 0$, the range $\mathcal{R}(\alpha I - \mathcal{A})$ is dense in $C_0\left(\overline{D} \setminus M\right)$.

Then the operator $\mathcal{A}$ is closable *in the space $C_0\left(\overline{D} \setminus M\right)$, and its minimal closed extension $\overline{\mathcal{A}}$ generates a Feller semigroup $\{T_t\}_{t \geq 0}$ on the state space $\overline{D} \setminus M$.*

Proof. We apply part (ii) of Theorem 3.36 (the Hille–Yosida–Ray theorem) to the linear operator $\mathcal{A}$. To do this, we have only to prove the following five assertions (1) through (5):

(1) The operator $\mathcal{A}$ is *closable* in the space $C_0\left(\overline{D} \setminus M\right)$.

(2) $\mathcal{R}(\alpha I - \overline{\mathcal{A}}) = C_0\left(\overline{D} \setminus M\right)$ for $\alpha > 0$.

(3) If $u \in \mathcal{D}\left(\overline{\mathcal{A}}\right)$ takes a positive maximum at a point x' of K, then we have the inequality

$$\overline{\mathcal{A}}u(x') \leq 0.$$

(4) $\widetilde{G}_\alpha := (\alpha I - \overline{\mathcal{A}})^{-1}$ for $\alpha > 0$.

(5) $f \in C_0\left(\overline{D} \setminus M\right)$, $f \geq 0$ on $\overline{D} \Longrightarrow 0 \leq \widetilde{G}_\alpha f \leq \dfrac{1}{\alpha}\max_{\overline{D}} f$ on $\overline{D}$.

The proof is divided into six steps.

Step 1: First, we prove that, for all $\alpha > 0$

$$u \in \mathcal{D}(\mathcal{A}), \ (\alpha - \mathcal{A})u = f \geq 0 \quad \text{on } \overline{D} \tag{14.31}$$

$$\Longrightarrow \ 0 \leq u \leq \frac{1}{\alpha}\max_{\overline{D}} f \quad \text{on } \overline{D}.$$

Our proof is based on a reduction to absurdity. If we assume, to the contrary, that

$$\min_{\overline{D}} u < 0,$$

then it follows from condition (b) with $u := -u$ that there exists a point $x_0 \in \overline{D} \setminus M$ such that

$$\begin{cases} -u(x_0) = \max_{\overline{D}}(-u) = -\min_{\overline{D}} u > 0, \\ -\mathcal{A}u(x_0) \leq 0. \end{cases}$$

Hence we have the assertion

$$0 \leq f(x_0) = \alpha u(x_0) - \mathcal{A}u(x_0) < 0.$$

This contradiction proves that

$$u \geq 0 \quad \text{on } \overline{D}.$$

Similarly, if we assume, to the contrary, that

$$\max_{\overline{D}} u > \frac{1}{\alpha}\max_{\overline{D}} f,$$

then we can find a point $x_1 \in \overline{D} \setminus M$ such that

$$\begin{cases} u(x_1) = \max_{\overline{D}} u > \dfrac{1}{\alpha}\max_{\overline{D}} f, \\ \mathcal{A}u(x_1) \leq 0. \end{cases}$$

Hence it follows that

$$f(x_1) = \alpha\,u(x_1) - \mathcal{A}u(x_1) \geq \alpha u(x_1) > \max_{\overline{D}} f.$$

This contradiction proves that

$$\max_{\overline{D}} u \leq \frac{1}{\alpha} \max_{\overline{D}} f.$$

Step 2: Secondly, by virtue of assertion (14.31) we can define an inverse

$$G_\alpha := (\alpha I - \mathcal{A})^{-1} : \mathcal{R}(\alpha I - \mathcal{A}) \longrightarrow \mathcal{D}(\mathcal{A})$$

which is non-negative and bounded on the range $\mathcal{R}(\alpha I - \mathcal{A})$:

$$f \in \mathcal{R}(\alpha I - \mathcal{A}), \ f \geq 0 \ \text{ on } \overline{D} \implies 0 \leq G_\alpha f \leq \frac{1}{\alpha} \max_{\overline{D}} f \ \text{ on } \overline{D}.$$

Furthermore, since the range $\mathcal{R}(\alpha I - \mathcal{A})$ is dense in $C_0\left(\overline{D} \setminus M\right)$, we find that the operator G_α can be uniquely extended to a non-negative, bounded linear operator on the whole space $C_0\left(\overline{D} \setminus M\right)$, denoted by $\widetilde{G}_\alpha$ (see Figure 14.1 below).

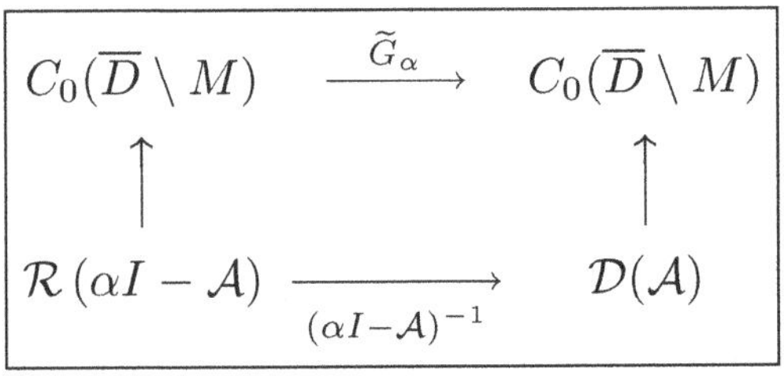

$$
\begin{array}{ccc}
C_0\left(\overline{D} \setminus M\right) & \xrightarrow{\ \widetilde{G}_\alpha\ } & C_0\left(\overline{D} \setminus M\right) \\
\Big\uparrow & & \Big\uparrow \\
\mathcal{R}(\alpha I - \mathcal{A}) & \xrightarrow[\ (\alpha I - \mathcal{A})^{-1}\ ]{} & \mathcal{D}(\mathcal{A})
\end{array}
$$

Fig. 14.1. The operators $(\alpha I - \mathcal{A})^{-1}$ and $\widetilde{G}_\alpha$

More precisely, it should be noticed that

$$f \in C_0\left(\overline{D} \setminus M\right), \ f(x) \geq 0 \ \text{ on } \overline{D} \tag{14.32}$$

$$\implies 0 \leq \widetilde{G}_\alpha f(x) \leq \frac{1}{\alpha} \max_{\overline{D}} f \ \text{ on } \overline{D},$$

and that

$$\left\|\widetilde{G}_\alpha\right\| \leq \frac{1}{\alpha} \ \text{ for all } \alpha > 0. \tag{14.33}$$

Step 3: Thirdly, we prove that

$$\lim_{\alpha \to \infty} \left\|\alpha \widetilde{G}_\alpha u - u\right\|_\infty = 0 \ \text{ for each } u \in C_0\left(\overline{D} \setminus M\right). \tag{14.34}$$

Here

$$\|u\|_\infty = \sup_{x \in D} |u(x)| = \max_{x \in \overline{D}} |u(x)|.$$

For any given $\varepsilon > 0$, we can find a function $v \in \mathcal{D}(\mathcal{A})$ such that

$$\|u - v\| < \varepsilon.$$

Then we have, by assertion (14.33),

$$
\begin{aligned}
\left\|\alpha \widetilde{G}_\alpha u - u\right\|_\infty &\leq \alpha \left\|\widetilde{G}_\alpha (u - v)\right\|_\infty + \left\|\alpha \widetilde{G}_\alpha v - v\right\|_\infty + \|v - u\|_\infty \quad (14.35) \\
&\leq 2 \|u - v\|_\infty + \left\|\alpha \widetilde{G}_\alpha v - v\right\|_\infty \\
&\leq 2\varepsilon + \left\|\alpha \widetilde{G}_\alpha v - v\right\|_\infty .
\end{aligned}
$$

However, it should be noticed that

$$v = \widetilde{G}_\alpha (\alpha I - \mathcal{A}) v = \alpha \widetilde{G}_\alpha v - \widetilde{G}_\alpha (\mathcal{A}v).$$

Hence we have, by assertion (14.33),

$$\left\|\alpha \widetilde{G}_\alpha v - v\right\|_\infty = \left\|\widetilde{G}_\alpha (\mathcal{A}v)\right\|_\infty \leq \frac{1}{\alpha} \|\mathcal{A}v\|_\infty . \quad (14.36)$$

Therefore, by combining inequalities (14.35) and (14.36) we obtain that

$$\left\|\alpha \widetilde{G}_\alpha u - u\right\|_\infty \leq 2\varepsilon + \frac{1}{\alpha} \|\mathcal{A}v\|_\infty ,$$

so that

$$\limsup_{\alpha \to \infty} \left\|\alpha \widetilde{G}_\alpha u - u\right\|_\infty \leq 2\varepsilon.$$

This proves the desired assertion (14.34), since ε is arbitrary.

Step 4: Now we can prove that the operator $\mathcal{A}$ is *closable* in the space $C_0 \left(\overline{D} \setminus M\right)$.

To do this, we assume that

$$
\begin{cases}
\{u_n\} \subset \mathcal{D}(\mathcal{A}), \ u_n \longrightarrow 0, \\
\mathcal{A}u_n \longrightarrow v \quad \text{in } C_0 \left(\overline{D} \setminus M\right).
\end{cases}
$$

Then it follows that, for all $\alpha > 0$,

$$(\alpha I - \mathcal{A}) u_n = \alpha u_n - \mathcal{A}u_n \longrightarrow -v \quad \text{in } C_0 \left(\overline{D} \setminus M\right).$$

Moreover, we have, by the boundedness of $\widetilde{G}_\alpha$,

$$u_n = G_\alpha (\alpha I - \mathcal{A}) u_n = \widetilde{G}_\alpha (\alpha I - \mathcal{A}) u_n \longrightarrow -G_\alpha v \quad \text{in } C_0 \left(\overline{D} \setminus M\right).$$

Hence we have the assertion

$$G_\alpha v = 0 \quad \text{for all } \alpha > 0.$$

Therefore, we obtain from assertion (14.34) that

$$v = \lim_{\alpha \to \infty} \alpha \widetilde{G}_\alpha v = 0.$$

This proves that $\mathcal{A}$ is closable in $C_0\left(\overline{D} \setminus M\right)$.

Step 5: Let $\overline{\mathcal{A}}$ be the minimal closed extension of $\mathcal{A}$ in the space $C_0\left(\overline{D} \setminus M\right)$. Finally, it remains to prove the formula

$$\widetilde{G}_\alpha = \left(\alpha I - \overline{\mathcal{A}}\right)^{-1} \quad \text{for all } \alpha > 0. \tag{14.37}$$

(a) First, we prove the formula

$$\widetilde{G}_\alpha \left(\alpha I - \overline{\mathcal{A}}\right) = I \quad \text{on } \mathcal{D}(\overline{\mathcal{A}}). \tag{14.38}$$

Let u be an arbitrary element of $\mathcal{D}(\overline{\mathcal{A}})$. Then we can find a sequence $\{u_n\}$ in $\mathcal{D}(\mathcal{A})$ such that

$$\begin{cases} u_n \longrightarrow u & \text{in } C_0\left(\overline{D} \setminus M\right), \\ \mathcal{A}u_n \longrightarrow \overline{\mathcal{A}}u & \text{in } C_0\left(\overline{D} \setminus M\right). \end{cases}$$

Since $\widetilde{G}_\alpha$ is bounded, it follows that

$$\begin{aligned} u_n &= \widetilde{G}_\alpha \left(\alpha I - \mathcal{A}\right) u_n = \alpha \widetilde{G}_\alpha u_n - \widetilde{G}_\alpha \left(\mathcal{A}u_n\right) \\ &\longrightarrow \alpha \widetilde{G}_\alpha u - \widetilde{G}_\alpha \left(\overline{\mathcal{A}}u\right) \quad \text{in } C_0\left(\overline{D} \setminus M\right). \end{aligned}$$

Hence we have the formula

$$u = \widetilde{G}_\alpha \left(\alpha I - \overline{\mathcal{A}}\right) u,$$

so that

$$\widetilde{G}_\alpha \left(\alpha I - \overline{\mathcal{A}}\right) u = u.$$

This proves formula (14.38).

(b) Next, we prove the formula

$$\left(\alpha I - \overline{\mathcal{A}}\right) \widetilde{G}_\alpha = I \quad \text{on } C_0\left(\overline{D} \setminus M\right). \tag{14.39}$$

Let f be an arbitrary element of the space $C_0\left(\overline{D} \setminus M\right)$. Since the range $\mathcal{R}\left(\alpha I - \mathcal{A}\right)$ is dense, we can find a sequence $\{u_n\}$ in $\mathcal{D}\left(\mathcal{A}\right)$ such that

$$\left(\alpha - \mathcal{A}\right) u_n \longrightarrow f \quad \text{in } C_0\left(\overline{D} \setminus M\right).$$

Then we have the assertions

$$u_n = \widetilde{G}_\alpha \left(\alpha - \mathcal{A}\right) u_n \longrightarrow \widetilde{G}_\alpha f \quad \text{in } C_0\left(\overline{D} \setminus M\right)$$

and

$$\mathcal{A}u_n = (\mathcal{A} - \alpha)\,u_n + \alpha\,u_n \longrightarrow -f + \alpha\widetilde{G}_\alpha f \quad \text{in } C_0\left(\overline{D} \setminus M\right).$$

Since $\overline{\mathcal{A}}$ is closed, it follows that

$$\begin{cases} \widetilde{G}_\alpha f \in \mathcal{D}(\overline{\mathcal{A}}), \\ \overline{\mathcal{A}}\left(\widetilde{G}_\alpha f\right) = -f + \alpha\widetilde{G}_\alpha f, \end{cases}$$

or equivalently,

$$\left(\alpha I - \overline{\mathcal{A}}\right)\widetilde{G}_\alpha f = f.$$

This proves formula (14.39).

Step 6: By virtue of assertions (14.37), (14.32) and (14.33), we obtain from part (ii) of Theorem 4.13 that the minimal closed extension $\overline{\mathcal{A}}$ is the infinitesimal generator of some Feller semigroup $\{T_t\}_{t\geq 0}$ on the state space $\overline{D} \setminus M$.

The proof of Theorem 14.16 is complete. $\quad\square$

End of Proof of Theorem 1.11: We have only to verify conditions (a), (b) and (c) in Theorem 14.16 for the operator $\mathcal{W}$ defined by formula (1.17).

Condition (b): We show that, for each $\alpha > 0$ the equation

$$(\alpha I - \mathcal{W})\,u = f$$

has a unique solution $u \in \mathcal{D}(\mathcal{W})$ for any $f \in C^{\theta_0}(\overline{D}) \cap C_0\left(\overline{D} \setminus M\right)$. Remark that the space

$$C^{\theta_0}(\overline{D}) \cap C_0\left(\overline{D} \setminus M\right)$$

is *dense* in $C_0(\overline{D} \setminus M)$.

Since we have the inequality

$$c(x) - \alpha \leq -\alpha \quad \text{for all } x \in D,$$

by applying Theorem 1.8 to the operator $\mathcal{W} - \alpha$ we obtain that the boundary value problem

$$\begin{cases} (\alpha - \mathcal{W})\,u = f & \text{in } D, \\ Lu = 0 & \text{on } \partial D \end{cases}$$

has a unique solution

$$u \in C^{2+\theta_0}(\overline{D}) \cap C_0\left(\overline{D} \setminus M\right)$$

for any $f \in C^{\theta_0}(\overline{D})$.

However, we have the assertion

$$Lu = \mu(x')\frac{\partial u}{\partial \mathbf{n}}(x') + \gamma(x')u(x') = 0 \quad \text{on } \partial D$$
$$\Longrightarrow u = 0 \text{ on } M = \{x' \in \partial D : \mu(x') = 0\}$$
$$\Longrightarrow u \in C_0\left(\overline{D} \setminus M\right).$$

Therefore, if $f \in C^{\theta_0}(\overline{D}) \cap C_0(\overline{D} \setminus M)$ then it follows that

$$W u = \alpha u - f \in C^{\theta_0}(\overline{D}) \cap C_0(\overline{D} \setminus M).$$

This proves that

$$\begin{cases} u \in \mathcal{D}(\mathcal{W}), \\ (\alpha I - \mathcal{W}) u = f. \end{cases}$$

Condition (c): We show that, for each $\alpha > 0$ the Green operator

$$G_\alpha := (\alpha I - \mathcal{W})^{-1} : C_0(\overline{D} \setminus M) \cap C^{\theta_0}(\overline{D}) \longrightarrow C_0(\overline{D} \setminus M) \cap C^{2+\theta_0}(\overline{D})$$

is non-negative:

$$f \in C^{\theta_0}(\overline{D}) \cap C_0(\overline{D} \setminus M), \ f(x) \geq 0 \quad \text{in } D$$
$$\implies u(x) = G_\alpha f(x) \geq 0 \quad \text{in } D.$$

Indeed, if we let

$$v(x) = -u(x) = -G_\alpha f(x),$$

then it follows that

$$\begin{cases} (W - \alpha) v = f \geq 0 & \text{in } D, \\ Lv = 0 & \text{on } \partial D. \end{cases}$$

Therefore, by applying Proposition 14.7 to the operator $W - \alpha$ we obtain that

$$v(x) \leq 0 \quad \text{in } D.$$

This proves that

$$u(x) = -v(x) \geq 0 \quad \text{in } D.$$

Condition (a): In order to verify condition (a), we show that, for each $\alpha > 0$ the Green operator

$$G_\alpha = (\alpha I - \mathcal{W})^{-1}$$

is bounded on the space $C^{\theta_0}(\overline{D}) \cap C_0(\overline{D} \setminus M)$ with norm $1/\alpha$:

$$\|G_\alpha\| \leq \frac{1}{\alpha} \quad \text{for all } \alpha > 0. \tag{14.40}$$

Let $f(x)$ be an arbitrary function in $C^{\theta_0}(\overline{D}) \cap C_0(\overline{D} \setminus M)$. If we let

$$u_\pm(x) := \pm \alpha G_\alpha f(x) - \|f\|_\infty \in C^{2+\theta_0}(\overline{D}),$$

then we have only to prove that

$$u_\pm(x) \leq 0 \quad \text{in } D. \tag{14.41}$$

Indeed, it follows that

$$(W - \alpha) u_{\pm}(x) = \pm \alpha f(x) + (\alpha - c(x)) \|f\|_{\infty}$$
$$= \alpha \left(\|f\|_{\infty} \pm f(x)\right) + (\alpha - c(x)) \|f\|_{\infty}$$
$$\geq 0 \quad \text{in } D,$$

and further that

$$L(u_{\pm}) = -L(\|f\|_{\infty}) = -\gamma(x') \|f\|_{\infty} \geq 0 \quad \text{on } \partial D.$$

Therefore, the desired assertion (14.41) follows by applying Proposition 14.7 to the operator $W - \alpha$.

Now we show that the domain $\mathcal{D}(\mathcal{W})$ is *dense* in $C_0\left(\overline{D} \setminus M\right)$. More precisely, we prove that, for each $u \in C^{2+\theta_0}(\overline{D}) \cap C_0\left(\overline{D} \setminus M\right)$ we have the assertion

$$\lim_{\alpha \to +\infty} \|\alpha G_{\alpha} u - u\|_{\infty} = 0. \tag{14.42}$$

Remark that the space

$$C^{2+\theta_0}(\overline{D}) \cap C_0\left(\overline{D} \setminus M\right)$$

is *dense* in $C_0(\overline{D} \setminus M)$.

First, by Lemma 14.5 it follows that

$$Su \in C^{\theta_0}(\overline{D}) \quad \text{for } u \in C^{2+\theta_0}(\overline{D}).$$

Hence we have the assertions

$$Wu = Au + Su \in C^{\theta_0}(\overline{D}) \quad \text{for } u \in C^{2+\theta_0}(\overline{D}),$$

and so

$$G_{\alpha}(Wu) \in C^{2+\theta_0}(\overline{D}) \cap C_0\left(\overline{D} \setminus M\right).$$

If we let

$$w := \alpha \, G_{\alpha} u - G_{\alpha}(Wu) \in C^{2+\theta_0}(\overline{D}) \cap C_0\left(\overline{D} \setminus M\right),$$

then it follows that

$$(W - \alpha) w = -\alpha u + Wu = (W - \alpha) u \quad \text{in } D.$$

Hence we have the assertions

$$\begin{cases} w - u \in C^{2+\theta_0}(\overline{D}) \cap C_0\left(\overline{D} \setminus M\right), \\ (W - \alpha)(w - u) = 0 \quad \text{in } D. \end{cases}$$

By applying Theorem 1.8 to the operator $W - \alpha$, we obtain that

$$w - u = 0 \quad \text{in } D,$$

that is,

$$u = w = \alpha G_\alpha u - G_\alpha(Wu).$$

Therefore, the desired assertion (14.42) follows from an application of inequality (14.40):

$$\limsup_{\alpha \to +\infty} \|u - \alpha G_\alpha u\|_\infty = \limsup_{\alpha \to +\infty} \|G_\alpha (Wu)\|_\infty \le \lim_{\alpha \to +\infty} \frac{1}{\alpha} \cdot \|Wu\|_\infty$$
$$= 0.$$

Now the proof of Theorem 1.11 is complete. □

14.6 Minimal Closed Extension $\overline{\mathcal{W}}$

By combining Theorems 1.11 and 1.10, we can prove that the operator $\mathfrak{W}$ coincides with the minimal closed extension $\overline{\mathcal{W}}$:

$$\mathfrak{W} = \overline{\mathcal{W}}. \tag{14.43}$$

Indeed, we recall that

- $\mathcal{D}(\mathcal{W})$
$= \left\{ u \in C^2(\overline{D}) \cap C_0 (\overline{D} \setminus M) : Wu \in C_0 (\overline{D} \setminus M) ,\ Lu = 0 \text{ on } \partial D \right\},$
- $\mathcal{D}(\mathfrak{W})$
$= \left\{ u \in H^{2,p}(D) \cap C_0 (\overline{D} \setminus M) : Wu \in C_0 (\overline{D} \setminus M) ,\ Lu = 0 \text{ on } \partial D \right\}.$

Then we have, by Theorems 1.11 and 1.10,

$$\overline{\mathcal{W}} \subset \mathfrak{W}. \tag{14.44}$$

Therefore, we have only to show that

$$\mathcal{D}(\overline{\mathcal{W}}) = \mathcal{D}(\mathfrak{W}). \tag{14.45}$$

Let u be an arbitrary element of $\mathcal{D}(\mathfrak{W})$. Since the operators $\overline{\mathcal{W}}$ and $\mathfrak{W}$ generate semigroups of class (C_0) on the space $C_0 (\overline{D} \setminus M)$, it follows from an application of part (i) of Theorem 3.34 (the Hille–Yosida theorem) with $\alpha := 1$ that the operators

$$I - \overline{\mathcal{W}} \colon \mathcal{D}(\overline{\mathcal{W}}) \longrightarrow C_0 (\overline{D} \setminus M)$$

and

$$I - \mathfrak{W} \colon \mathcal{D}(\mathfrak{W}) \longrightarrow C_0 (\overline{D} \setminus M)$$

are both *bijective*. Hence we can find an (unique) element $v \in \mathcal{D}(\overline{\mathcal{W}})$ such that

$$\left(I - \overline{\mathcal{W}}\right) v = (I - \mathfrak{W}) u.$$

However, by assertion (14.44) it follows that

$$\begin{cases} v - u \in \mathcal{D}(\mathfrak{W}), \\ (I - \mathcal{W})(v - u) = 0, \end{cases}$$

so that

$$u = v \in \mathcal{D}\left(\overline{\mathcal{W}}\right).$$

This proves assertion (14.45) and hence the desired assertion (14.43).

Summing up, we have proved the following theorem:

Theorem 14.17. *If conditions (A), (B) and (H) are satisfied, then the closed operator $\mathfrak{W}$ is the infinitesimal generator of some Feller semigroup*

$$\{e^{t\mathfrak{W}}\}_{t \geq 0}$$

on the state space $\overline{D} \setminus M$. Moreover, it generates a semigroup $e^{z\mathfrak{W}}$ on the Banach space $C_0\left(\overline{D} \setminus M\right)$ that is analytic *in the sector*

$$\Delta_\varepsilon = \{z = t + is : z \neq 0, |\arg z| < \pi/2 - \varepsilon\}$$

for any $0 < \varepsilon < \pi/2$ (see Figure 1.3).

14.7 Notes and Comments

The results discussed in this chapter are adapted from Taira [117], [121], [123], [127] and [128]. Gohberg–Kreĭn [51] is the classic for index theory of closed linear operators in Banach spaces.

15

Path Functions of Markov Processes via Semigroup Theory

In this book we have studied mainly Markov transition functions with only informal references to the random variables which actually form the Markov processes themselves (see Section 3.3). In this chapter we study this neglected side of our subject. The discussion will have a more measure-theoretical flavor than hitherto. Section 15.1 is devoted to a review of the basic definitions and properties of Markov processes. In Section 15.2 we consider when the paths of a Markov process are actually continuous, and prove Theorem 3.19 (Corollary 15.7). In Section 15.3 we give a useful criterion for path-continuity of a Markov process $\{x_t\}$ in terms of the infinitesimal generator $\mathfrak{A}$ of the associated Feller semigroup $\{T_t\}$ (Theorem 15.9). Section 15.4 is devoted to the study of three typical examples of multi-dimensional diffusion processes. More precisely, we prove that (1) the reflecting barrier Brownian motion (Theorem 15.11), (2) the reflecting and absorbing barrier Brownian motion (Theorem 15.14) and (3) the reflecting, absorbing and drifting barrier Brownian motion (Theorem 15.15) are multi-dimensional *diffusion processes*, namely, they are continuous strong Markov processes.

This chapter is taken from [122, Chapter 12].

15.1 Basic Definitions and Properties of Markov Processes

First, we recall the basic definitions of stochastic processes (see Subsection 3.2.1). Let K be a locally compact, separable metric space and $\mathcal{B}$ the σ-algebra of all Borel sets in K. Let $(\Omega, \mathcal{F}, P)$ be a probability space. A function X defined on Ω taking values in K is called a random variable if it satisfies the condition

$$X^{-1}(E) = \{X \in E\} \in \mathcal{F} \quad \text{for all } E \in \mathcal{B},$$

that is, X is $\mathcal{F}/\mathcal{B}$-measurable.

© Springer Nature Switzerland AG 2020
K. Taira, *Boundary Value Problems and Markov Processes*, Lecture Notes in Mathematics 1499,
https://doi.org/10.1007/978-3-030-48788-1_15

A family $\mathcal{X} = \{x(t, \omega)\}$, $t \in [0, \infty)$, $\omega \in \Omega$, of random variables is called a stochastic process. We regard the process $\mathcal{X}$ primarily as a function of t whose values $x(t, \cdot)$ for each t are random variables defined on Ω taking values in K. More precisely, we are dealing with one function of two variables, that is, for each fixed t the function $x_t(\cdot)$ is $\mathcal{F}/\mathcal{B}$-measurable. If, instead of t, we fix an $\omega \in \Omega$, then we obtain a function $x(\cdot, \omega) \colon [0, \infty) \to K$ which may be thought of as the motion in time of a physical particle. In this context, the space K is called the state space and Ω the sample space. The function $x_t(\omega) = x(t, \omega)$, $t \in [0, \infty)$, defines in the state space K a trajectory or a path of the process corresponding to the sample point ω.

Sometimes it is useful to think of a stochastic process $\mathcal{X}$ specifically as a function of two variables $x(t, \omega)$ where $t \in [0, \infty)$ and $\omega \in \Omega$. One powerful tool in this connection is Fubini's theorem. To do this, we introduce a class of stochastic processes which we will deal with in this appendix.

Definition 15.1. *A stochastic process $\mathcal{X} = \{x_t\}_{t \geq 0}$ is said to be* measurable *provided that the function $x(\cdot, \cdot) \colon [0, \infty) \times \Omega \to K$ is measurable with respect to the product σ-algebra $\mathcal{A} \times \mathcal{F}$, where $\mathcal{A}$ is the σ-algebra of all Borel sets in the interval $[0, \infty)$.*

We remark that the condition that the function $x_t(\cdot) = x(t, \cdot)$ is $\mathcal{F}/\mathcal{B}$-measurable for each t does not guarantee the measurability of the process $x(\cdot, \cdot)$.

Now let p_t be a Markov transition function on the metric space K (see Definition 3.17). The idea behind Definition 3.17 of a Markov transition function suggests the following definition (cf. formula (3.17)):

Definition 15.2. *A stochastic process $\mathcal{X} = \{x_t\}_{t \geq 0}$ is said to be* governed by *the transition function p_t provided that we have, for all $0 \leq t_1 < t_2 < \ldots < t_n < \infty$ and all Borel sets B_1, B_2, ..., $B_n \in \mathcal{B}$,*

$$P\left(\omega \in \Omega : x_{t_1}(\omega) \in B_1, x_{t_2}(\omega) \in B_2, \ldots, x_{t_n}(\omega) \in B_n\right) \tag{15.1}$$

$$= \int_{y_n \in B_n} \cdots \int_{y_1 \in B_1} \int_{y_1 \in B_1} \int_{x \in K} \mu(dx)\, p_{t_1}(x, dy_1)$$
$$p_{t_2-t_1}(y_1, dy_2) \cdots p_{t_n-t_{n-1}}(y_{n-1}, dy_n),$$

where μ is some probability measure on the measurable space $(K, \mathcal{B})$, and is called the initial distribution *of the process $\{x_t\}$.*

The notion of the Markov property is introduced and discussed in Chapter 3, Section 3.2: If $\mathcal{X} = \{x_t\}_{t \geq 0}$ is a stochastic process, we introduce three sub-σ-algebras of $\mathcal{F}$ as follows:

$$
\begin{cases}
\mathcal{F}_{\leq t} & := \sigma(x_s : 0 \leq s \leq t) \\
& = \text{the smallest } \sigma\text{-algebra, contained in } \mathcal{F} \\
& \quad \text{with respect to which all } x_s,\, 0 \leq s \leq t,\, \text{are measurable,} \\
\mathcal{F}_{=t} & := \sigma(x_t) \\
& = \text{the smallest } \sigma\text{-algebra, contained in } \mathcal{F} \\
& \quad \text{with respect to which } x_t \text{ is measurable,} \\
\mathcal{F}_{\geq t} & := \sigma(x_s : t \leq s < \infty) \\
& = \text{the smallest } \sigma\text{-algebra, contained in } \mathcal{F} \\
& \quad \text{with respect to which all } x_s,\, t \leq s < \infty,\, \text{are measurable.}
\end{cases}
$$

We recall that an event in $\mathcal{F}_{\leq t}$ is determined by the behavior of the process $\{x_s\}$ up to time t and an event in $\mathcal{F}_{\geq t}$ by its behavior after time t. Thus they represent respectively the "past" and "future" relative to the "present" moment.

A stochastic process $\mathcal{X} = \{x_t\}$ is called a Markov process if it satisfies the following condition:

$$
P(B \mid \mathcal{F}_{\leq t}) = P(B \mid \mathcal{F}_{=t}) \quad \text{for any "future" set } B \in \mathcal{F}_{\geq t}.
$$

More precisely, we have, for any "future" set $B \in \mathcal{F}_{\geq t}$,

$$
P(A \cap B) = \int_A P(B \mid \mathcal{F}_{=t})(\omega)\, dP(\omega) \quad \text{for every "past" set } A \in \mathcal{F}_{\leq t}.
$$

Intuitively, this means that the conditional probability of a "future" event B given the "present" is the same as the conditional probability of B given the "present" and "past".

The next theorem justifies Definition 15.2, and hence it is fundamental for our further study of Markov processes:

Theorem 15.3. *Let $\mathcal{X} = \{x_t\}_{t \geq 0}$ be any stochastic process with values in the metric space K which is governed by a Markov transition function p_t. Then it follows that $\{x_t\}$ is* a Markov process.

Our study of Markov processes is based on formula (15.1) that shows how the finite-dimensional distributions of the process $\mathcal{X} = \{x_t\}$ are calculated from the Markov transition function p_t. However, knowledge of all the finite-dimensional distributions may not be sufficient to determine precisely the path functions of a Markov process. Therefore, it is important to ask the following question:

Question 15.4. Given a Markov transition function p_t and an initial distribution μ, does there exist a Markov process $\mathcal{X} = \{x_t\}$ having the corresponding finite-dimensional distributions whose paths are almost surely "nice" in some sense ?

We say that two Markov processes $\mathcal{X} = \{x_t\}_{t\geq 0}$ and $\mathcal{Y} = \{y_t\}_{t\geq 0}$ defined on the same probability space $(\Omega, \mathcal{F}, P)$ is *equivalent* provided that we have, for all $t \in [0, \infty)$,

$$P\left(\{\omega \in \Omega : x_t(\omega) = y_t(\omega)\}\right) = 1.$$

The next theorem asserts that, under quite general conditions there does exist a Markov process with "nice" paths equivalent to any given process:

Theorem 15.5. *Let (K, ρ) be a* compact *metric space and let $\mathcal{X} = \{x_t\}_{t\geq 0}$ be a stochastic process with values in K which is governed by a* normal *Markov transition function p_t. Then there exists a Markov process $\mathcal{Y} = \{y_t\}$, equivalent to the process $\mathcal{X} = \{x_t\}$, such that*

$$P(\{\omega \in \Omega : \text{ the function } y_t(\omega) \text{ are right-continuous and}$$
$$\text{have left-hand limits for all } t \geq 0\}) = 1.$$

15.2 Path-Continuity of Markov Processes

It is naturally interesting and important to consider when the paths of a Markov process $\{x_t\}$ are actually continuous for all $t \geq 0$. The purpose of this section is to establish some useful sufficient conditions for path-continuity of the Markov process $\{x_t\}$.

First, we have the following theorem:

Theorem 15.6. *Let (K, ρ) be a* locally compact *metric space and let $\mathcal{X} = \{x_t\}_{t\geq 0}$ be a measurable stochastic process with values in K. Assume that, for each $\varepsilon > 0$ and each $M > 0$ the condition*

$$P(\{\omega \in \Omega : \rho(x_t(\omega), \rho_{t+h}(\omega)) \geq \varepsilon\}) = o(h) \tag{15.2}$$

holds true uniformly for all $t \in [0, M]$, as $h \downarrow 0$. Namely, we have, for all $t \in [0, M]$,

$$\lim_{h\downarrow 0} \frac{P(\{\omega \in \Omega : \rho(x_t(\omega), \rho_{t+h}(\omega)) \geq \varepsilon\})}{h} = 0.$$

Then it follows that

$$P(\{\omega \in \Omega : x_t(\omega) \text{ has a jump discontinuity somewhere}\}) = 0. \tag{15.3}$$

Proof. The proof is divided into two steps.

Step 1: First, we show that if we let

$$J_{\varepsilon,h}(\omega) = \{t \in [0, M] : \rho(x_t(\omega), x_{t+h}(\omega)) \geq \varepsilon\}, \quad \omega \in \Omega, \tag{15.4}$$

then it follows from condition (15.2) that

$$E\left[m(J_{\varepsilon,h})\right] = o(h) \quad \text{as } h \downarrow 0, \tag{15.5}$$

where $m = dt$ is the Lebesgue measure on $\mathbf{R}$.

To do this, we define the set (see Figure 15.1 below)

$$A_{\varepsilon,h} = \{(t,\omega) \in [0,M] \times \Omega : \rho(x_t(\omega), x_{t+h}(\omega)) \geq \varepsilon\}.$$

We remark that the set $A_{\varepsilon,h}$ is measurable with respect to the product σ-

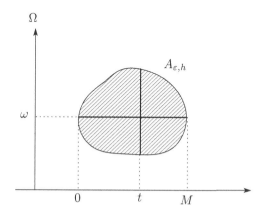

Fig. 15.1. The set $A_{\varepsilon,h}$

algebra $\mathcal{A} \times \mathcal{F}$. Indeed, it suffices to note the following two facts (a) and (b):

(a) The mapping $(t,\omega) \mapsto (x_t(\omega), x_{t+h}(\omega))$ is measurable from the product space $[0,M] \times \Omega$ into the product space $K \times K$, for each $h \geq 0$.
(b) The metric ρ is a continuous function on the product space $K \times K$.

By virtue of Fubini's theorem, we can compute the product measure $m \times P$ of $A_{\varepsilon,h}$ by integrating the measure of a cross section as follows:

$$(m \times P)(A_{\varepsilon,h}) = \int_0^M P(\{\omega \in \Omega : (t,\omega) \in A_{\varepsilon,h}\})\, dt.$$

By condition (15.2), it follows that

$$(m \times P)(A_{\varepsilon,h}) = o(h) \quad \text{as } h \downarrow 0. \tag{15.6}$$

Moreover, by integrating in the other order we obtain from definition (15.4) that

$$(m \times P)(A_{\varepsilon,h}) = \int_\Omega m(\{t \in \mathbf{R} : (t,\omega) \in A_{\varepsilon,h}\})\, dP(\omega) \tag{15.7}$$

$$= E\left[m(\{t \in [0, M] : \rho(x_t(\omega), x_{t+h}(\omega)) \geq \varepsilon\})\right]$$
$$= E\left[m(J_{\varepsilon,h})\right].$$

Therefore, the desired assertion (15.5) follows by combining assertions 15.6 and (15.7).

Step 2: Now we show that the existence of jumps in the trajectories of $\{x_t\}$ contradicts assertion (15.5).

Step 2-1: We assume that, for some $\varepsilon_0 > 0$,

$$P(\{\omega \in \Omega : x_t(\omega) \text{ has a jump with gap greater than } 2\varepsilon_0\}) > 0.$$

If $x_t(\omega)$ is a trajectory having such a jump at $t = t_0$ (see Figure 15.2 below), then we obtain that the two limits

$$x_{t_0^+}(\omega) = \lim_{h \downarrow 0} x_{t_0+h}(\omega), \quad x_{t_0^-}(\omega) = \lim_{h \downarrow 0} x_{t_0-h}(\omega)$$

exist and satisfy the condition

$$\rho(x_{t_0^-}(\omega), x_{t_0^+}(\omega)) \geq 2\varepsilon_0.$$

Then we have, for sufficiently small $h > 0$,

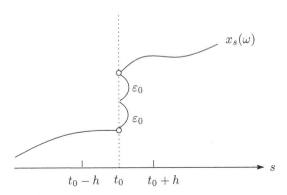

Fig. 15.2. The condition $\rho(x_{t_0^-}(\omega), x_{t_0^+}(\omega)) \geq 2\varepsilon_0$

$$\begin{cases} \rho(x_{t_0^-}(\omega), x_s(\omega)) \leq \frac{\varepsilon_0}{2} & \text{for all } s \in (t_0 - h, t_0), \\ \rho(x_{t_0^+}(\omega), x_s(\omega)) \leq \frac{\varepsilon_0}{2} & \text{for all } s \in (t_0, t_0 + h). \end{cases}$$

Moreover, by the triangle inequality it follows that

$$2\varepsilon_0 \leq \rho(x_{t_0^-}(\omega), x_{t_0^+}(\omega))$$
$$\leq \rho(x_{t_0^-}(\omega), x_t(\omega)) + \rho(x_t(\omega), x_{t+h}(\omega)) + \rho(x_{t+h}(\omega), x_{t_0^+}(\omega))$$

$$\leq \frac{\varepsilon_0}{2} + \rho(x_t(\omega), x_{t+h}(\omega)) + \frac{\varepsilon_0}{2} \quad \text{for all } t \in (t_0 - h, t_0),$$

so that

$$\rho(x_t(\omega), x_{t+h}(\omega)) \geq \varepsilon_0 \quad \text{for all } t \in (t_0 - h, t_0).$$

This implies that, for all sufficiently small $h > 0$,

$$m\left(\{t \in [0, M] : \rho(x_t(\omega), x_{t+h}(\omega)) \geq \varepsilon_0\}\right) \geq m\left((t_0 - h, t_0)\right) = h,$$

or equivalently,

$$\frac{m(J_{\varepsilon_0, h})(\omega)}{h} \geq 1 \quad \text{for all sufficiently small } h > 0. \tag{15.8}$$

Step 2-2: The proof is based on a reduction to absurdity. We assume, to the contrary, that

$$P(\{\omega \in \Omega : x_t(\omega) \text{ has a jump discontinuity somewhere}\}) > 0.$$

Then it follows from Step 2-1 that there exists a positive number ε_0 such that assertion (15.8) holds true.

Therefore, we can find a positive constant $\delta = \delta(\varepsilon_0)$ such that

$$E[m(J_{\varepsilon_0, h})] = \int_\Omega m(J_{\varepsilon_0, h})(\omega) \, dP(\omega) \tag{15.9}$$

$$\geq \int_{L_{\varepsilon_0}} m(J_{\varepsilon_0, h})(\omega) \, dP(\omega)$$

$$\geq h \delta \quad \text{for all sufficiently small } h > 0,$$

where

$$L_{\varepsilon_0} = \{\omega \in \Omega : x_t(\omega) \text{ has a jump with gap greater than } 2\varepsilon_0\}.$$

Assertion (15.9) contradicts condition (15.5).

The proof of Theorem 15.6 is complete. $\square$

The next corollary proves part (ii) of Theorem 3.19 under condition (N) with $E = K$:

Corollary 15.7. *Let (K, ρ) be a locally compact metric space and let $\mathcal{X} = \{x_t\}_{t\geq 0}$ be a right-continuous Markov process governed by a Markov transition function p_t. Assume that, for each $\varepsilon > 0$, the condition*

$$p_h\left(x, K \setminus U_\varepsilon(x)\right) = o(h) \tag{15.10}$$

holds true uniformly in $x \in K$ as $h \downarrow 0$. In other words, for each $\varepsilon > 0$ we have the condition

$$\lim_{h \downarrow 0} \frac{1}{h} \sup_{x \in K} p_t \left(x, K \setminus U_\varepsilon(x) \right) = 0.$$

Here $U_\varepsilon(x) = \{y \in K : \rho(y, x) < \varepsilon\}$ is an ε-neighborhood of x.
Then it follows that

$$P \left(\{\omega \in \Omega : x_t(\omega) \text{ is continuous for all } t \geq 0\} \right) = 1.$$

Proof. We shall apply Theorem 15.6 to our situation. It should be emphasized that the stochastic process $\{x_t\}$ is right-continuous and has limits from the left as well. The proof of Corollary 15.7 is divided into two steps.

Step 1: First, we prove the following lemma (see [33, Lemma 5.9]):

Lemma 15.8. *Every right-continuous stochastic process $\{x_t\}$ is measurable.*

Proof. If $\{t_n\}$ is an increasing sequence such that

$$0 = t_0 < t_1 < t_2 < \ldots \to \infty,$$

then we define a stochastic process $\{y_t\}$ by the formula (see Figure 15.3 below)

$$y_t(\omega) = \begin{cases} x_{t_0}(\omega) = x(t_0, \omega), \ \omega \in \Omega & \text{for } t_0 \leq t < t_1, \\ x_{t_1}(\omega) = x(t_1, \omega), \ \omega \in \Omega & \text{for } t_1 \leq t < t_2, \\ \quad \cdot \\ \quad \cdot \\ \quad \cdot \\ x_{t_k}(\omega) = x(t_k, \omega), \ \omega \in \Omega & \text{for } t_k \leq t < t_{k+1}, \\ \quad \cdot \\ \quad \cdot \\ \quad \cdot \end{cases}$$

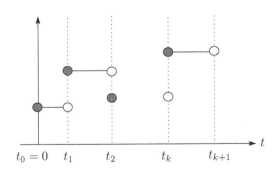

Fig. 15.3. The stochastic process $\{y_t\}$

Since we have, for any $a \in \mathbf{R}$,

$$\{(t,\omega) : y_t(\omega) < a\} = \bigcup_{i=0}^{\infty} [t_i, t_{i+1}) \times \{\omega : x_{t_i}(\omega) < a\},$$

it follows that the process $\{y_t\}$ is measurable.

For each integer $n \in \mathbf{N}$, we choose a non-negative integer k such that

$$\frac{k}{2^n} \le t < \frac{k+1}{2^n},$$

and let

$$\phi_n(t) = \frac{k+1}{2^n}.$$

It is clear that $\phi_n(t) \downarrow t$ as $n \to \infty$. Now, if we define a stochastic process $\{x_n\}$ by the formula

$$x_n(t, \omega) = x(\phi_n(t), \omega) \quad \text{for every } \omega \in \Omega,$$

then we have the following two assertions (a) and (b):

(a) The process $x_n(t)$ is measurable.
(b) $x_n(t, \omega) \to x(t, \omega)$ as $n \to \infty$.

Indeed, assertion (a) is proved just as in the case of the process $\{y_t\}$, while assertion (b) follows from the right-continuity of the process $\{x_t\}$.

Therefore, we obtain from assertions (a) and (b) that the process $\{x_t\}$ is measurable.

The proof of Lemma 15.8 is complete. $\square$

Step 2: Let μ be the initial distribution of the process $\{x_t\}$ in Definition 15.2. Namely, we have, for all $0 \le t_1 < t_2 < \ldots < t_n < \infty$ and all Borel sets $B_1, B_2, \ldots, B_n \in \mathcal{B}$,

$$P(\omega \in \Omega : x_{t_1}(\omega) \in B_1, x_{t_2}(\omega) \in B_2, \ldots, x_{t_n}(\omega) \in B_n)$$
$$= \int_{y_n \in B_n} \cdots \int_{y_1 \in B_1} \int_{y_1 \in B_1} \int_{x \in K} \mu(dx) p_{t_1}(x, dy_1)$$
$$p_{t_2-t_1}(y_1, dy_2) \cdots p_{t_n-t_{n-1}}(y_{n-1}, dy_n).$$

Then we have, by Fubini's theorem (see Figure 15.4 below),

$$P(\{\omega \in \Omega : \rho(x_t(\omega), x_{t+h}(\omega)) \ge \varepsilon\}) \tag{15.11}$$
$$= P(\{\omega \in \Omega : x_t(\omega) \in K, x_{t+h}(\omega) \in K \setminus U_\varepsilon(x_t(\omega))\})$$
$$= \int_K \int_{K \setminus U_{\varepsilon(y)}} \left(\int_K p_t(x, dy) \right) p_h(y, dz) \, \mu(dx)$$
$$= \int\int_K \left(\int_{K \setminus U_{\varepsilon(y)}} p_h(y, dz) \right) p_t(x, dy) \, \mu(dx)$$

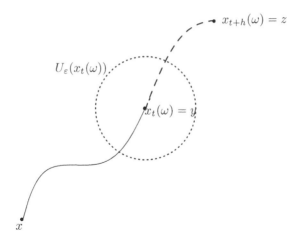

Fig. 15.4. The proof of formula (15.11)

$$= \iint_K p_t(x, dy) p_h\left(y, K \setminus U_\varepsilon(y)\right) \mu(dx).$$

In view of condition 15.10, we obtain from formula (15.11) that

$$\lim_{h \downarrow 0} \frac{P(\{\omega \in \Omega : \rho(x_t(\omega), x_{t+h}(\omega)) \geq \varepsilon\})}{h}$$

$$= \lim_{h \downarrow 0} \frac{\iint_K p_t(x, dy) p_h\left(y, K \setminus U_\varepsilon(y)\right) \mu(dx)}{h}$$

$$= \iint_K p_t(x, dy) \left(\lim_{h \downarrow 0} \frac{p_h\left(y, K \setminus U_\varepsilon(y)\right)}{h}\right) \mu(dx)$$

$$= 0$$

holds true uniformly for all $t \geq 0$. This assertion implies that condition (15.2) holds true uniformly for all $t \geq 0$, as $h \downarrow 0$. Hence it follows from an application of Theorem 15.6 that

$$P(\{\omega : x_t(\omega) \text{ has a jump discontinuity somewhere}\}) = 0. \qquad (15.12)$$

However, we recall that the stochastic process $\{x_t\}$ is right-continuous and has limits from the left as well.

Therefore, we obtain from assertion (15.12) that

$$P(\{\omega \in \Omega : x_t(\omega) \text{ is continuous for all } t \geq 0\}) = 1.$$

The proof of Corollary 15.7 is complete. $\quad\square$

15.3 Path-Continuity of Feller Semigroups

It is usually difficult to verify condition 15.10 directly, since it is rather exceptional when any simple formula for the transition probability function p_t is available. The purpose of this section is to give a useful criterion for path-continuity of the Markov process $\{x_t\}$ in terms of the infinitesimal generator $\mathfrak{A}$ of the associated Feller semigroup $\{T_t\}$.

Let (K, ρ) be a *compact* metric space and let $C(K)$ be the space of real-valued, bounded continuous functions on K; $C(K)$ is a Banach space with the maximum norm

$$\|f\|_\infty = \max_{x \in K} |f(x)|.$$

A strongly continuous semigroup $\{T_t\}_{t \geq 0}$ on the compact metric space $C(K)$ is called a Feller semigroup if it is non-negative and contractive on $C(K)$, that is,

$$f \in C(K),\ 0 \leq f(x) \leq 1 \quad \text{on } K \Longrightarrow 0 \leq T_t f(x) \leq 1 \quad \text{on } K.$$

We recall (see Theorem 3.31) that if p_t is a uniformly stochastically continuous, Feller function on K, then its associated operators $\{T_t\}_{t \geq 0}$, defined by formula

$$T_t f(x) = \int_K p_t(x, dy) f(y) \quad \text{for every } f \in C(K), \qquad (15.13)$$

form a Feller semigroup on K. Conversely, if $\{T_t\}_{t \geq 0}$ is a Feller semigroup on K, then there exists a uniformly stochastically continuous, Feller function p_t on K such that formula (15.13) holds true.

Indeed, it suffices to note that if K is compact, then condition (L) is trivially satisfied.

Furthermore, we know that the function p_t is the transition function of some *strong Markov process* $\mathcal{X} = \{x_t\}_{t \geq 0}$ whose paths are right-continuous and have no discontinuities other than jumps.

Our approach can be visualized as in Figure 15.5 below.

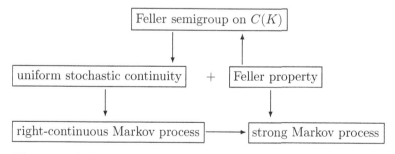

Fig. 15.5. A functional analytic approach to strong Markov processes

The next theorem gives some useful sufficient conditions for path-continuity of the Markov process $\{x_t\}$ in terms of the infinitesimal generator $\mathfrak{A}$ of the associated Feller semigroup $\{T_t\}$:

Theorem 15.9. *Let (K, ρ) be a compact metric space and let $\mathcal{X} = \{x_t\}_{t \geq 0}$ be a right-continuous Markov process governed by a uniformly stochastically continuous, Feller transition function p_t. Assume that the infinitesimal generator $\mathfrak{A}$ of the associated Feller semigroup $\{T_t\}_{t \geq 0}$, defined by formula (15.13), satisfies the following three conditions (15.14a), (15.14b) and (15.14c):*

For each $\varepsilon > 0$ and each point $x \in K$, there exists a function $f \in \mathcal{D}(\mathfrak{A})$ such that

$$f(x) \geq 0 \text{ on } K. \tag{15.14a}$$

$$f(y) > 0 \text{ for all } y \in K \setminus U_\varepsilon(x). \tag{15.14b}$$

$$f(y) = \mathfrak{A}f(y) = 0 \text{ in some neighborhood of } x. \tag{15.14c}$$

Here $U_\varepsilon(x) = \{z \in K : \rho(z, x) < \varepsilon\}$ is an ε-neighborhood of x.
Then it follows that

$$P(\{\omega \in \Omega : x_t(\omega) \text{ is continuous for all } t \geq 0\}) = 1.$$

Proof. We shall apply Corollary 15.7 to our situation.

The proof is based on a reduction to absurdity. We assume, to the contrary, that condition (15.10) does not hold true. Then we can find a positive number ε_0 such that

$$p_h\left(x, K \setminus U_{2\varepsilon_0}(x)\right) \quad \text{is not of order } o(h) \text{ uniformly in } x \text{ as } h \downarrow 0.$$

More precisely, for this ε_0 there exist a positive constant δ, a decreasing sequence $\{h_n\}$ of positive numbers, $h_n \downarrow 0$, and a sequence $\{x_n\}$ of points in K such that

$$p_{h_n}\left(x_n, K \setminus U_{2\varepsilon_0}(x_n)\right) \geq \delta\, h_n \quad \text{for all sufficiently large } n. \tag{15.15}$$

However, since K is compact, we may assume that the sequence $\{x_n\}$ itself converges to some point x_0 of K. If $\rho(x_n, x_0) < \varepsilon_0$, then it follows that (see Figure 15.6 below)

$$U_{\varepsilon_0}(x_n) \subset U_{2\varepsilon_0}(x_0).$$

Hence we have, by assertion (15.15),

$$p_{h_n}\left(x_n, K \setminus U_{\varepsilon_0}(x_0)\right) \geq p_{h_n}\left(x_n, K \setminus U_{2\varepsilon_0}(x_n)\right) \tag{15.16}$$
$$\geq \delta\, h_n \quad \text{for all sufficiently large } n.$$

Now, for these ε_0 and x_0 we can construct a function $f \in \mathcal{D}(\mathfrak{A})$ which satisfies

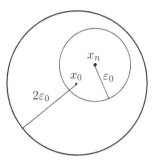

Fig. 15.6. The condition $U_{\varepsilon_0}(x_n) \subset U_{2\varepsilon_0}(x_0)$

conditions (15.14). Then it follows from condition (15.14b) that

$$c = \min_{x \in K \setminus U_{\varepsilon_0}} f(x) > 0.$$

Hence we have, by assertion (15.16),

$$T_{h_n} f(x_n) = \int_K p_{h_n}(x_n, dy)\, f(y) \tag{15.17}$$

$$\geq \int_{K \setminus U_{\varepsilon_0}(x_0)} p_{h_n}(x_n, dy)\, f(y)$$

$$\geq \left(\min_{K \setminus U_{\varepsilon_0}(x_0)} f \right) \cdot p_{h_n}(x_n, K \setminus U_{\varepsilon_0}(x_0))$$

$$\geq c\,\delta\, h_n \quad \text{for all sufficiently large } n.$$

However, since $f(x_n) = 0$ for x_n close to the limit point x_0, we obtain from assertion (15.16) that

$$\frac{T_{h_n} f(x_n) - f(x_n)}{h_n} = \frac{T_{h_n} f(x_n)}{h_n} \geq c\,\delta \quad \text{for all } x_n \in U(x_0). \tag{15.18}$$

On the other hand, since we have, for $f \in \mathcal{D}(\mathfrak{A})$,

$$\mathfrak{A}f(y) = \lim_{n \to \infty} \frac{T_{h_n} f(y) - f(y)}{h_n} \quad \text{uniformly in } y \in K,$$

it follows from condition (15.14c) that

$$\frac{T_{h_n} f(y) - f(y)}{h_n} \longrightarrow \mathfrak{A}f(y) = 0 \quad \text{uniformly in } y \in U(x_0).$$

Hence we have, for all $y \in U(x_0)$,

$$\left| \frac{T_{h_n} f(y) - f(y)}{h_n} \right| < \frac{c\,\delta}{2} \quad \text{for all sufficiently large } n. \tag{15.19}$$

However, since $f(x_n) = 0$ and $T_{h_n} f \geq 0$ on K, by letting $y := x_n$ in inequality (15.19) we obtain that

$$\frac{T_{h_n} f(x_n) - f(x_n)}{h_n} = \left| \frac{T_{h_n} f(x_n) - f(x_n)}{h_n} \right| < \frac{c\delta}{2} \quad \text{for all } x_n \in U(x_0).$$

This assertion contradicts assertion (15.18).

Therefore, we have proved that condition (15.10) holds true. Theorem 15.9 follows from an application of Corollary 15.7.

The proof of Theorem 15.9 is complete. □

15.4 Examples of Multi-dimensional Diffusion Processes

In this section we prove that (1) the reflecting barrier Brownian motion, (2) the reflecting and absorbing barrier Brownian motion, (3) the reflecting, absorbing and drifting barrier Brownian motion are typical examples of multi-dimensional diffusion processes, that is, examples of continuous strong Markov processes. It should be emphasized that these three Brownian motions correspond to the Neumann boundary value problem, the Robin boundary value problem and the oblique derivative boundary value problem for the Laplacian Δ in terms of elliptic boundary value problems, respectively.

Table 15.1 below gives a bird's eye view of diffusion processes and elliptic boundary value problems and how these relate to each other:

15.4.1 The Neumann Case

Let D be a bounded domain of Euclidean space $\mathbf{R}^N$ with smooth boundary ∂D; its closure $\overline{D} = D \cup \partial D$ is an N-dimensional, compact smooth manifold with boundary (see Figure 15.7 below).

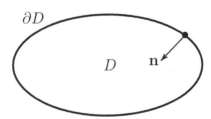

Fig. 15.7. The bounded domain D and the inward normal $\mathbf{n}$ to the boundary ∂D

First, we consider the *Neumann boundary condition*

$$\mathcal{B}_N u = \frac{\partial u}{\partial \mathbf{n}} = 0 \text{ on } \partial D,$$

Diffusion Processes (Microscope approach)	Elliptic Boundary Value Problems (Mesoscopic approach)
Brownian motion	Laplacian Δ
Reflecting barrier Brownian motion	Neumann boundary condition $\mathcal{B}_N$
Reflecting and absorbing barrier Brownian motion	Robin boundary condition $\mathcal{B}_R$
Reflecting, absorbing and drift barrier Brownian motion	Oblique derivative boundary condition $\mathcal{B}_o$

Table 15.1. A bird's-eye view of diffusion processes and elliptic boundary value problems

where $\mathbf{n} = (n_1, n_2, \ldots, n_N)$ is the unit *inward normal* to the boundary ∂D.

We introduce a linear operator $\mathfrak{A}_N$ as follows:

(a) The domain $D(\mathfrak{A}_N)$ is the space

$$D(\mathfrak{A}_N) = \left\{ u \in C(\overline{D}) : \Delta u \in C(\overline{D}), \frac{\partial u}{\partial \mathbf{n}} = 0 \text{ on } \partial D \right\}.$$

(b) $\mathfrak{A}_N u = \Delta u$ for every $u \in D(\mathfrak{A}_N)$.

Here Δu and $\frac{\partial u}{\partial \mathbf{n}}$ are taken in the sense of *distributions*.

Then it follows from an application of Theorem 1.6 with $L := \mathcal{B}_N$ that the operator $\mathfrak{A}_N$ is the infinitesimal generator of a Feller semigroup $\{S_t\}_{t \geq 0}$. Let $\{x_t(\omega)\}$ be the strong Markov process corresponding to the Feller semigroup $\{S_t\}_{t \geq 0}$ with Neumann boundary condition.

In this subsection we study the path-continuity of the Markov process $\{x_t(\omega)\}$. In order to make use of Theorem 15.9, we shall construct a function $f \in D(\mathfrak{A}_N)$ which satisfies conditions (15.14) of the same theorem. Our construction of the function $f(x)$ may be visualized as in Figure 15.8 below.

(I) The case where x_0 is an arbitrary (interior) point of D: By applying Theorem [121, Theorem 1.5] with $\mu(x') := 1$ and $\gamma(x') := 0$, we can find a function $\phi \in C^\infty(\overline{D})$ such that

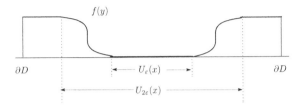

Fig. 15.8. The function $f \in D(\mathfrak{A}_N)$

$$\begin{cases} \Delta\phi = -1 & \text{in } D, \\ \frac{\partial\phi}{\partial \mathbf{n}} = 0 & \text{on } \partial D. \end{cases} \tag{15.20}$$

Then we have the following claim:

Claim 15.10. *The function satisfies the condition*

$$\phi(x) > 0 \quad \text{on } \overline{D}.$$

Proof. The proof is based on a reduction to absurdity. We assume, to the contrary, that

$$\min_{\overline{D}} \phi \le 0.$$

(i) First, we consider the case: There exists a point $x_0 \in D$ such that

$$\phi(x_0) = \min_{\overline{D}} \phi \le 0.$$

Since we have, by condition (15.20),

$$\Delta(-\phi) = 1 > 0 \quad \text{in } D, \tag{15.21}$$

it follows from an application of the strong maximum principle (Theorem 10.7) with $W := \Delta$ and $u := -\phi$ that

$$\phi(x) \equiv \phi(x_0) \quad \text{in } D.$$

Hence we have the assertion

$$\Delta(-\phi) = 0 \quad \text{in } D.$$

This contradicts condition (15.21).

(ii) Next we consider the case: There exists a point $x_0' \in \partial D$ such that

$$\phi(x_0') = \min_{\overline{D}} \phi \le 0.$$

Then, by applying the boundary point lemma (Lemma 10.11) with $W := \Delta$ to the function $u := -\phi$ we obtain from inequality (15.21) that

$$\frac{\partial \phi}{\partial \mathbf{n}}(x_0') > 0.$$

This contradicts the boundary condition

$$\frac{\partial \phi}{\partial \mathbf{n}} = 0 \quad \text{on } \partial D.$$

The proof of Claim 15.10 is complete. $\square$

Now we choose a real-valued, smooth function $\theta \in C^\infty(\mathbf{R})$ such that (see Figure 15.9 below)

$$\begin{cases} 0 \le \theta(t) \le 1 & \text{on } \mathbf{R}, \\ \operatorname{supp} \theta \subset (-1, 1), \\ \theta(t) = 1 & \text{for all } t \in \left[-\frac{1}{2}, \frac{1}{2}\right], \end{cases} \tag{15.22}$$

and define a function $f \in C^\infty(\overline{D})$ by the formula

$$f(x) = \left(1 - \theta\left(\frac{|x - x_0|}{\varepsilon}\right)\right) \phi(x),$$

where

$$0 < \varepsilon < \operatorname{dist}(x, \partial D) = \inf_{z \in \partial D} |x - z|.$$

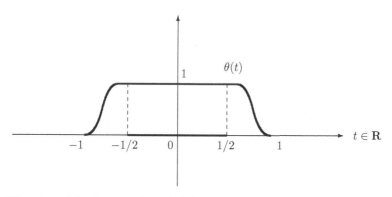

Fig. 15.9. The function $\theta \in C^\infty(\mathbf{R})$ that satisfies the conditions (15.22)

Then, in view of Claim 15.10 it is easy to verify that $f \in D(\mathfrak{A}_N)$ and satisfies conditions (15.14) of Theorem 15.9.

(II) The case where x_0 is an arbitrary (boundary) point of ∂D: By change of coordinates, we may assume that

$$\begin{cases} x_0 = (0', 0) = (0, \dots, 0, 0), \\ \mathbf{n} = (0', 1) = (0, \dots, 0, 1) \in \mathbf{R}^N. \end{cases} \tag{15.23}$$

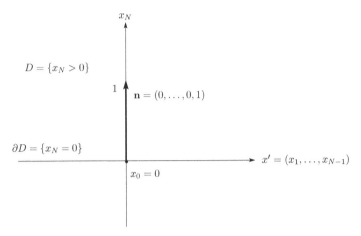

Fig. 15.10. The point x_0 and the unit inward normal $\mathbf{n}$ in the formula (15.23)

The situation may be represented schematically as in Figure 15.10 above.
If we define a function $f \in C^\infty(\overline{D})$ by the formula

$$f(x', x_N) = \left(1 - \theta\left(\frac{x_N}{\varepsilon}\right)\theta\left(\frac{|x'|}{\varepsilon}\right)\right)\phi(x', x_N),$$

$$x' = (x_1, x_2, \ldots, x_{N-1}) \in \mathbf{R}^{N-1}, \ x_N \in \mathbf{R},$$

then it follows from conditions (15.22) and (15.20) that

$$\frac{\partial f}{\partial \mathbf{n}} = \left.\frac{\partial f}{\partial x_N}\right|_{x_N=0}$$

$$= -\left.\left(\frac{\partial}{\partial x_N}\left(\theta\left(\frac{x_N}{\varepsilon}\right)\right)\right)\theta\left(\frac{|x'|}{\varepsilon}\right)\right|_{x_N=0} \cdot \phi(x', 0)$$

$$+ \left.\left(1 - \theta\left(\frac{x_N}{\varepsilon}\right)\theta\left(\frac{|x'|}{\varepsilon}\right)\right) \cdot \frac{\partial \phi}{\partial x_N}(x', x_N)\right|_{x_N=0}$$

$$= -\frac{1}{\varepsilon}\theta'(0)\theta\left(\frac{|x'|}{\varepsilon}\right) \cdot \phi(x', 0) + \left(1 - \phi\left(\frac{|x'|}{\varepsilon}\right)\right)\frac{\partial \phi}{\partial \mathbf{n}}$$

$$= 0 \quad \text{on } \partial D.$$

This proves that $f \in D(\mathfrak{A}_N)$.

Moreover, we have the following three assertions (α), (β) and (γ) (see Figure 15.11 below):

(α) Since $f(x) = 0$ in the $\varepsilon/2$ neighborhood $U_{\varepsilon/2}(x_0)$ of x_0, it follows that

$$f(y) = \mathfrak{A}_N f(y) = 0 \quad \text{for all } y \in U_{\varepsilon/2}(x_0).$$

This proves that condition (15.14c) in Theorem 15.9 is satisfied.

(β) It is clear that $f(x) \geq 0$ on $\overline{D}$, so that condition (15.14a) is satisfied.

(γ) Since $f(y) = \phi(y)$ outside the $\sqrt{N}\,\varepsilon$ neighborhood $U_{\sqrt{N}\,\varepsilon}(x_0)$ of x_0, it follows from Claim 15.10 that

$$f(y) > 0 \quad \text{for all } y \in \overline{D} \setminus U_{\sqrt{N}\,\varepsilon}(x_0).$$

This proves that condition (15.14b) is satisfied.

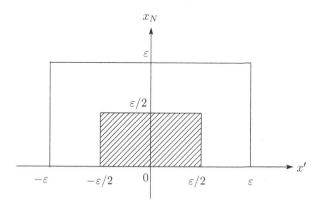

Fig. 15.11. The $\varepsilon/2$ neighborhood $U_{\varepsilon/2}(x_0)$ of $x_0 = 0$

Therefore, by applying Theorem 15.9 to the operator $\mathfrak{A}_N$ we obtain the following theorem:

Theorem 15.11 (the Neumann case). *The strong Markov process $\{x_t(\omega)\}$ associated with the Feller semigroup $\{S_t\}_{t \geq 0}$ enjoys the property*

$$P(\{\omega : x_t(\omega) \text{ is continuous for all } t \geq 0\}) = 1.$$

Since $\{S_t\}_{t \geq 0}$ is a Feller semigroup on the compact set $\overline{D}$, it follows from an application of Theorem 3.31 that there exists a uniformly stochastically continuous, Feller transition function $p_t(x, \cdot)$ on $\overline{D}$ such that the formula

$$S_t f(x) = \int_D p_t(x, dy) f(y) \tag{15.24}$$

holds true for all $f \in C(\overline{D})$.

Furthermore, we can prove the following important proposition:

Proposition 15.12. *The Feller transition function $p_t(x, \cdot)$ is conservative, that is, we have, for all $t > 0$,*

$$p_t(x, \overline{D}) = 1 \quad \text{for each } x \in \overline{D}. \tag{15.25}$$

Proof. First, we remark that

$$\begin{cases} 1 \in D(\mathfrak{A}_N), \\ \mathfrak{A}_N 1 = 0 \quad \text{in } D. \end{cases}$$

Hence, by applying the existence and uniqueness theorem for the initial-value problem with the semigroup S_t (cf. [115, Theorem 4.3]) we obtain that the function $u(t) = S_t 1$ is a unique solution of the initial-value problem

$$\begin{cases} \frac{du}{dt} = \mathfrak{A}_N u \quad \text{for all } t > 0, \\ u(0) = 1 \end{cases} \tag{$*$}$$

which satisfies the following three conditions (1), (2) and (3):

(1) The function $u(t)$ is continuously differentiable for all $t > 0$.
(2) $\|u(t)\| \leq 1$ for all $t \geq 0$.
(3) $u(t) \to 1$ as $t \downarrow 0$.

However, it is easy to see that the function $u(t) \equiv 1$ is also a solution of problem $(*)$ which satisfies three conditions (1), (2) and (3).

Therefore, it follows that

$$S_t 1 = 1 \quad \text{for all } t > 0.$$

In view of formula (15.24), we obtain the desired assertion (15.25) as follows:

$$1 = S_t 1(x) = \int_{\overline{D}} p_t(x, dy) = p_t(x, \overline{D}) \quad \text{for each } x \in \overline{D}.$$

The proof of Proposition 15.12 is complete. $\square$

15.4.2 The Robin Case

Secondly, we consider the *Robin boundary condition*

$$\mathcal{B}_R u = \mu(x') \frac{\partial u}{\partial \mathbf{n}} + \gamma(x') u = 0 \quad \text{on } \partial D, \tag{15.26}$$

where

$$\begin{cases} \mu \in C^\infty(\partial D), \quad \mu(x') > 0 \quad \text{on } \partial D, \\ \gamma \in C^\infty(\partial D), \quad \gamma(x') \leq 0 \quad \text{on } \partial D. \end{cases}$$

We introduce a linear operator $\mathfrak{A}_R$ as follows:

(a) The domain $D(\mathfrak{A}_R)$ is the space

$$D(\mathfrak{A}_R) = \left\{ u \in C(\overline{D}) : \Delta u \in C(\overline{D}), \; \mu(x') \frac{\partial u}{\partial \mathbf{n}} + \gamma(x') u = 0 \text{ on } \partial D \right\}.$$

(b) $\mathfrak{A}_R f = \Delta f$ for every $f \in D(\mathfrak{A}_R)$.

However, since $\mu(x') > 0$ on ∂D, by letting

$$\tilde{\gamma}(x') = \frac{\gamma(x')}{\mu(x')}$$

we find that the boundary condition 15.25 is equivalent to the following:

$$\frac{\partial f}{\partial \mathbf{n}} + \tilde{\gamma}(x')f = 0 \quad \text{on } \partial D.$$

In other words, without loss of generality we may assume that

$$\mu(x') = 1 \quad \text{on } \partial D,$$
$$\tilde{\gamma}(x') \leq 0 \quad \text{on } \partial D.$$

Then it follows from an application of Theorem 1.6 with $L := \mathcal{B}_R$ that the operator $\mathfrak{A}_R$ is the infinitesimal generator of a Feller semigroup $\{T_t\}_{t\geq 0}$ with Robin boundary condition. Let $\{y_t(\omega)\}$ be the strong Markov process corresponding to the Feller semigroup $\{T_t\}_{t\geq 0}$. In this subsection we study the path-continuity of the Markov process $\{y_t(\omega)\}$. To do this, we shall make use of Theorem 15.9.

(I) The case where x_0 is an arbitrary (interior) point of D: By applying Theorem [121, Theorem 1.5] with $\mu(x') := 1$ and $\gamma(x') := \tilde{\gamma}(x')$, we can find a function $\phi \in C^\infty(\overline{D})$ such that

$$\begin{cases} \Delta\psi = -1 & \text{in } D, \\ \frac{\partial\psi}{\partial\mathbf{n}} + \tilde{\gamma}(x')\psi = 0 & \text{on } \partial D. \end{cases} \tag{15.27}$$

Then we have the following claim:

Claim 15.13. *The function satisfies the condition*

$$\psi(x) > 0 \quad \text{on } \overline{D}.$$

Proof. The proof is based on a reduction to absurdity. We assume, to the contrary, that

$$\min_{\overline{D}} \psi \leq 0.$$

(1) First, we consider the case: There exists a point $x_0 \in D$ such that

$$\psi(x_0) = \min_{\overline{D}} \psi \leq 0.$$

Since we have, by condition (15.27),

$$\Delta(-\psi) = 1 > 0 \quad \text{in } D, \tag{15.28}$$

it follows from an application of the strong maximum principle (Theorem 10.7) with $W := \Delta$ and $u := -\psi$ that

$$\psi(x) \equiv \psi(x_0) \quad \text{in } D.$$

Hence we have the assertion

$$\Delta(-\psi) = 0 \quad \text{in } D.$$

This contradicts condition (15.28).

(2) Next we consider the case: There exists a point $x_0' \in \partial D$ such that

$$\psi(x_0') = \min_{\overline{D}} \psi \leq 0.$$

Then, by applying the boundary point lemma (Lemma 10.11) with $W := \Delta$ to the function $u := -\psi$ we obtain from inequality (15.28) that

$$\frac{\partial \psi}{\partial \mathbf{n}}(x_0') > 0.$$

This contradicts the boundary condition

$$0 = \frac{\partial \psi}{\partial \mathbf{n}}(x_0') + \widetilde{\gamma}(x_0')\psi(x_0') \geq \frac{\partial \psi}{\partial \mathbf{n}}(x_0') > 0 \quad \text{on } \partial D,$$

since $\widetilde{\gamma}(x_0') \leq 0$ on ∂D.

The proof of Claim 15.13 is complete. $\square$

Now we choose a real-valued, smooth function $\theta \in C^{\infty}(\mathbf{R})$ that satisfies condition (15.22), and define a function $f \in C^{\infty}(\overline{D})$ by the formula

$$f(x) = \left(1 - \theta\left(\frac{|x - x_0|}{\varepsilon}\right)\right)\psi(x),$$

where

$$0 < \varepsilon < \text{dist}\,(x, \partial D).$$

Then, in view of Claim 15.13 it is easy to verify that $f \in D(\mathfrak{A}_R)$ and satisfies conditions (15.14) of Theorem 15.9.

(II) The case where x_0 is an arbitrary (boundary) point of ∂D: By change of coordinates, we may assume that

$$\begin{cases} x_0 = (0', 0) = (0, \ldots, 0, 0), \\ \mathbf{n} = (0', 1) = (0, \ldots, 0, 1) \in \mathbf{R}^N. \end{cases} \tag{15.23}$$

If we define a function $f \in C^{\infty}(\overline{D})$ by the formula

$$f(x', x_N) = \left(1 - \theta\left(\frac{x_N}{\varepsilon}\right)\theta\left(\frac{|x'|}{\varepsilon}\right)\right)\psi(x', x_N),$$

it follows from conditions (15.22) and (15.27) that

$$
\begin{aligned}
&\frac{\partial f}{\partial \mathbf{n}} + \widetilde{\gamma}(x')f \\
&= \frac{\partial}{\partial x_N}\left(1 - \theta\left(\frac{x_N}{\varepsilon}\right)\theta\left(\frac{|x'|}{\varepsilon}\right)\right)\psi(x', x_N)\Big|_{x_N=0} \\
&\quad + \widetilde{\gamma}(x')\left(1 - \theta\left(\frac{x_N}{\varepsilon}\right)\theta\left(\frac{|x'|}{\varepsilon}\right)\right)\psi\Big|_{x_N=0} \\
&= -\frac{1}{\varepsilon}\theta'(0)\theta\left(\frac{|x'|}{\varepsilon}\right)\cdot\psi(x',0) + \left(1 - \theta(0)\theta\left(\frac{|x'|}{\varepsilon}\right)\right)\frac{\partial\psi}{\partial x_N}(x', x_N)\Big|_{x_N=0} \\
&\quad + \widetilde{\gamma}(x')\left(1 - \theta(0)\theta\left(\frac{|x'|}{\varepsilon}\right)\right)\psi(x',0) \\
&= \left(1 - \theta\left(\frac{|x'|}{\varepsilon}\right)\right)\frac{\partial\psi}{\partial x_N}(x', x_N)\Big|_{x_N=0} + \left(1 - \theta\left(\frac{|x'|}{\varepsilon}\right)\right)\widetilde{\gamma}(x')\psi(x',0) \\
&= \left(1 - \theta\left(\frac{|x'|}{\varepsilon}\right)\right)\left(\frac{\partial\psi}{\partial \mathbf{n}} + \widetilde{\gamma}(x')\psi\right) \\
&= 0 \quad \text{on } \partial D.
\end{aligned}
$$

This proves that $f \in D(\mathfrak{A}_R)$. Moreover, we have the following three assertions (1), (2) and (3) (see Figures 15.9 and 15.11):

(1) Since $f(x) = 0$ in the $\varepsilon/2$ neighborhood $U_{\varepsilon/2}(x_0)$ of x_0, it follows from Claim 15.13 that

$$f(y) = \mathfrak{A}_R f(y) = 0 \quad \text{for all } y \in U_{\varepsilon/2}(x_0).$$

This proves that condition (15.14c) in Theorem 15.9 is satisfied.

(2) It is clear that $f(x) \geq 0$ on $\overline{D}$, so that condition (15.14a) is satisfied.
(3) Since $f(y) = \psi(y)$ outside the $\sqrt{N}\,\varepsilon$ neighborhood $U_{\sqrt{N}\,\varepsilon}(x_0)$ of x_0, it follows that

$$f(y) > 0 \quad \text{for all } y \in \overline{D}\setminus U_{\sqrt{N}\,\varepsilon}(x_0).$$

This proves that condition (15.14b) is satisfied.

Therefore, by applying Theorem 15.9 to the operator $\mathfrak{A}_R$ we obtain the following theorem:

Theorem 15.14 (the Robin case). *The strong Markov process $\{y_t(\omega)\}$ associated with the Feller semigroup $\{T_t\}_{t\geq 0}$ enjoys the property*

$$P\left(\{\omega : y_t(\omega) \text{ is continuous for all } t \geq 0\}\right) = 1.$$

15.4.3 The Oblique Derivative Case

Finally, we consider the *oblique derivative boundary condition*

$$\mathcal{B}_O u = \frac{\partial u}{\partial \mathbf{n}} + \beta(x') \cdot u + \gamma(x')u = 0 \quad \text{on } \partial D.$$

where $\mathbf{n}$ is the unit inward normal vector to the boundary ∂D,

$$\gamma \in C^\infty(\partial D), \quad \gamma(x') \le 0 \text{ on } \partial D,$$

and

$$\beta(x') \cdot u := \sum_{i=1}^{N-1} \beta_i(x') \frac{\partial u}{\partial x_i}$$

is a tangent vector field of class C^1 on the boundary ∂D.

We introduce a linear operator $\mathfrak{A}_O$ as follows:

(a) The domain $D(\mathfrak{A}_O)$ is the space

$$D(\mathfrak{A}_O) = \left\{ u \in C(\overline{D}) : \Delta u \in C(\overline{D}), \frac{\partial u}{\partial \mathbf{n}} + \beta(x') \cdot u + \gamma(x')u = 0 \text{ on } \partial D \right\}.$$

(b) $\mathfrak{A}_O u = \Delta u$ for every $u \in D(\mathfrak{A}_O)$.

Then it follows from an application of Theorem 1.6 with $L := \mathcal{B}_O$ that the operator $\mathfrak{A}_O$ is the infinitesimal generator of a Feller semigroup $\{U_t\}_{t \ge 0}$ with oblique derivative boundary condition. Let $\{z_t(\omega)\}$ be the strong Markov process corresponding to the Feller semigroup $\{U_t\}_{t \ge 0}$. In this subsection we study the path-continuity of the Markov process $\{z_t(\omega)\}$. To do this, we shall make use of Theorem 15.9.

If we introduce a vector filed ℓ by the formula

$$\ell = \mathbf{n} + \beta,$$

then the oblique derivative boundary condition

$$\mathcal{B}_O u = \frac{\partial u}{\partial \mathbf{n}} + \beta(x') \cdot u + \gamma(x')u = 0 \quad \text{on } \partial D$$

can be written in the form (see Figure 15.12 below)

$$\frac{\partial u}{\partial \ell} + \gamma(x')u = 0 \quad \text{on } \partial D.$$

Furthermore, for each initial point $y' = (y_1, y_2, \ldots, y_{N-1}) \in \mathbf{R}^{N-1}$ we consider the following initial-value problem for ordinary differential equations:

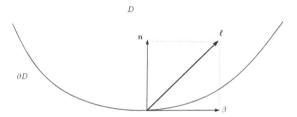

Fig. 15.12. The oblique derivative boundary condition $\frac{\partial u}{\partial \ell}$

$$
\begin{cases}
\frac{d\eta_1}{dt} = \beta_1(\eta(y',t)), & \eta_1(y',0) = y_1, \\
\vdots & \vdots \\
\frac{d\eta_{N-1}}{dt} = \beta_{N-1}(\eta(y',t)), & \eta_{N-1}(y',0) = y_{N-1}, \\
\frac{d\eta_N}{dt} = 1, & \eta_N(y',0) = 0.
\end{cases}
\tag{15.29}
$$

We remark that the initial-value problem (15.29) has a unique local solution

$$
\eta(y',t) = (\eta_1(y',t), \eta_2(y',t), \ldots, \eta_N(y',t)),
$$

since the vector filed $\beta(x')$ is Lipschitz continuous. Hence we can introduce a new change of variables by the formula

$$
\begin{aligned}
x = (x_1, x_2, \ldots, x_N) &= (\eta_1(y', y_N), \eta_2(y', y_N), \ldots, \eta_N(y', y_N)) \\
&= \eta(y) \quad \text{for } y = (y', y_N) \in \mathbf{R}^N.
\end{aligned}
$$

The situation may be represented schematically by Figure 15.13 below.

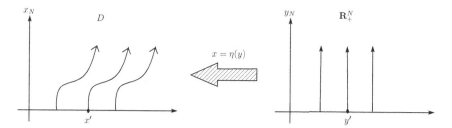

Fig. 15.13. The change of variables $x = \eta(y)$

Then it follows that the vector field ℓ can be simplified as follows:

$$
\frac{\partial}{\partial y_N} = \sum_{j=1}^{N-1} \frac{\partial x_j}{\partial y_N} \frac{\partial}{\partial x_j} + \frac{\partial x_N}{\partial y_N} \frac{\partial}{\partial x_N} = \sum_{j=1}^{N-1} \beta_j(x) \frac{\partial}{\partial x_j} + \frac{\partial}{\partial x_N} = \frac{\partial}{\partial \ell}.
$$

Hence we obtain that

$$\frac{\partial u}{\partial \mathbf{n}} + \beta(x') \cdot u + \gamma(x')u = \frac{\partial u}{\partial \ell} + \gamma(x')u = \left(\frac{\partial \widetilde{u}}{\partial y_N} + \widetilde{\gamma}(y')\widetilde{u} \right) \Bigg|_{y_N = 0},$$

where

$$\begin{cases} \widetilde{u}(y) = u(x), \\ \widetilde{\gamma}(y') = \gamma(x'). \end{cases}$$

In this way, we are reduced to the Robin boundary condition case in the half-space $\mathbf{R}_+^N$:

$$\mathcal{B}_R \widetilde{u} = \frac{\partial \widetilde{u}}{\partial y_N} + \widetilde{\gamma}(y')\widetilde{u} = 0 \quad \text{on } \partial \mathbf{R}_+^N.$$

Therefore, by applying Theorem 15.9 to the operator $\mathfrak{A}_O$ we obtain the following theorem:

Theorem 15.15 (the oblique derivative case). *The strong Markov process $\{z_t(\omega)\}$ associated with the Feller semigroup $\{U_t\}_{t \geq 0}$ enjoys the property*

$$P\left(\{\omega : \,_t(\omega) \text{ is continuous for all } t \geq 0\}\right) = 1.$$

15.5 Notes and Comments

This chapter is devoted to the study of path functions of Markov processes via the Hille-Yosida semigroup theory. The material is taken from Dynkin [34, Chapter V], Lamperti [74, Chapter 8] and Taira [122, Chapter 12].

Section 15.1: Theorem 15.3 is due to Lamperti [74, Chapter 8, Section 2] and Theorem 15.5 is due to Lamperti [74, Section 8.3, Theorem 1], respectively.

Section 15.2: Theorem 15.6 is adapted from Lamperti [74, Section 8.3, Theorem 2] and Corollary 15.7 is adapted from Dynkin [34, Theorem 3.5] (see also [74, Section 8.3, Corollary]).

Section 15.3: Theorem 15.9 is taken from Dynkin [34, Theorem 3.9] and Lamperti [74, Section 8.3, Theorem 3].

Section 15.4: Our functional analytic proof of Theorem 15.11 (the Neumann case), Theorem 15.14 (the Robin case) and Theorem 15.15 (the oblique derivative case) may be new.

Part VI

Concluding Remarks

16

The State-of-the-Art of Generation Theorems for Feller Semigroups

This book is devoted to a concise and accessible exposition of the functional analytic approach to the problem of construction of strong Markov processes with Ventcel' boundary conditions in probability. More precisely, we prove that there exists a Feller semigroup corresponding to such a diffusion phenomenon that a Markovian particle moves continuously in the state space $\overline{D} \setminus M$ until it "dies" at the time when it reaches the set M where the particle is definitely absorbed (see Figure 16.1 below). Our approach here is distinguished by the extensive use of the ideas and techniques characteristic of the recent developments in the theory of pseudo-differential operators which may be considered as a modern version of the classical potential theory.

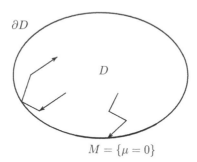

Fig. 16.1. A Markovian particle dies when it reaches the set M

More generally, it is known (see [15], [98], [114], [122], [143]) that the infinitesimal generator $\mathfrak{W}$ of a Feller semigroup $\{T_t\}_{t \geq 0}$ is described analytically by a Waldenfels integro-differential operator W and a Ventcel' boundary condition L, which we formulate precisely.

© Springer Nature Switzerland AG 2020
K. Taira, *Boundary Value Problems and Markov Processes*, Lecture Notes in Mathematics 1499,
https://doi.org/10.1007/978-3-030-48788-1_16

16.1 Formulation of the Problem

Now let D be a bounded domain of Euclidean space $\mathbf{R}^N$, with smooth boundary ∂D; its closure $\overline{D} = D \cup \partial D$ is an N-dimensional, compact smooth manifold with boundary. In this chapter we consider a second-order, *elliptic* integro-differential operator W with real coefficients such that

$$Wu(x) \tag{16.1}$$
$$= Au(x) + Su(x)$$

$$:= \left(\sum_{i,j=1}^N a^{ij}(x) \frac{\partial^2 u}{\partial x_i \partial x_j}(x) + \sum_{i=1}^N b^i(x) \frac{\partial u}{\partial x_i}(x) + c(x)u(x) \right)$$

$$+ \left(d(x)u(x) + \sum_{i=1}^N e^i(x) \frac{\partial u}{\partial x_i}(x) + \int_D s(x,y) \left[u(y) \right. \right.$$

$$\left. \left. - \sum_{\alpha=1}^\ell \sigma_\alpha(x,y) \left(u(x) + \sum_{j=1}^N (\chi_j^\alpha(y) - \chi_j^\alpha(x)) \frac{\partial u}{\partial \chi_j^\alpha}(x) \right) \right] dy \right).$$

Here:

(1) $a^{ij}(x) \in C^\infty(\mathbf{R}^N)$, $a^{ij}(x) = a^{ji}(x)$ for $x \in \mathbf{R}^N$ and $1 \le i, j \le N$, and there exists a constant $a_0 > 0$ such that

$$\sum_{i,j=1}^N a^{ij}(x)\xi_i\xi_j \ge a_0|\xi|^2 \quad \text{for all } (x,\xi) \in \mathbf{R}^N \times \mathbf{R}^N.$$

(2) $b^i(x) \in C^\infty(\mathbf{R}^N)$ for $1 \le i \le N$.
(3) $(A1)(x) = c(x) \in C^\infty(\mathbf{R}^N)$ and $c(x) \le 0$ and $c(x) \not\equiv 0$ in D.
(4) $d(x) \in C^\infty(\mathbf{R}^N)$.
(5) $e^i(x) \in C^\infty(\mathbf{R}^N)$ for $1 \le i \le N$.
(6) The integral kernel $s(x,y)$ is the distribution kernel of a properly supported, pseudo-differential operator $S \in L_{1,0}^{2-\kappa}(\mathbf{R}^N)$ with $\kappa > 0$, which has the *transmission property* with respect to ∂D (see Subsection 5.2), and $s(x,y) \ge 0$ off the diagonal $\{(x,x) : x \in \mathbf{R}^N\}$ in the product space $\mathbf{R}^N \times \mathbf{R}^N$. The measure dy is the Lebesgue measure on $\mathbf{R}^N$.
(7) Let $\{(U_\alpha, \chi_\alpha)\}_{\alpha=1}^\ell$ be a finite open covering of $\overline{D}$ by local charts, and let $\{\sigma_\alpha\}_{\alpha=1}^\ell$ be a family of functions in $C^\infty(\overline{D} \times \overline{D})$ (see [41, Proposition (8.15)]) such that

$$\operatorname{supp} \sigma_\alpha \subset U_\alpha \times U_\alpha,$$

and

$$\sum_{\alpha=1}^\ell \sigma_\alpha(x,y) = 1$$

in a neighborhood of the diagonal $\Delta_{\overline{D}} = \{(x,x) : x \in \overline{D}\}$.

(8) The function $(S1)(x) \in C^{\infty}(\mathbf{R}^N)$ satisfies the condition

$$(S1)(x) = d(x) + \int_D s(x,y)\left[1 - \sum_{\alpha=1}^{\ell} \sigma_\alpha(x,y)\right]dy \leq 0 \quad \text{in } D. \qquad (16.2)$$

The operator W given by formula (16.1) is called a second-order, *Waldenfels integro-differential operator* (cf. [140], [15]). The differential operator A is called a *diffusion operator* which describes analytically a strong Markov process with continuous paths (diffusion process) in the interior D. In fact, we remark that the differential operator A is *local*, that is, the value $Au(x^0)$ at an interior point $x^0 \in D$ is determined by the values of u in an arbitrary small neighborhood of x^0. Moreover, it is known from Peetre's theorem ([88]) that a linear operator is local if and only if it is a differential operator. Therefore, we have an *assurance* of the following assertion:

The infinitesimal generator $\mathfrak{A}$ of a Feller semigroup $\{T_t\}_{t\geq0}$ on the state space $\overline{D}$ is a differential operator in the interior D of $\overline{D}$ if the paths of its corresponding Markov process are continuous.

The operator S is called a second-order, *Lévy integro-differential operator* which is supposed to correspond to the jump phenomenon in the interior D; a Markovian particle moves by jumps to a random point, chosen with kernel $s(x,y)$, in the interior D. Therefore, the Waldenfels integro-differential operator W is supposed to correspond to such a diffusion phenomenon that a Markovian particle moves both by jumps and continuously in the state space D (see Figure 16.2 below).

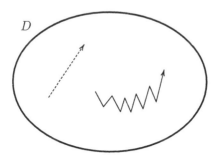

Fig. 16.2. A Markovian particle moves both by jumps and continuously in D

Intuitively, condition (16.2) implies that the jump phenomenon from a point $x \in D$ to the outside of a neighborhood of x in D is "dominated" by the absorption phenomenon at x.

We remark that the integral operator

$$S_r u(x) = \int_D s(x,y)\left[u(y)\right.$$

$$-\sum_{\alpha=1}^{\ell} \sigma_\alpha(x,y)\left(u(x) + \sum_{j=1}^{N}(\chi_j^\alpha(y) - \chi_j^\alpha(x))\frac{\partial u}{\partial \chi_j^\alpha}(x)\right)\Bigg] dy$$

is a "regularization" of S, since the integrand is absolutely convergent. Indeed, it suffices to note (see Theorem 4.51) that, for any compact $K \subset \mathbf{R}^N$, there exists a constant $C_K > 0$ such that the distribution kernel $s(x,y)$ of $S \in L_{1,0}^{2-\kappa}(\mathbf{R}^N)$ with $\kappa > 0$, satisfies the estimate

$$|s(x,y)| \le \frac{C_K}{|x-y|^{N+2-\kappa}} \quad \text{for all } x,\, y \in K \text{ and } x \ne y.$$

We give a simple example of S for $\kappa = 3$:

Example 16.1. The symbol $p(x,\xi)$ of $S \in L_{1,0}^{-1}(\mathbf{R}^N)$ is given by the formula

$$p(x,\xi) = \frac{1}{\langle\xi'\rangle} \exp\left[-\frac{\xi_N^2}{\langle\xi'\rangle^2}\right],$$

where

$$\xi = (\xi',\xi_N), \quad \xi' = (\xi_1,\xi_2,\dots,\xi_{N-1}), \quad \langle\xi'\rangle = \sqrt{1 + |\xi'|^2}.$$

It is easy to see that the distribution kernel $s(x,y)$ of S is equal to the following (see Example 4.1, (1)):

$$s(x,y) = \frac{1}{2\pi^{N/2}}\frac{1}{|x_N - y_N|^{N-1}} \exp\left[-\frac{|x_N - y_N|^2}{4} - \frac{|x' - y'|^2}{|x_N - y_N|^2}\right],$$

where

$$x = (x',x_N) \ne y = (y',y_N),$$
$$x' = (x_1,x_2,\dots,x_{N-1}), \quad y' = (y_1,y_2,\dots,y_{N-1}).$$

Let L be a second-order, boundary condition such that, in terms of local coordinates $(x_1,x_2,\dots,x_{N-1})$ on ∂D,

$$Lu(x') = \Gamma u(x') - \delta(x')\,Wu(x') = \Lambda u(x') + Tu(x') - \delta(x')\,Wu(x') \quad (16.3)$$
$$= Qu(x') + \mu(x')\frac{\partial u}{\partial \mathbf{n}}(x') - \delta(x')\,Wu(x') + Tu(x')$$
$$:= \left(\sum_{i,j=1}^{N-1} \alpha^{ij}(x')\frac{\partial^2 u}{\partial x_i \partial x_j}(x') + \sum_{i=1}^{N-1} \beta^i(x')\frac{\partial u}{\partial x_i}(x') + \gamma(x')u(x')\right)$$
$$+ \mu(x')\frac{\partial u}{\partial \mathbf{n}}(x') - \delta(x')\,Wu(x') + \left(\eta(x')u(x') + \sum_{i=1}^{N-1} \zeta^i(x')\frac{\partial u}{\partial x_i}(x')\right)$$
$$+ \int_{\partial D} r(x',y')\Big[u(y')$$

$$- \sum_{\beta=1}^{m} \tau_\beta(x',y') \left(u(x') + \sum_{j=1}^{N-1} (\chi_j^\beta(y) - \chi_j^\beta(x')) \frac{\partial u}{\partial x_j}(x')) \right) dy'$$

$$+ \int_D t(x',y) \left[u(y) \right.$$

$$\left. - \sum_{\beta=1}^{m} \tau_\beta(x',y)(u(x') + \sum_{j=1}^{N-1} (\chi_j^\beta(y) - \chi_j^\beta(x')) \frac{\partial u}{\partial x_j^\beta}(x')) \right] dy \right).$$

Here:

(1) The operator Q is a second-order, degenerate elliptic differential operator on ∂D with non-positive principal symbol. In other words, the a^{ij} are the components of a smooth symmetric contravariant tensor of type $\binom{2}{0}$ on ∂D satisfying the condition

$$\sum_{i,j=1}^{N-1} a^{ij}(x')\xi_i\xi_j \geq 0$$

for all $x' \in \partial D$ and $\xi' = \sum_{j=1}^{N-1} \xi_j dx_j \in T_{x'}^*(\partial D)$.

Here $T_{x'}^*(\partial D)$ is the cotangent space of ∂D at x'.

(2) $(Q1)(x') = \gamma(x') \in C^\infty(\partial D)$ and $\gamma(x') \leq 0$ on ∂D.

(3) $\mu(x') \in C^\infty(\partial D)$ and $\mu(x') \geq 0$ on ∂D.

(4) $\delta(x') \in C^\infty(\partial D)$ and $\delta(x') \geq 0$ on ∂D.

(5) $\mathbf{n} = (n_1, n_2, \ldots, n_N)$ is the unit inward normal to the boundary ∂D (see Figure 16.1).

(6) The integral kernel $r(x',y')$ is the distribution kernel of a pseudo-differential operator $R \in L_{1,0}^{2-\kappa_1}(\partial D)$ with $\kappa_1 > 0$, and $r(x',y') \geq 0$ off the diagonal $\Delta_{\partial D} = \{(x',x') : x' \in \partial D\}$ in the product space $\partial D \times \partial D$. The density dy' is a strictly positive density on ∂D.

(7) $\eta(x') \in C^\infty(\partial D)$.

(8) $\zeta(x') = \sum_{i=1}^{N-1} \zeta^i(x')\partial/(\partial x_i)$ is a smooth tangent vector field on ∂D.

(9) The integral kernel $t(x,y)$ is the distribution kernel of a properly supported, pseudo-differential operator $T \in L_{1,0}^{2-\kappa_2}(\mathbf{R}^N)$ with $\kappa_2 > 0$, which has the transmission property with respect to the boundary ∂D, and $t(x,y) \geq 0$ off the diagonal $\{(x,x) : x \in \mathbf{R}^N\}$ in the product space $\mathbf{R}^N \times \mathbf{R}^N$.

(10) Let $\{(U_\beta, \chi_\beta)\}_{\beta=1}^m$ be a finite open covering of $\overline{D}$ by local charts, and let $\{\tau_\beta\}_{\beta=1}^m$ be a family of functions in $C^\infty(\overline{D} \times \overline{D})$ (see [41, Proposition (8.15)]) such that

$$\operatorname{supp} \tau_\beta \subset U_\beta \times U_\beta,$$

and

$$\sum_{\beta=1}^{m} \tau_\beta(x,y) = 1$$

in a neighborhood of the diagonal $\Delta_{\overline{D}} = \{(x, x) : x \in \overline{D}\}$.

(11) The fundamental hypothesis concerning $T \in L_{1,0}^{2-\kappa_2}(\mathbf{R}^N)$ is formulated as follows (see [15, p. 436, Section II.2.3]): The function

$$\int_D t(x', y) \sum_{\beta=1}^{m} \tau_\beta(x', y) \left[\chi_N^\beta(y) + \sum_{j=1}^{N-1} (\chi_j^\beta(y) - \chi_j^\beta(x'))^2 \right] dy \qquad (16.4)$$

is continuous on ∂D.

(12) The function $(T1)(x') \in C^\infty(\partial D)$ satisfies the condition

$$(T1)(x') = \eta(x') + \int_{\partial D} r(x', y') \left(1 - \sum_{\beta=1}^{m} \tau_\beta(x', y') \right) dy' \qquad (16.5)$$

$$+ \int_D t(x', y) \left(1 - \sum_{\beta=1}^{m} \tau_\beta(x', y) \right) dy$$

$$\leq 0 \quad \text{on } \partial D.$$

We remark that the integral operator

$$R_r u(x') = \int_{\partial D} r(x', y') \left[u(y') \right.$$

$$\left. - \sum_{\beta=1}^{m} \tau_\beta(x', y') \left(u(x') + \sum_{j=1}^{N-1} (\chi_j^\beta(y) - \chi_j^\beta(x')) \frac{\partial u}{\partial \chi_j^\beta}(x') \right) \right] dy'$$

is a "regularization" of R, since the integrand is absolutely convergent. Indeed, it suffices to note (see Theorem 4.51) that, for any compact neighborhood $U_{x'}$ in ∂D there exists a constant $C_K' > 0$ such that the distribution kernel $r(x', y')$ of R satisfies the estimate

$$|r(x', y')| \leq \frac{C_K'}{|x' - y'|^{N+1-\kappa_1}} \quad \text{for all } y' \in U_{x'} \text{ and } y' \neq x',$$

where $|x' - y'|$ is the geodesic distance between x' and y' with respect to the Riemannian metric of ∂D induced by the natural metric of $\mathbf{R}^N$. The operator R_r is called a second-order, Ventcel'–Lévy boundary operator on the boundary.

We give a simple example of R for $\kappa_1 = 1$:

Example 16.2. Let $R = -\sqrt{-\Delta'} \in L_{1,0}^1(\partial D)$ where Δ' is the Laplace–Beltrami operator on the boundary ∂D. Then its principal symbol is equal to $-|\xi'|$ where $|\xi'|$ is the length of ξ' with respect to the Riemannian metric of ∂D. It is easy to see that the *distribution kernel* $r(x', y')$ of R is given by the formula

$$r(x', y') = \frac{\Gamma(N/2)}{\pi^{N/2}} \frac{1}{|x' - y'|^N} \quad \text{for all } x', y' \in \partial D \text{ and } x' \neq y'.$$

We give a simple example of T that satisfies condition (16.4) for $N = 3$ and $\kappa_2 = 5$:

Example 16.3. The symbol $q(x, \xi)$ of $T \in L_{1,0}^{-3}(\mathbf{R}^3)$ is given by the formula

$$q(x, \xi) = \frac{1}{\langle \xi' \rangle^3}\left(1 - \frac{2\xi_3^2}{\langle \xi' \rangle^2}\right) \exp\left[-\frac{\xi_3^2}{\langle \xi' \rangle^2}\right],$$

where

$$\xi = (\xi', \xi_3), \quad \xi' = (\xi_1, \xi_2), \quad \langle \xi' \rangle = \sqrt{1 + |\xi'|^2}.$$

It is easy to see that the *distribution kernel* $t(x, y)$ of T is equal to the following (see Example 4.1, (1)):

$$t(x, y) = \frac{1}{4\pi^{3/2}} \exp\left[-\frac{|x_3 - y_3|^2}{4} - \frac{|x' - y'|^2}{|x_3 - y_3|^2}\right].$$

Here

$$x = (x', x_3) \neq y = (y', y_3), \quad x' = (x_1, x_2), \quad y' = (y_1, y_2).$$

The boundary condition L given by formula (16.3) is called a second-order, *Ventcel' boundary condition* (cf. [143]). The six terms of L

$$\sum_{i,j=1}^{N-1} \alpha^{ij}(x') \frac{\partial^2 u}{\partial x_i \partial x_j}(x') + \sum_{i=1}^{N-1} \beta^i(x') \frac{\partial u}{\partial x_i}(x'),$$

$$\gamma(x')u(x'), \quad \mu(x')\frac{\partial u}{\partial \mathbf{n}}(x'), \quad \delta(x') \, W u(x'),$$

$$\int_{\partial D} r(x', y')\left[u(y') - \sum_{\beta=1}^{m} \tau_\beta(x, y)\left(u(x') + \sum_{j=1}^{N-1}(y_j - x_j)\frac{\partial u}{\partial x_j}(x')\right)\right] dy',$$

$$\int_{D} t(x', y)\left[u(y) - \sum_{\beta=1}^{m} \tau_\beta(x', y)\left(u(x') + \sum_{j=1}^{N-1}(y_j - x_j)\frac{\partial u}{\partial x_j}(x')\right)\right] dy$$

are supposed to correspond to the diffusion along the boundary, the absorption phenomenon, the reflection phenomenon, the viscosity phenomenon and the jump phenomenon on the boundary and the inward jump phenomenon from the boundary, respectively (see Figures 16.3 through 16.5 below).

Intuitively, condition (16.5) implies that the jump phenomenon from a point $x' \in \partial D$ to the outside of a neighborhood of x' in $\overline{D}$ is "dominated" by the absorption phenomenon at x'.

This chapter is devoted to a functional analytic approach to the problem of construction of Feller semigroups with Ventcel' boundary conditions. More precisely we consider the following problem (see [122, Problem 1.1]):

Problem. Conversely, given analytic data (W, L), can we construct a *Feller semigroup* $\{T_t\}_{t\geq 0}$ whose infinitesimal generator is characterized by (W, L) ?

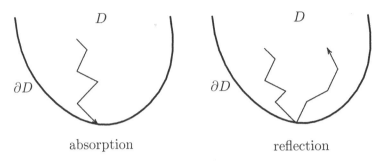

Fig. 16.3. Absorption and reflection phenomena

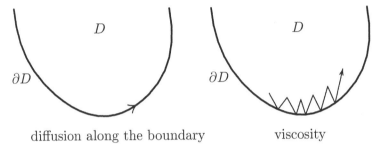

Fig. 16.4. Diffusion along ∂D and viscosity phenomenon

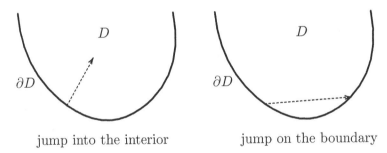

Fig. 16.5. Jump phenomena into D and on ∂D

Table 16.1 below gives a bird's-eye view of Markov processes, Feller semigroups and elliptic boundary value problems and how these relate to each other:

We shall only restrict ourselves to some aspects that have been discussed in the work [114] through [122]. Our approach is distinguished by the extensive use of the ideas and techniques characteristic of the recent developments in the theory of partial differential equations. It focuses on the relationship between two interrelated subjects in analysis; Feller semigroups and elliptic boundary value problems, providing powerful methods for future research.

Probability (Microscope)	Functional Analysis (Macroscope)	Elliptic Boundary Value Problems (Mesoscope)
Markov process $\mathcal{X} = (x_t)$	Feller semigroup $\{T_t\}_{t \geq 0}$	Infinitesimal generators $\mathfrak{W}_0$, $\mathfrak{W}$
Markov transition function $p_t(\cdot, dy)$	$T_t f = \int_{\overline{D}} p_t(\cdot, dy) \, f(y)$	$T_t = e^{t\,\mathfrak{W}_0}$ or $T_t = e^{t\,\mathfrak{W}}$
Chapman and Kolmogorov equation	Semigroup property $T_{t+s} = T_t \cdot T_s$	Waldenfels operator $W = A + S$
Various diffusion phenomena	Function spaces $C\left(\overline{D}\right)$ $C_0\left(\overline{D} \setminus M\right)$	Ventcel' (Wentzell) condition L

Table 16.1. A bird's-eye view of Markov processes, Feller semigroups and boundary value problems (Theorems 16.1 through 16.4)

16.2 Statement of Main Results

16.2.1 The Transversal Case

First, we generalize Theorem 14.1 to the transversal case. We say that the boundary condition L is *transversal* on the boundary ∂D if it satisfies the condition

$$\int_D t(x', y) \, dy = +\infty \quad \text{if } \mu(x') = \delta(x') = 0. \qquad (16.6)$$

The intuitive meaning of condition (16.6) is that a Markovian particle jumps away "instantaneously" from the points $x' \in \partial D$ where neither reflection nor viscosity phenomenon occurs (which is similar to the reflection phenomenon). Probabilistically, this means that every Markov process on the boundary ∂D is the "trace" on ∂D of trajectories of some Markov process on the closure $\overline{D} = D \cup \partial D$ (see Figure 16.6 below).

The next theorem asserts that there exists a Feller semigroup on the state space $\overline{D}$ corresponding to such a diffusion phenomenon that one of the re-

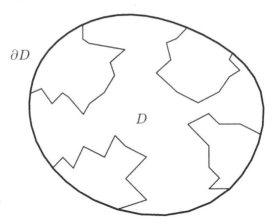

∂D

D

Fig. 16.6. The probabilistic meaning of transversality condition

flection phenomenon, the viscosity phenomenon and the inward jump phenomenon from the boundary occurs at each point of the boundary ∂D (see [122, Theorem 1.2]):

Theorem 16.1. *We define a linear operator*

$$\boxed{\mathfrak{W}_0 : C(\overline{D}) \longrightarrow C(\overline{D})}$$

in the space $C(\overline{D})$ as follows:

(a) The domain of definition $\mathcal{D}(\mathfrak{W}_0)$ is the space

$$\mathcal{D}(\mathfrak{W}_0) = \left\{ u \in C(\overline{D}) : Wu \in C(\overline{D}), \ Lu = 0 \ on \ \partial D \right\}. \qquad (16.7)$$

(b) $\mathfrak{W}_0 u = Wu$ for every $u \in \mathcal{D}(\mathfrak{W}_0)$.
   Here Wu and Lu are taken in the sense of distributions.

   Assume that the boundary condition L is transversal *on the boundary ∂D. Then the operator $\mathfrak{W}_0$ generates a Feller semigroup*

$$\left\{ e^{t\mathfrak{W}_0} \right\}_{t \geq 0}$$

on the state space $\overline{D}$.

   A probabilistic meaning of Theorem 16.1 may be represented schematically by Figure 16.7 below.
   We remark that Theorem 16.1 was proved by Taira [114, Theorem 10.1.3] under some additional conditions, and also by Cancelier [24, Théorème 3.2] and [116, Theorem 1]. On the other hand, Takanobu and Watanabe [130] proved a probabilistic version of Theorem 16.1 in the case where the domain D is the half space $\mathbf{R}_+^N$ (see [130, Corollary]).
   Our functional analytic approach to strong Markov processes in the transversal case may be visualized as in Figure 16.8 below.

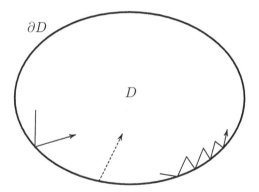

Fig. 16.7. A probabilistic meaning of Theorem 16.1 (transversal case)

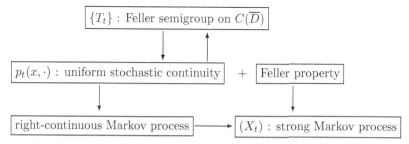

Fig. 16.8. The functional analytic approach to strong Markov processes in the transversal case

16.2.2 The Non-Transversal Case

Secondly, we generalize Theorem 14.8 to the *non-transversal case*. To do so, we assume that:

(G.1) There exists a second-order, Ventcel' boundary condition L_ν such that

$$Lu = m(x') L_\nu u + \gamma(x') u \quad \text{on } \partial D. \tag{16.8}$$

(G.2) $m(x') - \gamma(x') > 0$ on ∂D.

Here:

(2) $\gamma(x') \in C^\infty(\partial D)$ and $\gamma(x') \leq 0$ on ∂D.
(3') $m(x') \in C^\infty(\partial D)$ and $m(x') \geq 0$ on ∂D,
and L_ν is given, in terms of local coordinates $(x_1, x_2, \ldots, x_{N-1})$, by the formula

$$L_\nu u(x') = \overline{\Gamma} u(x') - \overline{\delta}(x') W u(x') = \overline{\Lambda} u(x') + \overline{T} u(x') - \overline{\delta}(x') W u(x')$$
$$= \overline{Q} u(x') + \overline{\mu}(x') \frac{\partial u}{\partial \mathbf{n}}(x') + \overline{T} u(x') - \overline{\delta}(x') W u(x')$$

$$:= \left(\sum_{i,j=1}^{N-1} \overline{\alpha}^{ij}(x') \frac{\partial^2 u}{\partial x_i \partial x_j}(x') + \sum_{i=1}^{N-1} \overline{\beta}^i(x') \frac{\partial u}{\partial x_i}(x') \right)$$

$$+ \overline{\mu}(x') \frac{\partial u}{\partial \mathbf{n}}(x') - \overline{\delta}(x') \, W u(x') + \left(\overline{\eta}(x') u(x') \right.$$

$$+ \sum_{i=1}^{N-1} \overline{\zeta}^i(x') \frac{\partial u}{\partial x_i}(x') + \int_{\partial D} \overline{r}(x', y') \Big[u(y')$$

$$- \sum_{\beta=1}^{m} \tau_\beta(x', y') \left(u(x') + \sum_{j=1}^{N-1} \left(y_j - x_j \right) \frac{\partial u}{\partial x_j}(x') \right) \Big] dy'$$

$$+ \int_D \overline{t}(x', y) \Big[u(y)$$

$$- \sum_{\beta=1}^{m} \tau_\beta(x', y') \left(u(x') + \sum_{j=1}^{N-1} \left(y_j - x_j \right) \frac{\partial u}{\partial x_j}(x') \right) \Big] dy \Big),$$

and it satisfies the condition

$$\overline{\Gamma} 1(x') = \overline{T} 1(x') \tag{16.5'}$$

$$= \overline{\eta}(x') + \int_{\partial D} \overline{r}(x', y') \left(1 - \sum_{\beta=1}^{m} \tau_\beta(x', y') \right) dy'$$

$$+ \int_D \overline{t}(x', y) \left(1 - \sum_{\beta=1}^{m} \tau_\beta(x', y) \right) dy$$

$$\leq 0 \quad \text{on } \partial D.$$

Moreover, we assume the *transversality* condition for L_ν

$$\int_D \overline{t}(x', y) \, dy = +\infty \quad \text{if } \overline{\mu}(x') = \overline{\delta}(x') = 0. \tag{16.6'}$$

We let

$$M = \left\{ x' \in \partial D : \mu(x') = \delta(x') = 0, \int_D t(x', y) \, dy < \infty \right\}.$$

Then, by condition (16.6') it follows that

$$M = \{ x' \in \partial D : m(x') = 0 \},$$

since we have the formulas

$$\mu(x') = m(x') \overline{\mu}(x'), \quad \delta(x') = m(x') \overline{\delta}(x'),$$
$$Q = m(x') \overline{Q}, \quad t(x', y) = m(x') \overline{t}(x', y).$$

Therefore, we find that the boundary condition L, defined by formula (16.8), is *not* transversal on ∂D.

The intuitive meaning of conditions (G.1) and (G.2) is that a Markovian particle does not stay on ∂D for any period of time until it "dies" at the time when it reaches the set M where the particle is definitely absorbed.

Now we introduce a subspace of $C(\overline{D})$ which is associated with the boundary condition L.

By condition (G.2), we find that the boundary condition

$$Lu = m(x')\, L_\nu u + \gamma(x')\, u = 0 \quad \text{on } \partial D$$

includes the condition

$$u = 0 \quad \text{on } M.$$

With this fact in mind, we let

$$C_0\left(\overline{D} \setminus M\right) = \left\{u \in C(\overline{D}) : u = 0 \text{ on } M\right\}.$$

The space $C_0\left(\overline{D} \setminus M\right)$ is a closed subspace of $C(\overline{D})$; hence it is a Banach space.

A strongly continuous semigroup $\{T_t\}_{t \geq 0}$ on the space $C_0\left(\overline{D} \setminus M\right)$ is called a *Feller semigroup* on the state space $\overline{D} \setminus M$ if it is non-negative and contractive on the Banach space $C_0\left(\overline{D} \setminus M\right)$:

$$f \in C_0\left(\overline{D} \setminus M\right) \quad \text{and} \quad 0 \leq f \leq 1 \text{ on } \overline{D} \setminus M$$
$$\implies 0 \leq T_t f \leq 1 \text{ on } \overline{D} \setminus M.$$

We define a linear operator

$$\boxed{\mathfrak{W} : C_0\left(\overline{D} \setminus M\right) \longrightarrow C_0\left(\overline{D} \setminus M\right)}$$

in the space $C_0\left(\overline{D} \setminus M\right)$ as follows:

(a) The domain of definition $\mathcal{D}(\mathfrak{W})$ is the space

$$\mathcal{D}(\mathfrak{W}) \tag{16.9}$$
$$= \left\{u \in C_0\left(\overline{D} \setminus M\right) : Wu \in C_0\left(\overline{D} \setminus M\right),\ Lu = 0 \text{ on } \partial D\right\}.$$

(b) $\mathfrak{W}u = Wu$ for every $u \in \mathcal{D}(\mathfrak{W})$.

The next theorem is a generalization of Theorem 16.1 to the *non-transversal case* (see [122, Theorem 1.3]):

Theorem 16.2. *Assume that conditions (G.1) and (G.2) are satisfied. Then the operator $\mathfrak{W}$ defined by formula (16.9) generates a Feller semigroup*

$$\left\{e^{t\mathfrak{W}}\right\}_{t \geq 0}$$

on the state space $\overline{D} \setminus M$.

Theorem 16.2 was proved by Taira [114, Theorem 2] under some additional conditions. We remark that Taira [116] has proved Theorem 16.2 in the case where $L_\nu = \partial/\partial\mathbf{n}$ and $\delta(x') \equiv 0$ on ∂D, by using the L^p theory of pseudo-differential operators (see [116, Theorem 4]).

If T_t is a Feller semigroup on the state space $\overline{D} \setminus M$, then there exists a unique Markov transition function $p_t(\cdot, dy)$ on $\overline{D} \setminus M$ such that

$$T_t f(x) = \int_{\overline{D}\setminus M} p_t(x, dy) f(y) \quad \text{for all } f \in C_0(\overline{D} \setminus M),$$

and that $p_t(\cdot, dy)$ is the transition function of some strong Markov process. On the other hand, the intuitive meaning of conditions (G.1) and (G.2) is that the absorption phenomenon occurs at each point of the set

$$M = \{x' \in \partial D : m(x') = 0\} \,.$$

Therefore, Theorem 16.2 asserts that there exists a Feller semigroup on the state space $\overline{D}\setminus M$ corresponding to such a diffusion phenomenon that a Markovian particle moves both by jumps and continuously in the state space $\overline{D} \setminus M$ until it "dies" at which time it reaches the set M.

The situation of Theorem 16.2 may be represented schematically by Figure 16.9 below.

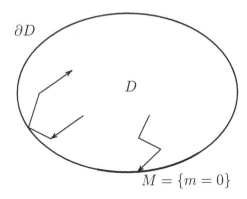

Fig. 16.9. A probabilistic meaning of Theorem 16.2 (the non-transversal case)

Our functional analytic approach to strong Markov processes in the non-transversal case may be visualized as in Figure 16.10 below.

16.2.3 The Lower Order Case

Finally, we consider the case where all the operators S, T and R are pseudo-differential operators of order *less than one*. Then we can take $\sigma(x, y) \equiv 1$ on $\overline{D} \times \overline{D}$, and write the operator W in the form (16.1'):

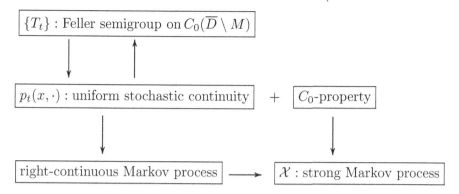

Fig. 16.10. The functional analytic approach to strong Markov processes in the non-transversal case

$$Wu(x) = Au(x) + Su(x) \tag{16.1'}$$

$$:= \left(\sum_{i,j=1}^{N} a^{ij}(x) \frac{\partial^2 u}{\partial x_i \partial x_j}(x) + \sum_{i=1}^{N} b^i(x) \frac{\partial u}{\partial x_i}(x) + c(x)u(x) \right)$$

$$+ \left(d(x)u(x) + \int_D s(x,y)[u(y) - u(x)]dy \right).$$

Here:

(3) $(A1)(x) = c(x) \in C^\infty(\mathbf{R}^N)$ and $c(x) \le 0$ and $c(x) \not\equiv 0$ in D.

(6') The integral kernel $s(x,y)$ is the distribution kernel of a properly supported, pseudo-differential operator $S \in L_{1,0}^{1-\kappa}(\mathbf{R}^N)$ with $\kappa > 0$, which has the transmission property with respect to the boundary ∂D, and $s(x,y) \ge 0$ off the diagonal $\{(x,x) : x \in \mathbf{R}^N\}$ in the product space $\mathbf{R}^N \times \mathbf{R}^N$.

(8') The function $(S1)(x) = d(x) \in C^\infty(\mathbf{R}^N)$ satisfies the condition

$$(S1)(x) = d(x) \le 0 \quad \text{in } D.$$

Similarly, the boundary condition L can be written in the form (16.3'):

$$Lu(x') = \Gamma u(x') - \delta(x') Wu(x') = \Lambda u(x') + Tu(x') - \delta(x') Wu(x') \tag{16.3'}$$

$$= Qu(x') + \mu(x') \frac{\partial u}{\partial \mathbf{n}}(x') - \delta(x') Wu(x') + Tu(x')$$

$$:= \left(\sum_{i,j=1}^{N-1} \alpha^{ij}(x') \frac{\partial^2 u}{\partial x_i \partial x_j}(x') + \sum_{i=1}^{N-1} \beta^i(x') \frac{\partial u}{\partial x_i}(x') + \gamma(x')u(x') \right)$$

$$+ \mu(x') \frac{\partial u}{\partial \mathbf{n}}(x') - \delta(x') Wu(x')$$

$$+ \left(\eta(x')u(x') \right.$$

$$+ \int_{\partial D} r(x', y')[u(y') - u(x')]dy' + \int_{D} t(x', y)[u(y) - u(x')]dy \Big).$$

Here:

(2) $(Q1)(x') = \gamma(x') \in C^\infty(\partial D)$ and $\gamma(x') \leq 0$ on ∂D.

(3) $\mu(x') \in C^\infty(\partial D)$ and $\mu(x') \geq 0$ on ∂D.

(4) $\delta(x') \in C^\infty(\partial D)$ and $\delta(x') \geq 0$ on ∂D.

(6') The integral kernel $r(x', y')$ is the distribution kernel of a pseudo-differential operator $R \in L_{1,0}^{1-\kappa_1}(\partial D)$ with $\kappa_1 > 0$, and $r(x', y') \geq 0$ off the diagonal $\{(x', x') : x' \in \partial D\}$ in the product space $\partial D \times \partial D$.

(9') The integral kernel $t(x, y)$ is the distribution kernel of a properly supported, pseudo-differential operator $T \in L_{1,0}^{1-\kappa_2}(\mathbf{R}^N)$ with $\kappa_2 > 0$, which has the transmission property with respect to the boundary ∂D, and $t(x, y) \geq 0$ off the diagonal $\{(x, x) : x \in \mathbf{R}^N\}$ in the product space $\mathbf{R}^N \times \mathbf{R}^N$.

(12') The function $(T1)(x') = \eta(x') \in C^\infty(\partial D)$ satisfies the condition

$$(T1)(x') = \eta(x') \leq 0 \quad \text{on } \partial D.$$

Then Theorems 16.1 and 16.2 may be simplified as follows (see [122, Theorems 1.4 and 1.5]):

Theorem 16.3. *Assume that the operator W and the boundary condition L are of the forms (16.1') and (16.3'), respectively. If the boundary condition L is* transversal *on the boundary ∂D, then the operator $\mathfrak{W}_0$ defined by formula (16.7) generates a Feller semigroup $\{e^{t \mathfrak{W}_0}\}_{t \geq 0}$ on the state space $\overline{D}$.*

Theorem 16.4. *Assume that the operator W and the boundary condition L are of the forms (16.1') and (16.3'), respectively. If conditions (G.1) and (G.2) are satisfied, then the operator $\mathfrak{W}$ defined by formula (16.9) generates a Feller semigroup $\{e^{t W}\}_{t \geq 0}$ on the state space $\overline{D} \setminus M$.*

Theorems 16.1, 16.2, 16.3 and 16.4 solve from the viewpoint of functional analysis the problem of construction of Feller semigroups with Ventcel' boundary conditions for *elliptic*, Waldenfels integro-differential operators.

Finally, we give an overview of general results on generation theorems for Feller semigroups proved mainly by the author using the theory of pseudo-differential operators in Table 16.2 below ([57], [105], [106]) and the Calderón–Zygmund theory of singular integral operators ([23], [108]):

Here (see Definitions 11.1 and 11.5):

$$W = A + S, \tag{16.1}$$

$$L = \Gamma - \delta(x')W = (\Lambda + T) - \delta(x')W \tag{16.3}$$

$$= \left(\mu(x') \frac{\partial}{\partial \mathbf{n}} + Q + T \right) - \delta(x')W.$$

diffusion operator A	Lévy operator S	Ventcel' condition $L = \Gamma - \delta(x')\,W$	using the theory of	proved by
elliptic smooth case	$S \equiv 0$	$\Lambda = \mu(x')\frac{\partial}{\partial \mathbf{n}} + Q$ $T \equiv 0$	pseudo-differential operators	[114]
elliptic smooth case	convex domain	$\Lambda = \mu(x')\frac{\partial}{\partial \mathbf{n}} + Q$ $Q = \gamma(x')$ (Robin case) $T \equiv 0$	pseudo-differential operators	[117]
elliptic smooth case	general case	general case	pseudo-differential operators	[116] [122]
elliptic discontinuous case	general case	$\mu(x') \equiv 0$ $Q = \gamma(x') \equiv 1$ (Dirichlet case) $T \equiv 0$	singular integral operators	[118] [127]
elliptic discontinuous case	$S \equiv 0$	$\alpha^{ij} \equiv 0$ (oblique derivative) (case) $T \equiv 0$	singular integral operators	[120] [129]

Table 16.2. An overview of generation theorems for Feller semigroups

In [118], [120] and [127], we prove existence theorems for Feller semigroups with Dirichlet boundary condition, oblique derivative boundary condition and first-order Ventcel' boundary condition for second-order, uniformly elliptic differential operators with *discontinuous* coefficients. Our approach there is distinguished by the extensive use of the ideas and techniques characteristic of the recent developments in the Calderón–Zygmund theory of singular integral operators with non-smooth kernels ([28], [29], [79]).

It should be emphasized that the Calderón–Zygmund theory of singular integral operators with non-smooth kernels provides a powerful tool to deal with smoothness of solutions of elliptic boundary value problems, with minimal assumptions of regularity on the coefficients. The theory of singular integrals continues to be one of the most influential works in modern history of analysis, and is a very refined mathematical tool whose full power is yet to be exploited (see [23], [109]).

16.3 Notes and Comments

We state a brief history of the *stochastic analysis* methods for Ventcel' boundary value problems. More precisely, we remark that Ventcel' boundary value problems are studied by Anderson [9], [10], Cattiaux [25] and Takanobu–Watanabe [130] from the viewpoint of stochastic analysis (see also Ikeda–Watanabe [62, Chapter IV, Section 7]).

(I) Anderson [9] and [10] studies the non-degenerate case under low regularity in the framework of the submartingale problem and shows the existence and uniqueness of solutions to the considered submartingale problem.

(II) Takanobu–Watanabe [130] study certain cases of both degenerate interior and boundary operators under minimal assumptions of regularity based on the theory of stochastic differential equations, and they show the existence and uniqueness of solutions. Such existence and uniqueness results on the diffusion processes corresponding to the boundary value problems imply the existence and uniqueness of the associated Feller semigroups on the space of continuous functions.

(III) Cattiaux [25] studies the hypoellipticity for diffusions with Ventcel' boundary conditions. By making use of a variant of the Malliavin calculus under Hömander's type conditions ([58]), he proves that some laws and conditional laws of such diffusions have a smooth density with respect to the Lebesgue measure.

References

[1] Abraham, R., Marsden, J. E. and Ratiu, T.: Manifolds, tensor analysis, and applications, second edition. Applied Mathematical Sciences, Vol. 75, Springer-Verlag, New York (1988)

[2] Adams, R. A. and Fournier J. J.: Sobolev spaces, second edition. Pure and Applied Mathematics, Vol. 140, Elsevier/Academic Press, Amsterdam (2003)

[3] Agmon, S.: Lectures on elliptic boundary value problems. Van Nostrand, Princeton (1965)

[4] Agmon, S., Douglis, A. and Nirenberg, L.: Estimates near the boundary for solutions of elliptic partial differential equations satisfying general boundary conditions I. Comm. Pure Appl. Math., **12**, 623–727 (1959)

[5] Agranovič (Agranovich), M. S. and Dynin, A. S.: General boundary-value problems for elliptic systems in higher-dimensional regions (Russian). Dokl. Akad. Nauk SSSR, **146**, 511–514 (1962)

[6] Agranovich, M. S. and Vishik, M. I.: Elliptic problems with a parameter and parabolic problems of general type (Russian). Uspehi Mat. Nauk **19** (3)(117), 53–161 (1964). English translation: Russian Math. Surveys, **19** (3), 53–157 (1964)

[7] Altomare, F., Cappelletti Montano, M., Leonessa, V. and Raşa, I.: Markov operators, positive semigroups and approximation processes. De Gruyter Studies in Mathematics, vol. 61, Walter de Gruyter, Berlin Munich Boston (2014)

[8] Amann, H.: Fixed point equations and nonlinear eigenvalue problems in ordered Banach spaces. SIAM Rev., **18**, 620–709 (1976)

[9] Anderson, R. F.: Diffusions with second order boundary conditions I, Indiana Univ. Math. J., **25**, 367–397 (1976)

[10] Anderson, R. F.: Diffusions with second order boundary conditions II, Indiana Univ. Math. J., **25**, 403–441 (1976)

[11] Applebaum, D.: Lévy processes and stochastic calculus. Second edition, Cambridge Studies in Advanced Mathematics, No. 116, Cambridge University Press, Cambridge (2009)

[12] Aronszajn, N. and Smith, K. T.: Theory of Bessel potentials I. Ann. Inst. Fourier (Grenoble), **11**, 385–475 (1961)

[13] Bergh, J. and Löfström, J.: Interpolation spaces, an introduction. Springer-Verlag, Berlin Heidelberg New York (1976)

© Springer Nature Switzerland AG 2020
K. Taira, *Boundary Value Problems and Markov Processes*, Lecture Notes in Mathematics 1499,
https://doi.org/10.1007/978-3-030-48788-1

486   References

[14] Blumenthal, R. M. and Getoor, R. K.: Markov processes and potential theory. Academic Press, New York (1968)

[15] Bony, J.-M., Courrège, P. et Priouret, P.: Semigroupes de Feller sur une variété à bord compacte et problèmes aux limites intégro-différentiels du second ordre donnant lieu au principe du maximum. Ann. Inst. Fourier (Grenoble), **18**, 369–521 (1968)

[16] Bourdaud, G.: L^p-estimates for certain non-regular pseudo-differential operators. Comm. Partial Differential Equations, **7**, 1023–1033 (1982)

[17] Boutet de Monvel, L.: Comportement d'un opérateur pseudo-différentiel sur une variété à bord I, La propriété de transmission. J. Analyse Math., **17**, 241–253 (1966)

[18] Boutet de Monvel, L.: Comportement d'un opérateur pseudo-différentiel sur une variété à bord II, Pseudo-noyaux de Poisson. J. Analyse Math., **17**, 255–304 (1966)

[19] Boutet de Monvel, L.: Boundary problems for pseudo-differential operators. Acta Math., **126**, 11–51 (1971)

[20] Brezis, H.: Functional analysis, Sobolev spaces and partial differential equations. Universitext. Springer-Verlag, New York (2011)

[21] Calderón, A. P.: Lebesgue spaces of differentiable functions and distributions. In: Proc. Sym. Pure Math., **X**, Singular integrals, A. P. Calderón (ed.), pp. 33–49, American Mathematical Society, Providence, Rhode Island (1961)

[22] Calderón, A. P.: Boundary value problems for elliptic equations. In: Outlines Joint Sympos. Partial Differential Equations (Novosibirsk, 1963), pp. 303–304. Acad. Sci. USSR Siberian Branch, Moscow.

[23] Calderón, A. P. and Zygmund, A.: On the existence of certain singular integrals. Acta Math., **88**, 85–139 (1952)

[24] Cancelier, C.: Problèmes aux limites pseudo-différentiels donnant lieu au principe du maximum. Comm. Partial Differential Equations, **11**, 1677–1726 (1986)

[25] Cattiaux, P.: Hypoellipticité et hypoellipticité partielle pour les diffusions avec une condition frontière, Ann. Inst. H. Poincaré Probab. Statist., **22**, 67–112 (1986)

[26] Chazarain, J. et Piriou, A.: Introduction à la théorie des équations aux dérivées partielles linéaires. Gauthier-Villars, Paris (1981)

[27] Chen, Z.-Q. and Song, R.: Estimates on Green functions and Poisson kernels for symmetric stable processes, Math. Ann., **312**, 465–501 (1998)

[28] Chiarenza, F., Frasca, M. and Longo, P.: Interior $W^{2,p}$ estimates for nondivergence elliptic equations with discontinuous coefficients, Ricerche Mat., **60**, 149–168 (1991)

[29] Chiarenza, F., Frasca, M. and Longo, P.: $W^{2,p}$- solvability of the Dirichlet problem for nondivergence elliptic equations with VMO coefficients, Trans. Amer. Math. Soc., **336**, 841–853 (1993)

[30] Coifman, R. R. et Meyer, Y.: Au-delà des opérateurs pseudo-différentiels. Astérisque, No. 57, Société Mathématique de France, Paris (1978)

[31] Duistermaat. J. J.: Fourier integral operators. Courant Institute Lecture Notes, New York (1973)

[32] Duistermaat, J. J. and Hörmander, L.: Fourier integral operators II, Acta Math., **128**, 183–269 (1972)

[33] Dynkin, E. B.: Foundations of the theory of Markov processes (Russian). Fiz-matgiz, Moscow (1959). English translation: Pergamon Press, Oxford London New York Paris (1960). German translation: Springer-Verlag, Berlin Göttingen Heidelberg (1961). French translation: Dunod, Paris (1963)

[34] Dynkin, E. B.: Markov processes I, II. Springer-Verlag, Berlin Göttingen Heidelberg (1965)

[35] Dynkin, E. B. and Yushkevich, A. A.: Markov processes, theorems and problems. Plenum Press, New York (1969)

[36] Einstein, A.: Investigations on the theory of the Brownian movement. Dover, New York (1956)

[37] Engel, K.-J. and Nagel, R.: One-parameter semigroups for linear evolution equations. Graduate Texts in Mathematics, Vol. 194, Springer-Verlag, New York Berlin Heidelberg (2000)

[38] Ethier, S. N. and Kurtz, T. G.: Markov processes, characterization and convergence. Wiley series in probability and statistics John Wiley & Sons Inc., Hoboken, New Jersey (2005)

[39] Feller, W.: The parabolic differential equations and the associated semigroups of transformations. Ann. of Math., (2) **55**, 468–519 (1952)

[40] Feller, W.: On second order differential equations. Ann. of Math., (2) **61**, 90–105 (1955)

[41] Folland, G. B.: Introduction to partial differential equations, second edition. Princeton University Press, Princeton, New Jersey (1995)

[42] Folland, G. B.: Real analysis, second edition. John Wiley & Sons, New York Chichester Weinheim Brisbane Singapore Toronto (1999)

[43] Friedman, A.: Partial differential equations. Dover Publications Inc., Mineola, New York (1969/2008)

[44] Friedman, A.: Foundations of modern analysis. Dover Publications Inc., New York (1970/1982)

[45] Fujiwara, D.: On some homogeneous boundary value problems bounded below. J. Fac. Sci. Univ. Tokyo Sec. IA, **17**, 123–152 (1970)

[46] Fujiwara, D. and Uchiyama, K.: On some dissipative boundary value problems for the Laplacian. J. Math. Soc. Japan, **23**, 625–635 (1971)

[47] Gagliardo, E.: Proprietà di alcune classi di funzioni in più variabili. Ricerche Mat., **7**, 102–137 (1958)

[48] Garroni, M. G. and Menaldi, J. L.: Green functions for second order integro-differential problems. Pitman Research Notes in Mathematics Series No. 275, Longman Scientific & Technical, Harlow (1992)

[49] Gel'fand, I. M. and Shilov, G. E.: Generalized functions (Russian), Vols. I–III, Moscow (1958). English translation: Academic Press, New York (1964, 1967, 1968)

[50] Gilbarg, D. and Trudinger, N. S.: Elliptic partial differential equations of second order, reprint of the 1998 edition. Classics in Mathematics, Springer-Verlag, New York Berlin Heidelberg Tokyo (2001)

[51] Gohberg, I. C. and Kreĭn, M. G.: The basic propositions on defect numbers, root numbers and indices of linear operators (Russian). Uspehi Mat. Nauk., **12**, 43–118 (1957). English translation: Amer. Math. Soc. Transl. (2) **13**, 185–264 (1960)

[52] Goldstein, J. A.: Semigroups of linear operators and applications. Oxford Mathematical Monographs, Clarendon Press, Oxford University Press, New York (1985)

[53] Grubb, G. and Hörmander, L.: The transmission property. Math. Scand., **67**, 273–289 (1990)

[54] Hille, E. and Phillips, R. S.: Functional analysis and semi-groups, American Mathematical Society Colloquium Publications, 1957 edition. American Mathematical Society, Providence, Rhode Island (1957)

[55] Hopf, E.: A remark on linear elliptic differential equations of second order. Proc. Amer. Math. Soc., **3**, 791–793 (1952)

[56] Hörmander, L.: Linear partial differential operators. Springer-Verlag, Berlin, Gottingen, Heidelberg (1963)

[57] Hörmander, L.: Pseudodifferential operators and non-elliptic boundary problems. Ann. of Math., (2) **83**, 129–209 (1966)

[58] Hörmander, L.: Hypoelliptic second order differential equations. Acta Math., **119**, 147–171 (1967)

[59] Hörmander, L.: Pseudo-differential operators and hypoelliptic equations. In: Proc. Sym. Pure Math., **X**, Singular integrals, A. P. Calderón (ed.), pp. 138–183, American Mathematical Society, Providence, Rhode Island (1967)

[60] Hörmander, L.: Fourier integral operators I. Acta Math., **127**, 79–183 (1971)

[61] Hörmander, L.: The analysis of linear partial differential operators III, Pseudo-differential operators, reprint of the 1994 edition, Classics in Mathematics Springer-Verlag, Berlin Heidelberg New York Tokyo (2007)

[62] Ikeda, N. and Watanabe, S.: Stochastic differential equations and diffusion processes, second edition. North-Holland Publishing Co., Amsterdam; Kodansha Ltd., Tokyo (1989)

[63] Ishikawa, Y.: Stochastic calculus of variations for jump processes, second edition. De Gruyter Studies in Mathematics, Vol. 54, De Gruyter, Berlin Boston (2016)

[64] Itô, K.: Stochastic processes (in Japanese). Iwanami Shoten, Tokyo (1957)

[65] Itô, K. and McKean, H. P., Jr.: Diffusion processes and their sample paths, reprint of the 1974 edition, Classics in Mathematics, Springer-Verlag, Berlin Heidelberg New York (1996)

[66] Iwasaki, N.: A sufficient condition for the existence and the uniqueness of smooth solutions to boundary value problems for elliptic systems. Publ. Res. Inst. Math. Sci., **11**, 559–634 (1975/76)

[67] Juhl, A.: An index formula of Agranovič–Dynin-type for interior boundary value problems for degenerate elliptic pseudo-differential operators. Math. Nachr., **126**, 101–170 (1986)

[68] Kannai, Y.: Hypoellipticity of certain degenerate elliptic boundary value problems. Trans. Amer. Math. Soc., **217**, 311–328 (1976)

[69] Knight, F. B.: Essentials of Brownian motion and diffusion. American Mathematical Society, Providence, Rhode Island (1981)

[70] Kolmogorov, A. N.: Uber die analytischen Methoden in der Wahrscheinlichkeitsrechnung. Math. Ann., **104**, 415–458 (1931)

[71] Kolmogorov, A. N. and Fomin, S. V.: Introductory real analysis, Translated from the second Russian edition and edited by R. A. Silverman. Dover Publications Inc., New York (1975)

[72] Komatsu, T.: Markov processes associated with certain integro-differential operators. Osaka J. Math., **10**, 271–303 (1973)

[73] Kumano-go, H.: Pseudodifferential operators. MIT Press, Cambridge, Massachusetts (1981)

[74] Lamperti, J.: Stochastic processes. Springer-Verlag, New York Heidelberg Berlin (1977)

[75] Lax, P. D.: Asymptotic solutions of oscillatory initial value problems. Duke Math. J., **24**, 627–646 (1957)

[76] Lévy, P.: Processus stochastiques et mouvement brownien. Gauthier-Villars, Paris (1948)

[77] Lions, J.-L. et Magenes, E.: Problèmes aux limites non-homogènes et applications 1, 2. Dunod, Paris (1968). English translation: Non-homogeneous boundary value problems and applications 1, 2. Springer-Verlag, Berlin Heidelberg New York (1972)

[78] Masuda, K.: Evolution equations (in Japanese). Kinokuniya Shoten, Tokyo (1975)

[79] Maugeri, A., Palagachev, D. K. and Softova, L. G.: Elliptic and parabolic equations with discontinuous coefficients Mathematical Research, **109**. Wiley-VCH, Berlin (2000)

[80] McLean, W.: Strongly elliptic systems and boundary integral equations. Cambridge University Press, Cambridge (2000)

[81] Meyers, N. and Serrin, J.: $W = H$. Proc. Nat. Akad. Sci. USA, **51**, 1055–1056 (1964)

[82] Mizohata, S.: The theory of partial differential equations. Cambridge University Press, London New York (1973)

[83] Munkres, J. R.: Elementary differential topology. Ann. of Math. Studies, No. 54, Princeton University Press, Princeton, New Jersey (1966)

[84] Noble, B.: Methods based on the Wiener–Hopf technique for the solution of partial differential equations, second edition. Chelsea Publishing Company, New York (1988)

[85] Oleĭnik, O. A.: On properties of solutions of certain boundary problems for equations of elliptic type (Russian). Mat. Sbornik, **30**, 595–702 (1952)

[86] Palais, R.: Seminar on the Atiyah–Singer index theorem. Ann. of Math. Studies, No. 57, Princeton University Press, Princeton, New Jersey (1965)

[87] Pazy, A.: Semigroups of linear operators and applications to partial differential equations. Springer-Verlag, New York Berlin Heidelberg Tokyo (1983)

[88] Peetre, J.: Rectification à l'article "Une caractérisation des opérateurs différentiels". Math. Scand., **8**, 116–120 (1960)

[89] Peetre, J.: Another approach to elliptic boundary problems. Comm. Pure Appl. Math., **14**, 711–731 (1961)

[90] Perrin, J.: Les atomes. Gallimard, Paris (1970)

[91] Protter, M. H. and Weinberger, H. F.: Maximum principles in differential equations, corrected second printing. Springer-Verlag, New York (1999)

[92] Ray, D.: Stationary Markov processes with continuous paths. Trans. Amer. Math. Soc., **82**, 452–493 (1956)

[93] Rempel, S. and Schulze, B.-W.: Index theory of elliptic boundary problems. Akademie-Verlag, Berlin (1982)

[94] Revuz, D. and Yor, M.: Continuous martingales and Brownian motion, third edition. Springer-Verlag, Berlin New York Heidelberg (1999)

[95] Rudin, W.: Real and complex analysis, third edition. McGraw-Hill, New York (1987)

[96] Sato, K.: Time change and killing for multi-dimensional reflecting diffusion. Proc. Japan Acad., **39**, 69–73 (1963)

[97] Sato, K. and Tanaka, H.: Local times on the boundary for multi-dimensional reflecting diffusion, Proc. Japan Acad., **38**, 699–702 (1962)

[98] Sato, K. and Ueno, T.: Multi-dimensional diffusion and the Markov process on the boundary. J. Math. Kyoto Univ., **4**, 529–605 (1965)

[99] Schaefer, H. H.: Topological vector spaces, third printing. Graduate Texts in Mathematics, Vol. 3, Springer-Verlag, New York Berlin (1971)

[100] Schechter, M.: Principles of functional analysis, second edition. Graduate Studies in Mathematics, Vol. 36, American Mathematical Society, Providence, Rhode Island (2002)

[101] Schrohe, E.: A short introduction to Boutet de Monvel's calculus. In: Approaches to Singular Analysis, J. Gil, D. Grieser and M. Lesch (eds), pp. 85–116, Oper. Theory Adv. Appl., **125**, Birkhäuser, Basel (2001)

[102] Schwartz, L.: Théorie des distributions, Hermann, Paris (1966)

[103] Seeley, R. T.: Extension of C^∞ functions defined in a half-space. Proc. Amer. Math. Soc., **15**, 625–626 (1964)

[104] Seeley, R. T.: Refinement of the functional calculus of Calderón and Zygmund. Proc. Nederl. Akad. Wetensch., Ser. A, **68**, 521–531 (1965)

[105] Seeley, R. T.: Singular integrals and boundary value problems. Amer. J. Math., **88**, 781–809 (1966)

[106] Seeley, R. T.: Topics in pseudo-differential operators. In: Pseudo-differential operators (C.I.M.E., Stresa, 1968), L. Nirenberg (ed.), pp. 167–305. Edizioni Cremonese, Roma (1969). Reprint of the first edition, Springer-Verlag, Berlin Heidelberg (2010)

[107] Stein, E. M.: The characterization of functions arising as potentials II. Bull. Amer. Math. Soc., **68**, 577–582 (1962)

[108] Stein, E. M.: Singular integrals and differentiability properties of functions. Princeton Mathematical Series, Princeton University Press, Princeton, New Jersey (1970)

[109] Stein, E. M.: Harmonic analysis: real-variable methods, orthogonality, and oscillatory integrals. Princeton University Press, Princeton, New Jersey (1993)

[110] Stewart, H. B.: Generation of analytic semigroups by strongly elliptic operators under general boundary conditions. Trans. Amer. Math. Soc., **259**, 299–310 (1980)

[111] Stroock, D. W.: Diffusion processes associated with Lévy generators. Z. Wahrscheinlichkeitstheorie verw. Gebiete, **32**, 209–244 (1975)

[112] Taibleson, M. H.: On the theory of Lipschitz spaces of distributions on Euclidean n-space I. J. Math. Mech., **13**, 407–479 (1964)

[113] Taira, K.: Un théorème d'existence et d'unicité des solutions pour des problèmes aux limites non-elliptiques. J. Functional Analysis, **43**, 166–192 (1981)

[114] Taira, K.: Diffusion processes and partial differential equations. Academic Press, San Diego New York London Tokyo (1988)

[115] Taira, K.: The theory of semigroups with weak singularity and its applications to partial differential equations. Tsukuba J. Math., **13**, 513–562 (1989)

[116] Taira, K.: On the existence of Feller semigroups with boundary conditions. Memoirs Amer. Math. Soc., No. 475, American Mathematical Society, Providence, Rhode Island (1992)

[117] Taira, K.: Boundary value problems for elliptic integro-differential operators. Math. Z., **222**, 305–327 (1996)

[118] Taira, K.: On the existence of Feller semigroups with discontinuous coefficients. Acta Math. Sinica (English Series), **22**, 595–606 (2006)

[119] Taira, K.: Introduction to boundary value problems of nonlinear elastostatics. Tsukuba J. Math., **32**, 67–138 (2008)

[120] Taira, K.: On the existence of Feller semigroups with discontinuous coefficients II. Acta Math. Sinica (English Series), **25**, 715–740 (2009)

[121] Taira, K.: Boundary value problems and Markov processes, second edition. Lecture Notes in Mathematics, No. 1499, Springer-Verlag, Berlin Heidelberg New York (2009)

[122] Taira, K.: Semigroups, boundary value problems and Markov processes, second edition. Springer Monographs in Mathematics series, Springer-Verlag, Berlin Heidelberg New York (2014)

[123] Taira, K.: Analytic semigroups and semilinear initial boundary value problems, second edition. London Mathematical Society Lecture Note Series, No. 434, Cambridge University Press, London New York (2016)

[124] Taira, K.: Spectral analysis of the subelliptic oblique derivative problem. Ark. Mat. **55**, 243–270 (2017)

[125] Taira, K.: A strong maximum principle for globally hypoelliptic operators. Rend. Circ. Mat. Palermo, II. Ser. **68**, 193–217 (2019)

[126] Taira, K.: Spectral analysis of the hypoelliptic Robin problem, Ann. Univ. Ferrara Sez. VII Sci. Mat., **65**, 171–199 (2019)

[127] Taira, K.: Dirichlet problems with discontinuous coefficients and Feller semigroups. Rend. Circ. Mat. Palermo, II. Ser. **69**, 287–323 (2020)

[128] Taira, K.: Ventcel' boundary value problems for elliptic Waldenfels operators. Boll. Unione Mat. Ital., **13**, 213–256 (2020)

[129] Taira, K.: Oblique derivative problems and Feller semigroups with discontinuous coefficients. Ricerche Mat.
Advance online publication. https://doi.org/10.1007/s11587-020-00509-5

[130] Takanobu, S. and Watanabe, S.: On the existence and uniqueness of diffusion processes with Wentzell's boundary conditions. J. Math. Kyoto Univ., **28**, 71–80 (1988)

[131] Tanabe, H.: Equations of evolution (in Japanese), Iwanami Shoten, Tokyo (1975). English translation: Monographs and Studies in Mathematics, Vol. 6, Pitman, Boston London (1979)

[132] Tanabe, H.: Functional analytic methods for partial differential equations. Marcel Dekker, New York Basel (1997)

[133] Taylor, M. E.: Pseudodifferential operators. Princeton Mathematical Series, Vol. 34. Princeton University Press, Princeton, New Jersey (1981)

[134] Treves, F.: Topological vector spaces, distributions and kernels. Academic Press, New York London (1967)

[135] Triebel, H.: Theory of function spaces. Monographs in Mathematics, Vol. 78. Birkhäuser-Verlag, Basel Boston Stuttgart (1983)

[136] Troianiello, G. M.: Elliptic differential equations and obstacle problems. The University Series in Mathematics, Plenum Press, New York London (1987)

[137] Ueno, T.: The diffusion satisfying Wentzell's boundary condition and the Markov process on the boundary II, Proc. Japan Acad., **36**, 625–629 (1960)

[138] Višik (Vishik), M. I.: On general boundary problems for elliptic differential equations (Russian). Trudy Moskov. Mat. Obšč. **1**, 187–246 (1952). English translation: Amer. Math. Soc. Transl. (2) **24**, 107–172 (1963)

[139] Volkonskiĭ, V. A.: Additive functionals of Markov processes (Russian). Trudy Moskov. Mat. Obšč., **9**, 143–189 (1960)

[140] von Waldenfels, W.: Positive Halbgruppen auf einem n-dimensionalen Torus. Archiv der Math., **15**, 191–203 (1964)

[141] Watanabe, S.: Construction of diffusion processes with Wentzell's boundary conditions by means of Poisson point processes of Brownian excursions. In: Probability Theory, 255–271. Banach Center Publications, Vol. 5, PWN-Polish Scientific Publishers, Warsaw (1979)

[142] Watson, G. N.: A treatise on the theory of Bessel functions, reprint of the second (1944) edition, Cambridge Mathematical Library, Cambridge University Press, Cambridge (1995)

[143] Wentzell (Ventcel'), A. D.: On boundary conditions for multidimensional diffusion processes (Russian). Teoriya Veroyat. i ee Primen. **4**, 172–185 (1959). English translation: Theory Prob. and its Appl. **4**, 164–177 (1959)

[144] Wiener, N.: Differential space. J. Math. Phys., **2**, 131–174 (1923)

[145] Wiener, N. and Hopf, E.: Über eine Klasse singulärer Integralgleichungen. Sitzungsberichte Preußsche Akademie, Math. Phys. Kl., 696–706 (1931)

[146] Wloka, J.: Partial differential equations. Cambridge University Press, Cambridge (1987)

[147] Yosida, K.: Functional analysis, reprint of the sixth (1980) edition. Classics in Mathematics, Springer-Verlag, Berlin Heidelberg New York (1995)

Index

© Springer Nature Switzerland AG 2020

K. Taira, *Boundary Value Problems and Markov Processes*, Lecture Notes in Mathematics 1499, https://doi.org/10.1007/978-3-030-48788-1

LECTURE NOTES IN MATHEMATICS

Editors in Chief: J.-M. Morel, B. Teissier;

Editorial Policy

1. Lecture Notes aim to report new developments in all areas of mathematics and their applications – quickly, informally and at a high level. Mathematical texts analysing new developments in modelling and numerical simulation are welcome.

   Manuscripts should be reasonably self-contained and rounded off. Thus they may, and often will, present not only results of the author but also related work by other people. They may be based on specialised lecture courses. Furthermore, the manuscripts should provide sufficient motivation, examples and applications. This clearly distinguishes Lecture Notes from journal articles or technical reports which normally are very concise. Articles intended for a journal but too long to be accepted by most journals, usually do not have this "lecture notes" character. For similar reasons it is unusual for doctoral theses to be accepted for the Lecture Notes series, though habilitation theses may be appropriate.

2. Besides monographs, multi-author manuscripts resulting from SUMMER SCHOOLS or similar INTENSIVE COURSES are welcome, provided their objective was held to present an active mathematical topic to an audience at the beginning or intermediate graduate level (a list of participants should be provided).

   The resulting manuscript should not be just a collection of course notes, but should require advance planning and coordination among the main lecturers. The subject matter should dictate the structure of the book. This structure should be motivated and explained in a scientific introduction, and the notation, references, index and formulation of results should be, if possible, unified by the editors. Each contribution should have an abstract and an introduction referring to the other contributions. In other words, more preparatory work must go into a multi-authored volume than simply assembling a disparate collection of papers, communicated at the event.

3. Manuscripts should be submitted either online at www.editorialmanager.com/lnm to Springer's mathematics editorial in Heidelberg, or electronically to one of the series editors. Authors should be aware that incomplete or insufficiently close-to-final manuscripts almost always result in longer refereeing times and nevertheless unclear referees' recommendations, making further refereeing of a final draft necessary. The strict minimum amount of material that will be considered should include a detailed outline describing the planned contents of each chapter, a bibliography and several sample chapters. Parallel submission of a manuscript to another publisher while under consideration for LNM is not acceptable and can lead to rejection.

4. In general, **monographs** will be sent out to at least 2 external referees for evaluation.

   A final decision to publish can be made only on the basis of the complete manuscript, however a refereeing process leading to a preliminary decision can be based on a pre-final or incomplete manuscript.

   Volume Editors of **multi-author works** are expected to arrange for the refereeing, to the usual scientific standards, of the individual contributions. If the resulting reports can be forwarded to the LNM Editorial Board, this is very helpful. If no reports are forwarded or if other questions remain unclear in respect of homogeneity etc, the series editors may wish to consult external referees for an overall evaluation of the volume.

5. Manuscripts should in general be submitted in English. Final manuscripts should contain at least 100 pages of mathematical text and should always include
   - a table of contents;
   - an informative introduction, with adequate motivation and perhaps some historical remarks: it should be accessible to a reader not intimately familiar with the topic treated;
   - a subject index: as a rule this is genuinely helpful for the reader.
   - For evaluation purposes, manuscripts should be submitted as pdf files.

6. Careful preparation of the manuscripts will help keep production time short besides ensuring satisfactory appearance of the finished book in print and online. After acceptance of the manuscript authors will be asked to prepare the final LaTeX source files (see LaTeX templates online: https://www.springer.com/gb/authors-editors/book-authors-editors/manuscriptpreparation/ 5636) plus the corresponding pdf- or zipped ps-file. The LaTeX source files are essential for producing the full-text online version of the book, see http://link.springer.com/bookseries/304 for the existing online volumes of LNM). The technical production of a Lecture Notes volume takes approximately 12 weeks. Additional instructions, if necessary, are available on request from lnm@springer.com.

7. Authors receive a total of 30 free copies of their volume and free access to their book on SpringerLink, but no royalties. They are entitled to a discount of 33.3 % on the price of Springer books purchased for their personal use, if ordering directly from Springer.

8. Commitment to publish is made by a *Publishing Agreement*; contributing authors of multiauthor books are requested to sign a *Consent to Publish form*. Springer-Verlag registers the copyright for each volume. Authors are free to reuse material contained in their LNM volumes in later publications: a brief written (or e-mail) request for formal permission is sufficient.

Addresses:

Professor Jean-Michel Morel, CMLA, École Normale Supérieure de Cachan, France
E-mail: moreljeanmichel@gmail.com

Professor Bernard Teissier, Equipe Géométrie et Dynamique,
Institut de Mathématiques de Jussieu – Paris Rive Gauche, Paris, France
E-mail: bernard.teissier@imj-prg.fr

Springer: Ute McCrory, Mathematics, Heidelberg, Germany,
E-mail: lnm@springer.com

Printed in the United States
By Bookmasters